Climate Change Management

Series Editor

Walter Leal Filho, International Climate Change Information and Research Programme, Hamburg University of Applied Sciences, Hamburg, Germany

The aim of this book series is to provide an authoritative source of information on climate change management, with an emphasis on projects, case studies and practical initiatives – all of which may help to address a problem with a global scope, but the impacts of which are mostly local. As the world actively seeks ways to cope with the effects of climate change and global warming, such as floods, droughts, rising sea levels and landscape changes, there is a vital need for reliable information and data to support the efforts pursued by local governments, NGOs and other organizations to address the problems associated with climate change. This series welcomes monographs and contributed volumes written for an academic and professional audience, as well as peer-reviewed conference proceedings. Relevant topics include but are not limited to water conservation, disaster prevention and management, and agriculture, as well as regional studies and documentation of trends. Thanks to its interdisciplinary focus, the series aims to concretely contribute to a better understanding of the state-of-the-art of climate change adaptation, and of the tools with which it can be implemented on the ground.

Notes on the quality assurance and peer review of this publication

Prior to publication, the quality of the works published in this series is double blind reviewed by external referees appointed by the editor. The referees are not aware of the author's name when performing the review; the referees' names are not disclosed.

Satyabrata Mandal
Muhammad Mubashar Ahmad Dogar
Burhan Uddin Choudhury
Editors

Climate Change Vulnerability

Measuring Impact and Developing Climate-Resilient Adaptation Options for Rural Livelihoods

Springer

Editors
Satyabrata Mandal
Department of Geography
Banwarilal Bhalotia College
Asansol, West Bengal, India

Burhan Uddin Choudhury
ICAR (Indian Council of Agricultural Research) Research Complex for Northeastern Hill Region Tripura Center
Lembucherra, Tripura, India

Muhammad Mubashar Ahmad Dogar
Center for Environmental Modeling and Application (CEMA), Research Institute for Global Change (RIGC), Japan Agency for Marine-Earth Science and Technology (JAMSTEC)
Yokohama, Kanagawa, Japan

ISSN 1610-2002 ISSN 1610-2010 (electronic)
Climate Change Management
ISBN 978-3-032-16782-8 ISBN 978-3-032-16783-5 (eBook)
https://doi.org/10.1007/978-3-032-16783-5

This Springer imprint is published by the registered company Springer Nature Switzerland AG
The registered company address is: Gewerbestrasse 11, 6330 Cham, Switzerland

Preface

Climate change is one of the most significant challenges facing humanity today, with its effects felt most acutely in rural regions where livelihoods depend heavily on natural resources, climate-sensitive agriculture, and fragile ecosystems. Rural households across the Global South are facing increased pressures due to increasing climate variability, recurring extreme weather events, land degradation, water scarcity, and socioeconomic instability. Rural communities are vulnerable due to the combination of environmental stresses and deep-rooted structural factors, such as limited adaptive capacity, poor infrastructure, social marginalization, and resource constraints. Understanding the intricate relationships between climate risks and livelihood systems is now a top priority for science, development, and policy. This edited volume, *Climate Change Vulnerability: Measuring Impact and Developing Climate-Resilient Adaptation Options for Rural Livelihoods*, is conceived as a comprehensive response to this need.

The presentation integrates empirical evidence, innovative methodologies, theoretical insights, and field-based experiences to demonstrate how climate change is affecting rural lives and how communities are coping with these challenges. Contributions from scholars across disciplines and regions form this book, offering diverse yet convergent viewpoints on rural vulnerability and resilience. The structure of this volume is based on eight themes that are interconnected, highlighting the multidimensional nature of climate-livelihood interactions. The progression from conceptual foundations to practical, policy-oriented insights is represented by these themes. In the opening themes, key concepts, approaches, and indicators are introduced for assessing rural vulnerability, and regional case studies are presented to reveal how climatic stresses vary across landscapes and livelihood systems. Afterward, the themes address community and household adaptation strategies, emphasizing both indigenous knowledge and science-based innovations, as well as the importance of participatory governance in boosting grassroots resilience. This volume deals with the broader socioeconomic implications of climate change, including livelihood insecurity, diversification challenges, and gendered vulnerabilities, and presents paths for sustainable and climate-resilient rural development. The final themes place these discussions within policy and global frameworks that offer futuristic perspectives on the future of rural livelihoods in the face of increasing climate uncertainty.

This volume aims to be an essential resource for academics, researchers, development practitioners, planners, and policymakers who are studying climate change, rural development, and sustainability transitions by integrating scientific analysis, regional insights, and policy-oriented discussions. It encourages a greater comprehension of climate vulnerability by highlighting practical adaptation alternatives that can guide rural communities toward more resilient futures. All contributing authors deserve our sincere gratitude for their insightful research and dedicated efforts that have made this book even better. Springer Nature's support for the publication of this volume under the Springer Climate Change Management Series is greatly appreciated. We express our gratitude for the encouragement and institutional support we received from our respective organizations during the development of this work. We are confident that this volume will encourage further research, guide effective policy action, and make a significant contribution to the ongoing global efforts to safeguard rural livelihoods in an age of rapid climate change.

Asansol, West Bengal, India — Satyabrata Mandal
Yokohama, Japan — Muhammad Mubashar Ahmad Dogar
Lembucherra, Tripura, India — Burhan Uddin Choudhury
November 30, 2025

Contents

Editors and Contributors

About the Editors

Satyabrata Mandal is a geographer by training and a passionate researcher dedicated to understanding climate impacts on human-environment systems. A Ph.D. graduate from the University of Calcutta, he has authored numerous high-quality research papers in international journals and book chapters that contribute to emerging debates in climate science. Dr. Mandal has published extensively in leading scientific journals, including *Applied Geography*, *Ecological Indicators*, the *Journal of Environmental Management*, *Earth System and Environment, Land Degradation and Development, International Journal of Bioclimatology, Theoretical and Applied Climatology, Current Science,* and many more. His research spans critical areas such as agroclimatology, extreme weather events, and climate adaptation strategies for vulnerable communities. His work has earned him several prestigious recognitions, including the Young Scientist Award, Best Paper award, and Best Research Award from nationally reputed scientific organizations. An alumnus of Ramakrishna Mission, Presidency College, and the University of Calcutta, Dr. Mandal currently serves as an Assistant Professor of Geography at Banwarilal Bhalotia College, Asansol, India, where he continues to mentor young geographers and advance impactful climate studies.

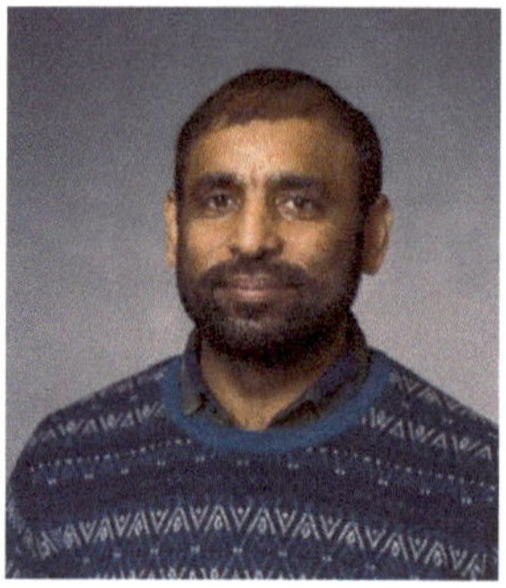

Muhammad Mubashar Ahmad Dogar is an accomplished climate scientist with extensive expertise in global and regional climate variability, climate modeling, and climate change impacts on socioecological systems. He earned his Ph.D. in Climate Modeling from Hokkaido University, Japan, with distinctions under the prestigious Japan Society for the Promotion of Science (JSPS) fellowship. His research has been recognized with the RONPAKU Medal Award for outstanding contributions to understanding natural climate variability and the impacts of external forcings, including anthropogenic influences and volcanic activity. Dr. Dogar has published extensively in leading scientific journals, including *Nature Scientific Reports*, npj Climate and Atmospheric Science, *Climate Dynamics*, and *Geophysical Research Letters* (*GRL*). His research spans critical topics such as climate teleconnections, extreme weather events, and adaptation strategies for vulnerable communities. He has collaborated with world-renowned institutions, including the International Centre for Theoretical Physics (ICTP, Italy), UK Met Office, the University of Tokyo (Japan), Hokkaido University (Japan), Texas A&M University (USA), and King Abdullah University of Science and Technology (KAUST, Saudi Arabia). Currently, Dr. Dogar is a Young Research Fellow at JAMSTEC, Japan, where he conducts research on coupled ocean–atmosphere processes, climate variability, Earth system predictability, and the response of the climate system to natural and anthropogenic forcings using observations, advanced climate models, and emerging AI-based approaches. He also has research interests in applying geospatial and bioclimatic analyses to assess the impacts of climate change on ecosystems and livelihoods, and to develop climate-resilient adaptation strategies, particularly for rural communities in regions like South Asia.

Burhan Uddin Choudhury obtained his Ph.D. in Agricultural Physics from the Indian Agricultural Research Institute (IARI), a national premier institute in Pusa, New Delhi. He completed a year-long postdoctoral research program in Soil Biophysics at Rothamsted Research in Harpenden, United Kingdom, as an RI Fellow. For the past 20 years, he has been working as a scientist (8 years as a Principal Scientist) at the national premier organization for agricultural research, Indian Council of Agricultural Research (ICAR). Currently, he is serving as the Regional Head, ICAR Research Complex for NEH Region, Tripura Center, Tripura, India. His research work entails mapping soil and crop resources through remote sensing and GIS, as well as evaluating soil carbon, land degradation, and adaptation strategies using simulation models at a regional level. Dr. Choudhury led a team of 35 multidisciplinary scientists as the Principal Investigator to develop adaptive agriculture and livestock technologies, specifically tailored to hill ecosystems, dispensing them to end-users in the ICAR-funded mega project, National Innovations in Climate Resilient Agriculture (NICRA). He conducted detailed investigations on the climate resiliency of hilly micro-watershed-based integrated farming system (IFS) models regarding soil erosion and adaptation strategies using hydrological and crop simulation models. He has authored research peer-reviewed in 65 international and 75 national journals, alongside 90 other publications, and has over 5100 citations on Google Scholar, with an i10-index of 90 and an h-index of 37. Dr. Choudhury has been listed among the top 2% of global scientists in agriculture by Stanford University, USA, for two consecutive years (2023–2024 and 2024–2025). He has been honored with the ICAR Award for Outstanding Interdisciplinary Team Research in NRM, the Rothamsted International Fellowship, the IARI Merit Medal, the IAWSC Gold Medal, the IASWC Dr. K. G. Tejwani Award, the Fellow of the Indian Hill Farming Association, and many other professional society awards. For 9 years, he taught postgraduate students at Central Agricultural University and supervised 23 M.Sc. students, 2 Ph.D. candidates, and 1 overseas postdoctoral scholar as a major advisor, co-guide, or advisory committee member.

Contributors

M. Ahmed India Meteorological Department, Meteorological Centre, MoES, Srinagar, India

Mst Sanjida Alam Department of Sustainability and Social Justice, Clark University, Worcester, MA, USA

S. K. Md. Asif Department of Agriculture, Brainware University, Barasat, Kolkata, West Bengal, India

Muhammad Umer Farooq Awan Dendrochronology Lab., Department of Botany, Government College University, Lahore, Pakistan

Md. Abu Bakkar Department of Geography and Environmental Studies, University of Chittagong, Chattogram, Bangladesh

Rupa Banerjee Department of Oceanography, Techno India University, Kolkata, West Bengal, India

Jayanta Barman Department of Physics, ADP College, Nagaon, Assam, India
Institutional Biotech Hub, ADP College, Nagaon, Assam, India

Prananath Barman ICAR Research Complex for NEH Region, Tripura Centre, Lembucherra, West Tripura, India

Gupinath Bhandari Department of Civil Engineering, Jadavpur University, Kolkata, India

Huirem Bharati ICAR Research Complex for NEH Region, Tripura Centre, Lembucherra, West Tripura, India

Sumana Bhattacharyya Department of Oceanography, Techno India University, Kolkata, West Bengal, India

Debasish Chakraborty ICAR Research Complex for North Eastern Hill Region, Umiam, Meghalaya, India

Asit Chakrabarti ICAR Research Complex for NEH Region, Tripura Centre, Lembucherra, Tripura, India

Saptadipa Chakraborty Department of Geography, Vedanta College, Kolkata, West Bengal, India

Shilpa Chakraborty Department of Geography, Bankura University, Bankura, West Bengal, India

Burhan Uddin Choudhury ICAR Research Complex for North Eastern Hill Region, Umiam, Meghalaya, India
ICAR Research Complex for North Eastern Hill Region, Tripura Centre, Lembucherra, Tripura, India

Samik Choudhury ICAR Research Complex for NEH Region, Mizoram Centre, Kolasib, Mizoram, India

Bapi Das ICAR Research Complex for NEH Region, Tripura Centre, Lembucherra, West Tripura, India

Bappa Das ICAR-Central Coastal Agricultural Research Institute, Old Goa, Goa, India
ICAR Research Complex for North Eastern Hill Region, Tripura Center, Lembucherra, Tripura, India

Biswajit Das ICAR Research Complex for NEH Region, Tripura Centre, Lembucherra, West Tripura, India

Ganesh Das ICAR-Krishi Vigyan Kendra West Tripura, ICAR RC for NEH Region, Belbari, West Tripura, India

Soumik Das ICAR-Indian Institute of Soil Science, Nabi Bagh, Bhopal, Madhya Pradesh, India

Sujit Das ICAR Research Complex for NEH Region, Tripura Centre, Lembucherra, West Tripura, India

Abhijit Debnath Krishi Vigyan Kendra, Dhalai, Tripura, India

Chandan Debnath ICAR Research Complex for NEH Region, Umiam, Meghalaya, India

Supriya Debnath ICAR Research Complex for the North Eastern Hill Region, Umiam, Meghalaya, India

Bijoy Krishna Deka Institutional Biotech Hub, ADP College, Nagaon, Assam, India

Centre for Multidisciplinary Research, Tezpur University, Sonitpur, Assam, India

Sourabh Deori ICAR Research Complex for NEH Region, Umiam, Meghalaya, India

R. Devdas ICAR Research Complex for North Eastern Hill Region, Sikkim Center, Tadong, Sikkim, India

P. Dhanya Department of Geography, Government College, Chittur, Palakkad, Kerala, India

S. Dharumarajan ICAR-National Bureau of Soil Survey and Land Use Planning, Regional Centre, Hebbal, Bangalore, Karnataka, India

Sunil Doley Department of Forestry, Mizoram University, Aizawl, Mizoram, India

A. Gangarani ICAR Research Complex for North Eastern Hill Region, Tripura Centre, Lembucherra, Tripura, India

Lopamudra Ganguly Department of Geography, Vedanta College, Kolkata, West Bengal, India

Subhadwip Ghorai Department of Agriculture, Brainware University, Barasat, Kolkata, West Bengal, India

Pabitra Kumar Ghosh Department of Agriculture, Brainware University, Barasat, Kolkata, West Bengal, India

Rajrupa Ghosh Department of Zoology, Banwarilal Bhalotia College, Asansol, India

Abid Hassan Department of Geography and Environmental Studies, University of Chittagong, Chattogram, Bangladesh

Ramchhanliana Hauchhum Department of Forestry, Mizoram University, Aizawl, Mizoram, India

S. Hazarika ICAR Research Complex for North Eastern Hill Region, Umiam, Meghalaya, India

K. Jayarajan Department of Geography, Government College, Chittur, Palakkad, Kerala, India

Himashree Kalita Institutional Biotech Hub, ADP College, Nagaon, Assam, India

Centre for Multidisciplinary Research, Tezpur University, Sonitpur, Assam, India

Sahely Kanthal Department of Agriculture, Brainware University, Barasat, Kolkata, West Bengal, India

Afifa Kausar Institutional Biotech Hub, ADP College, Nagaon, Assam, India

Department of Zoology, ADP College, Nagaon, Assam, India

Tahira Khalid Dendrochronology Lab., Department of Botany, Government College University, Lahore, Pakistan

Zaheer-ud-din Khan Dendrochronology Lab., Department of Botany, Government College University, Lahore, Pakistan

R. Krishnappa ICAR Research Complex for the North Eastern Hill Region, Umiam, Meghalaya, India

Amit Kumar ICAR Research Complex for North Eastern Hill Region, Sikkim Center, Tadong, Sikkim, India

Ashok Kumar ICAR Research Complex for North Eastern Hill Region, Sikkim Center, Tadong, Sikkim, India

Rahul Kumar ICAR Research Complex for NEH Region, Lembucherra, Tripura Centre, West Tripura, Tripura, India

Sananda Kundu Department of Geography, Manipur University, Imphal, Manipur, India

Lalhruaipuii ICAR RC NEH Region, Mizoram Centre, Kolasib, Mizoram, India

Evelyn Lalparmawii Department of Forestry, Mizoram University, Aizawl, Mizoram, India

P. L. Lalrinsanga ICAR RC NEH Region, Mizoram Centre, Kolasib, Mizoram, India

Lungmuana ICAR RC NEH Region, Mizoram Centre, Kolasib, Mizoram, India

Hossain Al Mahbub Department of Geography and Environmental Studies, University of Chittagong, Chattogram, Bangladesh

Parag Majumder ICAR Research Complex for NEH Region, Tripura Centre, Lembucherra, West Tripura, India

Ishanu Mandal Faculty of Agriculture, JIS University, Kolkata, West Bengal, India

Hanumantappa Meti Department of Soil Science and Agricultural Chemistry, UAS, GKVK, Bangalore, Karnataka, India

Abhijit Mitra Department of Oceanography, Techno India University, Kolkata, West Bengal, India

Rajdeep Mohanta Department of Agriculture, Brainware University, Barasat, Kolkata, West Bengal, India

Arun Mondal Department of Geography, University of Allahabad, Prayagraj, India

Souvik Mondal Department of Geography, University of Allahabad, Prayagraj, India

Subhadip Mondal Department of Geography, Vedanta College, Kolkata, West Bengal, India

Sohaib Muhammad Dendrochronology Lab., Department of Botany, Government College University, Lahore, Pakistan

Muhammad Muhibbullah Department of Geography and Environmental Studies, University of Chittagong, Chattogram, Bangladesh

Swarnali Mukhhopadhyay Department of Geography, Baruipur College, Kolkata, India

Department of Geography, Amity University, Kolkata, India

Rumi Narzari ICAR Research Complex for the North Eastern Hill Region, Umiam, Meghalaya, India

Lakshman Chandra Pal Department of Geography, Bidhan Chandra College, Hooghly, West Bengal, India

Tapas Pal Department of Geography, Raiganj University, Raiganj, West Bengal, India

Anupam Pandey Department of Geography, University of Allahabad, Prayagraj, India

Prashant Pandey ICAR Research Complex for North Eastern Hill Region, Sikkim Center, Tadong, Sikkim, India

Subrata Pan Department of Geography, Bankura Christian College, Bankura, West Bengal, India

Janmejay Parhi ICAR-CIFA, Bhubaneswar, India

Partha Paul Department of Geography, University of Allahabad, Prayagraj, India

Rubai Podder ICAR Research Complex for NEH Region, Tripura Centre, Lembucherra, West Tripura, India

Prosenjit Pramanick Department of Oceanography, Techno India University, Kolkata, West Bengal, India

K. Prasad Department of Geography, Government College, Chittur, Palakkad, Kerala, India

Mijanur Rahman Department of Geography and Environmental Studies, University of Chittagong, Chattogram, Bangladesh

T. Ramesh ICAR Research Complex for the North Eastern Hill Region, Umiam, Meghalaya, India

Soumik Dey Roy Department of Agriculture, Brainware University, Barasat, Kolkata, West Bengal, India

Arghya Saha Department of Geography, Bidhan Chandra College, Hooghly, West Bengal, India

Bipasha Saha ICAR Research Complex for NEH Region, Tripura Centre, Lembucherra, West Tripura, India

Saurav Saha ICAR Research Complex for North Eastern Hill Region, Sikkim Center, Tadong, Sikkim, India

Lopamudra Sahoo ICAR Research Complex for NEH Region, Tripura Centre, Lembucherra, Tripura, India

Blessa Sailo ICAR RC NEH Region, Mizoram Centre, Kolasib, Mizoram, India

H. Samarendra ICAR Research Complex for the North Eastern Hill Region, Umiam, Meghalaya, India

Pradip Kumar Sarkar ICAR Research Complex for NEH Region, Tripura Centre, Lembucherra, West Tripura, India

Md. Iqbal Sarwar Department of Geography and Environmental Studies, University of Chittagong, Chattogram, Bangladesh

T. S. Sindhushree Department of Soil Science and Agricultural Chemistry, UAS, GKVK, Bangalore, Karnataka, India

Abhijit Singha ICAR Research Complex for NEH Region, Tripura Centre, Lembucherra, West Tripura, India

Mahak Singh ICAR Research Complex for NEH Region, Nagaland Centre, Medziphema, Nagaland, India

Vinay Singh ICAR Research Complex for NEH Region, Tripura Centre, Lembucherra, Tripura, India

N. K. Sinha ICAR-Indian Institute of Soil Science, Nabi Bagh, Bhopal, Madhya Pradesh, India

Suman Sinha Department of Geography, Amity University, Kolkata, India

Jeetendra Kumar Soni ICAR-Directorate of Weed Research, Jabalpur, Madhya Pradesh, India

T. Sreelesh Department of Geography, Government College, Nilambur, Kerala, India

Sreelusreepadi HSST Geography, Kozhikkode, Kerala, India

Ashima Suklabaidya ICAR Research Complex for NEH Region, Tripura Centre, Lembucherra, West Tripura, India

Muhammad Tayyab Dendrochronology Lab., Department of Botany, Government College University, Lahore, Pakistan

Vanlalruati Department of Forestry, Mizoram University, Aizawl, Mizoram, India

Muhammad Zahid Dendrochronology Lab., Department of Botany, Government College University, Lahore, Pakistan

Sufia Zaman Department of Oceanography, Techno India University, Kolkata, West Bengal, India

Reviewers

Rabin Chakrabortty, Ph.D., Postdoc. Department of Civil Engineering, American University of Sharjah, Sharjah, United Arab Emirates

Burhan Uddin Choudhury, Ph.D., Postdoc. ICAR Research Complex for NEH Region, Tripura Centre, Lembucherra, West Tripura, India

Debanjana Das, Ph.D. NOAA Center for Atmospheric Sciences and Meteorology, Howard University, and Atmospheric, Oceanic & Earth Sciences, George Mason University, Washington, DC, USA

Muhammad Mubashar Ahmad Dogar, Ph.D., Postdoc. Center for Environmental Modeling and Application (CEMA), Research Institute for Global Change (RIGC), Japan Agency for Marine-Earth Science and Technology (JAMSTEC), Yokohama, Kanagawa, Japan

Amiya Gayen, Ph.D., Postdoc. Department of Geography, Midnapore College, Midnapore, West Bengal, India

Pritam Ghosh, Ph.D. Department of Geography, Hijli College, Paschim Medinipur, West Bengal, India

Syed Zohaib Hassan, Ph.D., Postdoc. Department of Civil, Environmental, and Construction Engineering, The University of Central Florida, Orlando, FL, USA

Mir Zaman Hussain, Ph.D., Postdoc. Michigan State University, East Lansing, MI, USA

Satyabrata Mandal, Ph.D. Department of Geography, Banwarilal Bhalotia College, Asansol, West Bengal, India

Shibu Ibrahim Mohammad, Ph.D., Postdoc. Sustainable Soils and Grassland Systems Department, Rothamsted Research, West Common, Harpenden, UK

Arun Mondal, Ph.D., Postdoc. Department of Geography, University of Allahabad, Prayagraj, India

Joy Rajbanshi, Ph.D. Council on Energy, Environment and Water, ISID Campus, 4, Vasant Kunj Institutional Area, New Delhi, India

S. K. Reza, Ph.D. ICAR-NBSS&LUP, Salt Lake City, D. K. Block, Sector-II, Bidhan Nagar, Kolkata, India

Priyabrata Santra, Ph.D. Division of Natural Resources, ICAR-Central Arid Zone Research Institute, Jodhpur, Rajasthan, India

Raj Kumar Setia, Ph.D., Postdoc. Punjab Remote Sensing Centre (PRSC), PAU Campus, Ludhiana, Punjab, India

Muhammad Taimur Siddique Graduate School of Environmental Science, Hokkaido University, Sapporo, Hokkaido, Japan

Environmental Consultant at HT Consultant, Karachi, Pakistan

Part I
Understanding the Vulnerability to Climate

Chapter 1
Rainfall Trend Assessment Using Non-Parametric and Innovative Trend Method During 1981–2022 in Bankura District, West Bengal

Shilpa Chakraborty and Subrata Pan

Abstract This study investigates the variability and trends in annual, seasonal, and monthly rainfall in the Bankura district of West Bengal, India, over the period 1981–2022. The analysis employs three statistical techniques: the Mann-Kendall (MK) test, the Modified Mann-Kendall (MMK) test, and the Innovative Trend Analysis (ITA) method. Two types of data were used: observed data sourced from NASA and station data retrieved via Google Earth Engine. The ITA method detected both monotonic and non-monotonic trends, independent of serial correlation, dataset size, and distribution. Results from the MK test revealed a statistically significant positive trend in annual rainfall, with an average increase of approximately 3.2 mm per annum. Seasonal analysis showed notable increases during the monsoon and post-monsoon periods, while winter rainfall exhibited a declining trend. Monthly trends varied, with June and December showing minimal autocorrelation at lag-1.

These findings offer valuable insights for planners and policymakers, particularly in the domains of irrigation planning, climate adaptation, and sustainable water resource management in the Bankura district.

Introduction

Climate change refers to long-term alterations in Earth's climate system, characterized by persistent changes in the mean state and variability of climatic variables over decades or longer. These changes can result from internal climate variability and external forcings, including both natural and anthropogenic influences (Dogar et al. 2017b, 2023, 2024, 2025a, 2025b); however, the observed global warming since the

S. Chakraborty (✉)
Department of Geography, Bankura University, Bankura, West Bengal, India

S. Pan
Department of Geography, Bankura Christian College, Bankura, West Bengal, India

S. Mandal et al. (eds.), *Climate Change Vulnerability*, Climate Change Management, https://doi.org/10.1007/978-3-032-16783-5_1

mid-twentieth century has been primarily attributed to anthropogenic activities (Dogar 2019, 2020). Among the sectors affected, agriculture remains the most vulnerable. In India, agricultural output and crop yields are closely tied to the amount and distribution of rainfall (Chakraborty et al. 2025). Precipitation plays a pivotal role in shaping water resources and influencing socioeconomic conditions (Islam and van Amstel 2021; Tabari and Talaee 2011; Khan et al. 2025; Dogar and Almzroui 2022). Its variability contributes to extreme events such as floods, droughts, and landslides, thereby increasing societal vulnerability and economic losses (Roxy et al. 2017; Mandal et al. 2018). According to Gajbhiye et al. (2016), both the distribution and volume of rainfall are key factors shaping agricultural productivity. Given agriculture's central role in India's economy and its importance for the livelihood of its population, these climatic variables hold significant influence (Kumar and Gautam 2014). The southwest monsoon contributes nearly 80% of India's total annual rainfall. Shifts in its distribution, timing, and intensity (Guhathakurta et al. 2015; Ummenhofer et al. 2017) have significant implications for agricultural output, water resource planning, and the broader national economy (Saha et al. 1979; Mandal et al. 2015). Recognising these impacts, a wide array of research has focused on analysing climatic trends relevant to the Indian context (e.g., Mandal et al. 2013; Mandal and Choudhury 2014; Halder et al. 2023; Dogar et al. 2017a; Dogar and Sato 2018, 2019). A 2024 study conducted in the Nanded district of Maharashtra analysed annual rainfall trends across multiple tehsils using the Mann-Kendall and Sen's slope tests. The findings revealed both rising and falling trends in average annual rainfall, with several tehsils showing statistically significant increases. The study emphasised the implications of these shifts for rain-fed agriculture and regional water resource planning, underscoring climate change's localized impact on precipitation patterns (Singh et al. 2024). A study by Subrahmanyam et al. (2025) analyzed spatial and temporal shifts in Indian summer monsoon rainfall between 2001 and 2020. Using GSMaP_ISRO satellite data, the researchers identified a notable increase in rainfall over west-central India during 2011–2020 compared to the previous decade. They also observed a shift in the timing of peak rainfall, with the Indo-Gangetic Plain experiencing earlier peaks and central India showing delayed ones. These changes are attributed to evolving climate dynamics, including aerosol loading and warming trends, and have direct implications for agriculture, hydrology, and disaster preparedness (Subrahmanyam et al. 2025). The predominant methodologies employed for the analysis of trends encompass the Mann-Kendall (MK) test (Mann 1945), Sen's slope estimator (Sen 1968), and linear regression trend analysis (Hann 1977). Recent work has advanced trend detection in hydrology using refined statistical tools. Deng et al. (2024) integrated machine learning for lake water quality; Amirthanathan et al. (2023) used MK and Pettitt tests to assess streamflow changes in Australia; and Mao et al. (2025) applied MK and wavelet analysis to detect runoff trends in China.

The Innovative Trend Analysis (ITA) method, originally proposed by Şen (2012) as a graphical, non-parametric approach delineated by a 1:1 (45°) reference line, has gained prominence for detecting trends in hydroclimatic time series without assumptions of normality or independence. Yu et al. (2024) applied ITA alongside seasonal MK tests to assess groundwater variability across Japan, demonstrating its effectiveness in identifying subtle seasonal shifts. More recently, Abu Arra et al. (2025) introduced an enhanced framework—ITA-NF—that integrates extreme precipitation indices, standardization, and classification techniques to improve interpretability and climate impact assessment. Despite receiving approximately 1400 mm/year of annual precipitation, the region exhibits a marked propensity for drought. The agricultural sector in Bankura is susceptible to risks stemming from recurrent arid intervals throughout the cultivation periods. The patterns of precipitation exert a significant influence on hydrological dynamics, encompassing phenomena such as flooding and drought. The availability and usability of irrigation infrastructure are influenced—both directly and indirectly—by adequate rainfall, which remains a key factor in shaping agricultural productivity. Hence, examining rainfall patterns in the Bankura district is critically important for shaping resilient water resource strategies and guiding long-term agricultural planning.

The Bankura district, located in West Bengal, is marked by its vulnerability to drought and predominantly engages in agrarian activities. As a result, the prompt onset of adequate precipitation is of paramount importance for the local populace, given that the successful growth of crops, especially rice, is entirely reliant on rainfall. However, the agricultural paradigm of the district is often disrupted due to erratic precipitation patterns. The availability of irrigation facilities is also affected, either directly or indirectly, by adequate rainfall, which exerts a considerable influence on agricultural yields. Consequently, the ongoing examination of rainfall trends in the Bankura district is critically important for forecasting climatological challenges such as floods and droughts, in addition to facilitating efficient water resource management (Ghosh and Mujumdar 2009). However, despite extensive national-level studies, there remains a lack of focused research on precipitation trends in semi-arid districts like Bankura—especially using updated datasets and advanced methodologies. Existing literature often overlooks localized hydrological dynamics and their implications for drought resilience and irrigation planning. This study addresses that gap by analyzing seasonal and annual rainfall trends in Bankura using recent data and robust statistical techniques.

Objective

The main objectives of this study are:

- To analyse seasonal and annual rainfall patterns in the Bankura district of West Bengal using long-term observational data.
- To detect trends and variability in precipitation through statistical methods such as the Mann-Kendall test, Sen's slope estimator, and Innovative Trend Analysis (ITA).

Consequently, the objective of the present research is to examine the fluctuations in precipitation within the Bankura districts, recognized as one of the semi-arid zones in the state of West Bengal, India. An analysis of both seasonal and annual trends has been conducted. This involves gaining a thorough understanding of the region's rainfall characteristics and their changing patterns. Gaining insight into the uncertainties linked to precipitation trends will establish a foundational knowledge for enhanced management of agricultural practices, irrigation systems, and other water-related endeavours within the designated area.

Materials and Methods

Study Area

Bankura, located in the Indian state of West Bengal, functions as both a city and a municipal region. Geographically (Fig. 1.1), it lies between latitudes 22°38′ to 23°38′ north and longitudes 86°36′ to 87°46′ east. The municipality spans a substantial area of approximately 6882 sq. km. (2657 sq. mi.). The Damodar River demarcates the region from the Burdwan district to the north. Generally, the quality of soil in the upland regions is suboptimal for the cultivation of cash crops. Conversely, the lowland areas exhibit fertile soil, thereby providing significant

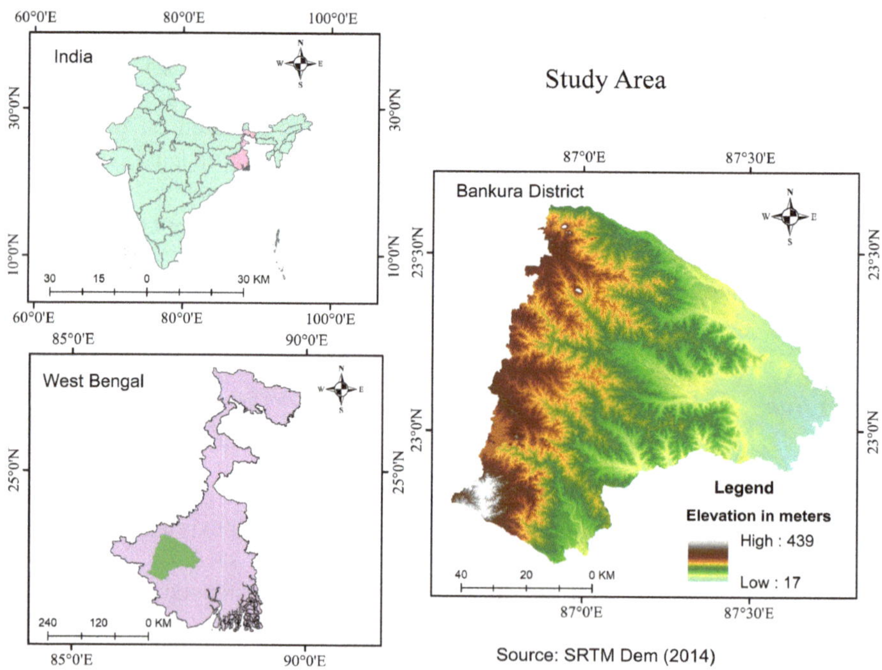

Fig. 1.1 Map depicting the research focus within Bankura District

opportunities for agricultural practices. Indus and Kotulpur blocks are characterized by the presence of loam and clay-based soils. The predominant landscape of the district comprises rolling terrain characterized by laterite and alluvial deposits. To the east, one can observe an extensive plain composed of recent alluvium, while the western extremity reveals gneisses and schists originating from the Archaean age, which delineate the eastern boundary of analogous geological formations in Chhotanagpur. Additionally, sedimentary rocks associated with the Gondwana system, which constitute the southern segment of the Raniganj coalfield, can be found in the northernmost district's land coverage, precisely located between Mejia and the Biharinath Hill region, where several economically viable coal seams are present. Despite Bankura being classified as a rain-fed district, it is widely recognized as a drought-prone region within the state. Although this district receives a substantial amount of precipitation, totalling approximately 1400 mm annually, agricultural cultivation and crop production are largely contingent upon a limited period of unpredictable rainfall. Typically, between 80% and 90% of the annual rainfall is received from June to September, depending on when the monsoon begins.

The climatic conditions in the northwestern upland regions are considerably more arid than those in the eastern and southern parts of Bengal. From March until early June, before the onset of the monsoon, the atmosphere is predominantly influenced by hot westerly winds. During the protracted dry season, extensive areas of red soil, virtually devoid of arboreal vegetation, contribute to a scorched and desolate landscape. In the eastern section, the observer's gaze is continuously met with vast expanses of rice paddies, verdant during the monsoon yet arid and parched in the summer. Monsoonal phase, spanning during the monsoon period, are relatively more temperate, as the climatic conditions are notably less oppressive than those experienced in other regions of Bengal.

Data Sources

This study applied three well-established statistical techniques for analyzing precipitation trends—namely, the Mann-Kendall (MK) test, its modified version (mMK), and Sen's slope estimator—along with the autocorrelation function to identify variations in rainfall patterns from 1981 to 2022. In addition, the Innovative Trend Analysis (ITA) approach was employed to further investigate rainfall fluctuations within the Bankura district. For the aforementioned analyses, the requisite data was procured from NASA's POWER project (https://power.larc.nasa.gov/data-access-viewer/), which provides climate data, as well as CHIRPS data (https://www.chc.ucsb.edu/data/chirps), which was also instrumental for conducting spatial analysis.

Autocorrelation Function

In line with Von Storch and Navarra (1999), the dataset was initially prewhitened to eliminate serial dependence before applying the Mann-Kendall (MK) test. This step ensures the validity of trend detection in autocorrelated time series. Autocorrelation was assessed using the lag-k correlation function: $r(k) = \frac{\sum_{t=1}^{N-k}(X_t - \bar{X})(X_{t+k} - \bar{X})}{\sum_{t=1}^{N}(X_t - \bar{X})^{2.}}$.

Here, r(k) denotes the autocorrelation at lag k, while X_t refers to the value of the time series at time t; the mean represents the average of the series, and N stands for the total number of data points. When the computed r(k) indicates no statistically significant trend at the 5% level, the Mann-Kendall (MK) test and Sen's slope estimator are then applied for further analysis. Conversely, if the r(k) outcome reveals a significant trend, the Hamed and Rao Modified MK test becomes applicable.

Innovative Trend Analysis (ITA)

Şen's (2012) Innovative Trend Analysis (ITA), a non-parametric approach, was applied to detect monotonic trends in the time series. The method involves splitting the dataset into two equal halves, sorting each independently, and plotting them against each other in a scatter diagram. A 1:1 alignment indicates no trend, while deviations below or above this line suggest decreasing or increasing trends, respectively. This visual diagnostic complements the Mann-Kendall test by offering intuitive insight into trend direction and strength (Alifujiang et al. 2020).

Mann-Kendall (MK) Test

The Mann-Kendall test is a rank-based statistical approach that does not rely on data distribution assumptions, invented by Kendall (1938), and Mann (1945). This test is widely used in environmental science, hydrology, and other fields where detecting trends in temporal data is important. The value of *S* in the Mann-Kendall test is defined by the following formula:

$$S = \sum_{k=1}^{n-1} \sum_{j=k+1}^{n} sign(x_j - x_k)$$

Where n denotes the number of data points $x_{k,}$ and x_j represents the data point of time j.

$$sign\left(x_j - x_k\right) = 1\, if\ x_j - x_k > 0$$
$$= 0\, if\ x_j - x_k = 0$$
$$= -1\, if x_j - x_k < 0$$

The variance of S is calculated by the following equation:

$$VAR\left(S\right) = \frac{1}{18} n\left(n-1\right)\left(2n+5\right) - \sum_{p=1}^{g} t_p\left(t_p - 1\right)\left(2t_p - 5\right)$$

Here, n refers to the total number of observations, g indicates the count of tied groups, and t_p denotes the number of ties within each group of size p. The test statistic S shares a close mathematical relationship with Kendall's tau (τ).

$$\tau = \frac{S}{D}$$

Where,

$$D = \sqrt{\frac{1}{2} n\left(n-1\right) - \frac{1}{2} \sum_{p=1}^{g} t_p\left(t_p - 1\right) \sqrt{\frac{1}{2} n\left(n-1\right)}}$$

If sample size n > 10, S and VAR(S) are used to compute the normalised test statistic Z by the following equation:

$$Z = \begin{cases} \frac{S-1}{\sqrt{VAR\left(S\right)}} & if\ S > 0 \\ 0 & if\ S = 0 \\ \frac{S+1}{\sqrt{VAR\left(S\right)}} & if\ S < 0 \end{cases}$$

A positive Z value reflects an upward trend, whereas a negative Z value signifies a downward trend. This calculation was performed using a Microsoft Excel function.

Modified Mann-Kendall Test

The revised MK (mMK) test is implemented when the autocorrelation coefficient of the specified dataset exhibits statistical significance at lag-k to mitigate the serial dependence present in the analysed time series (Hamed and Rao, 1998; Basak et al. 2021; Das et al. 2021). The VAR(S) can be assessed utilising the following equation:

$$VAR(S)=\left(\frac{N(N-1)(2N+5)}{18}\right),\left(\frac{N}{N_e^*}\right)$$

For datasets exhibiting autocorrelation, the correlation factor $\left(\frac{N}{N_e^*}\right)$ is fixed as follows:

$$\left(\frac{N}{N_e^*}\right)=1+\left(\frac{2}{N^3-3N^2+2N}\right)$$

$$\sum_{j=i}^{N-1}(N-f-1)(N-f-2)p_e(f)$$

$p_e(f)$ signifies the autocorrelation between observational ranks, which can be accessed as follows:

$$p_e(f)=2\sin\left(\frac{\pi}{6}p_e(\mathrm{f})\right)$$

Sen's Slope Estimator

Sen's slope offers a non-parametric approach for accurately estimating the true rate of change over time. This method captures the variability in trend slopes, which often fluctuate across seasonal and yearly time series datasets (Das and Bhattacharya 2018; Mandal et al. 2013). The slope is computed using the following formula:

$$Q_{i=\frac{x_k-x_j}{k-j},\quad i=1,\ldots,N}$$

Here, x_k and x_j represent the data points recorded at time instances *k* and *j*. Sen's slope estimator is defined as the median of the *N* computed Q_i values.

Result and Discussion

EDA Method

The research area has undergone a fluctuating mean annual precipitation, exhibiting an average value of 1250 ± 280 mm throughout the hydrological period spanning from 1981 to 2022. The most significant annual precipitation was observed in the year 2017, amounting to 1951 mm. The temporal variability of precipitation is most

Table 1.1 Monthly rainfall statistics for Bankura district (1981–2022)

Months	Minimum rainfall (mm)	Maximum rainfall (mm)	Mean rainfall (mm)	Standard deviation (mm)	Coefficient of variation (%)	Skewness	Kurtosis
January	0.00	47.46	10.95	13.49	123.15	1.36	1.17
February	0.00	58.01	15.05	17.62	117.04	1.07	0.09
March	0.00	94.92	17.84	20.56	115.27	2.14	5.79
April	0.00	126.56	32.13	28.61	89.04	1.13	1.62
May	5.27	219.13	96.50	56.80	58.86	0.55	−0.44
June	52.73	458.79	205.43	99.22	48.30	0.76	0.01
July	144.97	659.18	287.17	118.34	41.21	1.65	3.02
August	100.20	442.97	272.66	80.90	29.67	−0.28	−0.30
September	79.10	400.78	214.85	78.31	36.45	0.70	0.09
October	0.00	321.78	79.30	70.70	89.15	1.65	2.98
November	0.00	79.10	10.04	16.58	165.14	2.50	7.09
December	0.00	42.19	8.42	13.55	160.87	1.56	1.16

Source: NASA Power Data

Table 1.2 Seasonal rainfall statistics in Bankura district (1981–2022)

Statistic	Annual rainfall (mm)	Pre-monsoon rainfall (mm)	Monsoon rainfall (mm)	Post-monsoon rainfall (mm)	Winter rainfall (mm)
Minimum	743.55	10.54	527.35	5.27	0.00
Maximum	1951.17	305.85	1545.12	321.68	110.75
Mean	1250.36	146.47	980.13	89.34	34.43
Standard deviation	280.83	68.61	231.16	72.25	25.87
Coefficient of variation (%)	22.46	46.85	23.58	80.86	75.16
Skewness	0.78	0.52	0.53	1.48	0.98
Kurtosis	0.50	0.06	0.18	2.12	1.13

Source: NASA Power Data

pronounced in November, while rainfall stability is notably elevated during August (Table 1.1).

The highest volume of rainfall was recorded during the monsoon season, amounting to approximately 1545 ± 231 mm. This was followed by the post-monsoon season with 321 ± 72 mm, the pre-monsoon season with 305 ± 68 mm, and the lowest precipitation occurred in the winter months, totalling 110 ± 25 mm. For this study, the pre-monsoon or summer period includes March, April, and May; the monsoon season spans June through September; October and November are categorized as the post-monsoon or autumn phase; and December through February constitute the winter season. The seasonal distribution of rainfall as a percentage of the annual total was calculated as 11.53% for pre-monsoon, 78.48% for monsoon, 7.12% for post-monsoon, and 2.87% for winter (refer to Table 1.2). However, a higher coefficient of variation (CV) suggests greater inconsistency and lower predictability in rainfall patterns.

This investigation reveals that the coefficient of variation (CV) is at its lowest during the monsoon season and at its highest during the post-monsoon and winter seasons, indicating a heightened variability in precipitation within the Bankura district. In other terms, it can be articulated that the precipitation patterns in the analysed region exhibit a symmetric distribution throughout the monsoon months. An elevated CV suggests that agricultural practices in the Rabi and autumn seasons are heavily reliant on residual soil moisture or supplemental irrigation, owing to the unpredictability of rainfall. The average monthly precipitation has shown a progressive increase commencing in January, culminating in a peak of 287 mm in July, followed by a subsequent decline, with the lowest recorded precipitation being 8.42 mm in December. Analysing the long-term data spanning from 1981 to 2022, it is evident that July stands as the month with the highest precipitation, accounting for 22.73% of the total rainfall, closely followed by August, which accounts for 22.07%. Conversely, December is identified as the month with the least precipitation, contributing a mere 0.74% of the total rainfall, succeeded by November at 0.82%. Nonetheless, the exceedingly high CV observed from November to March signifies a substantial risk associated with the cultivation of Rabi crops in the lack of irrigation. Spatial distribution shows that the rainfall is gradually decreasing in the central part of Bankura district, and this region is the urban area of the city.

Influence of Autocorrelation on Rainfall Patterns

Figure 1.2 presents autocorrelation diagrams illustrating the distribution of precipitation across monthly, seasonal, and annual timescales within the Bankura district. This analysis aims to evaluate the degree of randomness and recurring patterns within the rainfall time series. Results indicate the presence of both short-term and long-term serial correlations across all temporal scales—monthly, seasonal, and annual. However, the annual totals, as well as rainfall during the monsoon, pre-monsoon, and post-monsoon seasons, demonstrated relatively weak autocorrelation. Similarly, the months of June and December showed minimal autocorrelation at the first lag, suggesting limited temporal dependence in rainfall during these periods, indicating that rainfall variability is largely stochastic rather than cyclic. This randomness implies greater uncertainty in forecasting rainfall and poses a challenge for sustainable crop planning.

The spatial maps (Fig. 1.3) reveal a persistent heterogeneity in rainfall distribution within the Bankura district. Higher mean daily rainfall is consistently observed in the southern and eastern blocks (e.g., Ranibandh, Sarenga, Raipur, and Patrasayer), while the northern blocks such as Mejhia, Saltora, and Barjora, record relatively lower values. This spatial contrast indicates that physiographic and orographic

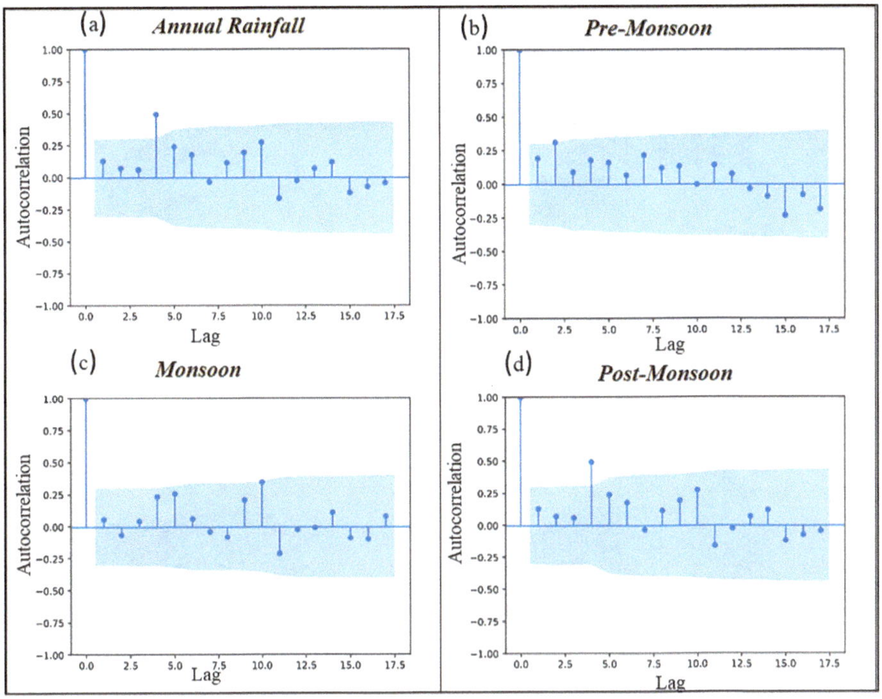

Fig. 1.2 Autocorrelation analysis of monthly rainfall in Bankura district (1981–2022) across four seasonal periods: (**a**) Annual, (**b**) Pre-Monsoon (March–May), (**c**) Monsoon (June–September), (**d**) Post-Monsoon (Oct-Nov)

factors, along with monsoon circulation, continue to influence rainfall patterns. Compared to the 1980s, the 2010s and 2020s have shown an overall decline in rainfall intensity, particularly in the north-central region, which is crucial for rain-fed agriculture.

Annual, Seasonal, and Monthly Trend by ITA

The assessment of annual and seasonal rainfall trends utilizing the ITA methodology revealed an upward trajectory in 9.5% of the years characterized as low, 67% classified as medium, and the remaining years exhibiting a high trend. Remarkably, rainfall has shown an increasing trend even during the monsoon and post-monsoon seasons. However, a detailed analysis of the ITA (Innovative Trend Analysis) graph reveals a downward trend in winter precipitation, with a notable 15% of the years exhibiting a significant decline (Fig. 1.4).

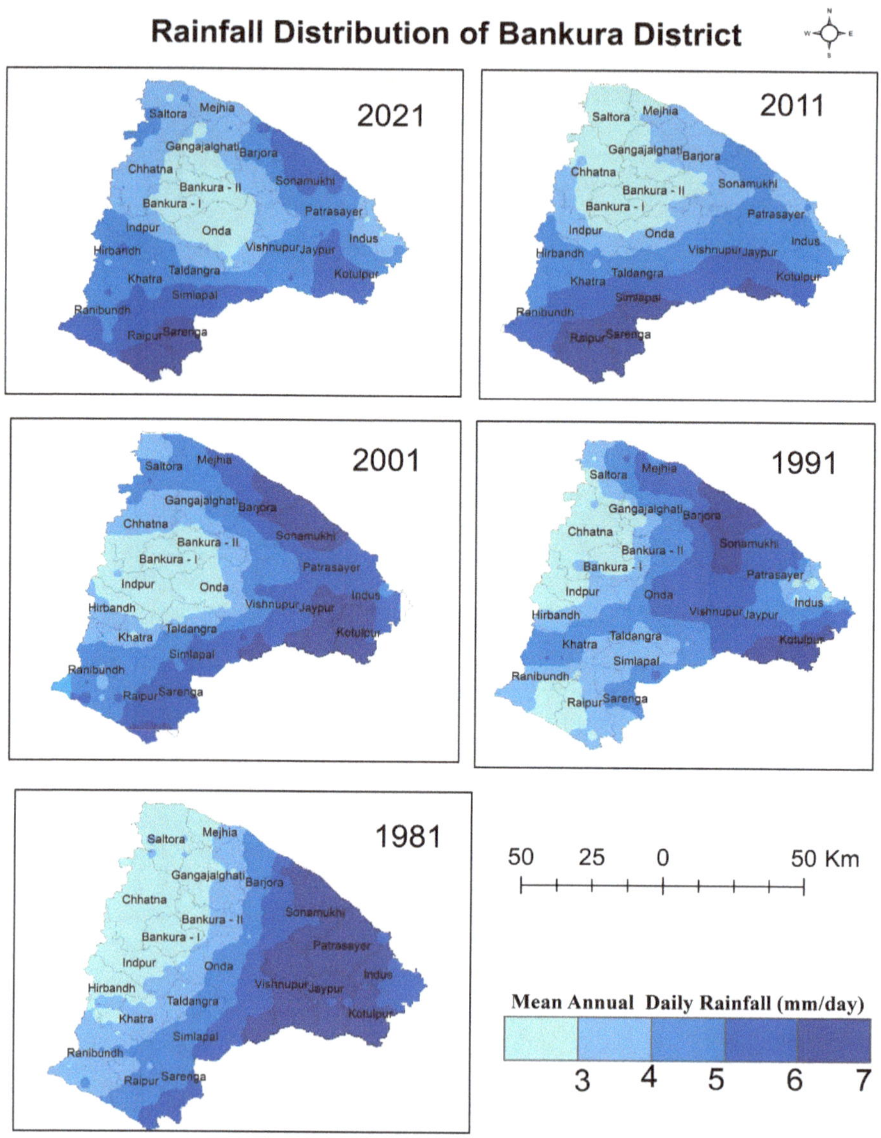

Fig. 1.3 Spatial distribution of mean annual daily rainfall in Bankura district across five time slices: 1981, 1991, 2001, 2011, and 2021

Annual and Seasonal Trend by MK and mMK Test

The distributional characteristics of precipitation across various seasons have demonstrated considerable variability A clear upward trend was detected using the Mann-Kendall method (z = 2.85) in the mean annual precipitation of the Bankura

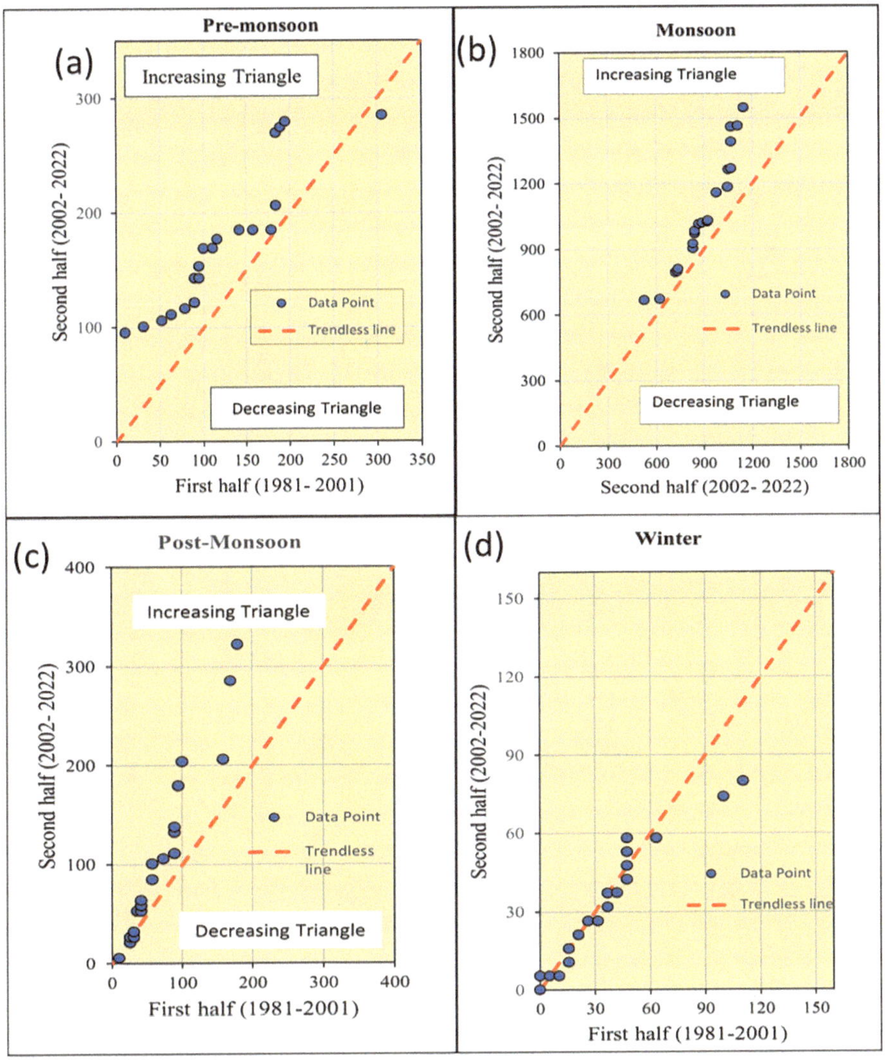

Fig. 1.4 ITA of seasonal rainfall across four periods (**a**) Pre-monsoon, (**b**) Monsoon, (**c**) post-monsoon, and (**d**) Winter. (Source: NASA Power Data)

district at a rate of 9.42 mm per annum. Furthermore, the Mann-Kendall test indicated an increasing trend during the pre-monsoon ($z = 3.10$) and monsoon periods ($z = 1.83$), with respective rates of 2.64 mm per annum and 6.1 mm per annum. Conversely, the post-monsoon period exhibited an upward trend ($z = 1.66$) at a rate of 1.11 mm per annum, while the winter seasonal precipitation demonstrated a non-significant declining trend (Fig. 1.5).

The months of January ($z = -0.37$) and February ($z = -0.64$) reflected a decreasing trend in precipitation (Fig. 1.5). In contrast, the months of March ($z = 0.68$),

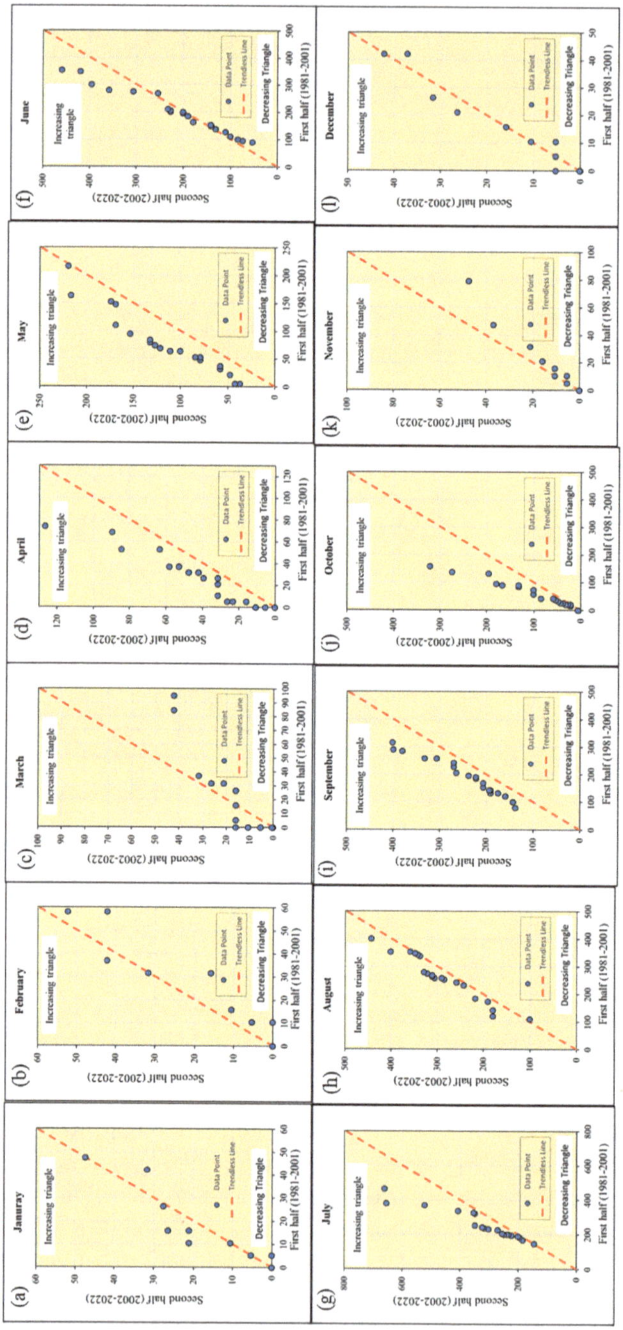

Fig. 1.5 ITA of monthly rainfall across 12-months **(a)** January, **(b)** February, **(c)** March, **(d)** April, **(e)** May, **(f)** June, **(g)** July, **(h)** August, **(i)** September, **(j)** October, **(k)** November, **(l)** December. (Source: NASA Power Data)

Table 1.3 Seasonal and monthly rainfall trends based on Mann–Kendall (MK) test, Sen's slope estimator, and innovative trend analysis (ITA) slope for Bankura District

Season and month	MK (Z value)	Sen's slope (Q)	ITA slope
Annual	2.85	9.42	11.15
Pre-monsoon	3.10	2.64	2.22
Monsoon	1.83	6.10	7.42
Post-monsoon	1.66	1.11	1.66
Winter	0.71	0.00	−0.16
January	−0.37	0.00	0.04
February	−0.64	0.00	−0.21
March	0.68	0.00	−0.16
April	1.14	0.33	0.68
May	2.93	2.10	1.70
June	0.57	0.82	0.73
July	1.87	2.19	3.05
August	1.15	1.58	1.23
September	1.71	1.62	2.41
October	1.93	1.13	1.82
November	−0.76	0.00	−0.16
December	0.34	0.00	0.01

Source: NASA Power Data

April (z = 1.14), May (z = 2.93), June (z = 0.57), July (z = 1.87), August (z = 1.15), September (z = 1.71), October (z = 1.93), November (z = −0.76), and December (z = 0.34) displayed an increasing trend in rainfall within the Bankura district (Table 1.3).

The reliability of the newly implemented ITA methodology is assessed through a comparative analysis of its outcomes against those of MK and mMK. The observed correlation coefficient of 0.99 and R^2 value of 0.98 indicates a strong and significant association between the two methodologies. Consequently, the dependability of these two approaches to trend detection substantiates that ITA is more effective and reliable, as it provides a comprehensive graphical representation of trend analysis across various value ranges. To contextualize the rainfall trends observed in Bankura, we incorporated findings from adjacent districts such as Purulia and Birbhum, which share similar semi-arid characteristics. A 35-year trend analysis in Purulia (1979–2013) revealed a significant shift in rainfall patterns around 1987–88, with contrasting changes between northern and southern sub-regions. The study employed Mann-Kendall and Sen's slope methods to detect seasonal variability and change points, highlighting increased rainfall variability and implications for agricultural planning. Additionally, research on Birbhum and Purulia has documented the socio-economic impacts of irregular rainfall, prolonged dry spells, and rising temperatures. These climatic stressors have led to declining agricultural productivity and increased seasonal migration, particularly among younger populations. The study emphasises the need for climate-resilient infrastructure and inclusive planning to support rural livelihoods.

These comparative insights reinforce the broader climatic instability across the western districts of West Bengal and underscore the urgency of localised adaptation strategies tailored to semi-arid zones. The district exhibits a notable upward trend in annual precipitation during the pre-monsoon and monsoon periods, while the diminished rainfall throughout the winter season adversely impacts the harvest of 'Ravi' crops, resulting in substantial yield losses. Thus, it is evident that Bankura district is more conducive to the cultivation of 'Aus' and 'Aman' crops, whereas the cultivation of 'Rabi' crops without adequate irrigation poses a considerable risk. Similarly, the escalating trend in rainfall during the months of June to September raises concerns regarding the potential for flooding and crop failures. Consequently, agricultural land is experiencing waterlogging, which leads to the deterioration of pre-harvested crops. Such erratic trends warrant significant attention due to their implications for the district's economic stability.

Together, these findings suggest a temporal redistribution of rainfall, where total annual amounts may not drastically decline, but the seasonal and monthly shifts create new agricultural risks. Reduced early monsoon rainfall threatens sowing schedules, while increased late monsoon and post-monsoon rainfall risks crop damage at harvesting. Weak persistence also implies that long-term prediction models must integrate stochastic approaches to improve reliability.

Conclusion

This study provides a detailed assessment of rainfall behaviour in Bankura district, analyzing annual, seasonal, and monthly patterns using ITA, Mann-Kendall variants, and Sen's slope techniques. The consistency across these methods confirms their reliability in detecting significant precipitation trends. Notably, the influence of climate change is evident in the shifting rainfall patterns observed across all temporal scales. As water demand continues to rise—driven by population growth, agricultural intensification, and industrial expansion—it becomes increasingly important to evaluate future water availability across both spatial and temporal dimensions. The insights from this research offer a valuable foundation for designing infrastructure and planning interventions to manage the risks associated with both droughts and floods. To translate these findings into practical action, the following localized strategies are recommended for Bankura:

- Smart Irrigation Planning: Adjust irrigation schedules to align with changing seasonal rainfall, improving water efficiency and reducing stress on groundwater.
- Drought Risk Management: Develop early warning systems and community-level contingency plans based on monthly rainfall variability to support farmers and vulnerable populations.
- Rainwater Harvesting Initiatives: Encourage the construction of decentralized rainwater storage systems, especially in rural and semi-urban areas, to mitigate erratic rainfall impacts.

- Climate-Resilient Farming: Promote drought-resistant crops and agroforestry practices to enhance agricultural resilience.
- Collaborative Water Governance: Foster coordination among local agencies, farmers, and community groups to implement integrated water resource management (IWRM) tailored to Bankura's unique climatic challenges.

By linking scientific analysis with actionable policy measures, this study contributes to the broader goal of climate adaptation and sustainable water management in semi-arid regions like Bankura.

Acknowledgments The authors express their sincere gratitude to Bankura University for providing the necessary academic support and facilities. The first author also extends heartfelt thanks to Dr. Subrata Pan for his valuable guidance and encouragement throughout the course of this research.

Funding This study did not receive any financial support from funding agencies, commercial, or not-for-profit sectors.

Competing Interests The authors declare that they have no known financial or personal conflicts of interest that could have influenced the work reported in this paper.

Informed Consent All participants involved in the study provided informed consent.

Authors' Contributions All authors contributed equally to the conception, analysis, and writing of this manuscript.

Data Availability The datasets generated and/or analyzed during the current study are available from the corresponding author upon reasonable request.

References

Abu Arra A, Alashan S, Şişman E (2025) A new framework for innovative trend analysis: integrating extreme precipitation indices, standardization, enhanced visualization, and novel classification approaches (ITA-NF). Nat Hazards 121:13543–13575. https://doi.org/10.1007/s11069-025-07334-9

Alifujiang Y, Abuduwaili J, Maihemuti B, Emin B, Groll M (2020) Innovative trend analysis of precipitation in the Lake Issyk-Kul Basin, Kyrgyzstan. Atmosphere 11(4):332

Amirthanathan GE, Bari MA, Woldemeskel FM, Tuteja NK, Feikema PM (2023) Regional significance of historical trends and step changes in Australian streamflow. Hydrol Earth Syst Sci 27(1):229–254

Basak A, Das J, Rahman AS, Pham QB (2021) An integrated approach for delineating and characterizing groundwater depletion hotspots in a coastal state of India. J Geol Soc India 97(11):1429–1440

Chakraborty S, Pan S, Pal SC et al. (2025) Evaluating the impact of rainfall patterns on sustainable agricultural crop production. Discov Sustain 6:516. https://doi.org/10.1007/s43621-025-01392-6

Das J, Bhattacharya SK (2018) Trend analysis of long-term climatic parameters in Dinhata of Koch Bihar district, West Bengal. Spat Inf Res 26:271–280

Das J, Mandal T, Rahman AS, Saha P (2021) Spatio-temporal characterization of rainfall in Bangladesh: an innovative trend and discrete wavelet transformation approaches. Theor Appl Climatol 143(3):1557–1579

Deng et al. (2024) Remote Sens
Dogar MM, Kucharski F, & Azharuddin S (2017a) Study of the global and regional climatic impacts of ENSO magnitude using SPEEDY AGCM. Journal of Earth System Science, 126(2):30. https://doi.org/10.1007/s12040-017-0804-4
Dogar MM, Stenchikov D,Osipov S, Wyman B, and Zhao M (2017b), Sensitivity of the regional climate in the Middle East and North Africa to volcanic perturbations, J. Geophys. Res. Atmos., 122, 7922–7948. https://doi.org/10.1002/2017JD026783.
Dogar MM, & Sato T (2018). Analysis of climate trends and leading modes of climate variability for MENA region. Journal of Geophysical Research: Atmospheres, 123, 13,074–13,091. https://doi.org/10.1029/2018JD029003
Dogar MMA (2019) The Sensitivity of DTR to the increased CO2 over Mid-latitude Semi-dry Regions. JOJ Hortic Arboric, 2(2):555584
Dogar MM, & Sato, T. (2019) Regional climate response of Middle Eastern, African, and South Asian monsoon regions to explosive volcanism and ENSO forcing. Journal of Geophysical Research: Atmospheres, 124(14):7580–7598.
Dogar MMA (2020) Study of the regional climatic impacts of tropical explosive volcanism in the Middle East and North Africa region (Doctoral dissertation, Hokkaido University). https://doi.org/10.14943/doctoral.r7091
Dogar MMA & Almazroui M (2022). Revisiting the strong and weak ENSO teleconnection impacts using a high-resolution atmospheric model. Atmospheric Environment, 270, 118866.
Dogar MM, Hermanson L, Scaife AA et al. (2023). A Review of El Niño Southern Oscillation Linkage to Strong Volcanic Eruptions and Post-Volcanic Winter Warming. Earth Syst Environ 7, 15–42. https://doi.org/10.1007/s41748-022-00331-z
Dogar MM, Fujiwara M, Zhao M, Ohba M, and Kosaka Y (2024). ENSO and NAO linkage to strong volcanism and associated post-volcanic high-latitude winter warming. Geophysical Research Letters, 51, e2023GL106114. https://doi.org/10.1029/2023GL106114
Dogar MM, Fujiwara M, Ohba M, and Almazroui M (2025a). Investigating post-eruption amplified winter cooling in the Middle East and North Africa—unraveling ENSO and NAO dynamics. Environmental Research Communications, 7(4), 045012. https://doi.org/10.1088/2515-7620/adc675
Dogar MM, Watanabe S, Fujiwara M et al. (2025b). Tropical volcanic impacts on MENA climate via ENSO and NAO dynamics in a high-top model. npj Clim Atmos Sci 8, 330. https://doi.org/10.1038/s41612-025-01201-x
Gajbhiye S, Meshram C, Singh SK, Srivastava PK, Islam T (2016) Precipitation trend analysis of Sindh River basin, India, from 102-year record (1901–2002). Atmos Sci Lett 17(1):71–77
Ghosh S, Mujumdar PP (2009) Climate change impact assessment: uncertainty modeling with imprecise probability. J Geophys Res Atmos 114(D18)
Guhathakurta P, Rajeevan M, Sikka DR, Tyagi A (2015) Observed changes in southwest monsoon rainfall over India during 1901–2011. Int J Climatol 35(8):1881–1898
Halder P, Dey RK, Mandal S (2023) Long-period trend analysis of annual and seasonal rainfall in West Bengal, India (1901–2020). Theor Appl Climatol 154(1–2):685–703. https://doi.org/10.1007/s00704-023-04577-z
Hamed KH, Rao AR (1998) A modified Mann-Kendall trend test for autocorrelated data. J Hydrol 204(1–4):182–196
Hann CT (1977) Statistical methods in hydrology. Lowa State University Press
Islam MN, van Amstel A (2021) India: climate change impacts, mitigation and adaptation in developing countries. Springer. https://doi.org/10.1007/978-3-030-67865-4
Kendall MG, (1938) A new measure of rank correlation. Biometrika, 30(1–2):81–93. https://doi.org/10.1093/biomet/30.1-2.81
Khan A, Mandal S, Satpati L, Dogar MM, Hasanean H, Salah AM (2025) Analyzing drought: the impact of climate change on precipitation patterns in the Shilabati River basin, West Bengal, India. Earth Syst Environ. https://doi.org/10.1007/s41748-025-00679-y

Kumar R, Gautam HR (2014) Climate change and its impact on agricultural productivity in India. J Climatol Weather Forecast 2(1):1–3

Mandal S, Choudhury BU (2014) Estimation and prediction of maximum daily rainfall at Sagar Island using best fit probability models. Theor Appl Climatol 117:87–97

Mandal S, Choudhury BU, Mondal M, Bej S (2013) Trend analysis of weather variables in Sagar Island, West Bengal, India: a long-term perspective (1982–2010). Curr Sci 105:947–953

Mandal S, Choudhury BU, Satpati LN (2015) Monsoon variability, crop water requirement, and crop planning for kharif rice in Sagar Island, India. Int J Biometeorol 59:1891–1903. https://doi.org/10.1007/s00484-0150995-9

Mandal S, Satpati LN, Choudhury BU, Sadhu S (2018) Climate change vulnerability to the agrarian ecosystem of the small Island: evidence from Sagar Island, India. Theor Appl Climatol 132:451–464. https://doi.org/10.1007/s00704-017-2098-5

Mann HB (1945) Nonparametric tests against trend. Econometrica 13:245–259

Mao et al. (2025) Water Resour Manage

Roxy MK, Ghosh S, Pathak A, Athulya R, Mujumdar M, Murtugudde R et al. (2017) A threefold rise in widespread extreme rain events over Central India. Nat Commun 8(1):1–11

Saha KR, Mooley DA, Saha S (1979) The Indian monsoon and its economic impact. GeoJournal 3:171–178

Sen PK (1968) Estimates of the regression coefficient based on Kendall's tau. J Am Stat Assoc 63(324):1379–1389

Şen Z (2012) Innovative trend analysis methodology. J Hydrol Eng 17(9):1042–1046

Singh H, Jadhav, M. G., Khobragade, A. M., Perke DS, Kumar P (2024) Rainfall trend analysis using Mann-Kendall and Sen's slope test method in Nanded District of Maharashtra State

Subrahmanyam KV, Kumar P, Nelli NR, Anoop S, Ramana MV, Rajeevan M, Chauhan P (2025) Regional shift in the peak time of maximum Indian summer monsoon rainfall in recent decades. Geophys Res Lett 52:e2024GL112697. https://doi.org/10.1029/2024GL112697

Tabari H, Talaee PH (2011) Analysis of trends in temperature data in arid and semi-arid regions of Iran. Glob Planet Chang 79(1–2):1–10

Ummenhofer CC, Seo H, Kwon YO, Parfitt R, Brands S, Joyce TM (2017) Emerging European winter precipitation pattern linked to atmospheric circulation changes over the North Atlantic region in recent decades. Geophys Res Lett 44(16):8557–8566

Von Storch H, Navarra A (eds) (1999) Analysis of climate variability: applications of statistical techniques. Springer Science & Business Media

Yu et al. (2024) Groundwater trend analysis in Japan. Water Resour Manage

Chapter 2
Real-Time Flood Monitoring of the Ganga River Basin Using SAR Data: A Spatial Approach to Climate Risk Mapping

Anupam Pandey, Arun Mondal, and Sananda Kundu

Abstract Synthetic Aperture Radar (SAR) technology is a widely used technique in real-time flood monitoring. The objective of this study is to demarcate the flood-affected area using SAR images. The Doab region of the Ganga-Ramganga River Basin is selected for the study, because this area is experienced several flood events. The study also emphasizes the detection of river flow variations during flood events. The flood event that occurred between August and September 2018 is selected for the study. Satellite images of Sentinel-1 and Sentinel-2 are used in this study to identify the temporal change of flood extent. Whereas, TRIMM (Tropical Rainfall Measuring Mission) precipitation data are employed to assess the impact of precipitation during the flood. Sentinel-1 supports real-time flood monitoring, and Sentinel-2 is used to identify the permanent waterbodies of the basin. 'Modified Normalized Differential Water Index' (MNDWI) and threshold methods are used to distinguish water and non-water features. The results indicate that the peak flood conditions occurred on 2nd September 2018 (28.27 km^2) due to heavy precipitation on 24th August 2018 (140 mm) and 25th August 2018 (55 mm). The result also shows that the river's course is changed during the peak flood time, devastating surrounding areas. The floodwater stays for a long time in places where the river bends or forms an abandoned channel. The results help to attain sustainable goals to make the human and human settlements more resilient, sustainable, inclusive, and safe.

A. Pandey · A. Mondal (✉)
Department of Geography, University of Allahabad, Prayagraj, Imphal, India
e-mail: amondal@allduniv.ac.in

S. Kundu
Department of Geography, Manipur University, Manipur, Imphal, India

S. Mandal et al. (eds.), *Climate Change Vulnerability*, Climate Change Management, https://doi.org/10.1007/978-3-032-16783-5_2

Introduction

Climate change, driven by both natural variability and anthropogenic influences (Dogar et al. 2017a, 2017b, 2019, 2023, 2024, 2025; Dogar and Almazroui 2022), has contributed to changes in the frequency and intensity of many extreme weather and hydrological events, including floods. Among natural disasters worldwide, floods are one of the most common and destructive hazards. It occurs frequently and damages the property and resources over a vast area (Hamidi et al. 2023; Tellman et al. 2021). The severity of floods has increased in the last sixty years due to the changing behavior of climatic parameters and land-use change (Wing et al. 2022; Kundzewicz et al. 2018). Floods and their impacts on human society are very concerning, particularly in South and Southeast Asia, where population density is high, people engage in primary activities, and structural development is unregulated. According to the report of NDMA (2008), India is a highly flood-affected area, where more than 40 mha area is highly vulnerable to floods. As India experiences orographic rainfall, heavy rainfall is very common in the Western Ghats and Himalayan regions (Mondal et al. 2026; Thirumurugan and Krishnaveni 2019; Balasch et al. 2019; Kale et al. 1994). The Ganga-Brahmaputra-Sindh basin has experienced several flood events in the past and the present also. A similarly high flood vulnerability is also observed in Bangladesh, where around 20,000 deaths were reported between 1954 and 2007 (Mirza 2011). Hence, the importance of flood inundation mapping becomes even more significant. It provides valuable support in emergencies, damage assessment, and post-disaster reconstruction (Bhatt and Rao 2016). Not only for disaster-related activities, monitoring surface water is also essential for achieving the UN Sustainable Development Goals (SDGs). It protects food security, water-related ecosystem management, and the interests of the poor and marginal people. In this context, the importance of real-time flood mapping is necessary to monitor the dynamics of the floodwater and take proper measures in an emergency.

The collection of real-time ground information is essential and typically involves multiple processes, including field data acquisition and remote sensing methods (aerial photography and satellite-based observations) (Cian et al. 2018; Yousefi et al. 2018; Burn and Milan 2012). Remote sensing technology is more convenient and flexible than field study, because it offers the advantages of a short time interval, low cost, and large coverage in obtaining valuable information about flood-affected regions. Sometimes flood-affected areas are not physically accessible due to the breakdown of communication networks (the destruction of roads, bridges, and other structures) (Muñoz et al. 2021; Belenguer-Plomer et al. 2019). Weather conditions in the area became highly unstable during the flood, with heavy precipitation and dense cloud cover. In such a situation, optical satellite images cannot obtain the ground information properly, because they can't penetrate the cloud cover (DeVries et al. 2020; Liang and Liu 2020). On the contrary, synthetic aperture radar (SAR) imagery can capture the ground information at all times (day-night) and all climatic conditions (cloudy, rainy, and lightning), because the bands (especially bands C and L) of the SAR sensor can penetrate the climatic obstacles. Hence, SAR images are an important data source for real-time flood monitoring (Tupas et al. 2023; Zhao et al. 2023; Hamidi et al. 2023; Tupas et al. 2023).

For preparing the flood inundation map from SAR images, the method is mainly categorized into two parts, named the post-classification comparison (PCC) method and change detection in multi-temporal SAR data (Zhao et al. 2023; Roo et al. 1999). In the PCC method, categories of ground objects (water & non-water) are analysed in multi-temporal periods. Whereas, categorisation of ground objects is not needed for change detection in the multi-temporal SAR data method. The flooded area is identified directly using multi-temporal data. Sentinel-1, RADARSAT, and ALOS PALSAR are some examples of active sensors. Among the above, Sentinel-1 is widely used by researchers in their work for mapping flooded and inundated areas (Zhao et al. 2023; Tupas et al. 2023; Hamidi et al. 2023). For categorizing flooded and non-flooded areas, several researchers employed the threshold method in their studies. Tupas et al. (2023) used four different methods, namely the normalized difference scattering index (NDSI), Bayesian inference, standardized residuals, and Shannon's entropy of NDSI (SNDSI) to separate water pixels from non-water pixels from the Sentinel-1 dataset. In addition, the polarisation mode of SAR images (HH, VV, HV, and VH) is an important factor in determining the flooded area. Among them, the VH polarisation mode is widely used by researchers due to its high sensitivity to water. McCormack et al. (2022) and Zhuang et al. (2022) conducted flood mapping using VH and VV polarisation modes, and concluded that in the less noisy environment, both modes can provide better results.

In India, several places face flood conditions every year. Bihar is one of the flood zones, where the Kosi River creates a flood-like situation and inundates surrounding areas. Tripathi et al. (2020) analysed the flood in North Bihar during August and September 2017 and reported that approximately 392 km^2 area was submerged. In addition, several other successive floods hit the land of Bihar in 2019, July–September 2020 (Gandak and Koshi River basin), and 2017 (Koshi River basin, north Bihar), etc. Borah et al. (2018) mapped the flood inundation area for the Kaziranga National Park (KNP) during the monsoon of 2017. Katiyar et al. (2021) identified the most flood-affected districts in Kerala during the 2018 flood with the help of SAR images. In northern India, especially in Punjab and Haryana, heavy precipitation devastated huge areas in 2019, and SAR imagery was used by Sahoo et al. (2021) and Tripathi et al. (2021) to assess the damage to croplands, property, and infrastructure from that flood. In the middle and lower Gangetic plain, several studies were done to identify flood-affected areas using SAR images. The analysis of Bajracharya et al. (2007) indicated that the maximum number of floods occurred in the north, south, and peninsular India. SAR imagery is also helpful for mapping the flood-affected area in the cloudburst conditions. For the monitoring of flood conditions in Mumbai (June 2015) (Lal et al. 2020; Gupta and Nair 2011) and Chennai (2005, 2010, and 2015), SAR imagery played an important role in providing updated information about the flood.

However, previous studies on the Ganga river basin focused on the zonation of flooded area due to several hazard-forming factors (Mukherjee and Deb 2024), sustainable use of flooded water through groundwater recharge (Reddy et al. 2020) and also analyse the nature of flood frequency of that area (Mukherjee and Bilas 2019). In the doab region of the Ganga-Ramganga river basin, Agnihotri et al. (2019) observed the changing course of the Ramganga river basin during the flood in 2018 and also mapped the damaged area due to the flood. All the above-mentioned studies indicate that there is a scope for further research. In this study, it is attempted (1) to use the SAR data for flood inundation mapping. Secondly, (2) to understand the extent of flood water during the flood and after flood conditions, and (3) to analyse the change of the river path during flood time.

Study Area

Figure 2.1 shows the location map of the study area, situated in the south-western part of Uttar Pradesh. The total area of the Ganga basin is about 861,404 km^2 and the total length of the river is about 2525 km (Mukherjee and Deb 2024; Singh 2022). The area was chosen to analyse the extent and depth of flood duration in the Ganga River basin, considering the higher frequency of flooding. Geographically, the area is extended from 27°23′36″ N to 27°43′26″ N and from 79°17′33″ E to 79°38′39″ E, covering an area of over 559.32 km^2. The total length of the Ganga river basin, considered in this study, is about 46.56 km long, and spreads over 3 districts of Uttar Pradesh, namely Farrukhabad, Budaun, and Shahjahanpur. It is a part of the large Gangetic plain, where the Ganga and Ramganga rivers deposited silt to form the plain land. The amount of river water discharge and sediment load at the Anikinghat is about 406.73 m^3/s and 15,689.14 metric tons/day from 2009 to 2011, respectively (Khan et al. 2016). Significant movement of channel shifting is observed in its floodplain area. Due to the plain surface area, the region experiences several flood events every year. The region experiences monsoonal climatic conditions, receiving maximum rainfall in the monsoonal season. The region receives precipitation of 90–100 cm/year in the monsoonal season (July–September) of each year (Agnihotri et al. 2019). Whereas dry climatic conditions are observed throughout the rest of the year. Although the yearly average temperature is about 25 °C, the temperature varies from 13 °C to 32 °C (Agnihotri et al. 2019).

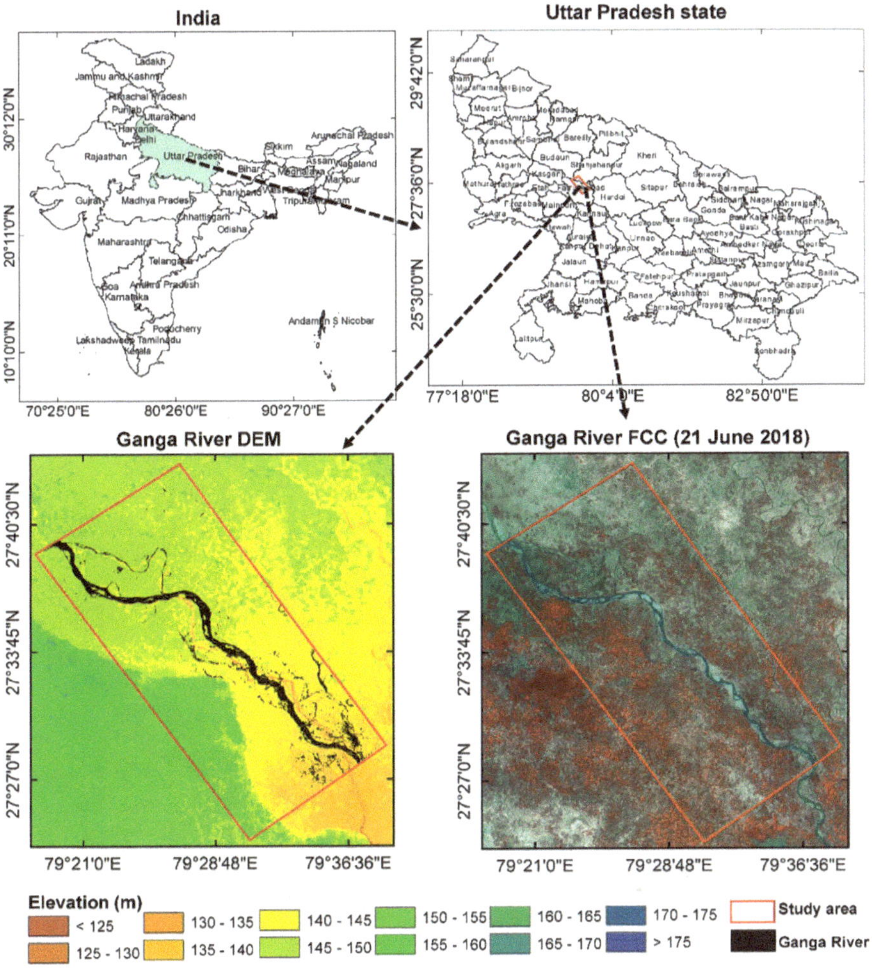

Fig. 2.1 Study area map showing the study region in the Uttar Pradesh state and Ganga River course in the study region over the Shuttle Radar Topography Mission (SRTM) DEM and Sentinel-2 FCC colour image (21 June 2018)

Materials and Methods

Throughout the study, satellite images of Sentinel-1 and Sentinel-2 played an important role in monitoring the flood conditions over time. These datasets were collected from the Sentinel data hub. The information about the dataset is given in Table 2.1. In addition, daily precipitation data were collected from the Tropical Rainfall Measuring Mission (TRMM) from 18th August to 27 September 2018.

Sentinel-1 is a SAR sensor that can be used for flood monitoring and is suitable for collecting real-time flood data, even in cloudy conditions (Borah et al. 2018;

Table 2.1 Data source

Sr. no.	Image ID	Date	Track ID
Sentinel-1 image 1	S1A_IW_GRDH_1SDV_20180809T003608_20180809T003633_023162_02842B_A8C2	09-Aug-2018	165
2	S1A_IW_GRDH_1SDV_20180818T124618_20180818T124647_023301_0288BA_2216	18-Aug-2018	129
3	S1A_IW_GRDH_1SDV_20180821T003609_20180821T003634_023337_0289D5_4EF2	21-Aug-2018	165
4	S1A_IW_GRDH_1SDV_20180830T124619_20180830T124648_023476_028E47_FE33	30-Aug-2018	129
5	S1A_IW_GRDH_1SDV_20180902T003610_20180902T003635_023512_028F64_63D3	02-Sep-2018	165
6	S1A_IW_GRDH_1SDV_20180911T124619_20180911T124648_023651_0293E3_B406	11-Sep-2018	129
7	S1A_IW_GRDH_1SDV_20180914T003610_20180914T003635_023687_029500_54F5	14-Sep-2018	165
8	S1A_IW_GRDH_1SDV_20180923T124619_20180923T124648_023826_029995_DEBC	23-Sep-2018	129
Sentinel-2 image 9	S2A_MSIL1C_20180621T051651_N0206_R062_T44RLR_20180621T081647	21-June-2018	99,999
10	S2B_MSIL1C_20180914T051639_N0206_R062_T44RLR_20180914T091348	14-Sep-2018	99,999

Vishnu et al. 2019). Hence, satellite images of Sentinel-1 from 21st August 2019 to 23rd September 2018 were used to detect the changes in the waterlogging area on different temporal bases. In this process, satellite images of VV polarization were used, because VV polarization provides more accurate results in flood monitoring (Clement et al. 2018; Twele et al. 2016). Therefore, data was processed through several steps, including geometric and spectral noise. Then, the corrected images were used to categorise flooded and non-flooded areas using the threshold method.

In this regard, Sentinel-2 satellite images of pre-flood (21st June 2018) and post-flood (14th September 2018) situations were also collected. The optical image of 21st June 2018 was used to mark it as permanent water. Similarly, the optical images from 14 September 2018 were used as reference data to validate the flood extent derived from the SAR images taken on the same day. However, to separate the water surface area from its surroundings, several algorithms, such as supervised and unsupervised classification, density slicing, and spectral water index, were used for Sentinel-2 satellite images (Rahman and Thakur 2018; Huang et al. 2014; Li et al. 2013). Among all algorithms, the spectral water index is more reliable and user-friendly. Hence, mapping of the flood-affected area was identified using Modified Normalized Differential Water Index (MNDWI). In the pre-processing stage, MNDWI is calculated using the green band (band 3) and the SWIR band (band 11) with a spatial resolution of 10 m and 20 m, respectively. Then, two bands were upscaled to 20 m (Du et al. 2016). The equation for calculating the MNDWI is given below.

$$\mathrm{MNDWI}_{20\mathrm{m}} = \frac{\rho_3^{20\mathrm{m}} - \rho_{11}}{\rho_3^{20\mathrm{m}} + \rho_{11}} \tag{2.1}$$

Where, ρ_{11} indicates the Top of the Atmosphere (TOA) reflection of band 11. $?_3^{20\mathrm{m}}$ is for TOA reflectance of the 20 m upscaled band 3 of the Sentinel-2A.

After the MNDWI classification, the water and non-water areas were separated using the threshold value. The resulting output indicated permanent water bodies before the flood, and was also used to check the result, derived from Sentinel-1 images on 14th September 2018.

However, a total of nine Sentinel-1 satellite images were used to monitor the flood conditions. After subtracting the total permanent waterbodies from the flooded image of Sentinel 1, the actual flooded area was computed. Considering this process, the changes in the flooded area over time were observed. The flood duration map was also prepared using the Sentinel-1 data. Based on the duration of flood water, the extent of flood water was categorized as 1-day, 2 to 5 days, 6 to 10 days, and 11 to 16 days of flood. The framework of the study is given in Fig. 2.2.

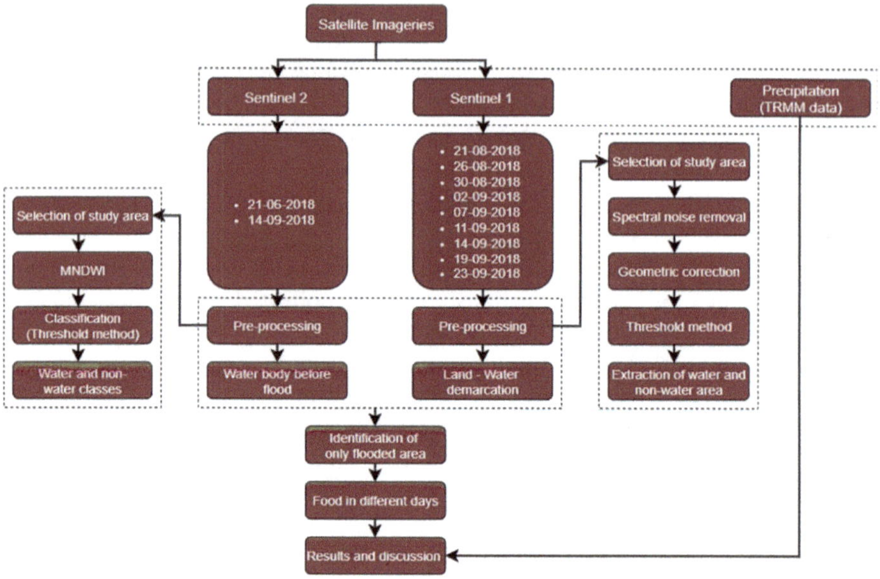

Fig. 2.2 Methodological flow chart of entire study

Table 2.2 Backscatter values of Sentinel-1 images from 21 August 2018 to 23 September 2018

Date	Min	Max	Mean	SD
21 August 2018	−47.01	10.75	−10.47	4.02
26 August 2018	−42.19	11.16	−9.87	4.00
30 August 2018	−43.30	12.20	−10.69	4.14
02 September 2018	−33.46	11.33	−10.43	4.02
07 September 2018	−38.25	13.22	−9.82	4.43
11 September 2018	−33.40	12.13	−10.14	3.27
14 September 2018	−31.22	9.36	−10.58	3.30
19 September 2018	−44.21	10.31	−9.22	3.52
23 September 2018	−30.30	10.68	−10.14	2.71

Result

The analysis and findings of the study is discussed below-.

Analysis of Extent of Flood Using Temporal SAR Images

Sentinel-1 images are very useful in mapping floods due to their capability to capture data in all weather conditions and their uniqueness in having short revisit frequency in the same place (Kundu et al. 2022). The Sentinel-1 images consist of backscatter values, and on the basis of these backscatter values, a threshold method was used to identify the flooded area in the study region. Table 2.2 gives a detailed description of the backscatter values (minimum, maximum, mean, and standard deviation) of Sentinel −1 images used in the present study. The backscatter values are not uniform in all the images. It varies from one image to another. The minimum, maximum, and mean backscatter value of the 21 August 2018 image were −47.01, 10.75, and −10.47, 26 August 2018 were −42.19, 11.16, and −9.87, 30 August 2018 were −43.30, 12.20, and −10.69. On the other hand, the 02 September image had a minimum, maximum, and mean value of −33.46, 11.33, and −10.43, 07 September image had −38.25, 13.22, and −9.82, 11 September image had −33.40, 12.13, and −10.14, 14 September image had −31.22, 9.36, and −10.58, and 19 September image had −44.21, 10.31, and −9.22. In last, the minimum, maximum, and mean value of 23 September 2018 image were −30.30, 10.68, and −10.14. The highest minimum backscatter value was observed on 21 August 2018 (−47.01) and the highest maximum backscatter value was observed on 07 September 2018 (13.22). Meanwhile, the highest mean backscatter value was observed in the 19 September 2018 image (−9.22). The threshold method is a very unique and very useful for determining flood and non-flood areas using Sentinel-1 images. This method uses a fixed value based on the backscatter value to demarcate flood and non-flood areas. Image pixel values that were less than or equal to the threshold limit were identified as pixels with water (value 1) and pixel values that were more than the threshold value were marked as land pixels (value 2). Table 2.3 shows the threshold limit of Sentinel-1 images that were used in this present study to demarcate the land and water areas in the present study region.

Figure 2.3 and Table 2.4 show the variation in flood inundation on different dates (21 August 2018 to 23 September 2018) in the study region. The peak flood was observed on 30 August, 02 September and 07 September 2018, with an actual flooded area of 24.55 km^2 (4.39%), 28.27 km^2 (5.05%), and 21.70 km^2 (3.88%) respectively. Meanwhile, the least area (5.91 km^2) under flood was observed on 19 September 2018. On 14 September 2018, there was an increase in the actual flooded area (1.16 km^2) compared to 11 September 2018 actual flooded area (6.31 km^2). Meanwhile, the actual flooded area on 23 September 2018 was 5.97 km^2 (1.07%).

Table 2.3 Threshold values for water and non-water area demarcation of Sentinel-1 images from 21 August 2018 to 23 September 2018

Date	Min	Max	Threshold for water demarcation
21 August 2018	−47.01	10.75	−18.5
26 August 2018	−42.19	11.16	−17.5
30 August 2018	−43.30	12.20	−17.5
02 September 2018	−33.46	11.33	−16.0
07 September 2018	−38.25	13.22	−18.0
11 September 2018	−33.40	12.13	−18.0
14 September 2018	−31.22	9.36	−17.5
19 September 2018	−44.21	10.31	−16.5
23 September 2018	−30.30	10.68	−16.0

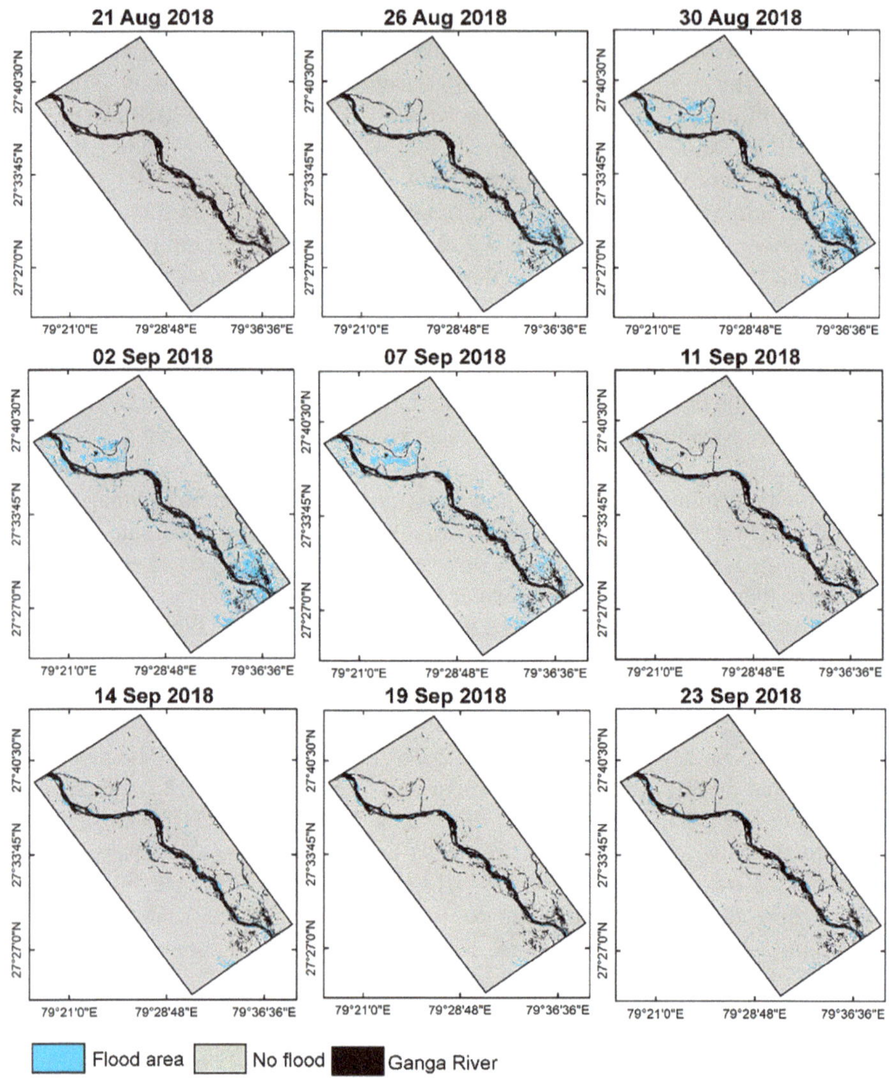

Fig. 2.3 Spatial-temporal distribution of the flooded area in the study region is prepared based on Sentinel-1 images (21 August 2018–23 September 2018). The Ganga River course is represented by black colour, and the flooded area is shown by blue colour

Figure 2.4 displays the temporal variations of flood water on 21 June 2018 and 14 September 2018. The Sentinel-2 images were used for the mentioned dates to calculate MNDWI and identify areas under flood. The flood water and non-flood water regions were extracted using the threshold method on MNDWI. The minimum MNDWI value for 21 June 2018 was −0.68, and the maximum MNDWI value

Table 2.4 Actual flooded area excluding permanent water body from 21 August 2018 to 23 September 2018

Date	Actual flooded area (sq. km)
21 August 2018	0
26 August 2018	15.18
30 August 2018	24.55
02 September 2018	28.27
07 September 2018	21.70
11 September 2018	6.31
14 September 2018	7.47
19 September 2018	5.91
23 September 2018	5.97

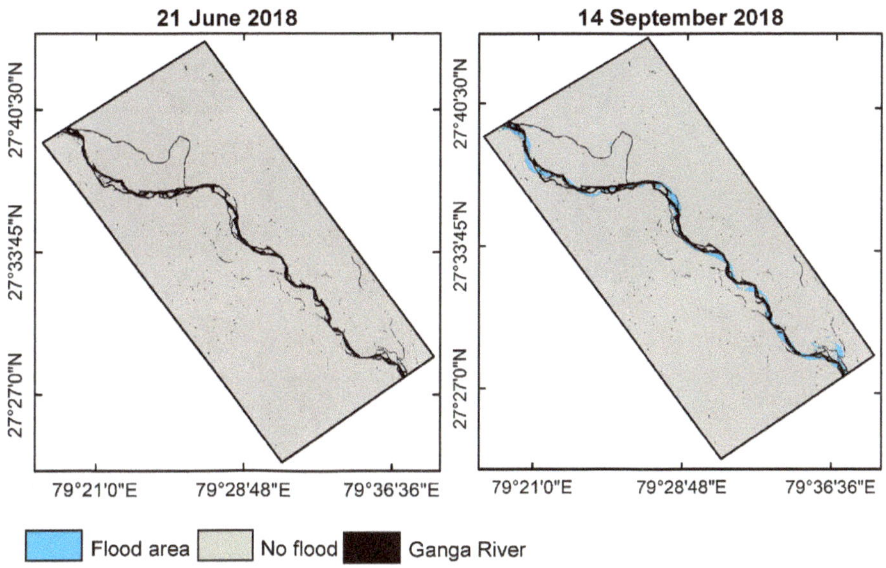

Fig. 2.4 Spatio-temporal distribution of flood area based on Sentinel-2 data of 21 June 2018 and 14 September 2018 (Blue colour representing flood inundated area)

for the same date was 0.23. A threshold of −0.15 was used to demarcate water and non-water areas on 21 June 2018. On the other hand, the minimum MNDWI value for 14 September 2018 image was −0.50 and maximum MNDWI value for same date was 0.63. A threshold value of +0.50 was used on this image to demarcate water and non-water areas (Table 2.5). There was no flood on 21 June 2018, whereas on 14 September 2018 flood water spread along the course of the Ganga River in the study area and small patches of flood water were also observed in the north-eastern, south, and south-eastern parts of the study region with a flooded area of 7.87 km^2 (Table 2.6).

Figure 2.5 shows the course of the Ganga River in the study area on pre-flood (21 June 2018) and post-flood day (14 September 2018) using Sentinel-2 False Color

Table 2.5 Threshold values for water and non-water area demarcation of Sentinel-2 images of 21 June 2018 and 14 September 2018

Date	MNDWI min	MNDWI max	Threshold for water demarcation
21 June 2018	−0.68	0.23	−0.15
14 September 2018	−0.50	0.63	+0.50

Table 2.6 Actual flooded area excluding permanent water body of 21 June 2018 and 14 September 2018 using Sentinel-2 images

Date	Actual flooded area (km^2)
21 June 2018	0
14 September 2018	7.87

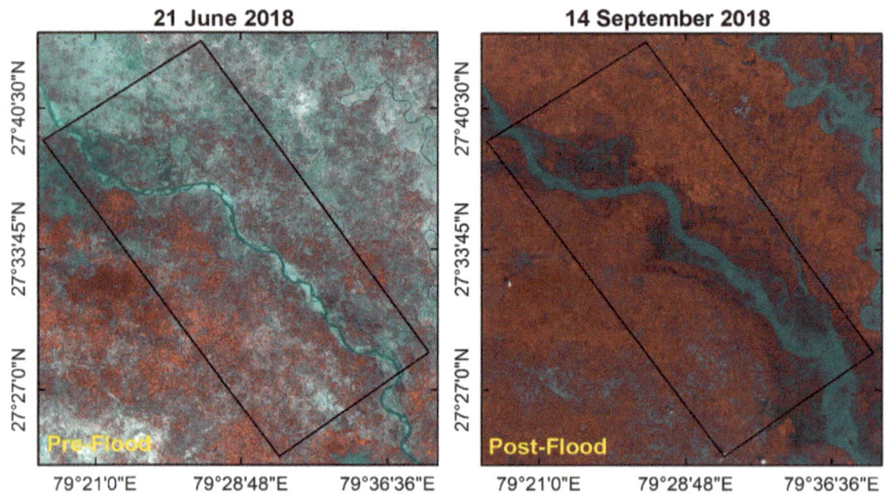

Fig. 2.5 Sentinel-2 images (21 June 2018—Pre-flood and 14 September 2018—Post-flood) of the Ganga River

Composite (FCC) images. From the images, it's clearly visible that on 14 September 2018, the width of the river was expanded greatly compared to the river of 21 June 2018. The areas close to the Ganga River covered with agricultural land vegetation were inundated on 14 September 2018.

Figure 2.6 and Table 2.7 display the duration of the flood in the study region. The duration of the flood in the study region ranges from 1 day to 17–22 days. Most of the areas in the study region remained underwater for 1 day duration (22.97 km^2) and 6–10 days duration (13.62 km^2). Whereas, very less area (0.80 km^2) remained underwater for 17–22 days. Areas that remained underwater for 1 day duration were primarily observed away from the main river channel. On the other hand, areas that remained underwater for 17–22 days were mainly observed along the main channel of the river.

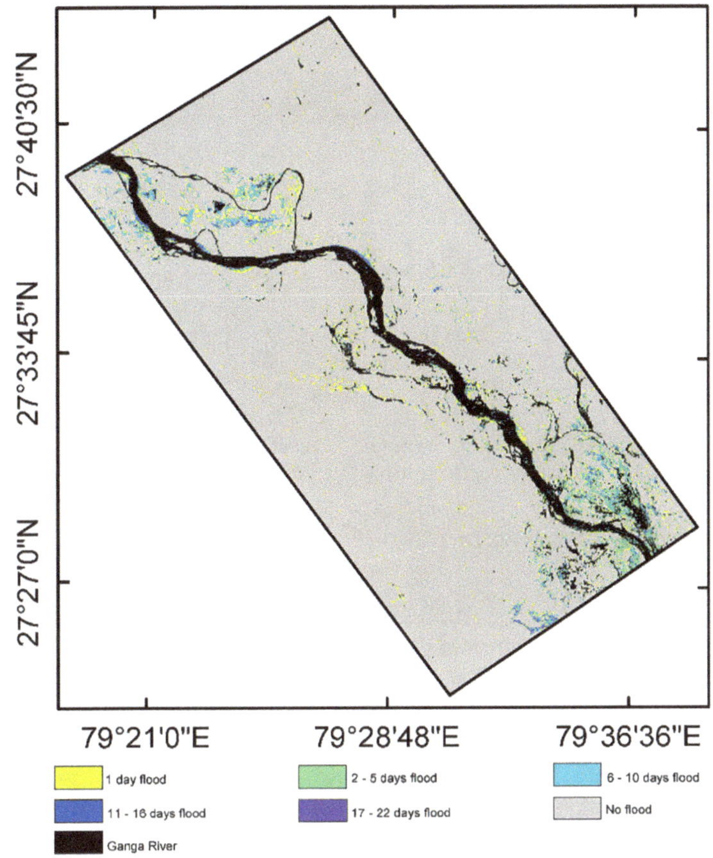

Fig. 2.6 Duration (1 day, 2–5 days, 6–10 days, 11–16 days, and 17–22 days) of flood in study region from 21 August 2018 to 23 September 2018

Table 2.7 Actual flooded area excluding permanent water body in different days

No. of days of flood	Actual flooded area (km^2)
1 day water	22.97
2–5 days water	6.68
6 to 10 days water	13.62
11–16 days water	3.10
17–22 days water	0.80
No flood	476.16

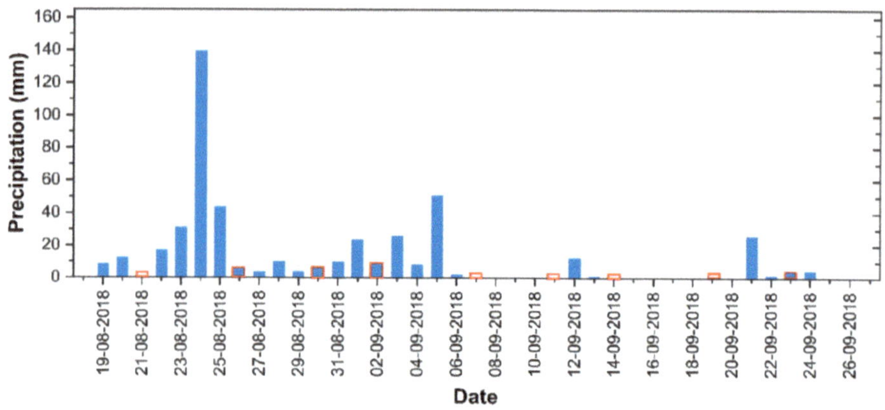

Fig. 2.7 Amount of daily precipitation in the Ganga River from 18 August 2018 to 27 September 2018. The red mark indicates flood extents using Sentinel-1 data

Meteorological Observation

Figure 2.7 displays the daily rainfall in the study region from 18 August 2018 to 27 September 2018. The highest rainfall (139.63 mm) in the study area occurred on 24 August 2018. The majority of the rainfall in the study region occurred from 22 August 2018 to 5 September 2018, and in this period more of the dates experienced rainfall >5 mm. The only exception to this was 27 and 29 August 2018, when the rainfall amounts were 3.28 mm and 3.79 mm. This was the period when the study region experienced major flooding along the course of the Ganga River due to the rise in water in the river channel.

Figure 2.8 shows the distribution of daily rainfall over the Ganga River and its adjacent areas from 21 August 2018 to 23 September 2018. The rainfall in the study region varied from 0 to >15 mm. On most of the dates, rainfall mainly varied from 0 to 3 mm. The only exceptions to this were 26 August 2018, 30 August 2018, 02 September 2018, and 23 September 2018. From 30 August 2018 to 02 September 2018, the major portion of the study region experienced rainfall >15 mm and notably this was the reason that during this time period the region experienced the flood event. Meanwhile, on 23 September the study region also experienced rainfall over >15 mm in north-eastern portions of the study area.

Conclusion

In this study, an attempt is made to identify the flood-inundated area during the flood that occurred in August–September 2018 using SAR images. As the basin is located in the plain land of the Daob region (Ganga-Ramganga River), it experiences severe consequences of floods every year. This study provides detailed

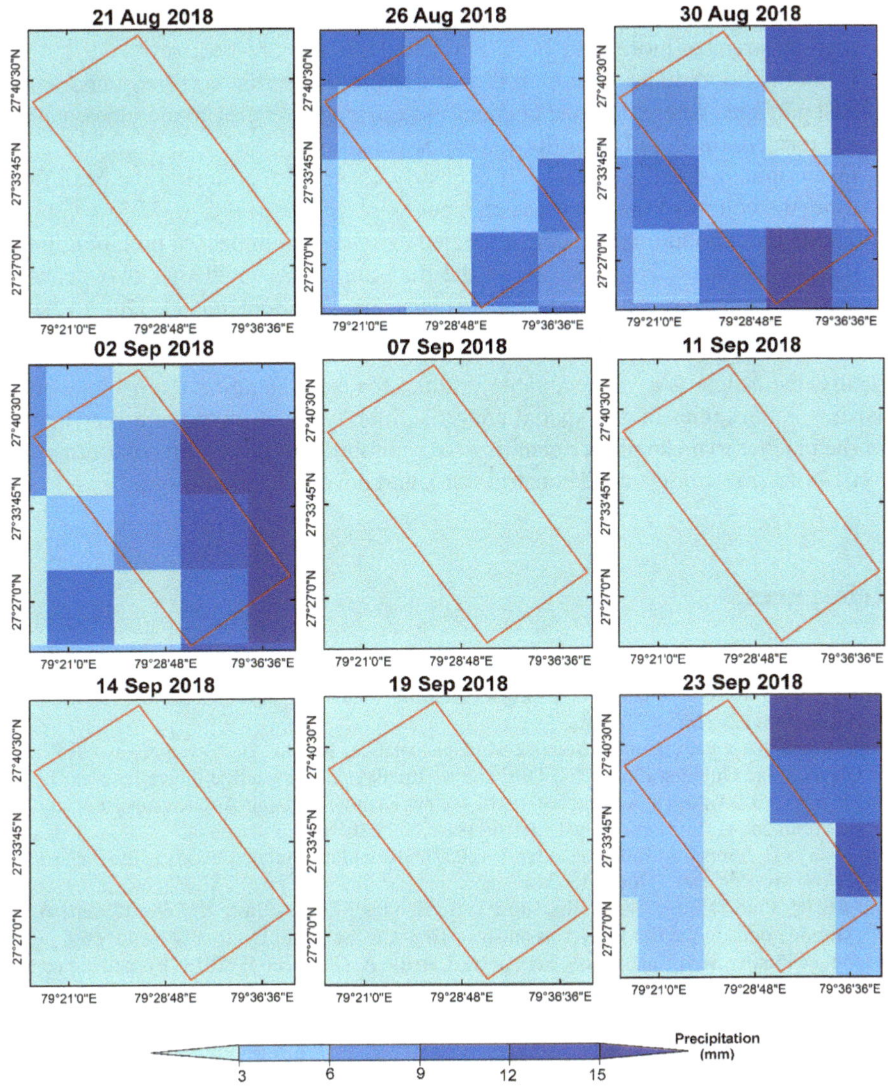

Fig. 2.8 Spatio-temporal distribution of daily precipitation over the Ganga River and adjacent areas from TRMM observations (21 August–23 September 2018)

information regarding the pre-during-post flood conditions of the study area. Some important observations of the study are discussed below.

1. The threshold value of SAR images is quite low to separate water and non-water areas, and it ranges between −18.5 and −16 dB in VV polarization.
2. The flood happened after 21st August. Later, due to heavy precipitation and other external factors, the flood was at its peak on 2nd September, devastating

the area of 28.28 km^2. On 30th August and 23rd September, flooded water covered the areas of more than 24.55 km^2 and 21.70 km^2, respectively.
3. From the flood duration map, it is also notable that in the north-east and south-east portions, where the river changes its course and several abandoned channels are formed, the duration of flood water was higher.
4. From optical satellite imagery, it is clearly observed that during the peak flood time, the course of the river was changed.
5. TRMM precipitation data were collected to assess the impact of precipitation on the extension of the flood. From the diagram, it is stated that two peaks of precipitation were observed (24th August and 5th September). Due to heavy precipitation, the flooded area suddenly increased to 24.55 km^2.

Finally, the article is a very valuable product for understanding the use of remote sensing technologies in geo-spatial analysis. Such types of images are very helpful for the planner to make proper planning for economic (crop type, crop pattern selection), structural (bridge, road construction), and development purposes.

References

Agnihotri AK, Ohri A, Gaur S, Shivam, Das N, Mishra S (2019) Flood inundation mapping and monitoring using SAR data and its impact on Ramganga River in Ganga basin. Environ Monit Assess 191(12):760

Arun, Mondal Anupam, Pandey Souvik, Mondal Sananda, Kundu (2026) Impact assessment of Glacial Lake Outburst Floods (GLOFs) in the Himalaya using satellite imageries over the last 25 years (2000–2024): A comprehensive review Remote Sensing Applications: Society and Environment 41. https://doi.org/10.1016/j.rsase.2025.101845

Bajracharya B, Shrestha AB, Rajbhandari L (2007) Glacial lake outburst floods in the Sagarmatha region. Mt. Res. Dev. 27(4):336–344

Balasch JC, Pino D, Ruiz-Bellet JL, Tuset J, Barriendos M, Castelltort X, Peña JC (2019) The extreme floods in the Ebro River basin since 1600 CE. Sci Total Environ 646:645–660

Belenguer-Plomer MA, Tanase MA, Fernandez-Carrillo A, Chuvieco E (2019) Burned area detection and mapping using Sentinel-1 backscatter coefficient and thermal anomalies. Remote Sens Environ 233:111345

Bhatt CM, Rao GS (2016) Ganga floods of 2010 in Uttar Pradesh, North India: a perspective analysis using satellite remote sensing data. Geomat Nat Haz Risk 7(2):747–763

Borah SB, Sivasankar T, Ramya MNS, Raju PLN (2018) Flood inundation mapping and monitoring in Kaziranga National Park, Assam using Sentinel-1 SAR data. Environ Monit Assess 190(9):520

Cian F, Marconcini M, Ceccato P (2018) Normalized difference flood index for rapid flood mapping: taking advantage of EO big data. Remote Sens Environ 209:712–730

Clement MA, Kilsby CG, Moore P (2018) Multi-temporal synthetic aperture radar flood mapping using change detection. J Flood Risk Manag 11(2):152–168

De Roo A, Van Der Knijff J, Horritt M, Schmuck G, De Jong S (1999, August) Assessing flood damages of the 1997 Oder flood and the 1995 Meuse flood. In: Proceedings of the Second International ITC Symposium on Operationalization of Remote Sensing, Enschede. The Netherlands, pp 16–20

DeVries B, Huang C, Armston J, Huang W, Jones JW, Lang MW (2020) Rapid and robust monitoring of flood events using Sentinel-1 and Landsat data on the Google Earth Engine. Remote Sens Environ 240:111664

Dogar MM, Kucharski F, Azharuddin S (2017a). Study of the global and regional climatic impacts of ENSO magnitude using SPEEDY AGCM. J Earth Syst Sci 126, 30. https://doi.org/10.1007/s12040-017-0804-4

Dogar MM, Stenchikov G, Osipov S, Wyman B, Zhao M (2017b). Sensitivity of the regional climate in the Middle East and North Africa to volcanic perturbations, J. Geophys. Res. Atmos., 122, 7922–7948, https://doi.org/10.1002/2017JD026783.

Dogar MM, Hermanson L, Scaife AA, et al. (2023). A Review of El Niño Southern Oscillation Linkage to Strong Volcanic Eruptions and Post-Volcanic Winter Warming. Earth Syst Environ 7, 15–42. https://doi.org/10.1007/s41748-022-00331-z

Dogar MMA, Almazroui M (2022). Revisiting the strong and weak ENSO teleconnection impacts using a high-resolution atmospheric model. Atmospheric Environment, 270, 118866. https://doi.org/10.1016/j.atmosenv.2021.118866

Dogar MM, Fujiwara M, Zhao M, Ohba M, Kosaka Y (2024). ENSO and NAO linkage to strong volcanism and associated post-volcanic high-latitude winter warming. Geophysical Research Letters, 51, e2023GL106114. https://doi.org/10.1029/2023GL106114

Dogar MM, Kucharsk, F, Sato T, Mehmood S, Ali S, Gong Z, ... & Arraut J (2019). Towards understanding the global and regional climatic impacts of Modoki magnitude. Global and Planetary Change, 172, 223–241. https://doi.org/10.1016/j.gloplacha.2018.10.004

Dogar MM, Watanabe S, Fujiwara M, et al. (2025). Tropical volcanic impacts on MENA climate via ENSO and NAO dynamics in a high-top model. npj Clim Atmos Sci 8, 330. https://doi.org/10.1038/s41612-025-01201-x

Du Y, Zhang Y, Ling F, Wang Q, Li W, Li X (2016) Water bodies' mapping from Sentinel-2 imagery with modified normalized difference water index at 10-m spatial resolution produced by sharpening the SWIR band. Remote Sens 8(4):354

Gupta AK, Nair SS (2011) Urban floods in Bangalore and Chennai: risk management challenges and lessons for sustainable urban ecology. Curr Sci:1638–1645

Hamidi E, Peter BG, Muñoz DF, Moftakhari H, Moradkhani H (2023) Fast flood extent monitoring with SAR change detection using google earth engine. IEEE Trans Geosci Remote Sens 61:1–19

Huang C, Chen Y, Wu J (2014) Mapping spatio-temporal flood inundation dynamics at large river basin scale using time-series flow data and MODIS imagery. Int J Appl Earth Obs Geoinf 26:350–362

Kale VS, Ely LL, Enzel Y, Baker VR (1994) Geomorphic and hydrologic aspects of monsoon floods on the Narmada and Tapi Rivers in central India. In: Geomorphology and natural hazards. Elsevier, pp 157–168

Katiyar V, Tamkuan N, Nagai M (2021) Near-real-time flood mapping using off-the-shelf models with SAR imagery and deep learning. Remote Sens 13(12):2334

Khan MYA, Daityari S, Chakrapani GJ (2016) Factors responsible for temporal and spatial variations in water and sediment discharge in Ramganga River, Ganga Basin, India. Environ Earth Sci 75(4):283

Kundu S, Lakshmi V, Torres R (2022) Flood depth estimation during hurricane Harvey using Sentinel-1 and UAVSAR data. Remote Sens 14(6):1450. https://doi.org/10.3390/rs14061450

Kundzewicz ZW, Pińskwar I, Brakenridge GR (2018) Changes in river flood hazard in Europe: a review. Hydrol Res 49(2):294–302

Lal P, Prakash A, Kumar A, Srivastava PK, Saikia P, Pandey AC et al. (2020) Evaluating the 2018 extreme flood hazard events in Kerala, India. Remote Sens Lett 11(5):436–445

Li W, Du Z, Ling F, Zhou D, Wang H, Gui Y et al. (2013) A comparison of land surface water mapping using the normalized difference water index from TM, ETM+ and ALI. Remote Sens 5(11):5530–5549

Liang J, Liu D (2020) A local thresholding approach to flood water delineation using Sentinel-1 SAR imagery. ISPRS J Photogramm Remote Sens 159:53–62
McCormack T, Campanyà J, Naughton O (2022) A methodology for mapping annual flood extent using multi-temporal Sentinel-1 imagery. Remote Sens Environ 282:113273
Milan DJ (2012) Geomorphic impact and system recovery following an extreme flood in an upland stream: Thinhope Burn, Northern England, UK. Geomorphology 138(1):319–328
Mirza MMQ (2011) Climate change, flooding in South Asia and implications. Reg Environ Chang 11(Suppl 1):95–107
Mukherjee R, Bilas R (2019) Flood frequency analysis of Ramganga river basin in western Gangetic Plain, India. Natl Geogr J India 35(3):286–299
Mukherjee R, Deb P (2024) Application of GIS-based analytical hierarchy process for assessment and mapping of flood risk zone in the lower Ramganga River Basin, Western Gangetic Plain, India. Environ Dev Sustain 26(3):6163–6193
Muñoz DF, Muñoz P, Moftakhari H, Moradkhani H (2021) From local to regional compound flood mapping with deep learning and data fusion techniques. Sci Total Environ 782:146927
NDMA (2008) National Disaster Management Guidelines: Management of Floods. National Disaster Management Authority, Ministry of Home Affairs, Government of India, New Delhi.
Rahman MR, Thakur PK (2018) Detecting, mapping and analysing of flood water propagation using synthetic aperture radar (SAR) satellite data and GIS: a case study from the Kendrapara District of Orissa State of India. Egypt J Remote Sens Space Sci 21:S37–S41
Reddy VR, Rout SK, Shalsi S, Pavelic P, Ross A (2020) Managing underground transfer of floods for irrigation: a case study from the Ramganga basin, India. J Hydrol 583:124518
Sahoo S, Arora M, Pateriya B (2021) Mapping of flood inundated area and impact assessment using Sar data over Ghaggar river basin of Punjab, India
Singh IB (2022) The Ganga River. Large Rivers: Geomorphology and Management, 2nd edn, pp 521–550
Tellman B, Sullivan JA, Kuhn C, Kettner AJ, Doyle CS, Brakenridge GR et al. (2021) Satellite imaging reveals increased proportion of population exposed to floods. Nature 596(7870):80–86
Thirumurugan P, Krishnaveni M (2019) Flood hazard mapping using geospatial techniques and satellite images—a case study of coastal district of Tamil Nadu. Environ Monit Assess 191(3):193
Tripathi A, Attri L, Tiwari RK (2021) Spaceborne C-band SAR remote sensing–based flood mapping and runoff estimation for 2019 flood scenario in Rupnagar, Punjab, India. Environ Monit Assess 193(3):110
Tripathi G, Pandey AC, Parida BR, Kumar A (2020) Flood inundation mapping and impact assessment using multi-temporal optical and SAR satellite data: a case study of 2017 flood in Darbhanga district, Bihar, India. Water Resour Manage 34(6):1871–1892
Tupas ME, Roth F, Bauer-Marschallinger B, Wagner W (2023) An intercomparison of Sentinel-1 based change detection algorithms for flood mapping. Remote Sens 15(5):1200
Twele A, Cao W, Plank S, Martinis S (2016) Sentinel-1-based flood mapping: a fully automated processing chain. Int J Remote Sens 37(13):2990–3004
Vishnu CL, Sajinkumar KS, Oommen T, Coffman RA, Thrivikramji KP, Rani VR, Keerthy S (2019) Satellite-based assessment of the august 2018 flood in parts of Kerala, India. Geomat Nat Hazards Risk 10:758–767
Wing OE, Lehman W, Bates PD, Sampson CC, Quinn N, Smith AM et al. (2022) Inequitable patterns of US flood risk in the anthropocene. Nat Clim Chang 12(2):156–162
Yousefi S, Mirzaee S, Keesstra S, Surian N, Pourghasemi HR, Zakizadeh HR, Tabibian S (2018) Effects of an extreme flood on river morphology (case study: Karoon River, Iran). Geomorphology 304:30–39
Zhao B, Sui H, Liu J (2023) Siam-DWENet: flood inundation detection for SAR imagery using a cross-task transfer siamese network. Int J Appl Earth Obs Geoinf 116:103132
Zhuang H, Fan H, Deng K, Zhang K, Wang X, Wang M (2022) Change detection in SAR images based on progressive nonlocal theory. IEEE Trans Geosci Remote Sens 60:1–13

Part II
Regional Climate Change Impacts on Rural Livelihood: Case Studies from South Asia and Beyond

Chapter 3
Geo-Spatial Exploration of Climate Change Vulnerability: Impacts on Agriculture and Livelihoods in Kapasia Upazila of Gazipur District, Bangladesh

Hossain Al Mahbub, Mst Sanjida Alam, Muhammad Muhibbullah, Md. Iqbal Sarwar, Md. Abu Bakkar, Abid Hassan, and Mijanur Rahman

Abstract A major worldwide concern, climate change disproportionately impacts vulnerable areas like Bangladesh. Climate variability has progressively harmed local livelihoods and agricultural systems in Kapasia Upazila of Gazipur, Bangladesh. Assessing climate change vulnerability from 2010 to 2024 and analyzing its effects on agriculture (2014 to 2024) and livelihoods in the study region were the goals of the current study. A mixed-method approach has been used to incorporate both socioeconomic and spatial data. Primary data was collected through 248 household questionnaires, 10 focus group discussions, and 5 experts' interviews. To monitor environmental changes, spatial data, including temperature, precipitation, land use, and land cover (LULC), and normalized difference vegetation index (NDVI) maps, were created and analyzed using ArcGIS 10.8. The results showed that a slow increase in temperature and erratic precipitation patterns made the region more vulnerable to droughts, floods, and seasonal instability. While LULC results showed significant loss of cropland and green cover along with urban expansion, NDVI analysis revealed deteriorating vegetation health. Reduced crop yields, changed cropping patterns, increased soil degradation, and an increase in pest incidence were all consequences of these environmental changes. According to household surveys, there were notable impacts on rainfall predictability (69%), soil quality (59%), and food security (85%). Reduced water availability, biodiversity loss, and deteriorating ecosystem services all had an impact on livelihoods. The study came to the conclusion that livelihoods and agriculture had suffered as a result of climate change vulnerability. The findings offered community-based and spatially explicit evidence to guide resource allocation for resilience building in

H. A. Mahbub (✉) · M. Muhibbullah · M. I. Sarwar · M. A. Bakkar · A. Hassan · M. Rahman
Department of Geography and Environmental Studies, University of Chittagong, Chattogram, Bangladesh
e-mail: almahbub.ges@cu.ac.bd

M. S. Alam
Department of Sustainability and Social Justice, Clark University, Worcester, MA, USA

S. Mandal et al. (eds.), *Climate Change Vulnerability*, Climate Change Management, https://doi.org/10.1007/978-3-032-16783-5_3

comparable high-risk areas, targeted climate adaptation policies, and sustainable land management techniques.

Abbreviations

AOI	Area of interest
APA	American Psychological Association
FGDs	Focus Group Discussions
GOs	Government Organizations
LULC	Land use and land cover
NDVI	Normalized difference vegetation index
NGOs	Non-Government Organizations
SPSS	Statistical Package for the Social Sciences
UNO	Upazila *Nirbahi* Officer

Introduction

Bangladesh is a resourceful and densely populated country that has been experiencing frequent disasters, which have increased significantly in recent decades due to climate change and global warming (Hossain et al. 2025; Dastagir 2015; Dogar et al. 2017, 2019; Dogar and Almazroui 2022). According to Dey (2023), extreme weather events like heatwaves, droughts, cyclones, and floods frequently affect the country, causing substantial impacts on biodiversity, agriculture and socioeconomic systems. Climate change's effects on agriculture are a global concern (Mandal et al. 2015, 2018; Dey et al. 2025), but for Bangladesh, where agriculture is the foundation of lives and livelihoods, it poses a danger to the country's food security (Bosello and Zhang 2005). Gazipur is seeing alarming rates of urbanization as more people move into metropolitan areas, which greatly contributes to the metropolitan Heat Island (UHI) phenomenon (Shima and Hasan 2024). With forecasts showing an increase in the frequency and severity of extreme weather events, Bangladesh has been named one of the nations most vulnerable to climate change by the Intergovernmental Panel on Climate Change (IPCC) (2021). According to studies by Hossain et al. (2023), the nation's high population density, reliance on agriculture, and little capability for adaptation make it even more vulnerable. Additionally, the nation's deltaic terrain renders it especially vulnerable to saline intrusion and sea level rise, which impacts both agriculture and biodiversity (Morshed et al. 2025).

Extreme weather events have become more frequent and intense in Bangladesh due to substantial temperature increases and changes in precipitation patterns over the past few decades (Rozario et al. 2024). The nation is among the most climate-vulnerable in the world, with 185 extreme weather events recorded between 2000 and 2019 (ICCCAD 2024). The problems already facing agriculture and biodiversity protection have been made worse by these climate shifts. According to recent studies, Bangladesh's coastal storm tide threats have increased due to an increase in the frequency of tropical cyclones and storm surges (Qiu et al. 2025). Furthermore, it has been noted that sea level rise has caused saline intrusion into freshwater systems, which affects agricultural and aquatic ecosystems (Farha et al. 2024). Climate change has a significant impact on Bangladesh's economy, which is based primarily on agriculture. Bangladesh is a country mostly focused on an agri-economy (Hoque et al. 2019). In recent years, cropping patterns have changed, and agricultural yields have declined (Afjal Hossain et al. 2012). The topography, water availability, and subtropical climate of Bangladesh make it a great place to grow rice. Currently, varieties created by the Bangladesh Rice Research Institute (BRRI) are grown on 80% of the nation's rice-growing territory (Shapla et al. 2015).

Bangladesh is susceptible to more than just the effects of climate change. With one of the greatest population densities in the world (950 individuals per sq. km.) and an increasing population (139 million in 2011), a significant section of its population (about 43%) lives in poverty, and it faces significant challenges in providing for its people (Lázár et al. 2015). According to studies, crop yields have decreased as a result of rising temperatures and unpredictable rainfall patterns, particularly in the production of vegetables and rice. It was discovered that a 4 °C temperature increase will significantly affect the production of food grains, particularly wheat (Karim et al. 1999). Droughts and floods have become more frequent, causing nutrient depletion and soil erosion, reducing agricultural production even further. Because of the interruption of traditional farming practices caused by these climate pressures, rural residents in Kapasia are experiencing economic insecurity. In 2024, floods in Bangladesh ruined around 1.1 million metric tons of rice (Urvashi and Sergio 2010).

Habitat loss and ecological degradation brought on by climate change pose a threat to Bangladesh's biodiversity. Freshwater fish populations are declining, and plant and animal species are disappearing as a result of altered hydrological cycles and rising salinity (Islam 2022). Deforestation and water shortages in areas like Kapasia have made biodiversity declines worse, impacting both ecological balance and livelihoods that rely on natural resources. Communities that depend on these systems are more vulnerable as a result of climate change-induced ecological deterioration (Rozario et al. 2024). Furthermore, ecological services that are essential to human well-being, such as pollination and water purification, are impacted by biodiversity loss (Islam et al. 2021). Numerous adaptation techniques have been put forth and put into practice to address these issues. Resilience can be increased by community-based strategies, including using sustainable farming methods and climate-resilient crop types (UNDP 2024). Furthermore, programs that emphasize biodiversity conservation education, water resource management, and reforestation are essential for reducing the negative consequences of climate change (Uddin and Hoque 2025).

Several studies in Bangladesh have investigated the effects of climate change on agriculture and biodiversity. The majority of existing research focuses on national or coastal contexts, failing to consider the links between agriculture and biodiversity at the local level. Furthermore, empirical data do not support the effectiveness of adaption strategies or the disparity between national climate policies and their local implementation in Kapasia Upazila of Gazipur district in Bangladesh are not well supported by empirical data.

Aims and Objectives

The purpose of this study is to fill in the knowledge gaps regarding the socioeconomic and spatial effects of climate change in the region under investigation. The particular aims and objectives are:

(a) To use spatial analysis of temperature and precipitation to evaluate the vulnerability of climate change during the last 14 years (2010–2024). In order to capture recent trends in climate variability over a long enough time span for accurate analysis, the year 2010 was chosen as the starting point.
(b) To assess how vulnerability to climate change has affected agriculture over the past 10 years (2014–2024), with an emphasis on changes in land use and land cover (LULC), vegetation health (NDVI), and farmer-reported experiences. The 2014–2024 timeframe was selected because it aligned with reliable agricultural records and high-resolution satellite data that were available.
(c) To investigate how climate change vulnerability affects livelihoods, including biodiversity loss, resource availability, and ecosystem services, using survey data.

Materials and Methods

Study Area

Gazipur district is in the middle part of Bangladesh. It is a very geographically important area for Bangladesh. Gazipur is famous for agricultural production. Rice, the staple food of Bangladesh, is abundant in this district. Besides rice, jute, seasonal fruits, vegetables, livestock farming, and so many other agricultural activities are seen in the area. But recently, agricultural production and livelihoods have been greatly hampered by climate vulnerability. There are 5 (five) Upazila in the Gazipur District. Kapasia Upazila is one of them. This has been chosen as the study area of this present study (Fig. 3.1). The study area was selected because of its unique topography, biology, and socioeconomic characteristics that make it particularly vulnerable to the consequences of climate sensitivity. The region of Kapasia, which is situated between rural and urban areas, is vulnerable to a variety of climate-related stressors, including rising temperatures, seasonal floods, droughts, and erratic rainfall. The region's rich biodiversity is at risk due to these climate

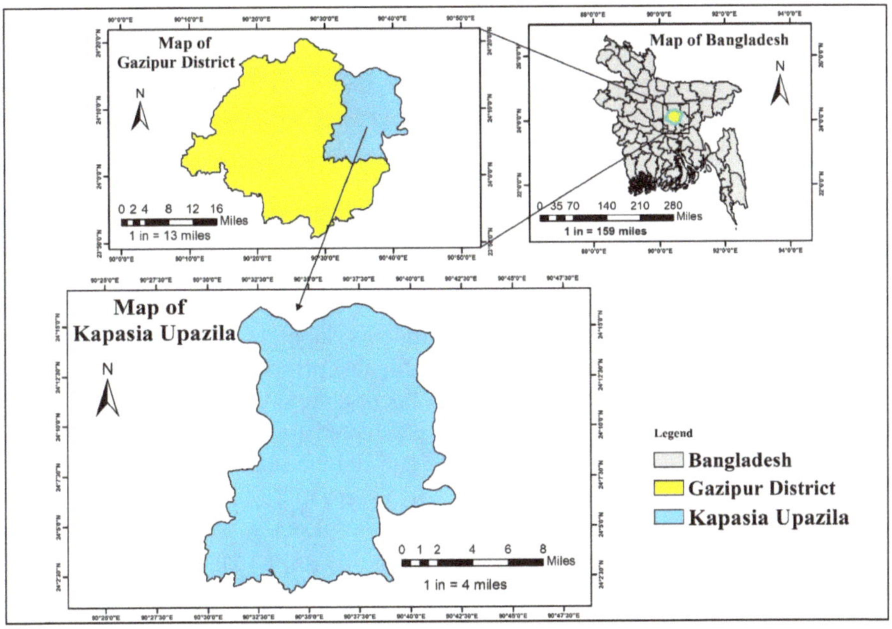

Fig. 3.1 Map of the study area

disruptions, which also have a direct impact on the agriculture-based economy, especially in wetlands, woodland patches, and agricultural areas. The region's reliance on natural resources for livelihoods and its comparatively low capacity for adaptation further emphasize how urgent it is to conduct a thorough assessment of vulnerabilities and impacts. Understanding how climate vulnerability affects rural populations in the peri-urban area of Bangladesh and putting forward well-informed resilience-building measures are two opportunities presented by studying about Kapasia Upazila of Gazipur District in Bangladesh.

Data Sources

To achieve the goals and objectives of this study, both primary and secondary data sources were used.

Primary Data Sources

In the present study, the primary data have been collected from several ways (Table 3.1). To fulfill the objectives of the present research, direct field observation surveys have been conducted from January 2024 to April 2025. These surveys were

Table 3.1 Sampling design of the present study

Sample type	Sample size
Questionnaire survey	248
Focus group discussions (FGDs)	10
Expert opinion survey	5

done to fulfill the aims and objectives of the present research regarding climate change vulnerabilities and their impacts. There are 248 (Two hundred forty-eight) questionnaires that have been conducted through face-to-face interviews from different types of professionals at the time of primary data collection. The stakeholders were chosen from the residents, farmers, students, laborers, school teachers, businessmen and so on. Farmers were given the highest priority while taking the questionnaire survey. During the primary data collection period, a total of ten (10) Focus Group Discussions (FGDs) surveys were conducted for the current study. These were carried out to get a better concept and to double-check the information that was collected from the respondents. Each focus group discussion included 10–15 participants from various occupational categories. Locals, farmers, civil society organizations, educators, and so on were among the groupings. Five (05) in-depth interviews with various professional personnel were conducted as part of the current study. These included the Chairman of Kapasia *Sadar* Union, two officials from the Upazila Agricultural and Livestock office, Upazila *Nirbahi* Officer (UNO), and one of the Upazila Administration Officials. Since they were in charge of the study area, their opinions, ideas, and experience have been adopted.

Secondary Data Sources

Secondary data was gathered from various publications, including books, journals, conference proceedings, websites, and more. Additionally, information from international organizations, non-governmental organizations, and government organizations (GOs) was gathered for this study. Landsat 8 images were downloaded from USGS Earth Explorer. Precipitation and temperature data were taken from online (https://crudata.uea.ac.uk/cru/data/hrg/).

Sampling Design

To gather representative household-level data on the effects of climate change, a total of 248 questionnaire surveys were administered using random sampling techniques. Five expert opinion surveys were conducted to include technical and policy insights, and ten Focus Group Discussions (FGDs) were arranged to confirm findings and gather group viewpoints. Data diversity, reliability, and thorough coverage of the study area were guaranteed by this mixed sampling strategy (Table 3.1).

Data Analysis

The data was analyzed with IBM SPSS 25.0. To accurately describe the data, the analysis employs the mean, frequency distribution, percentages and correlations. Similar to prior studies, SPSS has proven to be quite useful in managing large datasets while retaining statistical correctness in the current investigation. The research area maps were created using ArcGIS 10.8 software. Furthermore, it played an important role in creating temperature maps, precipitation maps, Land Use and Land Cover (LULC) maps, and Normalized Difference Vegetation Index (NDVI) maps. Microsoft Word was used to visualize the data, create the chart, and write the entire article for the current research. Microsoft Excel was utilized for data organization and other purposes. Utilizing sophisticated algorithms, conditional formatting, and data validation to effectively handle datasets. To effectively visualize the data, the software was also utilized to build interactive charts, figures, infographics, and dashboards. Additionally, this was utilized to gather information for creating maps. In this present research, the American Psychological Association (APA) 7th Edition referencing style was used. For that, Zotero Connectors software was used. Grammarly and QuillBot were used to paraphrase and correct grammatical mistakes.

Preparation of the Maps

In geospatial analysis of climate sensitivity in the study region, maps are essential. Several maps were made for this study, including-.

Land Use Land Cover (LULC) Map

Landsat 8 imagery data was used to create the Land Use Land Cover (LULC) map for the study area, and ArcGIS 10.8 software was used throughout the process. First, the USGS EarthExplorer portal was used to download the Landsat 8 image. After that, ArcGIS was updated with the downloaded data bands (Bands 1–7). Then used "Composite Bands" in the Data Management Toolbox to generate a composite image. The study area shapefile was then used to trim the composite image to the Area of Interest (AOI). Following the creation of AOI, the Image Classification Toolbar was used to manually digitize representative training samples and land cover type classifications for different land use classes, including water bodies, cropland, built-up, barren land, and healthy vegetation, based on the following Band Combinations: {Natural Color: 4 (Red), 3 (Green), 2 (Blue); False Color (Vegetation): 5 (NIR), 4 (Red), 3 (Green)}. Following image classification, select ArcToolbox → Spatial Analyst Tools → Maximum Likelihood Classification. Next, enter the composite picture and training sample signature file. The output in that instance was a LULC map classification. Essential map components, such as a scale bar, title, north arrow, and legend, were then included to create the final LULC map.

The Normalized Difference Vegetation Index (NDVI) Map

The spread of vegetation throughout the research region was evaluated using the Normalized Difference Vegetation Index (NDVI). First, the USGS Earth Explorer website was used to download Landsat 8 data. The image should then be entered into the ArcGIS program. For NDVI calculation, the following bands were used: Band 4 (Red) and Band 5 (Near-Infrared—NIR). Then to Spatial Analyst Tools > Map Algebra → Raster Calculator in that scenario. The NDVI calculation that was used was:

$$\begin{aligned} \text{NDVI} &= \frac{(NIR - RED)}{(NIR + RED)} \\ &= \frac{(Band\ 5 - Band\ 4)}{(Band\ 5 + Band\ 4)} (\text{for Landsat } 8) \end{aligned}$$

Here, **NIR** = Near Infrared band value, **RED** = Red band value.

Higher positive values from this equation denoted dense, healthy flora, while lower, zero, or negative values denoted water, populated regions, or bare terrain. The values range from -1 to $+1$. Following that, the "Clip" tool was used to clip the computed NDVI image to the research area shapefile. NDVI Categorization: Water ws represented by -1 to 0; barren land/built-up by 0 to 0.2; and vegetation by 0.2–1. The final NDVI map was created by incorporating the necessary map components, such as a scale bar, title, north arrow, and legend.

Temperature Map

The Climatic Research Unit's (CRU TS v4.07) data were obtained through the official website (https://crudata.uea.ac.uk/cru/data/hrg/). Next, selecting the "Multidimension" Tools toolbox, the following tool was used: Create the NetCDF Raster Layer by entering the downloaded NetCDF file, then choosing the values for the "tmp" variable and the time dimension. In that instance, the researcher ensured that the coordinate system was WGS 1984 UTM Zone 46 N and entered the study area shapefile. The researcher then exported the image in accordance with the shapefile and data frame, entered the final 12 bands from the projected raster image, and used "Composite Bands" in the Data Management Toolbox to generate a composite image.

Once the outcome was displayed, navigated to ArcToolbox → Spatial Analyst Tools → Local. The overlay and cell statistics were "mean." Go to Conversion tools → From Raster → Raster to point in that scenario. Next, navigate to Interpolation → Spatial Analyst Tools → IDW. The researchers then employed a color ramp and classified the image. The final Temperature map was created by including the necessary map components, such as a scale bar, north arrow, legend, and appropriate title.

Precipitation Map

Data were downloaded from the Climatic Research Unit (CRU TS v4.07) via the official website (https://crudata.uea.ac.uk/cru/data/hrg/). Then, going to the Multidimension Tools toolbox, the tool used: Make NetCDF Raster Layer, input downloaded NetCDF file, after that, "pre" variable, and specific time dimension values were selected. In that case, the researcher has input the study area shapefile and has made sure the coordinate system is WGS 1984 UTM Zone 46 N. After that, the researcher has exported the image according to shapefile and data frame, then input the last 12 bands from the projected raster image, then went to Data Management Toolbox and created a composite image using "Composite Bands." The overlay statistic was "sum" once the results were sent to ArcToolbox → Spatial Analyst Tools → Local → Cell statistics. The researcher has chosen Conversion tools → From Raster → Raster to point in that scenario. Next, Interpolation → Spatial Analyst Tools → IDW. The image was then categorized by the researcher using a color ramp. The final Temperature map was created by including the necessary map components, such as a scale bar, north arrow, legend, and appropriate title.

Result and Discussions

Climate Change Vulnerability in the Study Area

The study area has seen significant climatic shifts, such as increased temperatures and erratic rainfall patterns, which have exacerbated climate-related risks like droughts, floods, and soil erosion. These climate changes have had a significant impact on agriculture and livelihoods, affecting cropping cycles, livestock farming, and the production of staple crops like rice, fruits, and vegetables. According to the study, deforestation and deteriorating river health were the main causes of the sharp decline in fish populations, the substantial losses in plant and animal species, and the changes in forest cover. Water scarcity during the dry season and restricted access to natural resources also contributed to the region's environmental stress. The results also included local community perspectives gathered through focus groups and surveys. These have illustrated how people have experienced climate change and how traditional knowledge is being applied to these problems.

Geo-Spatial Analysis of Climate Change Vulnerability

Temperature Map of the Research Area During the Last 14 Years (Year 2010–2024)

The region's growing susceptibility to climate change is evident from the temperature analysis of the study area (Fig. 3.2) from 2010 to 2024, which showed a steady and gradual increase in average temperatures. A discernible change in thermal zones

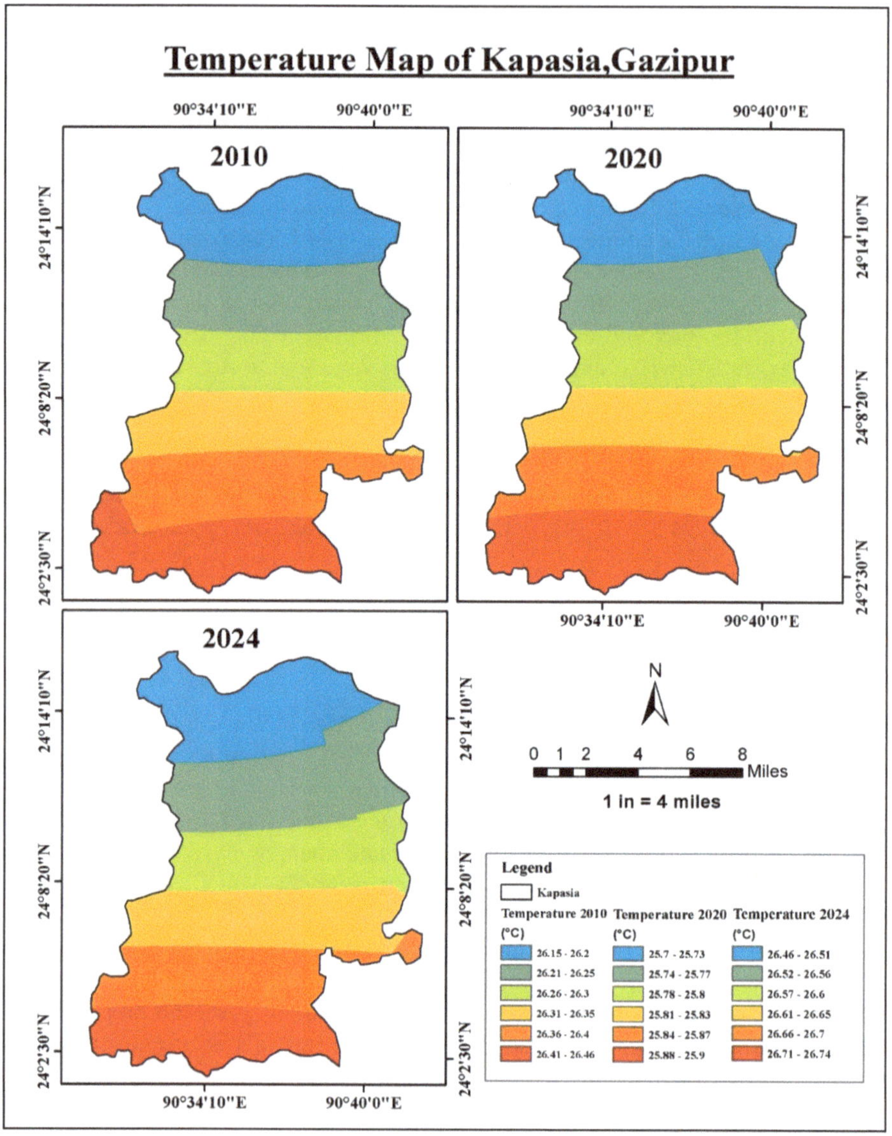

Fig. 3.2 Temperature map of the research area (in 2010, 2020 and 2024). Source: https://crudata.uea.ac.uk/cru/data/hrg/

has been noted throughout the study area, according to the comparative temperature maps for 2010, 2020, and 2024. The majority of the study area experienced temperatures between 26.15 °C and 26.46 °C in 2010, with the central and northern regions experiencing comparatively lower temperatures. However, by 2020, the warmer zones (defined as 26.61–26.65 °C and above) had significantly expanded in spatial extent, causing the cooler bands to shift further north. Warmer temperature zones have been moving northward over time, as evidenced by the acceleration of this trend by 2024, when notable temperatures above 26.71 °C were recorded in portions of the central and southern zones. This thermal shift indicates that there has been a noticeable warming trend in the study area over the last 14 years. The southern belt, in particular, has stayed consistently hotter throughout all three periods, but the increase in temperature values and spatial coverage throughout the rest of the area has suggested a wider climate vulnerability. The slow warming depicted in the temperature maps is in line with local perceptions gathered through field surveys, where participants have mentioned rising temperatures, shifting seasonal patterns, and declining agricultural productivity. These findings demonstrated how warming temperatures harmed the study area, making it more vulnerable to climate risks like heat stress, crop failure, and drought. In summary, the temperature map analysis has provided strong evidence that the study area has become increasingly vulnerable to climate change from 2010 to 2024.

Precipitation Map of the Research Area During the Last 14 Years

From 2010 to 2024, precipitation patterns of the research area displayed significant temporal and spatial variability, indicating the region's increasing susceptibility to climate change (Fig. 3.3). Comparing the precipitation maps in three different years (In the years 2010, 2020, and 2024) has shown that the distribution of rainfall in the area had not stayed constant over time, but had instead experienced significant changes and irregularities. The northern region of the study area received more precipitation in 2010 (above 2188 mm), while the southern and southeast regions received comparatively less (between 1993.9 and 2080 mm). But this trend hasn't continued. With the central-southern belt receiving over 2500 mm of rainfall, the zones of higher rainfall have significantly spread to the southern region by 2020, indicating increased rainfall concentration and potentially flood-prone conditions. The distribution will have changed once more by 2024. The central and southern regions have seen a decrease in precipitation, with many areas receiving less than 2000 mm, while the northern portion continues to receive comparatively high levels. An important indicator of climate variability, this fluctuation has shown that the rainfall regime has grown more unpredictable.

The availability of water resources and the agro-ecological balance have been upset by these precipitation variations. More frequent seasonal anomalies, such as extended dry spells and unexpected monsoon floods, have been reported by the respondents. Thus, the goal of determining the region's vulnerability to climate change has been achieved, as this pattern of rainfall fluctuation over the previous 14 years has verified that the study area has been exposed to climatic instability.

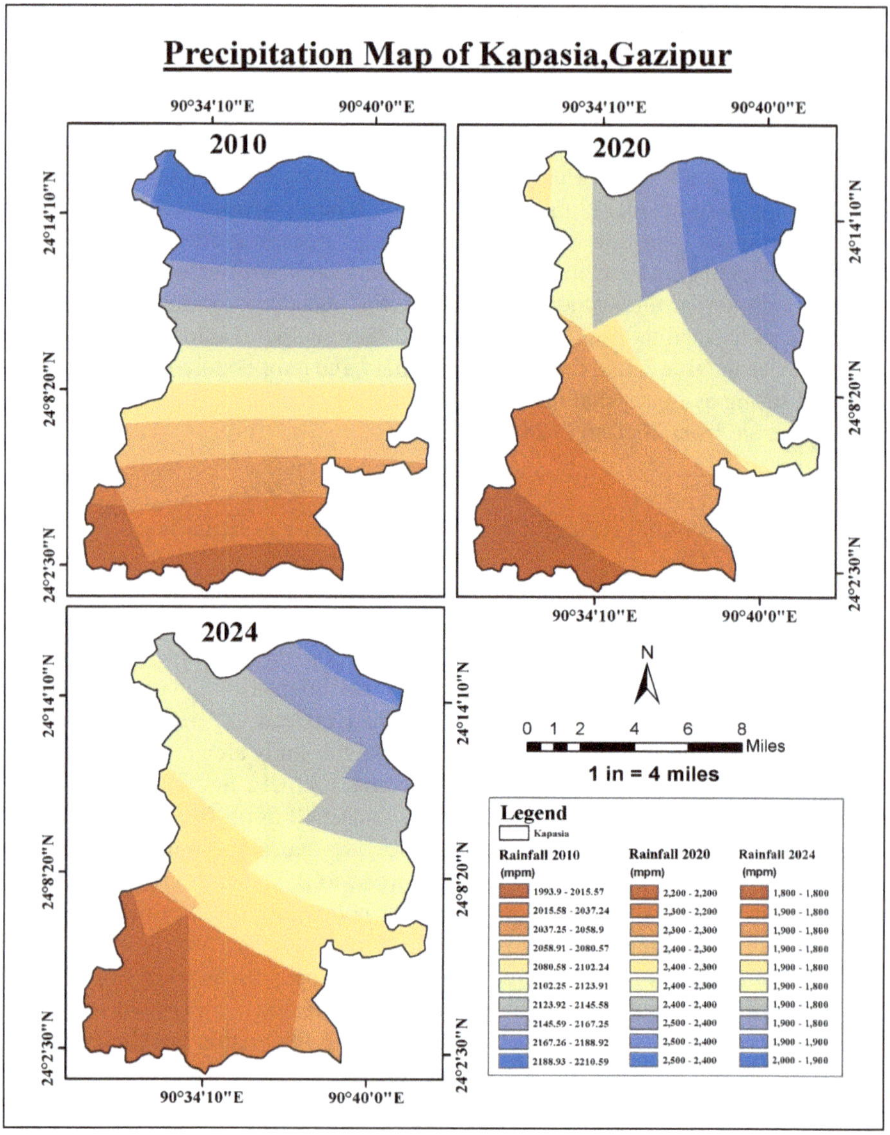

Fig. 3.3 Precipitation map of the research area (in 2010, 2020 and 2024). Source: https://crudata.uea.ac.uk/cru/data/hrg/

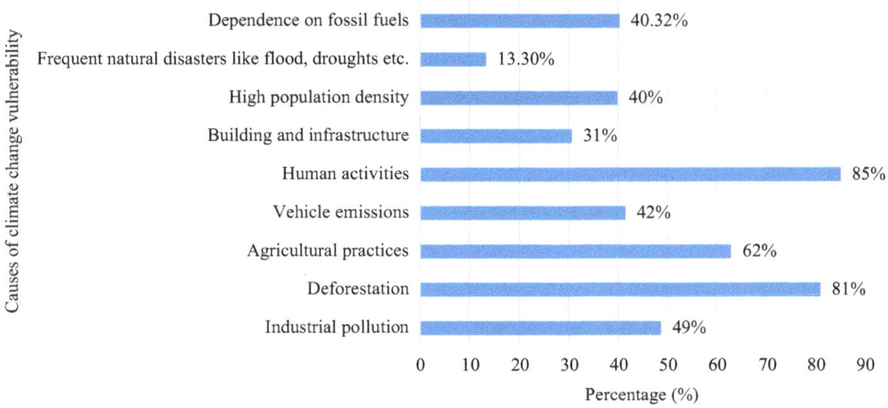

Fig. 3.4 Main Causes of climate change vulnerability according to the respondents

Main Causes of Climate Change Vulnerability

The graph (Fig. 3.4) shows that human activities have been identified as the most significant driver of climate change, with 85% of respondents accepting their contribution. This high percentage demonstrates public understanding of the overall impact of human-caused changes to the climate system. Deforestation came in second place, with 81% of respondents recognizing it. Forest clearing was widely acknowledged to contribute to global warming by diminishing carbon sinks and increasing $CO2$ levels. 62% of participants acknowledged agricultural practices. These activities, such as fertilizer use and livestock methane emissions, have increased greenhouse gas emissions. 49% of respondents identified industrial pollutants. This indicates that nearly half of the population understands the role of industrial waste and emissions in air degradation. 42% of respondents reported vehicle emissions as a source of climate change. The use of fossil fuels in transportation has significantly increased carbon emissions. Dependence on fossil fuels and high population density were noted by 40.32% and 40%, respectively. These causes have made a significant contribution to rising energy demand and environmental stress. Building and infrastructure development was cited by 31% of respondents. Urban growth and the construction industry have contributed to land use changes and increasing energy demand. Only 13.3% of respondents identified regular natural catastrophes such as floods and droughts as drivers of climate change. This has suggested a potential misconception, as these are usually impacts rather than direct causes of climate change.

Climatic Events Experienced by the Respondents in Recent Times

The chart (Fig. 3.5) illustrates the climatic events that respondents have experienced in recent times. A significant portion of the population has reported facing irregular rainfall patterns (67%) and temperature increases (62%), indicating that these have

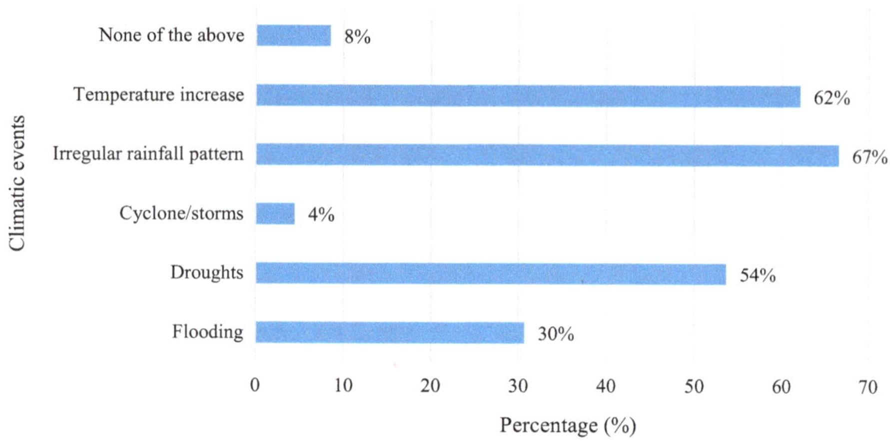

Fig. 3.5 Climatic events experienced by the respondents in recent times

been the most commonly perceived manifestations of climate change. Droughts have also been widely experienced, as mentioned by 54% of the respondents. In contrast, only 30% have reported flooding, suggesting either regional variation or possible mitigation measures in place. Notably, a very small percentage (4%) has encountered cyclones or storms, which may indicate a lower frequency or localized exposure in the surveyed area. Interestingly, 8% of the respondents have claimed to have experienced none of these climatic events. According to these findings, there was an urgent need for adaptation measures because changes in temperature and rainfall have already had an impact on the local ecosystem and people's perceptions of climate change.

Impacts of Climate Change Vulnerability on Agriculture

Over the past decade, agriculture in the study area has been greatly impacted by climate vulnerability. Crop yields have decreased, and seasonal farming cycles have been upset by changes in temperature and precipitation patterns. Maps from 2010 to 2024 have showed how evapotranspiration has grown due to rising temperatures, causing agricultural areas to experience moisture stress. At the same time, unpredictable heavy showers or protracted dry periods have hindered crop growth due to unpredictable rainfall patterns. Reduced agriculture and expanding urban areas were signs that land cover and land use have changed as a result of these climate changes. Additionally, NDVI data shows a drop in vegetation density and health, indicating stress on forest and farming areas. A rise in crop failures and pest infestations has led some farmers to scale back operations or transition to less climate-sensitive crops. These circumstances have decreased the region's total agricultural productivity and jeopardized food security. Therefore, in addition to altering the physical terrain, climate-related challenges have also threatened the agricultural system's stability. This research has indicated that the study area's agricultural dynamics have been significantly altered by climate sensitivity, necessitating the implementation of swift adaptation programs and sustainable land management techniques.

Geospatial Examination of how Agriculture Is Affected by Climate Change Vulnerability

LULC Map Analysis of the Study Area

Between 2014 and 2024, the study area's land use and land cover (LULC) experienced substantial changes that mirrored the increasing susceptibility of agriculture to climate change (Fig. 3.6). Spatial analysis of the LULC maps over the last 10 years has shown dynamic changes in the main land use categories, especially in vegetation, water bodies, built-up areas, and cropland.

In 2014, cropland dominated the landscape, but built-up areas have been encroaching more and more. Indicating continued urbanization and population pressure, the red zones denoting built-up land have notably grown, particularly in the central and southern regions. Due to this change, there was less agricultural land available, endangering rural livelihoods and food production. Healthy vegetation areas, denoted by dark green, have also become less dense and continuous. This suggests that there has been natural cover degradation and deforestation, possibly due to changing climate conditions and increased habitation. Over time, the blue zones that stand in for water bodies have changed; by 2024, some areas had shrunk. This implies water stress, which immediately affected irrigation and crop cycles. The appearance of beige barren land in isolated patches has further confirmed the degradation of once productive agricultural lands. Along with decreasing the amount of arable land, these changes have made already-existing agricultural zones more vulnerable to floods, droughts, and reduced soil fertility due to erratic weather patterns. In general, the LULC patterns from 2014 to 2024 have made it abundantly evident that climate vulnerability and human-induced pressure have significantly altered land use in the research area, endangering the region's agricultural sustainability.

NDVI Analysis of the Study Area

The NDVI maps of the study area from 2014 to 2024 have shown notable changes in vegetation density and health, offering crucial new insights into the climate change-induced vulnerability of agriculture (Fig. 3.7). The NDVI values have shown a slow decline in vegetative vigor over the last decade, underscoring the negative impacts of climate variability and land use changes on agriculture and regional ecosystems. In 2014, when the NDVI readings ranged from nearly −0.046

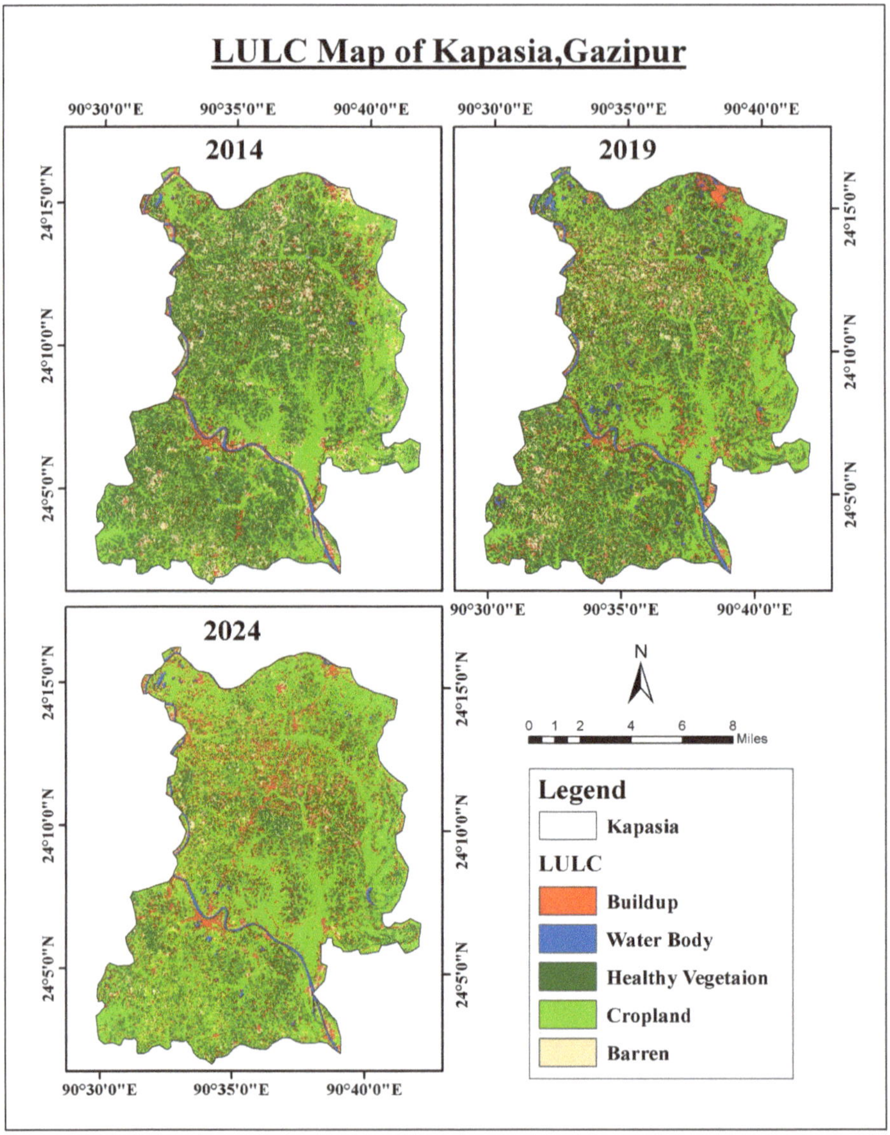

Fig. 3.6 LULC analysis of the study area (in 2014, 2019 and 2024). Source: Landsat 8 USGS Earth Explorer

to 0.420, a preponderance of green colors reflected the region's robust vegetation. This pattern indicated that at that time, a sizable the study area had maintained healthy crop cover and native vegetation. But by 2019, the NDVI values were almost between 0.026 and 0.441, showing a mixed pattern with some zones denoting a slight improvement and others reflecting a decline, especially in the southern and

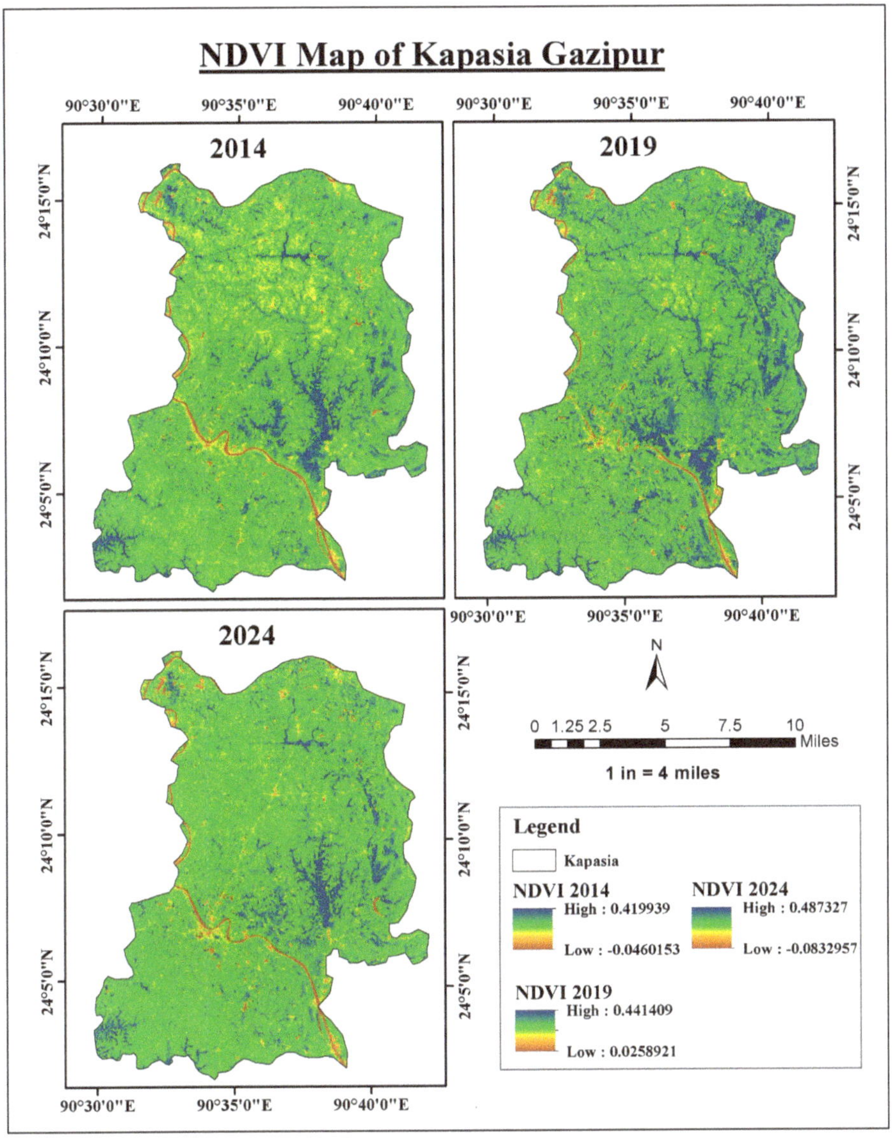

Fig. 3.7 NDVI analysis of the study area. Source: Landsat 8 USGS Earth Explorer

southeastern regions. Extreme weather, sporadic rainfall, and land conversion for non-agricultural purposes were probably the causes of these variations. With values ranging from −0.083 to 0.487, the 2024 NDVI map has shown even more decline in vegetative health. A larger area, particularly in the western and central regions, has shown yellow to reddish hues, indicating stressed or sparse vegetation, despite some pockets displaying high vegetation.

The LULC analysis has clearly shown that this degradation has been associated with rising temperatures, decreasing rainfall, and increasing urban pressure. Overall, by showing declining vegetation health in the study area, the NDVI analysis has aided in the goal of assessing how climate vulnerability affects agriculture. Crop productivity and ecological balance have been directly impacted by these changes in vegetative cover, which has emphasized the necessity of adaptive agricultural planning in the face of climate change.

Impacts of Climate Change Vulnerability on Agriculture

The respondents' perceptions of the various effects of climate change on agriculture are depicted in the graph (Fig. 3.8). A majority (85%) have reported that food security and livelihoods have been severely affected, reflecting the central role agriculture has played in sustaining rural communities. Decreased crop yield has also been a major concern, as indicated by 81% of the participants, followed by the altered rainfall pattern (69%), which has disrupted the agricultural calendar and water availability. Soil degradation (59%) and increased pests and crop diseases (40%) have further reduced agricultural productivity. Moreover, 36% of respondents have observed a change in cropping patterns, while 32% have identified adverse impacts on livestock. Interestingly, only 26% have mentioned a geographic shift in agriculture, suggesting that while such shifts may be occurring, they have not yet been widely experienced or recognized. These findings have confirmed that climate change has already posed serious threats to agricultural systems, compromising food production, land quality, and rural resilience.

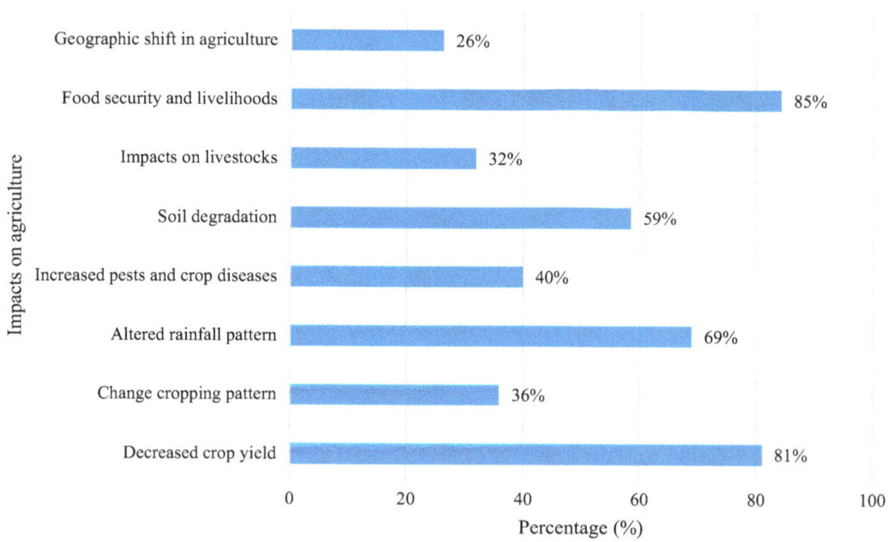

Fig. 3.8 Impacts of climate change vulnerability on agriculture according to the respondents

Impacts of Climate Change Vulnerability on Livelihoods

Nine effects of climate change on livelihoods were found in this study (Fig. 3.9) as, Reduced availability of natural resources like food and water (69%), impacts on human health (48%), habitat loss and degradation (72%), forest ecosystem under stress (80%), introduction of new crops and animal species (71%), migration of species to different areas (8%), loss of plant species (80%), and loss of animal species (26%).

Correlation Matrix Between Temperature, Precipitation, and NDVI

Statistically significant correlations were found between temperature, precipitation, and NDVI according to the Pearson correlation analysis (Table 3.2). Vegetation health decreased as average temperature rose, according to a strong negative correlation between temperature and NDVI ($r = -0.93$, $p < 0.05$). NDVI and precipitation showed a positive correlation ($r = 0.73$, $p < 0.05$), indicating that increased

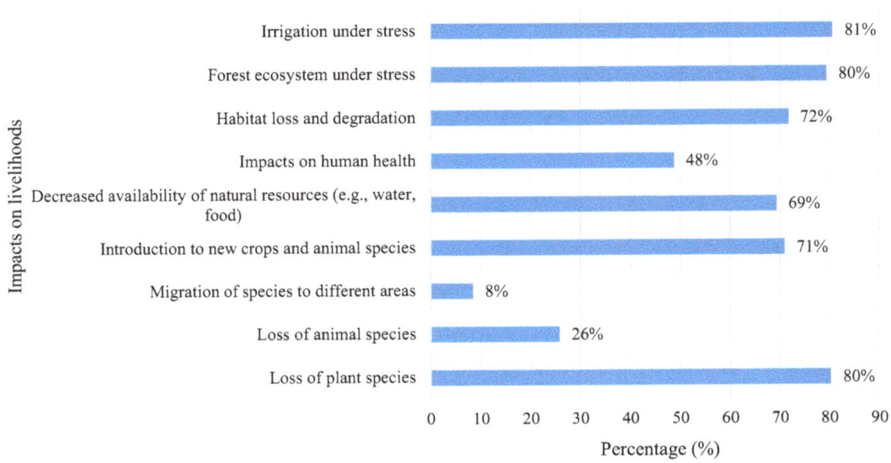

Fig. 3.9 Impacts of climate change vulnerability on livelihoods according to the respondents

Table 3.2 Pearson correlation between temperature, precipitation and NDVI

Variables	Temperature (°C)	Precipitation (mm)	NDVI
Temperature (°C)	**0.89**	**−0.67**	**−0.93**
Precipitation (mm)	**−0.67**	**0.89**	**0.73**
NDVI	**−0.93**	**0.73**	**0.89**

Values in bold indicate statistically significant correlations at $p < 0.05$

rainfall promoted agricultural productivity and vegetation health. Pointing climatic instability in the study area, a moderately negative correlation was also found between temperature and precipitation ($r = -0.67$, $p < 0.05$). These results demonstrated that during the study period, vegetation dynamics and agricultural vulnerability were greatly impacted by both warming temperatures and unpredictable rainfall patterns.

Limitations and Implications of the Study

The use of secondary climatic data, which might not adequately represent micro-scale changes in temperature, precipitation, and land use, has limited this study. Although the NDVI and LULC analyses have yielded insightful information, their temporal coverage and resolution may limit the accuracy of identifying localized changes. Moreover, only two climatic variables were given the main focus, which might not be impactful due to the time limitations. Furthermore, community perceptions, which can be arbitrary and impacted by transient events rather than long-term climate trends, have been reflected in questionnaire and focus group discussion responses. These restrictions suggest that rather than being absolute, the results should be regarded as indicative. To increase the robustness of findings, future studies should incorporate larger socioeconomic surveys, continuous ground-based climate monitoring, and higher-resolution remote sensing data. Notwithstanding these drawbacks, the study has offered a solid basis for comprehending climate vulnerability and how it affects local agriculture and livelihoods in the study area and as a whole in Bangladesh.

Conclusion

The study has shown how the study area's temperature, precipitation, and land use patterns have changed dramatically due to climate change, with direct effects on biodiversity and agriculture. Reduced crop yields, livelihood stress, and deteriorating vegetation health have all been exacerbated by rising temperatures, unpredictable rainfall, and shrinking cropland. The vulnerability of rural households is highlighted by the increased food insecurity, soil degradation, and loss of ecosystem services already experienced by local communities. To improve resilience in the future, adaptation plans must give community-based water management, biodiversity preservation, and sustainable farming methods top priority. To create localized adaptation policies, future studies should combine socioeconomic analyses with high-resolution climate models. Effective planning for climate resilience in vulnerable areas can be facilitated by this study's integration of science, policy, and community action.

References

Afjal Hossain M, Imran Reza M, Rahman S, Kayes I (2012) Climate change and its impacts on the livelihoods of the vulnerable people in the southwestern coastal zone in Bangladesh. In: Leal Filho W (ed) Climate change and the sustainable use of water resources. Springer, pp 237–259. https://doi.org/10.1007/978-3-642-22266-5_15

Bosello F, Zhang J (2005) Assessing climate change impacts: agriculture (SSRN Scholarly Paper No. 771245). Social Science Research Network. https://doi.org/10.2139/ssrn.771245

Dastagir MR (2015) Modeling recent climate change induced extreme events in Bangladesh: a review. Weather Clim Extrem 7:49–60. https://doi.org/10.1016/j.wace.2014.10.003

Dey S (2023) Climate change effects on livestock production in Bangladesh and its economic impacts. SAARC J Agric 21(2):227–238. https://doi.org/10.3329/sja.v21i2.68030

Dey RK, Ansary R, Mandal S (2025) Impact of climate variability on mango cultivation: A Case Study in Malda District of West Bengal, India. In: Livelihoods and well-being in the era of climate change: risk to resilience across India. Springer Nature Switzerland, Cham, pp 217–232. https://doi.org/10.1007/978-3-031-81132-6

Dogar MM, Kucharski F, & Azharuddin S (2017) Study of the global and regional climatic impacts of ENSO magnitude using SPEEDY AGCM. J Earth Syst Sci 126, 30. https://doi.org/10.1007/s12040-017-0804-4

Dogar MM, Kucharski F, Sato T, Mehmood S, Ali S, Gong Z, & Arraut J (2019). Towards understanding the global and regional climatic impacts of Modoki magnitude. Global and Planetary Change, 172, 223–241. https://doi.org/10.1016/j.gloplacha.2018.10.004

Dogar MMA, & Almazroui M. (2022). Revisiting the strong and weak ENSO teleconnection impacts using a high-resolution atmospheric model. Atmospheric Environment, 270, 118866. https://doi.org/10.1016/j.atmosenv.2021.118866

Farha UJ, Subah Z, Uddin MH, Rashid H (2024) Impacts of climate change-induced salinity intrusion on physiological parameters of aquatic hydrophytes from Coastal Rivers of Bangladesh. https://doi.org/10.13140/RG.2.2.31365.36324

Hoque MZ, Cui S, Xu L, Islam I, Jianxiong T, Ding S (2019) (PDF) assessing agricultural livelihood vulnerability to climate change in coastal Bangladesh. Int J Environ Res Public Health 16(22). https://doi.org/10.3390/ijerph16224552

Hossain SS, Cui Y, Delin H, Zhang X (2023) The economic influence of climate change on Bangladesh agriculture: application of a dynamic computable general equilibrium model. Int J Clim Change Strat Manag 15(3):353–370. https://doi.org/10.1108/IJCCSM-10-2021-0123

Hossain MA, Begum M, Akhtar MN. et al. (2025) Predicting water scarcity in northern Bangladesh using deep learning and climate data. npj Clim Atmos Sci 8, 348. https://doi.org/10.1038/s41612-025-01030-y

ICCCAD (2024, February 11) Climate change impacts in Bangladesh. What climate change means for a country and its people. International Centre for Climate Change and Development (ICCCAD). https://icccad.net/publications/climate_change_impacts_in_bangladesh_report_2024/

IPCC (2021) Five reflections on the IPCC climate change report. https://www.natureaustralia.org.au/what-we-do/our-insights/perspectives/five-reflections-on-the-ipcc-climate-change-report/

Islam R (2022, October 20) Habitat loss, climate change threaten Bangladesh's native freshwater fishes with extinction. Mongabay Environmental News. https://news.mongabay.com/2022/10/habitat-destruction-climate-change-threaten-bangladeshs-native-freshwater-fishes-with-extinction/

Islam S, Anzum F, Kamal AB, Akter M, Razie AM, Khan MR (2021) The integration of nature-based solutions into climate adaptation policy and planning in Bangladesh. J Sci Policy Gov 18(02). https://doi.org/10.38126/JSPG180204

Karim Z, Hussain SG, Ahmed AU (1999) Climate change vulnerability of crop agriculture. In: Huq S, Karim Z, Asaduzzaman M, Mahtab F (eds) Vulnerability and adaptation to climate change for Bangladesh. Springer Netherlands, pp 39–54. https://doi.org/10.1007/978-94-015-9325-0_4

Lázár AN, Clarke D, Adams H, Akanda AR, Szabo S, Nicholls RJ, Matthews Z, Begum D, Saleh AFM, Abedin MA, Payo A, Streatfield PK, Hutton C, Mondal MS, Moslehuddin AZM (2015) Agricultural livelihoods in coastal Bangladesh under climate and environmental change—a model framework. Environ Sci: Processes Impacts 17(6):1018–1031. https://doi.org/10.1039/C4EM00600C

Mandal S, Choudhury BU, Satpati LN (2015) Monsoon variability, crop water requirement, and crop planning for kharif rice in Sagar Island, India. Int J Biometeorol 59:1891–1903. https://doi.org/10.1007/s00484-0150995-9

Mandal S, Satpati LN, Choudhury BU, Sadhu S (2018) Climate change vulnerability to the agrarian ecosystem of the small Island: evidence from Sagar Island, India. Theor Appl Climatol 132:451–464. https://doi.org/10.1007/s00704-017-2098-5

Morshed G, Tortajada C, Hossain MS (2025) The state of climate change adaptation research in Bangladesh: A systematic literature review. Mitig Adapt Strateg Glob Chang 30. https://doi.org/10.1007/s11027-025-10219-8

Qiu J, Ravela S, Emanuel K (2025) Bangladesh's amplified coastal storm tide Hazard from tropical cyclones and rising sea levels in a warming climate (no. arXiv:2312.06051). arXiv. https://doi.org/10.48550/arXiv.2312.06051

Rozario S, Mirza AB, Ranon R (2024) Enhancing climate resilience of Bangladesh's food system: vulnerability, measures, and way forward

Shapla T, Park J, Hongo C, Kuze H (2015) Agricultural land cover change in Gazipur, Bangladesh, in relation to local economy studied using Landsat images. Adv Remote Sens 4(3):214–223. https://doi.org/10.4236/ars.2015.43017

Shima MUA, Hasan MM (2024) Temporal mapping of vegetation cover change in Gazipur District, Bangladesh: a framework for environmental sustainability. SSRN Electr J. https://doi.org/10.2139/ssrn.4711628

Uddin MM, Hoque MA (2025) Climate change trends and vulnerabilities in Bangladesh's crop sector: A review of crop production challenges and resilience strategies. Turk J Agric Food Sci Technol 13(2):258–272. https://doi.org/10.24925/turjaf.v13i2.258-272.6923

UNDP (2024) How green agriculture bolsters Bangladesh's climate resilience. UNDP. https://www.undp.org/bangladesh/blog/how-green-agriculture-bolsters-bangladeshs-climate-resilience

Urvashi M, Sergio N (2010) The costs to developing countries of adapting to climate change: new methods and estimates—the global report of the economics of adaptation to climate change study [Text/HTML]. World Bank. https://documents.worldbank.org/en/publication/documents-reports/documentdetail/en/667701468177537886

Chapter 4
Leveraging CMIP6 Climate Projections for Policy-Oriented Flood Management in the Beas River Basin, Himalaya

Arun Mondal, Anupam Pandey, Partha Paul, Souvik Mondal, and Sananda Kundu

Abstract Increasing temperature accelerates the depletion of glaciers, hydrological imbalance, destroying the natural flow of the ecosystem, and the pattern of precipitation in the north-western Himalaya. For this, the study focuses on the distribution of precipitation and its changes in different periods of time (historical, near-future, mid-future, and far future). The study is also used to analyse the trends of precipitation. The Geophysical Fluid Dynamics Laboratory's Earth System Model version 4.1 (GFDL-ESM4) model from the CMIP-6 dataset is used to collect the past and future precipitation data, used in this analysis. The precipitation data under the scenario of SSP5–8.5 is collected across the three periods, such as Historical (1990–2014), near-future (2026–2050), mid-future (2051–2075), and far future (2076–2100). The trend of precipitation is calculated using the parametric method. The data is analysed on an annual and seasonal basis. For the study, the Beas River basin is selected to calculate the future precipitation data. The change of precipitation distribution increases to the north-eastern side, indicating the decrease of the glaciated area of the basin. Along with spatial variation, the differences in precipitation are also analysed on an annual and season-wise. But the seasonal variation of precipitation is more prominent than the annual variation. The mean annual precipitation of different periods has increased in the monsoon and post-monsoon seasons, which may create severe extreme events in these periods in the future. It also shows the positive trend of precipitation in different periods. So, the information may help to take proper policy to protect the water balance of the basin.

A. Mondal · A. Pandey · P. Paul · S. Mondal
Department of Geography, University of Allahabad, Prayagraj, India

S. Kundu (✉)
Department of Geography, Manipur University, Manipur, India
e-mail: skundu@manipuruniv.ac.in

S. Mandal et al. (eds.), *Climate Change Vulnerability*, Climate Change Management, https://doi.org/10.1007/978-3-032-16783-5_4

Introduction

In the environment, the weather, climate, and hydrological cycle of the Earth operate in a systematic and interlinked way. Any change of one component can influence the functioning process of other components (Doulabian et al. 2021). It also disturbs the balance of the environment (Mandal et al. 2013). In the environment, due to anthropogenic and natural causes, the concentration of the greenhouse gases (CO_2, NH_4, CFC, CO, and others) is increasing, accelerating the heat-absorbing capacity of the Earth (Knutti and Hegerl 2008; Gao et al. 2015; Fanta et al. 2023). This process is called global warming. Consequently, the average temperature of the Earth is also increasing (Hegdahl et al. 2016). And its adverse impact is witnessed on the hydro-biological system of the Earth (Yin et al. 2016). Although the effects of global warming are universal, the hydrosphere of the Earth is highly impacted by it, especially the reservoir of fresh water, because water is more sensitive to heat (Hagg et al. 2021; Scapozza et al. 2019). Global warming alters the pattern of precipitation, the intensity and frequency of extreme events, the moisture content of soil and the atmosphere, the depletion of glaciers, and the evaporation of water (Mandal et al. 2018; Halder et al. 2023; Kumar et al. 2023; Rani et al. 2019; Aggarwal et al. 2016). Global warming not only decreases the freshwater reservoir of the Earth, but also hydrological hazards like floods and droughts become very unpredictable and frequent (Khwairakpam and Kundu 2025). In the humid and semi-arid region, temperature and precipitation are the two key components in the domain of climatic fluctuations and their impact on the hydrological cycle (Jaweso et al. 2019; Gao et al. 2015).

Past studies indicate how the temperature of the Earth increases over time and its influence on the hydrological process (Liu et al. 2024; Kumar et al. 2023). The report of the Intergovernmental Panel for Climate Change (IPCC) has indicated that the average surface temperature of the Earth has increased approximately 0.89 °C between 1901 and 2012 (Rani and Sreekesh 2021). However, its effect on natural and human systems is equivalent to the effects of a 1.5 °C increase in temperature (Masson-Delmotte 2018; IPCC 2013). It is also projected that the surface temperature of the Earth will rise to 2 °C by the end of the century (Van Vuuren et al. 2011). The warming rate of different places on Earth is not the same because it depends on the latitude, solar inclination, thickness of the ozone layer, presence of pollutants, the concentration of greenhouse gases in the air, and others (Barbhuiya et al. 2025; Scapozza et al. 2019; Agarwal et al. 2014). The average temperature of India has increased 0.44 °C/100 years (1941–2012). Although it varies between 0.51 °C/100 years (maximum) and 0.19 °C/100 years (minimum) (Saxena and Mathur 2019; Ray et al. 2019). Due to this increasing temperature, the minimum temperature of the Western Himalaya is rising faster than the rate of maximum temperature (Babar 2013). It hampers the balance of the cryosphere and hydrological cycle in the river basin (Barbhuiya et al. 2025; Lakhera et al. 2016; Agarwal et al. 2014; Xu et al. 2011; Middelkoop et al. 2001). The Indus River and its tributaries, in particular, are highly sensitive to climate change (Shah et al. 2013). So, to take the proper planning, management of those basins, planners need to understand the changing patterns of temperature and precipitation over the years (Minville et al. 2008; Immerzeel et al. 2012; Mandal et al., 2014; Mandal and Choudhury 2014; Mandal et al. 2015).

To know the past and future trends of precipitation, global climate models (GCMs) and regional climate models (RCMs) are mainly used to project the precipitation trends of any area (Dogar et al. 2017a, 2017b, 2020, 2025; Dogar and Sato 2018, 2019; Dogar 2020; Kundu et al. 2017, 2018; Jehanzaib et al. 2020). Due to coarser resolution, the GCM is not suitable for application at the regional level (Dogar et al. 2019). But, GCM data can be converted to RCM data after the downscaling of its spatial resolution (Tsegay et al. 2015). The past and future precipitation data of the research are collected from the Coupled Model Intercomparison Project Phase 6 (CMIP6). Various shared socioeconomic pathways (SSPs) scenarios (SSP1–2.6, SSP2–4.5, SSP3–7.0, and SSP5–8.5) have been introduced in this model (Gidden et al. 2019). SSP is the enhanced approach of the representative concentration pathway (RCP), which is used in the CMIP6 model. SSP is the pathway that combines several projections, such as economic development, emissions, land use change, use of technology, and others. Nepal and Shrestha (2015) used climate models to estimate the changes of climatic parameters in the Ganga, Indus, and Brahmaputra River basins and also analysed how those changes affect the hydrological regime of these selected basins. The findings of several studies have indicated the increasing tendency of climatic parameters in the Southeast Asian region (Prasad et al. 2016). Furthermore, the Beas River has experienced several instances of hydrological imbalances (flood and drought) in recent years, which indicates the change of climatic parameters of the basin (Prasad et al. 2016). So, the analysis of trends of the precipitation becomes necessary for the analysis of past-future trends for proper hydrological planning.

As the river basin is located in an area where several evidences of glacial retreats and hydrological imbalances have been observed due to the changes of climatic parameters, multiple researchers have given their contribution to understand the change of precipitation and hydraulic behaviour of the river (Mondal et al. 2022, 2026; Khare et al. 2015, 2017; Bhadwal et al. 2019; Sonali and Kumar 2013). In this attempt, Rani and Sreekesh (2021) have used the Soil and Water Assessment Tool (SWAT) model to simulate the river discharge data from the river channel, and the study has also proved that climate change is more responsible for the variation of precipitation and basin hydrology than land-use changes. A similar method has also been used by Rani et al. (2019) to estimate the changes of potential evapotranspiration of the basin. Some researchers have also noticed that the snow cover of the basin area has decreased, and consequently, the flow of the river has also been impacted.

Along with the depletion of glaciers, variation of precipitation is also responsible for changing the hydrodynamics of the river (Mondal et al. 2026; Singh et al. 2023; Li et al. 2019; Yin et al. 2016; Khan et al. 2025). Some extreme events of the basin have given evidence of a changing rainfall pattern of the basin. From this concern, researchers used various precipitation models to predict the future scenario of precipitation. Kant and Meena (2024) have used 3 models (ACCESS, CanESM, and NorESM) of precipitation in the SSP2–4.5 and SSP5–8.5 scenarios. Kant et al. (2024) have used a total of five precipitation models (BNU-ESM, Can-ESM2, CNRM CM5, MPI-ESM MR, and MPI-ESM LR) to compare the prediction of precipitation among models in two representative concentration pathway (RCP) scenarios (4.5, 8.5). Aggarwal et al. (2016) have used the WAS_GFDL_ESM2M_SMHI_RCAV2_DAILY data in RCP 8.5 and RCP 4.5 scenarios. In this research, precipitation data of GFDL-ESM4 have been used in the SSP5–8.5 scenario, which is not used by the others. Along with this, the spatial variation of precipitation is also analysed in different time periods (2026–2050, 2051–2075, 2076–2100) and seasonal basis (monsoon, pre-monsoon, post-monsoon).

The study mainly focused on the precipitation trend of future rainfall in the Beas River basin. The objective of the study is to understand the precipitation distribution of the basin area in the seasonal pattern. The study also analysis the changes of precipitation and its future scenario. The study is important to be aware of the future precipitation scenario of the basin, which is required for the proper management of the water resources.

Study Area

The Beas River, an important tributary of the Sutlej River, is selected for the study. It is a river in North-West India, contributing to the Indus River system through its flow into the Sutlej River. The area of the catchment is about 20,303 km^2, of which only 3.83 of % area (777 km^2) is permanent snow cover (Bhattacharya et al. 2019). The high mountainous area of Himachal Pradesh (The Beas Kund) is the origin place of the Beas River basin. From its origin, it flows around 470 km in a south-western direction to join the Sutlej River in Punjab. The average elevation of the basin is about 4361 m, which varies from 6600 m in the north-west to 802 m near Pandoh dam, covering the area from 32°21′59″N to 31°16′09″N latitude and from 77°05′08″E to 74°58′31″E longitude (Fig. 4.1) (Bhattacharya et al. 2019). The basin area is extended over eleven districts of Himachal Pradesh (Lahaul and Spiti, Kullu, Mandi, Hamirpur, Kangra) and Punjab (Gurdaspur, Pathankot, Hoshiarpur, Kapurthala, Amritsar, and Tarn Taran). The water of melting glaciers in summer is the main source of its flow (Kant and Meena 2024). But in winter, a significant portion of the basin is covered by the ice. Some important tributaries of the river are Tirthan, Sainj, Parvati, Larji, and others. In winter and summer, the mean monthly temperature of the basin is 5 °C and 25 °C, respectively (Rani and Sreekesh 2021). The average annual precipitation of the study area is 1217 mm, of which more than 70% precipitation occurs in monsoon (July–September) (Bajracharya et al. 2018). In the basin, deciduous and alpine forest is observed that depend on the elevation of the basin.

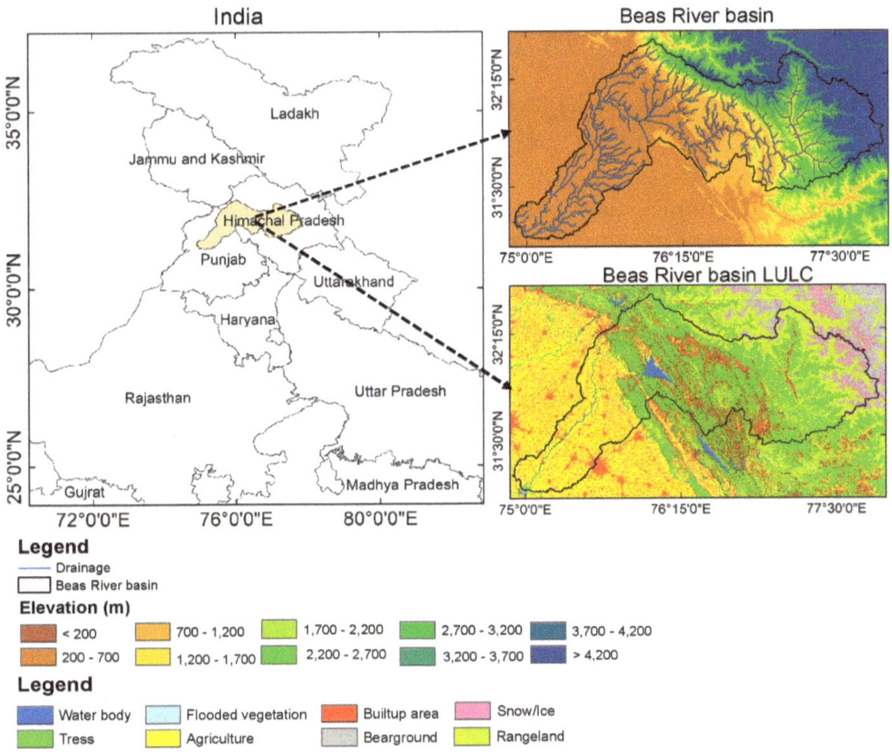

Fig. 4.1 Study area map shows the basin boundary of the Beas River basin in the Indian states of Himachal Pradesh and Punjab. The elevation distribution map was generated from the SRTM 30 m DEM (downloaded from USGS EarthExplorer), and the land use and land cover map of the Beas River basin was obtained from Environmental Systems Research Institute (ESRI)

Data

In this study, the CMIP6 dataset plays a crucial role in collecting past and future precipitation data of the study area. It is a scientific collaboration under the CMIP project designed to share and raise awareness of the data about global warming. In this project, several datasets of climatic parameters have been stored, including mean air temperature (Tas), precipitation (Pr), relative humidity (Hurs), air pressure (Ps), wind speed (Wind), surface downwelling shortwave radiation (Rsds), and surface downwelling longwave radiation (Rlds) (Liu et al. 2024). The data is available in two periods such as historical periods and future periods. In the historical periods, the data is available from the year 1990 to 2014. Similarly, the dataset provides future data from the year 2015 to 2100. The dataset is provided in 4 SSP scenarios: SSP1–2.6, SSP2–4.5, SSP3–7.0, and SSP5–8.5. Among those, SSP1–2.6 indicates the sustainable growth of the society and an eco-friendly approach to the environment. Contrarily, SSP5–8.5 indicates the highest level of challenges due to low economic conditions, high population growth rate, and intensive burning the fossil fuels. The SSP2–4.5, SSP3–7.0 indicate the intermediate pathway to predict the future weather conditions.

In the database of CMIP-6, several models of precipitation are available, such as ACCESS, CanESM, GFDL-ESM4, and NorESM (Kant and Meena 2024; Zhang et al. 2023). Among several precipitation models, GFDL-ESM4, developed by the USA, was used in this study in the SSP5–8.5 scenario. The dataset was used to understand the analysis of trends in past and future precipitation data. Past precipitation data was collected and analysed from the year 1990–2014, because the observed precipitation data is not available before the year 1990. In this study, the GFDL-ESM4 precipitation model was used in the SSP5–8.5 future emission scenario. The data was analysed during three future periods, such as 2026–2050, 2051–2075, and 2076–2100 (Liu et al. 2024).

Methodology

In the whole study, CMIP6 precipitation data played a significant role. For the analysis of future precipitation, several precipitation models are available in CMIP-6. However, our study is conducted using the GFDL-ESM4 precipitation data. The data is available for four different emission scenarios, such as SSP1–2.6, SSP2–4.5, SSP3–7.0, and SSP5–8.5, but the study only covered the SSP5–8.5 emission scenario. After selecting the data, the time scale of the study was confirmed. For the past, the precipitation trends were calculated from the year 1990 to 2014. As the range of time calculated for the computation of future precipitation is quite long (2025–2100), it is divided into 3 periods such as near future (2026–2050), middle future (2051–2075), and far future (2076–2100). Considering this time and data frame, past and future precipitation were analysed based on annual and seasonal (pre-monsoon March–May, Monsoon = June–September, Post-monsoon = October–November) precipitation. Then, the spatial variation of the precipitation data was analysed in different time frames and seasons to determine the total average precipitation over the years. Similarly, the changes in precipitation in the future were also analysed from the historical precipitation data. The parametric test was used to understand the trend of changes in precipitation on different time scales. Lastly, the validation of results was an important step to check the significance of the study. Therefore, the calculated p-value of the basin helps to check the validity of the results (0.05% level of significance). The methodology of the study is depicted in Fig. 4.2.

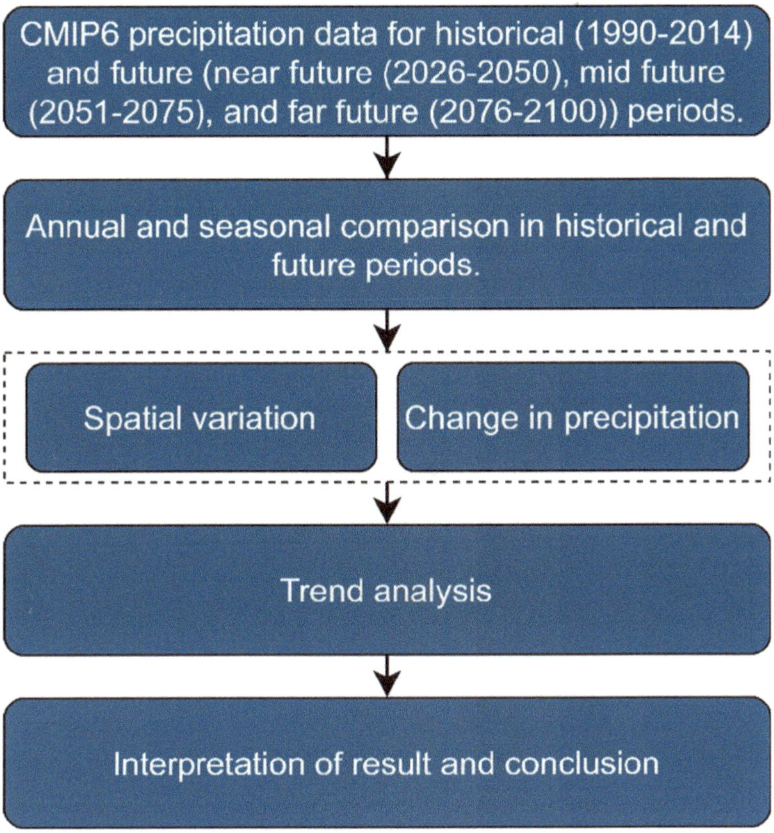

Fig. 4.2 Methodological flow chart of the entire study

Result

The result is analysed in annual, pre-monsoon, monsoon, post-monsoon, and winter seasons. The temporal changes of precipitation are also discussed in historical, near future, mid future, and far future. The results of the study are discussed below. Figure 4.3 illustrates the variation of the annual and seasonal distribution of precipitation over the Beas River basin in the past and future scenarios. The annual precipitation distribution data indicate that the central part of the basin received more precipitation than the other remaining parts in the eastern and western portions. However, over time, the precipitation has increased on the eastern and western side, receiving more rainfall than past. The average annual precipitation of different periods is given in Table 4.1. From the analysis, the seasonal variation of precipitation is more highlighted. In the post-monsoon and monsoon seasons, the precipitation of the basin has increased significantly. The data show that the basin will receive less

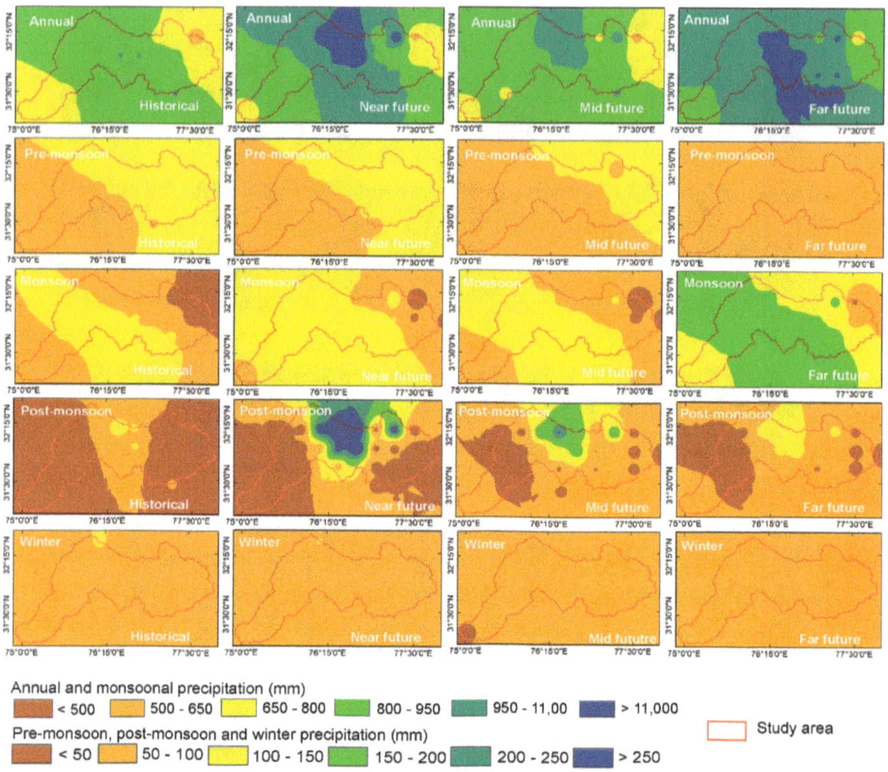

Fig. 4.3 Annual and seasonal (pre-monsoon, monsoon, post-monsoon, winter) precipitation distribution over Beas River basin (indicated by red line) across historical (1990–2014) and future (near future (2026–2050), mid future (2051–2075), and far future (2076–2100)) time period using CMIP6 SSP5–8.5 scenario precipitation data

Table 4.1 Average annual rainfall basic statistics of the Beas River basin

Periods	Mini	Max	Mean	Median
Historical (1990–2014)	490.43	1426.54	842.68	836.49
Near-future (2026–2050)	638.82	2023.68	963.73	873.71
Mid-future (2051–2075)	375.82	1536.73	884.32	861.77
Far-future (2076–2100)	559.78	1429.41	1027.49	1025.24

rainfall during the pre-monsoon season, and no significant changes will be observed in winter (Table 4.2).

Figure 4.4 shows the changes of precipitation (percentage) in different time and seasonal scales. From the historical dataset, the highest changes of annual precipitation have been observed in the near future and the far future. Major differences have been observed in the NE and SW directions. The change of annual precipitation in the mid-future does not significantly differ from the historical data (sometimes decreases). Whatever differences have been observed in the average annual

Table 4.2 Average seasonal rainfall basic statistics of the Beas River basin

	Periods	Mini	Max	Mean	Median	SD
Pre-monsoon	Historical (1990–2014)	16.40	211.65	101.41	94.65	58.65
	Near-future (2026–2050)	17.82	303.48	97.47	81.19	66.14
	Mid-future (2051–2075)	13.82	179.26	88.56	75.19	47.67
	Far-future (2076–2100)	3.26	269.86	80.71	66.30	60.65
Monsoon	Historical (1990–2014)	240.57	1225.88	609.89	575.13	235.86
	Near-future (2026–2050)	401.88	1339.55	680.53	651.82	221.98
	Mid-future (2051–2075)	232.71	1089.14	644.21	652.81	207.86
	Far-future (2076–2100)	301.26	1207.86	801.06	772.20	240.70
Post-monsoon	Historical (1990–2014)	0.00	205.86	44.60	26.51	56.76
	Near-future (2026–2050)	0.00	1123.56	96.90	19.53	241.64
	Mid-future (2051–2075)	0.23	544.68	82.99	35.94	123.88
	Far-future (2076–2100)	0.10	204.62	66.37	44.88	61.69
Winter	Historical (1990–2014)	12.75	170.23	86.78	77.90	39.60
	Near-future (2026–2050)	8.43	192.58	88.83	89.57	47.43
	Mid-future (2051–2075)	7.79	267.95	68.57	66.37	55.45
	Far-future (2076–2100)	5.97	231.03	79.35	86.43	53.81

precipitation, the difference is more prominent on a seasonal basis. The result indicates the shifting of precipitation from pre-monsoon to monsoon and post-monsoon periods. The precipitation of the future has increased from its historical time periods to the post-monsoon and monsoon season, whereas a negative change in precipitation is observed in the months of winter and monsoon. Although there are some small patches in the post-monsoon, where the precipitation has decreased. Similarly, patches of positive change in precipitation have been observed in the winter season (near future).

Figure 4.5 indicates the annual and seasonal slope of precipitation, indicating the trend of precipitation. The diagram clearly depicts that the average annual slope of the precipitation has increased from the past (historical) to the far future, except for the period of the near future. In the near future, the trend of precipitation is highly negative. Season-wise average slope of precipitation indicates the actual trend of precipitation over the years. In the monsoon and post-monsoon seasons, a positive trend of precipitation has been observed, but the trend is more prominent in mid-future and far-future of the monsoon season. At the same time, some small patches of negative trend of precipitation have been observed in those 2 months, especially in the near future period of the post-monsoon. The negative trends of precipitation have also been observed in the pre-monsoon and winter seasons, when the precipitation has decreased from 0 to 5 mm/year. Along with negative trends, some small patches of positive trends of precipitation are also observed.

Figure 4.6a depicts the monthly average precipitation over the Beas River basin in different periods (historical, near-future, mid-future, and far future). The differences in monthly precipitation among all periods do not show any significant differences from the month January to May. In the meantime, the mean precipitation data

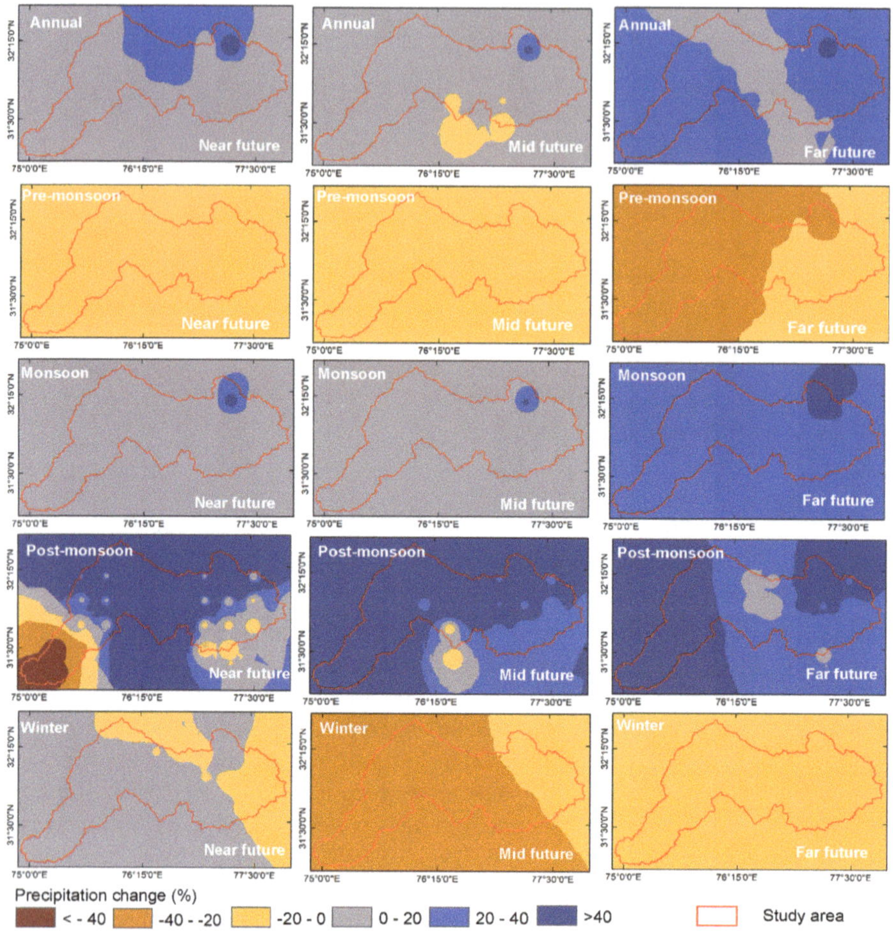

Fig. 4.4 Annual and seasonal (pre-monsoon, monsoon, post-monsoon, winter) percentage change in precipitation over the Beas River basin (indicated by red line) across three futures (near future (2026–2050), mid future (2051–2075), and far future (2076–2100)) time period compared to historical period (1990–2014) using CMIP6 SSP5–8.5 scenario precipitation data

of the near future (January and May) and far future (February) have exceeded the precipitation value of historical data. But in March and April, precipitation amounts have decreased in the future. Sharp changes in precipitation have been observed from the month of June and have continued till November. In June, the future precipitation is more than twice that of the past precipitation. Mean monthly precipitation of the far future has increased significantly from the month June to September. But after November, the differences in precipitation have decreased. The average daily precipitation is described in Table 4.3.

Figure 4.6b shows the changes in annual and seasonal precipitation with the changes in time periods over the Beas River basin. The average annual precipitation

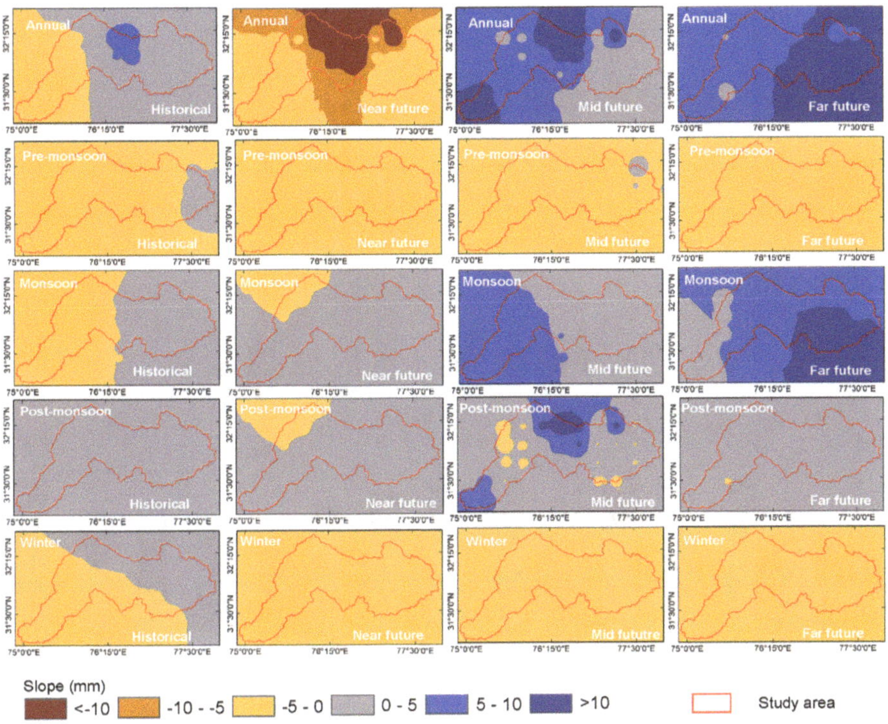

Fig. 4.5 Annual and seasonal (pre-monsoon, monsoon, post-monsoon, winter) precipitation trend over Beas River basin (indicated by red line) across historical (1990–2014) and future (near future (2026–2050), mid future (2051–2075), and far future (2076–2100)) time period using CMIP6 SSP5–8.5 scenario precipitation data

of the basin has increased from the past to the future. From the historical period (842.68 mm/year) to the far-future (1429.41 mm/year), the average annual precipitation has increased significantly, but in the mid-future, the precipitation value is slightly decreased (884.32 mm/day) than the near-future and far-future. Although no significant differences in annual precipitation have been observed, the variation is more prominent on a seasonal scale. Season-wise average annual precipitation shows that in the pre-monsoon and winter seasons. The total average precipitation values have decreased from the historical period to the far-future periods. In the post-monsoon and monsoon season, total mean precipitation has increased from the historic period to the far-future period. The future precipitation has increased more than twice from the historic to the far-future period.

Figure 4.7a indicates the trend of precipitation data according to time period and seasonal basis. Positive trend of annual precipitation has been observed in all periods from the historical to the near future (9.45–1.52 mm/year), except the period of the near future, where a negative trend (−7.01 mm/year) of average annual precipitation is observed. As more than 70% precipitation occurs in the monsoon season, the trend of annual average precipitation follows the trend of monsoonal

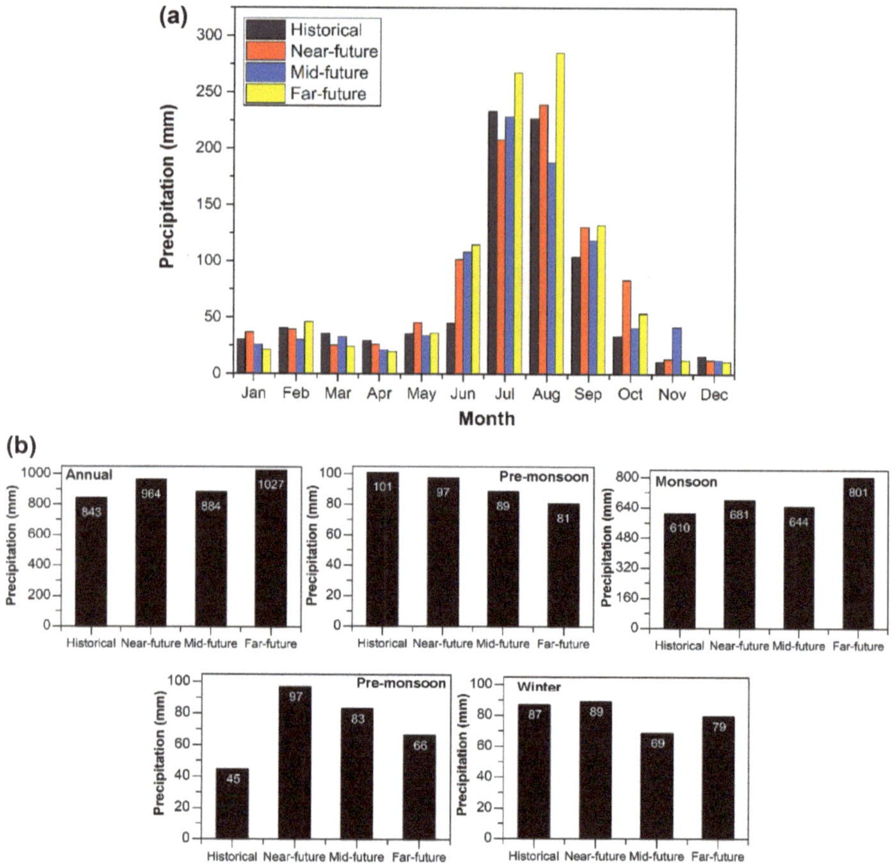

Fig. 4.6 (**a**) Monthly average precipitation in different time periods (Historical (1990–2014), near-future (2026–2050), mid-future (2051–2075), and far-future (2076–2100)) over the Beas River basin. (**b**) Average annual and seasonal (pre-monsoon, monsoon, post-monsoon, winter) precipitation in different periods (Historical (1990–2014), near-future (2026–2050), mid-future (2051–2075), and far-future (2076–2100)) over the Beas River basin

Table 4.3 Daily rainfall basic statistics of the Beas River basin

Periods	Mini	Max	Mean	Median	SD
Historical (1990–2014)	0.00	118.25	2.31	0.00	6.94
Near-future (2026–2050)	0.00	745.69	2.64	0.00	12.34
Mid-future (2051–2075)	0.00	443.63	2.42	0.00	9.07
Far-future (2076–2100)	0.00	157.36	2.81	0.00	8.11

precipitation. Although the precipitation trends have increased in the near future, the annual average precipitation has decreased due to highly negative trends of precipitation in other seasons. In the pre-monsoon and winter season, the trend of precipitation is negative. On the contrary, the trend of mean precipitation has increased in the season of monsoon and post-monsoon seasons, except for the monsoon in the historical period and the post-monsoon in the near future.

After analysing the trend, the data is used to check its significance level. While the result shows the positive and negative trends of precipitation, the calculated probability (p) value is higher than the alpha value (0.05), indicating no significant trend of precipitation (Fig. 4.7a). However, despite these highly insignificant trends of precipitation of the full basin, some significant patches of precipitation trends are also observed in the local scale (Fig. 4.7b). On the annual basis, significant precipitation trend is observed only at eastern side of the basin in far-future. But the difference is more acute at the seasonal level. Some patches of significant precipitation trends have also been observed at the eastern and north-eastern side of the basin in the monsoon and post-monsoon seasons in the far future. The precipitation of the post-monsoon season in the near-future period has shown the highest significant trend among others. However, in contrast, no significant precipitation trend is observed in the pre-monsoon and winter.

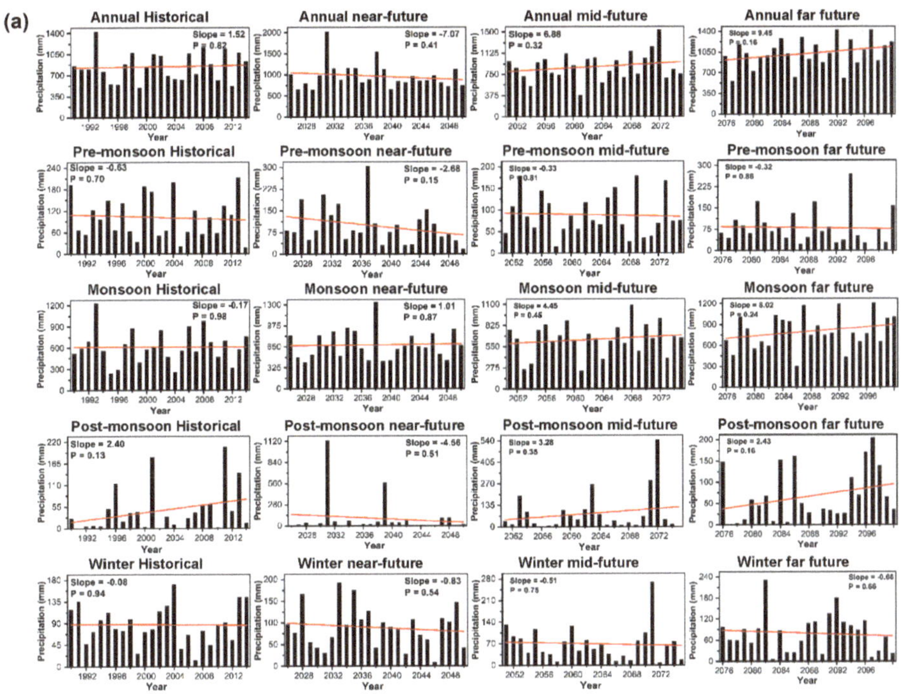

Fig. 4.7 (**a**) Average annual and seasonal (pre-monsoon, monsoon, post-monsoon, winter) precipitation trend over the Beas River basin in different time period (historical (1990–2014) and future (near future (2026–2050), mid future (2051–2075), and far future (2076–2100)). (**b**) Annual and seasonal (pre-monsoon, monsoon, post-monsoon, winter) p value over Beas River basin (indicated by red line) across historical (1990–2014) and future (near future (2026–2050), mid future (2051–2075), and far future (2076–2100)) time period using CMIP6 SSP5-8.5 scenario precipitation data

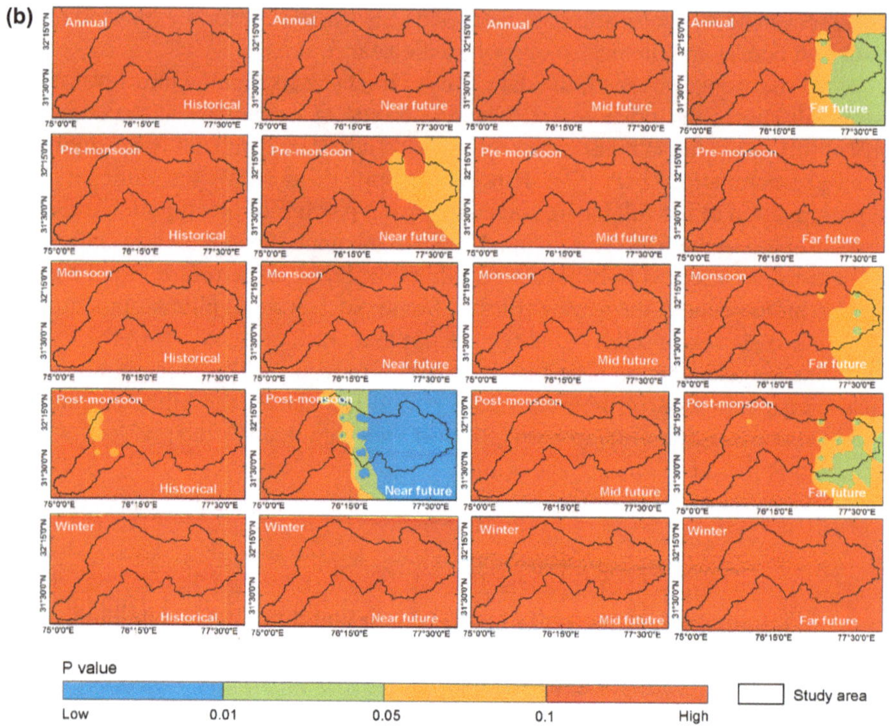

Fig. 4.7 (continued)

Conclusion

The Beas River is located in a place where climatic parameters are widely affected by climate change. The increasing temperature of the basin has a deep impact on the hydrological balance, increasing the depletion rate of glaciers, speeding up and rate of evapotranspiration, and changing the pattern of precipitation. Findings of several researchers indicate the changing precipitation pattern of the basin, affecting the water balance and flow of the river. For this purpose, past and future CMIP6 precipitation data are analysed to understand the nature of precipitation in the past and future. In the study area, the seasonal variation of precipitation is more prominent than the variation of annual precipitation. The average annual precipitation of the basin has increased, but the highest increase in precipitation is observed in the monsoon and post-monsoon seasons. It is also observed that the precipitation is also increased to the north-eastern side, indicating the reduction of the glaciated area of the basin. The change of mean annual precipitation from the past period to the future also denotes the positive changes of precipitation, but the decreasing rate of precipitation is observed in the pre-monsoon and winter seasons. Consequently, in the future, the basin will receive higher amounts of rainfall in very short periods of time (monsoon and post-monsoon), increasing the frequency of extreme events of the basin area. Along with the precipitation distribution data, a changing trend of precipitation over time is also needed to take and execute early, medium, and long-term plans. The increasing amount of average precipitation is very high in mid mid-future and far-future. The period-wise average precipitation data indicate the positive trend of precipitation in the future (except near-future). so, from the above discussion, it is noted that the precipitation pattern of the basin area will be changed in the future. And this change will be more acute on a seasonal basis. Therefore, a proper plan is needed to manage the water resources of the basin in a scientific way. Lastly, the study will help to understand the changes in precipitation scenarios over time, especially in the north-western Himalaya.

References

Arun, Mondal Manh-Hung, Le Venkataraman, Lakshmi (2022) Land use climate and water change in the Vietnamese Mekong Delta (VMD) using earth observation and hydrological modeling Journal of Hydrology: Regional Studies 42. https://doi.org/10.1016/j.ejrh.2022.101132

Arun, Mondal Anupam, Pandey Souvik, Mondal Sananda, Kundu (2026) Impact assessment of Glacial Lake Outburst Floods (GLOFs) in the Himalaya using satellite imageries over the last 25 years (2000–2024): A comprehensive review Remote Sensing Applications: Society and Environment 41. https://doi.org/10.1016/j.rsase.2025.101845

Agarwal A, Babel MS, Maskey S (2014) Analysis of future precipitation in the Koshi River Basin, Nepal. J Hydrol 513:422–434

Aggarwal SP, Thakur PK, Garg V, Nikam BR, Chouksey A, Dhote P, Bhattacharya T (2016) Water resources status and availability assessment in current and future climate change scenarios for Beas River Basin of north western Himalaya. Int Arch Photogramm Remote Sens Spat Inf Sci 41:1389–1396

Babar SF (2013) Analysis of southwest monsoon rainfall trend using statistical techniques over Nethravathi Basin. Int J Adv Technol Civil Eng 2(1):22–31

Bajracharya AR, Bajracharya SR, Shrestha AB, Maharjan SB (2018) Climate change impact assessment on the hydrological regime of the Kaligandaki Basin, Nepal. Sci Total Environ 625:837–848

Barbhuiya S, Sharma S, Pathania A, Gupta V (2025) Assessing the impacts of climate change on hydroclimatic regimes in Beas River Basin. In: Navigating the nexus: hydrology, agriculture, pollution and climate change, vol 1. Springer Nature Switzerland, Cham, pp 325–350

Bhadwal S, Sharma G, Gorti G, Sen SM (2019) Livelihoods, gender and climate change in the Eastern Himalayas. Environ Dev 31:68–77

Bhattacharya T, Khare D, Arora M (2019) A case study for the assessment of the suitability of gridded reanalysis weather data for hydrological simulation in Beas River Basin of North Western Himalaya. Appl Water Sci 9:1–15

Deepak, Khare Diptendu, Patra Arun, Mondal Sananda, Kundu (2015) Impact of landuse/land cover change on run-off in a catchment of Narmada river in India Applied Geomatics 7(1) 23–35. https://doi.org/10.1007/s12518-014-0148-6

Deepak, Khare Diptendu, Patra Arun, Mondal Sananda, Kundu (2017) Impact of landuse/land cover change on run-off in the catchment of a hydro power project Applied Water Science 7(2) 787–800. https://doi.org/10.1007/s13201-015-0292-0

Dogar MM, Stenchikov G, Osipov S, Wyman B, & Zhao M (2017a). Sensitivity of the regional climate in the Middle East and North Africa to volcanic perturbations. Journal of Geophysical Research: Atmospheres, 122(15), 7922–7948.

Dogar MM, Kucharski F, & Azharuddin S. (2017b). Study of the global and regional climatic impacts of ENSO magnitude using SPEEDY AGCM. J Earth Syst Sci 126, 30. https://doi.org/10.1007/s12040-017-0804-4

Dogar MM, Sato T (2018). Analysis of climate trends and leading modes of climate variability for MENA region. Journal of Geophysical Research: Atmospheres, 123(23), 13–074.

Dogar MM, Sato T (2019). Regional climate response of Middle Eastern, African, and South Asian monsoon regions to explosive volcanism and ENSO forcing. Journal of Geophysical Research: Atmospheres, 124(14), 7580–7598.

Dogar MM, Kucharski F, Sato T, Mehmood S, Ali S, Gong Z, ... & Arraut J (2019). Towards understanding the global and regional climatic impacts of Modoki magnitude. Global and Planetary Change, 172, 223–241.

Dogar MM (2020). Study of the regional climatic impacts of tropical explosive volcanism in the Middle East and North Africa region (Doctoral dissertation, Hokkaido University).

Dogar MM, Sato T, & Liu F (2020). Ocean sensitivity to periodic and constant volcanism. Scientific Reports, 10(1), 293.

Dogar MM, Watanabe S, Fujiwara M, Abid MA, Habeebullah TM, & Khan B (2025). Tropical volcanic impacts on MENA climate via ENSO and NAO dynamics in a high-top model. npj Climate and Atmospheric Science, 8(1), 330.

Doulabian S, Golian S, Toosi AS, Murphy C (2021) Evaluating the effects of climate change on precipitation and temperature for Iran using RCP scenarios. J Water Clim Change 12(1):166–184

Fanta SS, Yesuf MB, Saeed S, Bhattacharjee S, Hossain MS (2023) Analysis of precipitation and temperature trends under the impact of climate change over ten districts of Jimma Zone, Ethiopia. Earth Syst Environ 7(2):393–410

Gao Y, Li X, Leung LR, Chen D, Xu J (2015) Aridity changes in the Tibetan Plateau in a warming climate. Environ Res Lett 10(3):034013

Gidden MJ, Riahi K, Smith SJ, Fujimori S, Luderer G, Kriegler E et al (2019) Global emissions pathways under different socioeconomic scenarios for use in CMIP6. Geosci Model Dev 12(4):1443–1475

Hagg W, Ram S, Klaus A, Aschauer S, Babernits S, Brand D et al (2021) Hazard assessment for a glacier lake outburst flood in the Mo Chu River Basin, Bhutan. Appl Sci 11(20):9463

Halder P, Dey RK, Mandal S (2023) Long-period trend analysis of annual and seasonal rainfall in West Bengal, India (1901–2020). Theor Appl Climatol 154(1–2):685–703. https://doi.org/10.1007/s00704-023-04577-z

Hegdahl TJ, Tallaksen LM, Engeland K, Burkhart JF, Xu CY (2016) Discharge sensitivity to snowmelt parameterization: a case study for Upper Beas Basin, Himachal Pradesh, India. Hydrol Res 47(4):683–700

Immerzeel WW, Van Beek LPH, Konz M, Shrestha AB, Bierkens MFP (2012) Hydrological response to climate change in a glacierized catchment in the Himalayas. Clim Chang 110:721–736

IPCC (2013) Climate change 2013: the physical science basis. Cambridge University Press

Jaweso D, Abate B, Bauwe A, Lennartz B (2019) Hydro-meteorological trends in the Upper Omo-Ghibe River Basin, Ethiopia. Water 11(9):1951

Jehanzaib M, Sattar MN, Lee JH, Kim TW (2020) Investigating effect of climate change on drought propagation. Stoch Env Res Risk A 34:7–21

Kant C, Meena RS (2024) Projection of future rainfall events over the Beas River basin, Western Himalaya. J Water Clim Change 15(8):3536–3548

Kant C, Kumar A, Meena RS (2024) Exploring climate variables and drought patterns: a comprehensive trend analysis and evaluation of Beas Basin in Western Himalaya. Earth Syst Environ 8(2):449–464

Khan A, Mandal S, Satpati L, Dogar MM, Hasanean H, Salah AM (2025) Analyzing drought: the impact of climate change on precipitation patterns in the Shilabati River Basin, West Bengal, India. Earth Syst Environ. https://doi.org/10.1007/s41748-025-00679-y

Khwairakpam S.K, Robindro Singh (2025) Dealing with flood disaster– different techniques modelling and real-time flood analysis: a systematic review of different river basins in India Discover Civil Engineering 2(1). https://doi.org/10.1007/s44290-025-00284-y
Knutti R, Hegerl GC (2008) The equilibrium sensitivity of the Earth's temperature to radiation changes. Nat Geosci 1(11):735–743
Kumar M, Tiwari RK, Kumar K, Rautela KS (2023) Prioritization of sub-watersheds…. J Geol Soc India 99(4):545–553
Lakhera N, Kumar N, Shankar V (2016) Water balance study of Beas River using ArcGIS. In: CEC–2016 Proceedings, vol 9, p 10
Li L, Shen M, Hou Y, Xu CY, Lutz AF, Chen J et al (2019) Twenty-first-century glacio-hydrological changes in the Himalayan headwater Beas River Basin. Hydrol Earth Syst Sci 23(3):1483–1503
Liu C, Xu C, Zhang Z, Xiong S, Zhang W, Zhang B et al (2024) Modeling hydrological consequences…. Geosci Front 15(5):101819
Mandal S, Choudhury BU (2014) Estimation and prediction of maximum daily rainfall atSagar Island using best fit probability models. Theor Appl Climatol 117:87–97
Mandal S, Choudhury BU, Mondal M, Bej S (2013) Trend analysis of weather variables in Sagar Island, West Bengal, India: a long-term perspective (1982–2010). Curr Sci 105:947–953
Mandal S, Choudhury BU, Satpati LN (2015) Monsoon variability, crop water requirement, and crop planning for kharif rice in Sagar Island, India. Int J Biometeorol 59:1891–1903. https://doi.org/10.1007/s00484-0150995-9
Mandal S, Satpati LN, Choudhury BU, Sadhu S (2018) Climate change vulnerability to the agrarian ecosystem of the small Island: evidence from Sagar Island, India. Theor Appl Climatol 132:451–464. https://doi.org/10.1007/s00704-017-2098-5
Masson-Delmotte V (2018) IPCC special report: global warming of 1.5 °C
Middelkoop H, Daamen K, Gellens D, Grabs W, Kwadijk JC, Lang H et al (2001) Impact of climate change on hydrological regimes in the Rhine Basin. Clim Chang 49:105–128
Minville M, Brissette F, Leconte R (2008) Uncertainty of the impact of climate change on hydrology of a Nordic watershed. J Hydrol 358(1–2):70–83
Nepal S, Shrestha AB (2015) Impact of climate change on hydrological regimes of the Indus, Ganges and Brahmaputra basins. Int J Water Resour Dev 31(2):201–218
Prasad AS, Pandey BW, Leimgruber W, Kunwar RM (2016) Mountain hazard susceptibility…. Geoenviron Disasters 3:1–17
Rani S, Sreekesh S (2021) Flow regime changes under future climate scenarios in upper Beas Basin. Int J Environ Stud 78(3):382–397
Rani S, Sreekesh S, Krishnan P (2019) Effect of climate change on evapotranspiration…. ISPRS Arch 42:51–57
Ray LK, Goel NK, Arora M (2019) Trend analysis and change point detection of temperature over India. Theor Appl Climatol 138:153–167
Sananda, Kundu Deepak, Khare Arun, Mondal (2017) Future changes in rainfall temperature and reference evapotranspiration in the central India by least square support vector machine Geoscience Frontiers 8(3) 583–596. https://doi.org/10.1016/j.gsf.2016.06.002
Sananda, Kundu Arun, Mondal Deepak, Khare Christopher, Hain Venkat, Lakshmi (2018) Projecting Climate and Land Use Change Impacts on Actual Evapotranspiration for the Narmada River Basin in Central India in the Future Remote Sensing 10(4) 578. https://doi.org/10.3390/rs10040578
Saxena R, Mathur P (2019) Recent trends in rainfall and temperature over North West India (1871–2016). Theor Appl Climatol 135(3):1323–1338
Scapozza C, Ambrosi C, Cannata M, Strozzi T (2019) GLOF hazard assessment in the Himalayas. Geogr Helv 74(1):125–139
Shah SK, Bhattacharyya A, Shekhar M (2013) Reconstructing discharge of Beas River Basin using tree-ring data. Quat Int 286:138–147
Singh SK, Kanga S, Gulati B, Raič M, Sajan B, Đurin B, Singh S (2023) Hydrological modelling in the Beas Basin using SWAT+. Water 15(19):3338

Sonali P, Kumar DN (2013) Review of trend detection methods for temperature changes in India. J Hydrol 476:212–227
Tsegay D, Nugussie DHA, Hamdi R (2015) Analysis of precipitation based on regional climate models over Ethiopia (1989–2008). Int J Climatol 35:948–971
Van Vuuren DP, Stehfest E, den Elzen MG, Kram T, van Vliet J, Deetman S et al (2011) RCP2.6: exploring…. Clim Chang 109:95–116
Xu H, Taylor RG, Xu Y (2011) Uncertainty in climate change impacts on river discharge in China. Hydrol Earth Syst Sci 15(1):333–344
Yin Y, Xu CY, Chen H, Li L, Xu H, Li H, Jain SK (2016) Trend and concentration characteristics of precipitation in the Beas River Basin. Glob Planet Chang 145:116–129
Zhang Y, Knutson TR, Shevliakova E, Paynter D (2023) Long-term trends of global land precipitation in GFDL models. J Clim:6287–6307. https://doi.org/10.1175/JCLI-D-22-0764.1

Chapter 5
Climate Change Vulnerability Assessment of the North East Hill Regions of India

Prashant Pandey, Saurav Saha, Burhan Uddin Choudhury, Debasish Chakraborty, Amit Kumar, Samik Choudhury, Bappa Das, M. Ahmed, S. Hazarika, Ashok Kumar, and R. Devdas

Abstract The diverse terrain of north-east India bears fragile ecosystems, socio-economic variety, and is particularly sensitive to the adverse impact of climate change and environmental degradation. Pervasive regional warming together reduced monsoon showers, winter rainfall, and seasonal precipitation extremes were evident over the North Eastern India (NEI) region. Our vulnerability assessment study adapted the multidimensional framework of IPCC covering 661 sub-districts in seven north-eastern hill states—Arunachal Pradesh, Manipur, Meghalaya, Mizoram, Nagaland, Sikkim, and Tripura. About 34 indicators were divided into three primary categories: exposure, sensitivity, and adaptive capacity. Exposure indicators included climatic and environmental extremes such as yearly rainfall variability, variations in maximum (Tmax) and minimum (Tmin) temperatures throughout crop seasons, frequency of droughts and surplus rainfall years, landslides etc. Demographic and biophysical data were used to determine sensitivity, such as population density, soil properties, forest cover, slope etc. Adaptive capability was assessed using socioeconomic indicators such as literacy rates, sanitation, irrigation availability, livestock density, rural employment patterns etc. All indicators were standardized to calculate sub-district level climate change vulnerability using a weighted aggregation method. We identified the most vulnerable sub-istricts

P. Pandey · S. Saha (✉) · A. Kumar · A. Kumar · R. Devdas
ICAR Research Complex for North Eastern Hill Region, Sikkim Center, Tadong, Sikkim, India

B. U. Choudhury · D. Chakraborty · S. Hazarika
ICAR Research Complex for North Eastern Hill Region, Umiam, Meghalaya, India

S. Choudhury
ICAR Research Complex for NEH Region, Mizoram Centre, Kolasib, Mizoram, India

B. Das
ICAR-Central Coastal Agricultural Research Institute, Old Goa, Goa, India

M. Ahmed
India Meteorological Department, Meteorological Centre, MoES, Srinagar, India

S. Mandal et al. (eds.), *Climate Change Vulnerability*, Climate Change Management, https://doi.org/10.1007/978-3-032-16783-5_5

of Arunachal Pradesh (Doimukh SDO and Koyu EAC), Meghalaya (Shella Bholaganj, Mawsynram, and Amlarem), Manipur (Phungyar and Kasom Khullen), Mizoram (Chawngte and Lawngtlai), Nagaland (Kohima Sadar and Wokha Sadar), Sikkim (Gangtok and Rongli), and Tripura (Jirania, Sadar, Sabroom, and Mohanpur). Our findings showed substantial regional variation in vulnerability across the area, exhibiting the hotspots with high exposure and limited adaptive capability. Thus, our study offered vital insights for policymakers, planners, and local stakeholders as they construct focused adaptation policies and encourage climate-resilient development in the north-east Indian hill states.

Introduction

Earth's natural ecosystem will continue to function as it should have for all of recorded time with periodic regularity. Anthropogenic disturbances from developmental activities of progressive human civilization progressively compromised or often disrupted the natural pace and imbued the global climate change issue (IPCC 2007). Accelerated atmospheric warming of the land-ocean-atmosphere system and associated regional precipitation anomalies pose the risk of frequent natural disasters. This often resulted in extensive property damage and loss of life across the globe. The geographical perimeter of NEI hillside is its palpable blend of fragile Himalayan ecosystem, unique weather pattern, rich biodiversity, and diverse sociocultural environment (Choudhury et al. 2022), primarily outlining the distinctness of the region's climate change vulnerability from the rest of India (Das et al. 2009; Chakraborty et al. 2017). Climate change-driven rise in long-term local weather anomalies has increased the vulnerability of farming communities residing in the Eastern Himalayas. The impact of climate change on regional agricultural productivity, water, and forest resources is uncertain, with varying degrees of spatiotemporal variability across the NEI states (Ravindranath et al. 2011; Maiti et al. 2017).

An increase in surface air temperature and precipitation variability brought on by climate change has worsened natural disasters in recent years. This hurts crop production, food security, and socioeconomic development globally (Wassmann et al. 2009). The crop production sector, particularly in developing nations like India, is at risk due to more frequent and severe extreme weather events (Halder et al. 2023; Khan et al. 2025). The temperature on the global surface has already gone up by roughly 1.1 °C and is expected to go up by 1.4–4.4 °C by 2100 (IPCC 2023). In both developing and underdeveloped countries, agriculture is responsible for 63% of cumulative losses and about 26% of climate-related impacts (FAO 2021). Three prime factors influenced this agricultural vulnerability: sensitivity to damage, exposure to climate phenomena, and adaptive capacity for recovery or adjustment (IPCC 2022). As sensitivity and adaptive capacity determine the extent of impacts and coping ability (Brooks and Adger 2004; Mandal et al. 2018), high exposure raises the risks of crop failure and decreased productivity (Lobell and Gourdji 2012; Mandal et al. 2015). In order to measure these aspects and direct climate-resilient agricultural planning, indicator-based vulnerability assessments under the IPCC framework have become an essential tool.

Climate Change in North East India (NEI)

With a population of about 40 million spread over ~8% of India's geographical area, NEI became vulnerable to climate change because of its complex orography, reliance on *rainfed* agriculture, and delicate Eastern Himalayan ecosystem functioning often contrasted with the all-India long-term climate pattern (Jain et al. 2012; Saha et al. 2017). The NEI region experienced a significant warming trend between 1901 and 2009 (+0.77 °C) (Kumar et al. 2013). Localized rise in temperature (~0.06 °C year^{-1}) and precipitation (6.5 mm year^{-1}) have changed the traditional sowing seasons with increased climate risks (Shrestha et al. 2012). Divergent Tmax (+1.02 °C 100 year^{-1}) and Tmin (−0.32 °C 100 year^{-1}) trends (Deka et al. 2009), increasing atmospheric dryness, fewer sunshine hours, and shifting diurnal temperature range (Chakraborty et al. 2025; Saha et al. 2016; Jhajharia and Singh 2011; Saha et al. 2022 are examples of the significant spatiotemporal variability. Higher altitudes experience more warming, particularly during the winter and after the monsoon season (Gautam et al. 2024). Heatwaves, extended dry spells, and fewer cold extremes are examples of extreme events that are occurring more frequently (Saha et al. 2022; Kalita et al. 2023). By 2100, temperatures are predicted to rise by 3–5 °C under SSP585, with Tmax and Tmin surpassing 4 °C and 5.5 °C, respectively (Mishra et al. 2020; Paul and Maity 2023).

Even though NEI receives high rainfall annually, the region experiences significant spatial variability, 1577–6002 mm, and "winter water scarcity" as a result of dwindling winter precipitation (Jhajharia et al. 2009; Saha et al. 2021). About 60–70% of the yearly rainfall is attributed to the monsoon, but trends indicate a long-term decrease in summer monsoon rains across NEI (Guhathakurta and Revadekar 2016; Saha et al. 2024), with regional variability—increases in Meghalaya and Mizoram and decreases in Assam, Arunachal, and Nagaland (Saha et al. 2016; Bora et al. 2022). El Niño-Southern Oscillation (ENSO), Indian Ocean Dipole (IOD), Atlantic Sea Surface Temperature fluctuations, Eurasian snow cover, and human-generated aerosols are the main causes of shifts (Dogar et al. 2017, 2023, 2024, 2025; Azharuddin and Dogar 2016; Krishnan et al. 2016). A mixed trend pattern was recorded for rainfall extremes, while consecutive dry days are increasing and stressing crops. Higher one-day maximum rainfall in parts of Arunachal Pradesh and Meghalaya increases the risk of floods and landslides (Guhathakurta et al. 2011; Saha et al. 2022). At present, monsoon months often experience coincident drought and flood-like events (Dubey et al. 2022). Urbanization, deforestation, and anthropogenic land-use changes add to warming and change rainfall (Tahir et al. 2025; Roy et al. 2015; Pathak et al. 2024). The Coupled Model Intercomparison Project-6 (CMIP-6) generated future projections of Coordinated Regional Climate Downscaling Experiment—South Asia (CORDEX-SA) projections indicated similar warming trend will continue in coming days (0.23–0.52 °C decade^{-1}) along with changes in regional precipitation pattern particularly during post-monsoon (increasing) and winter (decreasing) months (Kumar and Dimri 2018; Paul and Maity 2023). Rising temperatures and hydrological extremes raise the risk of crop failure, glacial lake outburst floods, flash floods, and landslides, even though increased rainfall may benefit agriculture (Mandal and Choudhury 2014; Khan et al. 2025). To maintain agriculture and livelihoods of hill farmers in NEI, location-specific adaptive measures like water harvesting, climate-resilient crops, and better land-use management are essential to address future agricultural sustainability (Chakraborty et al. 2025; Das et al. 2021).

Climate Change Vulnerability of NEI States

Local weather anomalies associated with global climate change are making farming communities in NEI more vulnerable. In order to improve food security, disaster preparedness, and livelihood diversification, policymakers must conduct composite vulnerability assessments utilizing bio-physical and socio-economic indicators, given the region's reliance on rainfed agriculture (Brooks and Adger 2004; Saha et al. 2015; Mandal et al. 2018). Several earlier reported studies on state and district-level climate change vulnerability assessment gave out the localized variability in both bio-physical and socioeconomic indicators within the delineated district boundaries of NEI states. To enhance preparedness, local adaptability, and comprehensive disaster management strategy formulation, research that focuses on the human aspects of climate change impact studies should be the basis for targeted critical policy interventions (Brooks and Adger 2004).

Our present interest would be restricted mostly to the agriculture sector. Indicator-based assessment identified greater vulnerability of thickly populated southern slopes of the Eastern Himalayas (including Bhutan, Nepal, and NE India) over the sparsely populated Chinese Tibet regions (Sultan et al. 2022). Combined bio-physical and socio-economic vulnerability assessment of Indian states scored several NEI states within the top 10 most vulnerable category, viz. Nagaland (highest) > Sikkim > Meghalaya > Arunachal Pradesh > Manipur (Das 2013). Only Assam, followed by Tripura, remained moderately to less vulnerable. Yomcha et al. (2019) identified Manipur as the most vulnerable state for climate hazards with lower adaptive capacity and the least for Tripura with higher adaptive capacity across NEI. In contrast, 38 indicator-based analyses showed the least district-level agricultural vulnerability for all NE states among 572 districts of India (Rao et al. 2016). Futuristic district-level modeling studies projected the highest decrease in crop yield at North Sikkim from climate vagaries, limited agricultural infrastructural facilities (Ravindranath et al. 2011; Maiti et al. 2017). In contrast, the highest agricultural vulnerability from low input levels was identified for Arunachal Pradesh (Tirap, West Siang and Changlang) and Assam (Nalbari and Dibrugarh) under the

A1B scenario (Ravindranath et al. 2011). Agricultural vulnerability decreased from hill districts (Tamenglong, Churachandpur and Chandel) to the valleys regions of Manipur (Feroze et al. 2014). The higher degree of climate change-induced social vulnerability was sourced from the imbalanced distribution of infrastructural facilities across Anjaw, Upper Siang, West Siang, Lower Dibang Valley, East Siang, East Kameng and KurungKurmey districts of Arunachal Pradesh (Maiti et al. 2017). The gradient of spatial decline of agricultural vulnerability was extended from North to South and projected to reduce from high to moderate category across NEI in the near future.

The intricate linkage between poverty, environmental stressors, and vulnerability must be addressed in the vulnerability assessments of hill agriculture (Shukla et al. 2017). In general, the exposures of tribal communities towards climate change were much higher with greater sensitivity and lesser adaptation capacity than non-tribal residents of NEI (Roy et al. 2023). An individual household survey-based participatory community-based approach for climate change vulnerability assessment provided the baseline for framing policy formulation towards future livelihood development in a few villages of Arunachal Pradesh, experiencing frequent crop yield loss, livestock mortality, and landslide and flood events from the unpredictable weather conditions (Rehman et al. 2023). Sole dependency on shifting cultivation without practicing any adaptation option increased vulnerability of farm families at Aizawl and Champhai of Mizoram (Thong et al. 2022) and Chandal of Manipur (Devi and Choudhury 2013). The increased prevalence of crop diseases became a sequel of rising temperature along with decreased annual precipitation quantity with more advent of droughts in Nagaland (Schröder et al. 2024). The highest drought vulnerability of Zutovi village (Dimapur district) was contrasted with the highest overall livelihood vulnerability to climate change-induced natural disasters in the Niuland block of Dimapur district, Nagaland (Kumari et al. 2023). Rising temperatures coupled with increased precipitation quantity and intensity will amplify the need to adapt soil and water conservation practices through livestock ownership and participation in civil society organizations (Schröder et al. 2024). Agricultural extension services should facilitate better land management techniques, efficient decision making, crop diversification and viable market linkage. This will ensure indigenous communities have greater adaptability to climate change for farmers practicing shifting agriculture in the lower Assam region of NE India (Hazarika et al. 2024).

The NEI forests were resilient to climate extremes with low to medium inherent vulnerability (Sharma et al. 2017). The ecological vulnerability of low-density forest cover was higher than very dense or moderately dense forests in the eastern Himalayas (Gupta et al. 2024). The primary impact of climate change was more on the forest cover of the northern region, with a gradient of increasing vulnerability towards the southern region of NEI. This is due to a higher degree of human disturbances in the fragmented forest cover, which will decrease further in future climate change scenarios (Ravindranath et al. 2011). The climate change-driven weather extremes in the Hindu Kush Himalayas have a pernicious impact on the forest ecosystem, which is threatening the food and water security of indigenous populations (Wester et al. 2019).

The Brahmaputra basin, Nagaland, Manipur, Mizoram, Tripura, and Meghalaya often experience regions with high vulnerability to seasonal drought due to the increased flood vulnerability (Ravindranath et al. 2011). District level vulnerability based on unequal weights identified in Champhai, Mizoram (Lalthanpuia et al. 2022), Dhalai (Roy et al. 2023), Churachandpur and Senapati (Manipur) (Sanayanbi et al. 2021), as the most vulnerable districts in terms of water resource, livelihood, and flood risk management of NEI. However, the contributing location-specific indicators as top drivers of vulnerability assessment are not identical in regional climate change vulnerability assessment studies. In Mizoram, limited perennial spring availability per household, shrinking forest cover, and scarce groundwater resources became the top drivers of climate change vulnerability (Lalthanpuia et al. 2022). Therefore, localized vulnerability assessment methods should be redesigned in the heterogeneous Himalayan regions of NEI from project-type to process-type approaches to improve the livelihoods of the agriculture sector and the inhabiting communities. With this background, we have set up a methodological framework for mapping the agricultural vulnerability at the sub-district level in NE Indian states with the following objectives:

1. To determine the highly vulnerable sub-districts of NEI that require immediate attention from the policy makers.
2. To help stakeholders and policymakers devise adaptation strategies and climate risk management plans that are tailored to their respective sub-districts of NEI.

Methodology of Sub-District Level Vulnerability Assessment

Study Area

Our present study area for sub-level vulnerability assessment focuses on the North-Eastern Indian hill states, including Meghalaya, Nagaland, Manipur, Mizoram, Sikkim, Arunachal Pradesh, and Tripura. The study region collectively comprised 131 districts (~661sub-districts; Fig. 5.1), characterized by variable topography, socioeconomic disparities, abundant biological diversity, and frequent exposure to climatic extremes such as droughts, landslides, and floods, etc. of Eastern Himalaya, including Meghalaya Plateau and *Patkai* range at Indo Myanmar Border. Due to its ecological sensitivity and developmental challenges, this region presents a critical landscape for evaluating climate change susceptibility. We adapted the IPCC's conceptual framework, which defines vulnerability as a function of three components, namely: sensitivity, exposure adaptive capacity. According to this, vulnerability is a result of the amount and grade to which an entity is exposed, the entity's sensitivity

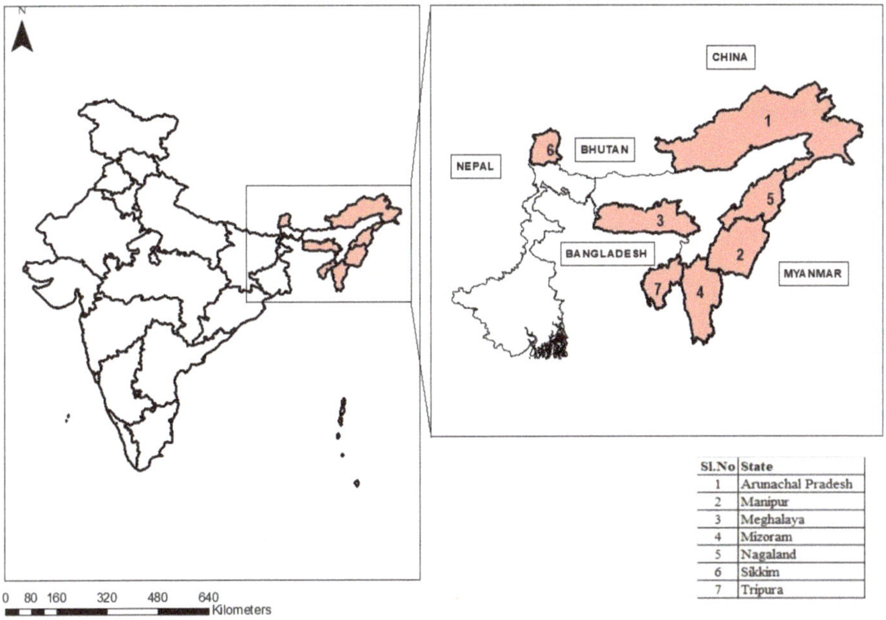

Sl.No	State
1	Arunachal Pradesh
2	Manipur
3	Meghalaya
4	Mizoram
5	Nagaland
6	Sikkim
7	Tripura

Fig. 5.1 Delineated boundary of North East Indian (NEI) hill region apprised for sub-district-level climate change vulnerability analysis

to climate change, and the adaptive capability to adjust and deal with changing climate.

Selection of Indicators

A total of 34 indicators covering all three components were selected based on data accessibility, regional relevance, broad reviews on previously published literature (O'Brien et al. 2004; Rao et al. 2016; Gbetibouo and Ringler 2009; Hiremath and Shiyani 2013; Monterroso et al. 2014; Newton and Weichselgartner 2014).

1. *Exposure indicators:* The daily subdistrict-wise rainfall and temperature data from 1981 to 2024 were collected from IMD gridded data (Pai et al. 2014). The daily data were converted to annual and monthly, and the Tmax and Tmin data were also distributed with respect to kharif (June to September) and rabi (October to March) seasons. Total 13 indicators were considered under exposure from climate variablity, that includes Variability in annual rainfall (CoV%), Rate of change in maximum temperature during *kharif* (ΔTmax_kharif; °C year^{-1}), Rate of change in maximum temperature during *rabi* (ΔTmax_rabi; °C year^{-1}), Rate of change in minimum temperature during *kharif* (ΔTmin_kharif; °C year^{-1}),

Rate of change in minimum temperature during *rabi* (ΔTmin_rabi; °C year^{-1}), Change in rainy days (RD; days year^{-1}), Frequency of excess rainfall years (ER; total number of years receiving annual rainfall ≥120% of normal rainfall), Frequency of moderate drought years (MDY; Total number of years receiving annual rainfall between 26–50% of normal rainfall), Frequency of severe drought years (SDY; Total number of years receiving annual rainfall <50% of normal rainfall), No of average landslides (LSD; Landslides Atlas and newspaper articles between 2011–2020), Heat wave frequency (HW; events year^{-1}) and Change in Monsoon intensity (MI; days year^{-1}), along with Average R factor (MJ mm ha^{-1} h^{-1} year^{-1}; https://esdac.jrc.ec.europa.eu/).

2. *Sensitivity indicators:* As these indicators show the degree to which a system is affected, a total of 11 indicators were selected. Human density (HD; Persons km^{-2}), illiterate population (ILP; %), accessibility of water sources (WR) all taken from Census 2011 data, while rest indicators organic C content of soil (SOC; g kg^{-1}), Water holding capacity of soil (TAWC; cm m^{-1}) and soil pH (source: https://soilgrids.org/), Soil erosion (EROS in kg ha^{-1} year^{-1}: https://esdac.jrc.ec.europa.eu/), Vegetation degradation (VEG DEG%; source: https://vedas.sac.gov.in/desertification/index.html), Normalized Difference Vegetation Index—NDVI profile (10 day max NDVI-10 m; https://vedas.sac.gov.in/vegetation-monitoring/index.html), Forest cover(FRC in km^{-2}; National Bureau of Soil Survey and Land Use Planning, National Bureau of Soil Survey and Land Use Planning) andmaximum slope (degrees; https://earthexplorer.usgs.gov/), soil pH (https://data.isric.org/geonetwork/srv/eng/catalog.search#/home, https://esdac.jrc.ec.europa.eu) having direct impact on agriculture system directly or indirectly.
3. *Adaptive capacity indicators:* This indicator should represents ability to cope/adjust with the changes, so a total of 10 indicators were selected including, Percentage of rural population (RURAL%), Literacy (LTRCY%), rural worker population (RW%), gender gap (GGAP), dependency on electricity (ELEC%), household with sanitation facilities (SANF%) derived from Census 2011 data, while indicators like large livestock density (LD_large; km^{-2}), small livestock density (LD_small; km^{-2}; https://www.data.gov.in/catalog, Livestock Census 2012), Net irrigated area (NIA%) and Groundwater availabilityfrom CGWB-Central Ground Water Board (GW in ha-mkm^{-2}; https://ingres.iith.ac.in and, nasagrace.unl.edu).

Data Normalization

Using min–max normalization, which scales the values of each parameter between 0.1 and 1, the indicator data were normalized independently for each NEI state. Through this transformation, indicators become directly comparable across various dimensions and independent of units. Following Rao et al. (2016), the following formulas were used to treat the positive and negative functional relationships between indicators and vulnerability separately.

$$X' = 0.1 + \frac{X - X_{\min}}{X_{\max} - X_{\min}} \times 0.9 \tag{5.1}$$

$$X' = 1.1 - 0.1 - \frac{X - X_{\min}}{X_{\max} - X_{\min}} \times 0.9 \tag{5.2}$$

Where X, X_{min}, and X $_{max}$ are the actual, minimum, and maximum values of the indicators.

For indicators that were positively correlated with vulnerability (higher value = greater vulnerability), Eq. (5.1) was utilized, whereas Eq. (5.2) was applied to indicators that were negatively correlated (higher value = lower vulnerability). Since no missing data were observed in the dataset, inconsistencies across indicators were handled entirely through normalization, ensuring comparability and uniform scaling of values.

Weight Calculation

To eliminate subjectivity, Principal Component Analysis (PCA) was used to calculate weights for each indicator in the three categories (exposure, sensitivity, and adaptive ability) separately for each NEI state. The PCA was carried out using correlation matrices, and indicators with substantial loadings in the first five principal components were kept. Each indicator was weighted according to its factor loading and eigenvalue contribution (Mandal et al. 2018). The PCA-derived weights represented the relative relevance of each indicator through offering a statistically solid alternative to subjective weighing.

Vuelnerability Index (VI) Calculation and Spatial Mapping

In order to calculate weighted indicator values, each indicator was multiplied by the corresponding PCA-derived weight and then summed up to derive the Sensitivity Index, Exposure Index, and Adaptive Capacity Index, respectively. Finally, the state-wise Vulnerability index (VI) was calculated as follows (Jakariya et al. 2020):

$$VI = (Exposure\,Index + Sesnsitivity\,Index) - Adavtive\,Capacity\,index \tag{5.3}$$

The inverse distance weighting (IDW) method was adapted for spatial interpolation and vulnerability mapping using ArcGIS 10.7.1 software.

Result and Discussion

Our present study on the indicator-based sub-district level vulnerability assessment involved all three components of climate change vulnerability under the IPCC-2007 framework, viz., sensitivity, and adaptive capacity in finer resolution, subjugating the limitations of past district level vulnerability studies of NE India (Ravindranath et al. 2011; Rao et al. 2016). Limited exposure, low sensitivity, and strong adaptive capacity aggregated into low district-level vulnerability. Even so, the relative contribution of driving indicators to sub-district level vulnerability was not identical among the administrative state boundaries of NEI hills (Fig. 5.1). Exposure frequency of moderate and severe drought occurrence acquired the nil weightage for Sikkim sub-districts, yet received higher loading for the sub-districts of Meghalaya, Manipur and Mizoram state (Fig. 5.2). The alteration in seasonal maximum and minimum temperature got moderate weightage for all the studied NEH hill states and the least for human density as sensitivity parameter for Meghalaya, Manipur, Mizoram and Nagaland sub districts. Manipur received the minimum weightage for soil erosion and population literacy factor. In order to prioritize location-specific policy formulation options, we set forth our outcome of vulnerability analysis study for each state boundary of NEI hill states, separately as summarized in Table 5.1.

1. **Arunachal Pradesh:** Substantial variability in district-level exposure, sensitivity, and adaptive capacity was evident across Arunachal Pradesh (Fig. 5.3a–c). Low exposure across both western and eastern high-altitude areas was at odds with higher exposure levels in foothill and central valley region including Gumto, Naharlagun, and Balijan sub-districts (Fig. 5.3a). Manmao, Soha, and Namtok sub-districts showed a high sensitivity level (Fig. 5.3b), with virtuous adaptive capacity for Lekang and Chongkham sub-districts (Fig. 5.3c). High exposure and poor adaptive capacity consequence higher sub-district level vulnerability at

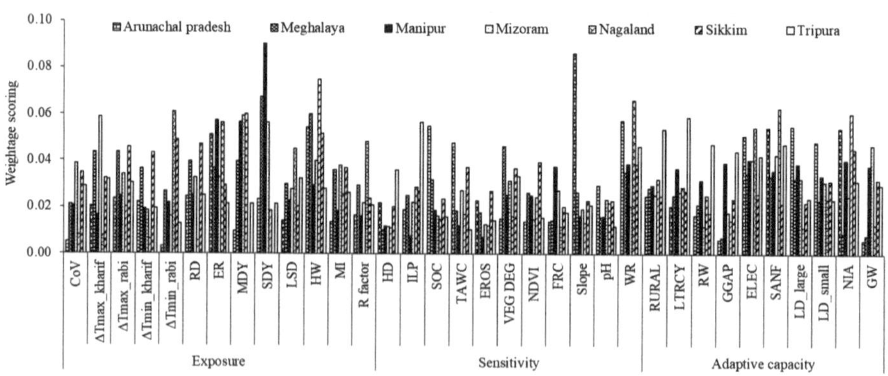

Fig. 5.2 Magnitude of driving indicators for overall sub-district level vulnerability assessment of the hill states in the NEI region

Abbreviation	Parameters
CoV	Variability in annual rainfall (coefficient of variation)
ΔTmax_kharif	Rate of change in maximum temperature during the kharif season
ΔTmax_rabi	Rate of change in maximum temperature during the rainy season
ΔTmin_kharif	Rate of change in minimum temperature during the kharif season
ΔTmin_rabi	Rate of change in minimum temperature during the rabi season
RD	Change in annual rainy days
ER	Frequency of excess rainfall years
MDY	Frequency of moderate drought years
SDY	Frequency of severe drought years
LSD	Number of average landslides
HW	Heat wave frequency
MI	Change in monsoon intensity
R factor	Rainfall erosivity factor
HD	Human density
ILP	Illiterate population
SOC	Organic C content of soil
TAWC	Water holding capacity of soil
EROS	Soil erosion
VEG DEG	Vegetation degradation
NDVI	NDVI profile
FRC	Forest cover
Slope	Maximum slope
pH	Soil pH
WR	Accessibility of water sources
RURAL	Percentage of rural population
LTRCY	Literacy percentage
RW	Percentage of rural worker population
GGAP	Gender gap
ELEC	Dependency on electricity
SANF	Sanitation facilities
LD_large	Livestock density (large)
LD_small	Livestock density (small)
NIA	Net irrigated area
GW	Groundwater availability

Doimukh, Koyu, Taraso, and Banderdewa (Fig. 5.3d). Low exposure and strong adaptive capacity manifested low vulnerability for Seppa, Dirang, and Bomdila. Overall, vulnerable hotspots were concentrated in the eastern and foothills, and very minimal over the central and western upland region of Arunachal Pradesh.

2. **Meghalaya:** Sub-district level analysis identified a wide range of exposure with variable degrees of sensitivity and adaptive capacity over Meghalaya (Fig. 5.4a–c). Low exposure level was observed in northern Meghalaya that increased

Table 5.1 Sub-district vulnerability hotspots, low-risk zones, and key adaptation needs in NEI hill states

State	High vulnerability sub-district[a]	Low vulnerability sub-district[a]	Key drivers	Driving indicators	Adaptation strategy (for highly vulnerable sub-districts)
Arunachal Pradesh	Doimukh (Papum Pare), Koyu (East Siang), Taraso (Papum Pare), Banderdewa (Papum Pare)	Seppa (East Kameng), Dirang (West Kameng), Bomdila (West Kameng)	High exposure in foothills; weak adaptive capacity	(i) Heat wave frequency	(i) Cultivation of short duration/heat-tolerant crop varieties, optimization of sowing window, crop rotation, precision water management (mulching, drip/sprinkler), crop insurance, nursery management/protected cultivation, foliar application of water, KNO_3, melatonin, or micronutrients (Zn, Fe and B), soil health improvement through diversified intercropping and agroforestry systems, real time dissemination of weather based advisory services
				(ii) Change in minimum temperature	(ii) Avoid cultivation of temperature-sensitive crop varieties/breeding lines, shifting sowing window, crop diversification, precision water and shade management in protected cultivation, strengthen capacity building, and improve extension services
				(iii) Rainfall erosivity (R-factor)	(iii) Afforestation and establish vegetative barriers, soil conservation measures viz. cover cropping, reduced tillage, contour ploughing, soil mulching, legume-based crop diversification, contour cultivation, run off farming, agroforestry-based farming systems establishment, and efficient drainage management through watershed management programme
Meghalaya	Shella Bholaganj (East Khasi Hills), Mawsynram (East Khasi Hills), Amlarem (West Jaintia Hills)	Kharkutta (North Garo Hills), Bajengdoba (North Garo Hills)	Extreme rainfall; high sensitivity in western Meghalaya	(i) Severe drought frequency	(i) Drought resilient crop varieties, crop insurance, soil mulching, improving soil health management/organic matter application, community rainwater harvesting, precision water management/conjunctive water use, crop diversification, foliar application of plant hormone (salicylic acid), periodic dissemination of medium-range weather forecasting-based agro advisory services
				(ii) Heat wave frequency	(ii) Heat-tolerant crop varieties, poly-house cultivation, sowing window adjustment, crop diversification and crop insurance, foliar application of water, KNO_3, melatonin, or micronutrients (Zn, Fe and B), and real-time dissemination of weather-based agro-advisory services

State	High vulnerability sub-district[a]	Low vulnerability sub-district[a]	Key drivers	Driving indicators	Adaptation strategy (for highly vulnerable sub-districts)
Manipur	Phungyar (Kamjong), Kasom Khullen (Kamjong)	Patsoi (Imphal West)	Rising exposure from NE–SE; weak adaptive capacity	(i) Severe to moderate drought years frequency	Drought-tolerant variety cultivation, crop insurance, establishment of community water harvesting and storage facility, precision irrigation tools crop diversification conservation agriculture, foliar application of plant hormone (salicylic acid), promotion of fruit crop/agroforestry based farming systems and further strengthening of extension services on drought contingency planning and integrated watershed development, periodic dissemination of medium range weather forecasting based agro advisory services
				(ii) Frequency of excess rainfall years	Cultivation of short duration flood tolerant crop varieties, construction of ridges and drainage channels/farm dams or on-farm reservoirs, crop insurance, protected cultivation, soil mulching, repositioning planting dates, cover crop cultivation, foliar application of K salts/salicylic acid/ silicon compounds/brassinosteroid analogues, crop diversification, integrated farming system establishment, diversifying rural livelihood beyond agriculture, resource sharing and timely meteorological information dissemination network
Mizoram	Chawngte (Lawngtlai), Lawngtlai (Lawngtlai)	Aibawk (Aizawl), Champhai (Champhai)	High exposure, low adaptive capacity	(i) Severe to moderate drought frequency	Creation of community water harvesting and storage, precision irrigation tools, crop insurance, cultivation of drought-tolerant crop variety, crop diversification, surface mulching, soil health improvement through organic matter application, foliar application of plant hormone (salicylic acid), promotion of fruit crop/agroforestry based farming systems and further strengthening of extension services on drought contingency planning and integrated watershed development, periodic dissemination of medium range weather forecasting based agro advisory services
				(ii) Rate of change in Tmax	Cultivation of short-duration/heat-tolerant crop varieties, precision water management, sowing window adjustment, intercropping, crop insurance, livelihood diversification, protected cultivation/shade management, strengthen capacity building, and improve extension services

(continued)

Table 5.1 (continued)

State	High vulnerability sub-district[a]	Low vulnerability sub-district[a]	Key drivers	Driving indicators	Adaptation strategy (for highly vulnerable sub-districts)
Nagaland	Shamator (Shamator)	Jalukie (Peren)	Eastward exposure rise; sensitivity in southern Nagaland	(i) Heat wave frequency	Cultivation of short-duration/heat-tolerant crop varieties, sowing window adjustment, precision water management, crop diversification, crop insurance, foliar application of water, KNO_3, melatonin, or micronutrients (Zn, Fe and B), soil health management through promotion of alder-based agroforestry systems and polyhouse cultivation, and real-time dissemination of weather-based advisory services
				(ii) Accessibility of water sources	Boost community-based rainwater harvesting along with the roof water harvesting technique for domestic use, precision irrigation management, establishment of a water storage network coupled with gravity-based irrigation lining
Sikkim	Gangtok (Gangtok), Rongli (Pakyong)	Chungthang (Mangan), Mangan (Mangan)	High exposure and sensitivity, low adaptive capacity	(i) Rate of change in Tmin	Avoid cultivation of sensitive crop varieties/breeding lines, adjust sowing window, crop diversification, shade management in protected cultivation, and temperature moderation through precision water management, strengthen capacity building, and improve extension services
				(ii) Change in rainy days	Cultivation of drought/flood-tolerant crop varieties, crop diversification, rainwater harvesting, and facilitate micro-irrigation, practicing conservation agriculture, spring-shed management, crop insurance, promoting climate-smart urban-agriculture, fruit crop/agroforestry-based farming system establishment in an integrated organic farming system mode, and diversifying rural livelihood beyond agriculture
Tripura	Jirania (Sipahijala), Sadar (West Tripura), Sabroom (South Tripura), Mohanpur (West Tripura)	Bishalgarh (Sipahijala), Sonamura (Sipahijala)	Mixed patterns; exposure in southern sub-districts	(i) Literacy percentage	(i) Boost skill development; improve extension services. Expand climate literacy and awareness initiatives, community campaign initiatives on digital literacy promotion, ICT tools usage, etc.
				(ii) Percentage of rural population	(ii) Management of rural workforce and employment generation schemes, strengthening rural health service, community participation in targeted rural enrolment schemes, and increasing accessibility of financial services
				(iii) Accessibility for electricity	(iii) Encourage rural electrification, renewable energy use, solar home systems, solar farming, and capacity building activities

[a] Note: District name mentioned in the parentheses

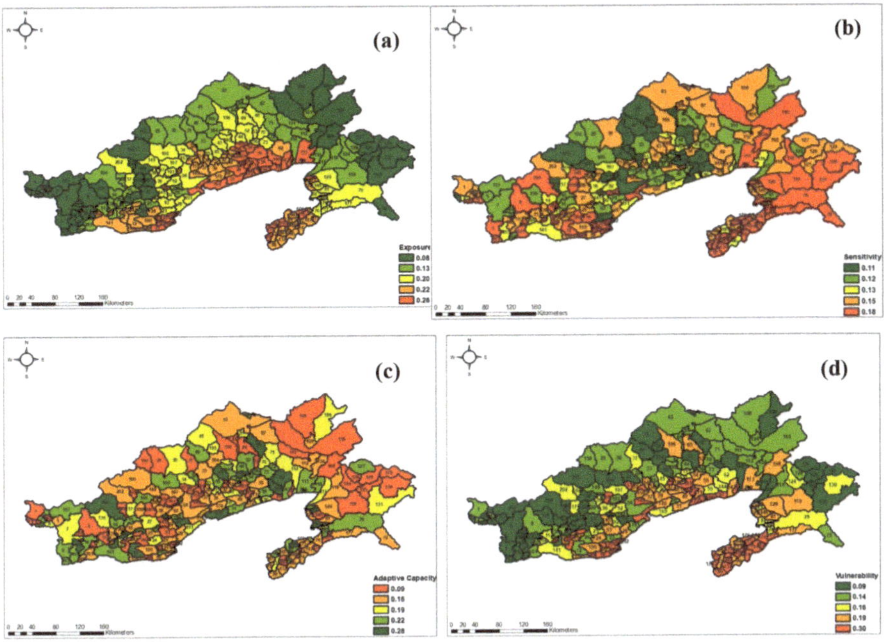

Fig. 5.3 Sub-district level (**a**) exposure, (**b**) sensitivity, (**c**) adaptive capacity, and (**d**) vulnerability profile of Arunachal Pradesh towards climate change
(Sub district notations: (1) Zemithang Circle, (2) Lumla ADC, (3) Tawang HQ, (4) Kitpi Circle, (5) Jang ADC, (6) Mukto Circle, (7) Dirang ADC, (8) Nafra EAC, (9) Kalaktang ADC, (10) Singchung ADC, (11) Thrizino ADC, (12) Pakke Kessang EAC, (13) Seppa HQ, (14) Bameng ADC, (15) PipuDipu Circle, (16) GyawePurang Circle, (17) Chayangtajo ADC, (18) Sawa Circle, (19) Balijan ADC, (20) Taraso Circle, (21) Naharlagun EAC, (22) Doimukh SDO, (23) Toru Circle, (24) Sagalee ADC, (25) Parang Circle, (26) Leporiang Circle, (27) Mengio EAC, (28) Kimin SDO, (29) Limeking Circle, (30) Nacho SDO, (31) Siyum ADC, (32) Taliha ADC, (33) Payeng Circle, (34) Giba Circle, (35) Chetam Peer Yapu Circle, (36) Daporijo HQ, (37) Puchi Geko Circle, (38) Dumporijo ADC, (39) Baririjo Circle, (40) Mechuka ADC, (41) Monigong EAC, (42) Kaying EAC, (43) Darak Circle, (44) Rumgong ADC, (45) Liromoba EAC, (46) Yomcha ADC, (47) Tirbin EAC, (48) Basar ADC, (49) Daring Circle, (50) Gensi EAC, (51) Likabali SDO, (52) Boleng ADC, (53) Riga EAC, (54) Pangin EAC, (55) Kebang Circle, (56) Rebo Perging Circle, (57) Koyu EAC, (58) Nari ADC, (59) Bilat Circle, (60) Ruksin ADC, (61) Oyan Circle, (62) Mebo ADC, (63) Tuting ADC, (64) Migging Circle, (65) Palling Circle, (66) Gelling Circle, (67) Singa Circle, (68) Yingkiong Circle, (69) Jengging Circle, (70) Geku EAC, (71) Mariyang ADC, (72) MopomAdipasi Circle, (73) Katan Circle, (74) Khimiyang EAC, (75) Changlang HQ, (76) Manmao EAC, (77) Nampong SDO, (78) Vijoynagar EAC, (79) Miao ADC, (80) Kharsang Circle, (81) Diyun EAC, (82) Bordumsa ADC, (83) Deomali ADC, (84) Soha Circle, (85) Khonsa HQ, (86) Dadam Circle, (87) Kanubari ADC, (88) Lawnu Circle, (89) Pangchao SDO, (90) Laju SDO, (91) Ziro HQ, (92) Yachuli ADC, (93) Yazali Circle, (94) Kamporijo Circle, (95) Dollungmukh Circle, (96) Palin Circle, (97) Yangte Circle, (98) Sangram SDO, (99) Nyapin ADC, (100) Phassang Circle, (101) Chambang Circle, (102) Gangte Circle, (103) Tarak Lengdi Cicle, (104) Sarli Circle, (105) Damin Circle, (106) Pipsorang Circle, (107) Tali ADC, (108) Mipi, (109) Anini, (110) Etalin, (111) Anelih, (112) Kronli Arzoo Circle, (113) Roing HQ, (114) Dambuk ADC, (115) Tinali Paglam Circle, (116) Koronu Circle, (117) Sunpura Circle, (118) Tezu HQ, (119) Wakro Circle, (120) Chongkham EAC, (121) Namsai HQ, (122) Lathao Circle, (123) LekangMahadevpur Circle, (124) Hayuliang ADC, (125) Manchal Circle, (126) Goiliang Circle, (127) Chaglagam Circle, (128)
(continued)

towards the south southwest and southeast sub-districts, including Mawsynram, as the wettest sub-district of the earth. A higher sensitivity level of western Meghalaya was coupled with better adaptive capacity. Shella Bholaganj, Mawsynram, and Amlarem exhibited higher vulnerability, as Kharkutta and Bajengdoba sub-districts remained under the very low vulnerability category (Fig. 5.4d).

3. **Manipur:** The Higher sensitivity level of the south-southwestern sub-districts of Manipur was contrasted with the higher adaptive capacity of central Manipur (Fig. 5.5a, b). Evitable rising gradient of exposure extended from north eastern to southeastern Manipur was dominated by a low sensitivity level flexed higher sub-district level vulnerability (Fig. 5.5c). Phungyar and Kasom Khullen sub-districts of north eastern Manipur both exhibited the highest vulnerability, whereas Patsoi sub-districts of Imphal West district were the least vulnerable (Fig. 5.5d).
4. **Mizoram:** Higher exposure level together with low adaptive capacity exhibited higher sub district level vulnerability irrespective of marked variation in sensitivity (Fig. 5.6a–c). Chawngte and Lawngtlai sub-districts topped as highly vulnerable (Fig. 5.6d). Conversely, Aibawk and Champhai sub districts exhibited the lower vulnerability amongst all sub-districts of Mizoram.
5. **Nagaland:** Increased sub-district level exposure eastward was evident with a higher sensitivity level of southern Nagaland sub-districts (Fig. 5.7a, b). Wokha

Fig. 5.3 (continued)

Metengliang Circle, (129) Kibithoo Circle, (130) Walong Circle, (131) Hawai HQ, (132) Wakka SDO, (133) Longding HQ, (134) Hunli SDO, (135) Aalo HQ, (136) Lada Circle, (137) Dissing Passo Circle, (138) Pizirang Veo Circle, (139) Richukrong Circle, (140) Bana EAC, (141) Seijosa ADC, (142) Old Ziro SDO, (143) Namsing Circle, (144) Pasighat HQ, (145) Pistana Circle, (146) Piyong Circle, (147) Raga ADC, (148) Koloriang HQ, (149) Patuk ADC, (150) Taksing Circle, (151) Dudunghar Circle, (152) Gite Ripa Circle, (153) Nilling Circle, (154) Segi Gusar Circle, (155) Maro Circle, (156) Lhou Circle, (157) Thingbu Circle, (158) Bongkhar Circle, (159) Gepen Circle, (160) Rupa SDO, (161) Bhalukpong EAC, (162) Jamiri Circle, (163) Balemu Circle, (164) Thembang Circle, (165) Shergaon Circle, (166) Kamengbari Doimara Circle, (167) Bomdila HQ, (168) Desali Circle, (169) Kangku Circle, (170) Sibe Circle, (171) Kora Circle, (172) New Seren Circle, (173) Renuk Circle, (174) LyngokLongtoi Circle, (175) Yatdam Circle, (176) Namtok EAC, (177) Tikhak Rima Putok Circle, (178) Jairampur ADC, (179) Pumao Circle, (180) Kamba ADC, (181) Kombo Circle, (182) Nikte Kodum Circle, (183) Bagra Circle, (184) Tato EAC, (185) Pidi Circle, (186) Dambuen Circle, (187) Borduria Circle, (188) Katang Circle, (189) Longo Circle, (190) Sangdupota Basar Nello Circle, (191) Kakoi Circle, (192) Gumto Circle, (193) Banderdewa Circle, (194) Khenawa Circle, (195) JomloMobuk Circle, (196) Payum Circle, (197) Kantang Circle, (198) Sago, (199) Deed Circle, (200) Silsango Circle, (201) Nyobia Circle, (202) Polosang Circle, (203) Passa Valley Circle, (204) Buragaon EAC, (205) Pakoti Circle, (206) Debeyar Co Circle, (207) Borum, (208) Itanagar EAC, (209) Parbuk SDO

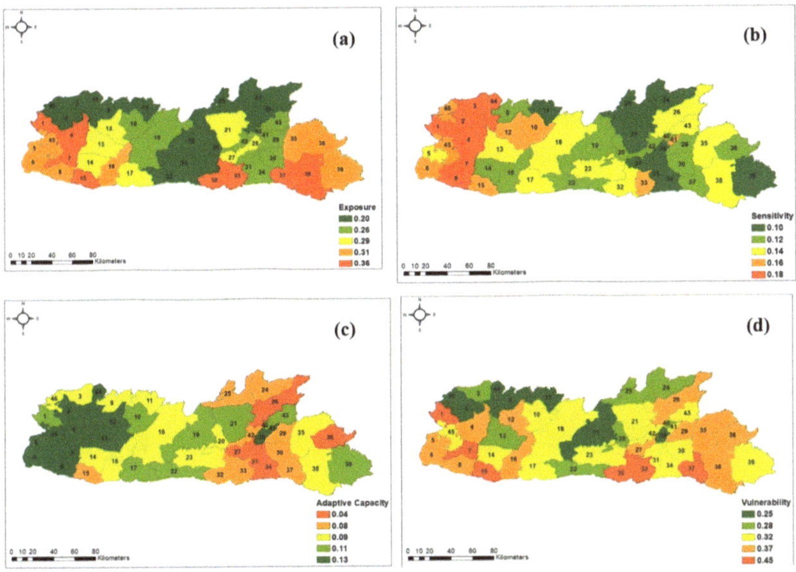

Fig. 5.4 Sub-district level (**a**) exposure, (**b**) sensitivity, (**c**) adaptive capacity, and (**d**) vulnerability profile of Meghalaya towards climate change
(Sub district notations:(1) Selsella, (2) Dadenggiri, (3) Tikrikilla, (4) Rongram, (5) Betasing, (6) Zikzak, (7) Gambegre, (8) Dalu, (9) Resubelpara, (10) Dambo Rongjeng, (11) Kharkutta, (12) Songsak, (13) Samanda, (14) Chokpot, (15) Gasuapara, (16) Baghmara, (17) Rongara, (18) Mawshynrut, (19) Nongstoin, (20) Mawthadraishan, (21) Mairang, (22) Ranikor, (23) Mawkyrwat, (24) Umling, (25) Jirang, (26) Umsning, (27) Mawphlang, (28) Mylliem, (29) Mawryngkneng, (30) Mawkynrew, (31) KhatarshnongLaitkroh, (32) Mawsynram, (33) Shella Bholaganj, (34) Pynursla, (35) Thadlaskein, (36) Laskein, (37) Amlarem, (38) Khliehriat, (39) Saipung, (40) Mawlai, (41) Mawpat, (42) Sohiong, (43) Bhoirymbong, (44) Bajengdoba, (45) Rerapara, (46) Demdema)

Sadar sub-district evinced the highest adaptive capacity, followed by Kohima Sadar sub-district, and the least for Monyakshu sub-districts (Fig. 5.6c). Eventually, Shamator and Jalukie became the most and least vulnerable sub-districts of Nagaland (Fig. 5.7d).

6. **Sikkim: A** Higher degree of sensitivity and exposure level with moderately low adaptive capacity warranted the highest vulnerability for Gangtok, followed by the Rongli sub-district (Fig. 5.8a–d). In contrast, Chungthang and Manganwere were categorized under the low agricultural vulnerability category.

Our present findings referred to substantial spatial variability in climate change vulnerability calculated based on various exposure, sensitivity, and adaptive capacity indicators across the hill states of NEI. Nevertheless, the research outcome will be the building blocks for designing precise, site-specific, efficient resource management planning for ensuring high farmers' profitability through future climate change adaptation strategies in the NEI hill region (Table 5.1).

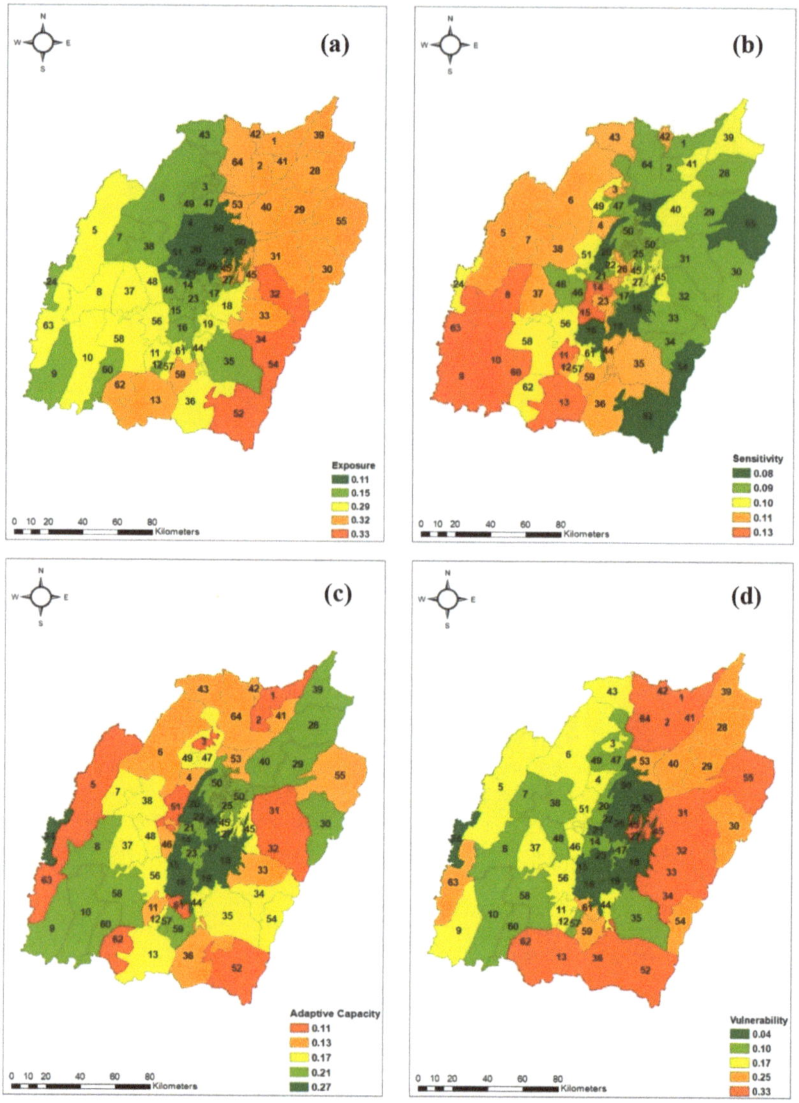

Fig. 5.5 Sub-district level (**a**) exposure (**b**) sensitivity (**c**) adaptive capacity and (**d**) vulnerability profile of Manipur towards climate change
(Sub district notations: (1) Paomata, (2) Purul, (3) Kangpokpi, (4) Saitu Gamphazol, (5) Tamenglong West, (6) Tamenglong North, (7) Tamenglong, (8) Nungba, (9) Tipaimukh, (10) Thanlon, 11) Tuibong, (12) Churachandpur, (13) Singngat, (14) Nambol, (15) Bishnupur, (16) Moirang, (17) Lilong, (18) Thoubal, (19) Kakching, (20) Lamsang, (21) Patsoi, (22) Lamphelpat, (23) Wangoi, (24) Jiribam, (25) Sawombung, (26) Porompat, (27) Keirao Bitra, (28) Chingai, (29) Ukhrul, (30) Kamjong, (31) Phungyar, (32) KasomKhullen, (33) Machi, (34) Tengnoupal, (35) Chandel, (36) Chakpikarong, (37) Khoupum, (38) Haochong, (39) Jessami, (40) Lm, (41) ChilivaiPhaibung, (42) Song Song, (43) Willong, (44) Waikhong, (45) Island, (46) Bungte Chiru, (47) Champhai, (48) Longmai, (49) T Vaichong, (50) Saikul, (51) Kangchup Geljang, (52) Khengjoy, (53) Lhungtin, (54) Moreh, (55) Sahamphung, (56) Kangvai, (57) Saikot, (58) Henglep, (59) Sangaikot, (60) Mualnuam, (61) Samulamlan, (62) Suangdoh, (63) Vangai Range, (64) Tadubi)

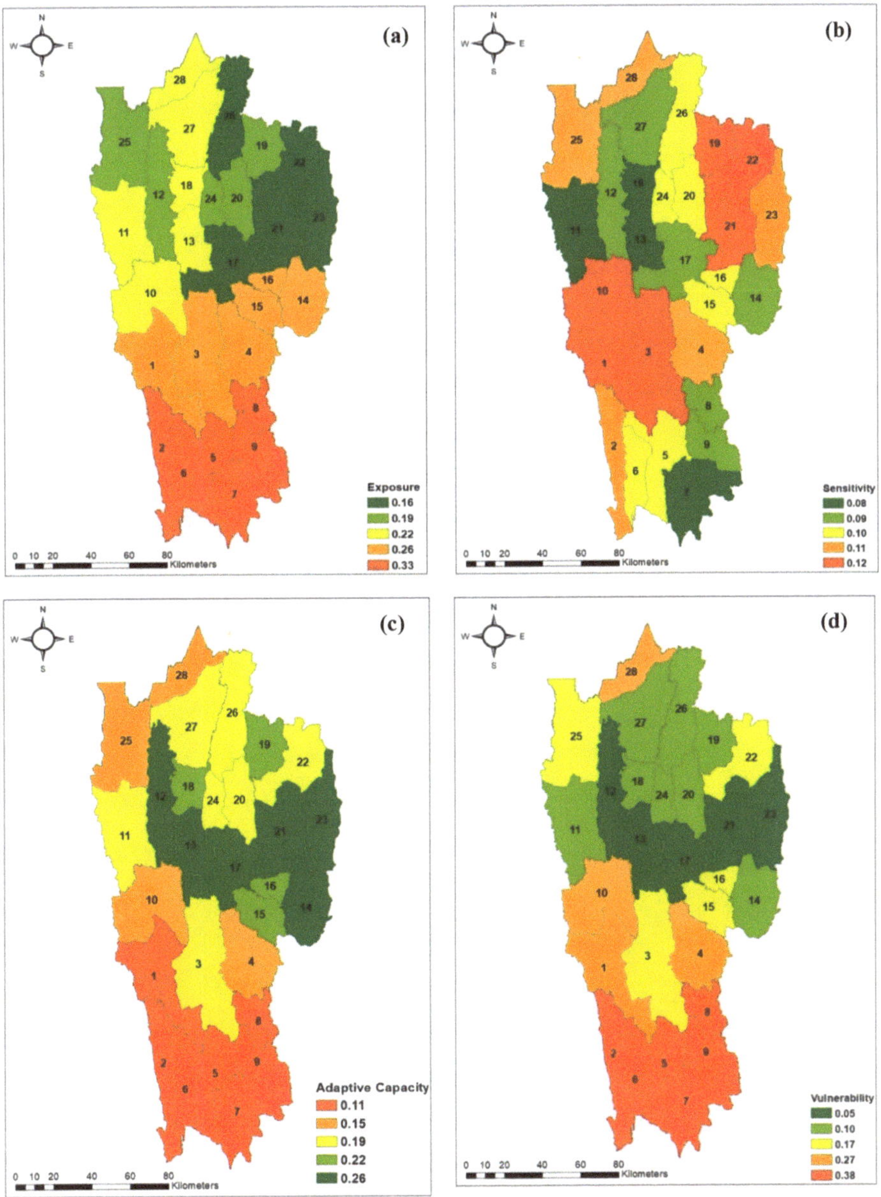

Fig. 5.6 Sub-district level (**a**) exposure, (**b**) sensitivity, (**c**) adaptive capacity, and (**d**) vulnerability profile of Mizoram towards climate change
(Sub district notations: (1) Lungsen, (2) Chawngte, (3) Lunglei, (4) Hnahthial, (5) Lawngtlai, (6) S′ Bungtlang, (7) Tuipang, (8) Sangau, (9) Saiha, (10) West Bunghmun, (11) West Phaileng, (12) Reiek, (13) Aibawk, (14) Khawbung, (15 and 16) East Lungdar, (17) Serchhip, (18) Tlangnuam, (19) Phullen, (20) Thingsulthliah, (21) Khawzawl, (22) Ngopa, (23) Champhai, (24) Thingsulthliah, (25) Zawlnuam, (26) Darlawn, (27) ‘N’ Thingdawl, (28) Bilkhawthlir)

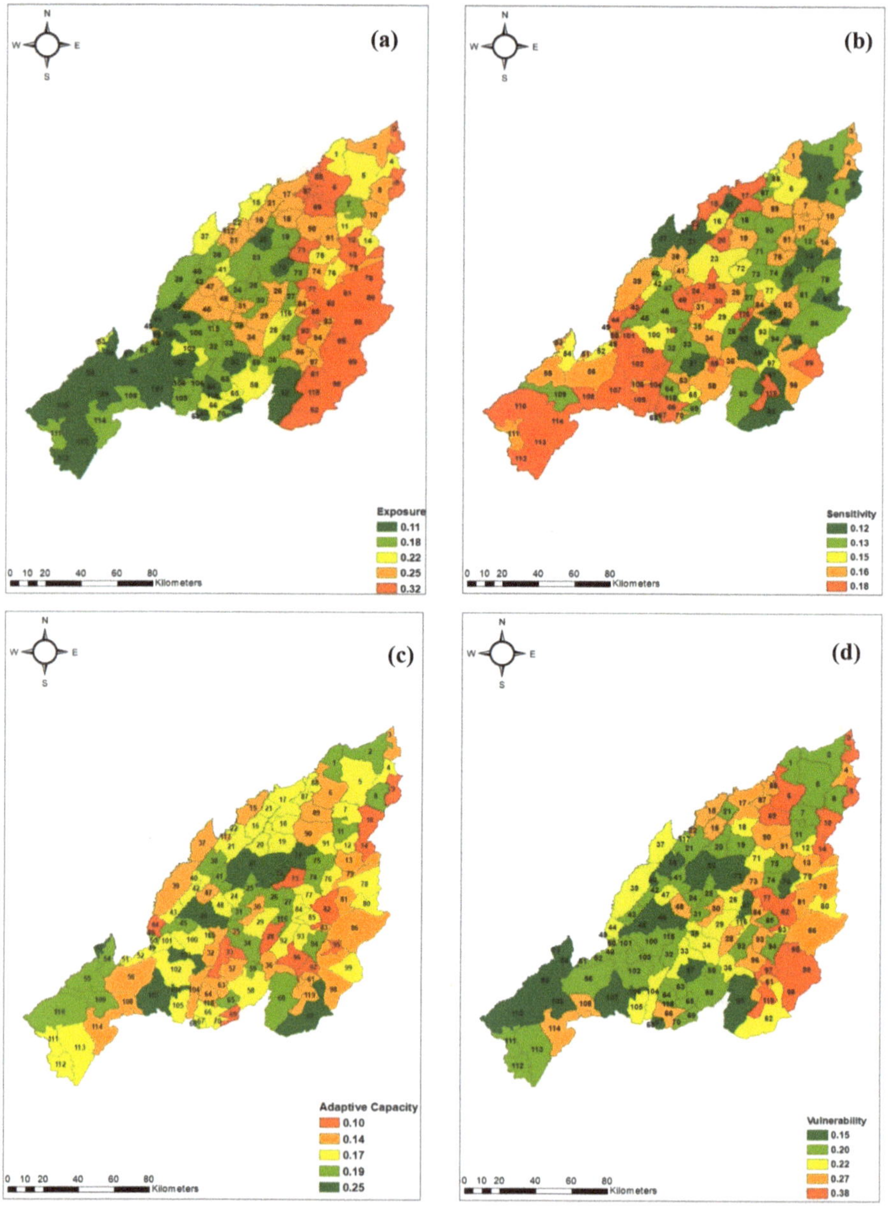

Fig. 5.7 Sub-district level (**a**) exposure, (**b**) sensitivity, (**c**) adaptive capacity, and (**d**) vulnerability profile of Nagaland towards climate change

(Sub district notations: (1) Naginimora, (2) Tizit, (3) Hunta, (4) Shangnyu, (5) Mon Sadar, 6) Wakching, (7) Aboi, (8) Longshen, (9) Phomching, (10) Chen, (11) Angjangyang, (12) Mopong, (13) Tobu, (14) Monyakshu, (15) Longchem, (16) Alongkima, (17) Tuli, (18) Changtongya, (19) Chuchuyimlang, (20) Kubolong, (21) Mangkolemba, (22) Merangmen, (23) Ongpangkong, (24) V K, (25) Akuluto, (26) Suruhuto, (27) Asuto, (28) Aghunato, (29) Zunheboto Sadar, (30) Atoizu,

The significance of customized, site-specific adaptation strategies is underscored by the observed spatial heterogeneity in vulnerability across NEI. Integrated flood-drought management strategies are necessary in foothill and valley regions (such as Arunachal Pradesh and Tripura). Rainwater harvesting and soil and water conservation are essential in high-rainfall areas, such as Mawsynram in Meghalaya. Agroforestry promotion, livelihood diversification, and improved extension services are necessary in shifting cultivation-dominated regions (like Mizoram and Nagaland). High sensitivity for the urbanizing hubs, like Gangtok in Sikkim, requires better market connections and climate-smart urban-agriculture planning. In order to prioritize areas for crucial policy intervention, effective resource allocation, and farmer-centric adaptation planning, the vulnerability assessment offers a framework for decision-making. The scaling of community-based adaptation strategies and their integration into current government initiatives can be based on these findings.

7. **Tripura:** A Limited synchronized spatial pattern was evident for all three indicators of sub-district vulnerability assessment (Fig. 5.9a–c). The upper level of exposure was sequenced as Sabroom > Kailasahar > Dharmanagar. The high sensitivity level of Bishalgarh and Sonamura sub-district experienced was countered with high adaptive capacity; thereby categorized as less vulnerable (Fig. 5.9d). Jirania, Sadar, Sabroomand Mohanpur became the four preeminent vulnerable sub-districts in Tripura, respectively, as revealed in our study.

Fig. 5.7 (continued)

(31) Akuhaito, (32) Pughoboto, (33) Ghathashi, (34) Satakha, (35) Saptiqa, (36) Satoi, (37) Changpang, (38) Aitepyong, (39) Bhandari, (40) Baghty, (41) Sungro, (42) Sanis, (43) Lotsu, (44) Ralan, (45) Wozhuro, (46) Wokha Sadar, (47) Englan, (48) Chukitong, (49) Niuland, (50) Aquqhnaqua, (51) Kuhoboto, (52) Nihokhu, (53) Dimapur Sadar, (54) Chumukedima, (55) Dhansiripar, (56) Medziphema, (57) Sekruzu, (58) Phek Sadar, (59) Khuza, (60) Meluri, (61) Phor, (62) Phokhungri, (63) Chozuba, (64) Chetheba, (65) Sakraba, (66) Pfutsero, (67) Zuketsa, (68) Khezhakeno, (69) Chizami, (70) Razieba, (71) Noksen, (72) Chare, (73) Longkhim, (74) Tuensang Sadar, (75) Ngoungchung, (76) Sangsangnyu, (77) Sotokur, (78) Noklak, (79) Chingmei, (80) Nokhu, (81) Panso, (82) Shamator, (83) Tsurungto, (84) Chessore, (85) Mangko, (86) Thonoknyu, (87) Tamlu, (88) Namsang, (89) Yongnyah, (90) Longleng, (91) Sakshi, (92) Seyochung, (93) Amahator, (94) Kiphire Sadar, (95) Kiusam, (96) Sitimi, (97) Longmatra, (98) Pungro, (99) Khongsa, (100) Tseminyu, (101) Tsogin, (102) Chiephobozou, (103) Botsa, (104) Kezocha, (105) Jakhama, (106) Kohima Sadar, (107) Sechu Zubza, (108) Pedi Ngwalwa, (109) Jalukie, (110) Athibung, (111) KebaiKhelma, (112) Nsong, (113) Tening, (114) Peren, (115) Chunlikha, (116) Tokiye, (117) Tsurangkong, (118) Kikruma, (119) Weziho)

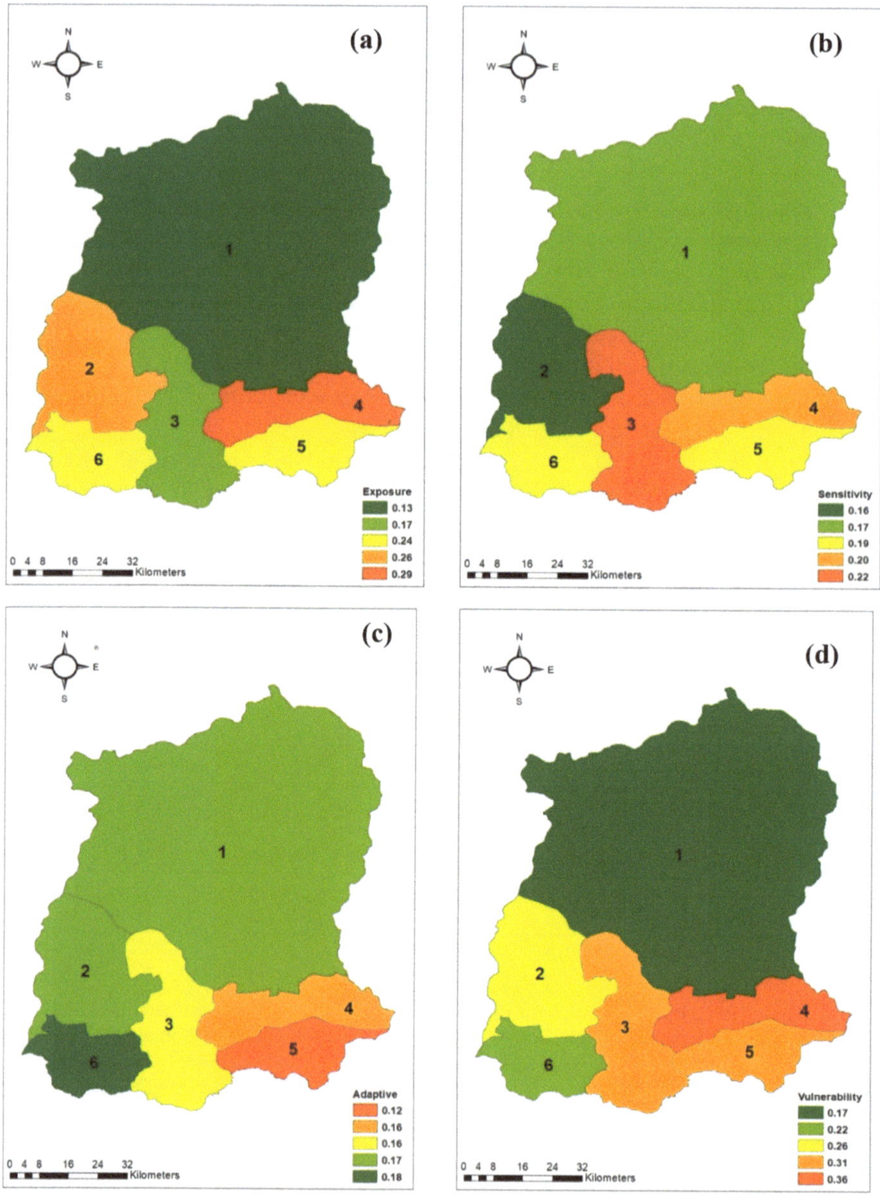

Fig. 5.8 Sub-district level (**a**) exposure, (**b**) sensitivity, (**c**) adaptive capacity, and (**d**) vulnerability profile of Sikkim towards climate change
(Sub district notations: (1) Mangan, (2) Gyalshing, (3) Namchi, (4) Gangtok, (5) Pakyong, (6) Soreng)

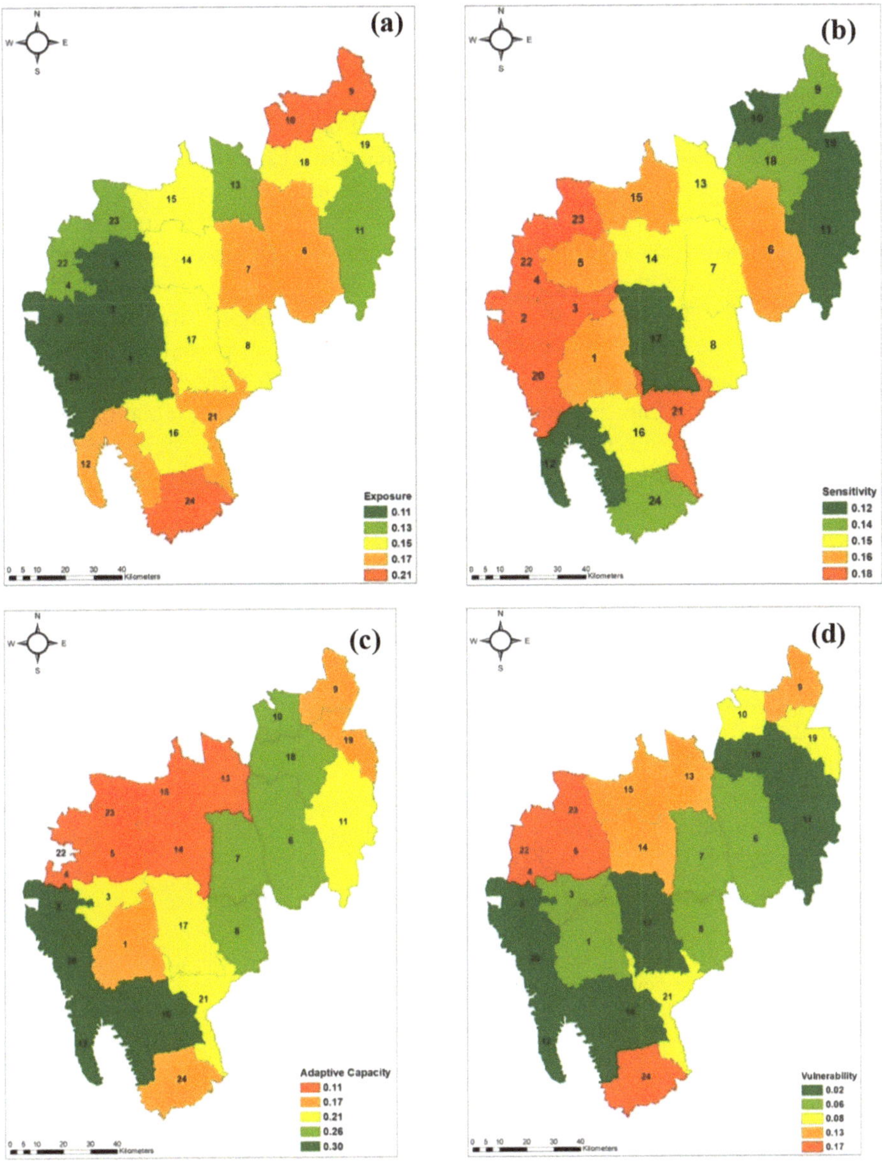

Fig. 5.9 Sub-district level (**a**) exposure, (**b**) sensitivity, (**c**) adaptive capacity, and (**d**) vulnerability profile of Tripura towards climate change
(Sub district notations) (1) Udaipur, (2) Bishalgarh, (3) Jampuijala, (4) Sadar, (5) Jirania, (6) Longtharai Valley, (7) Ambassa, (8) Gandachhara, (9) Dharmanagar, (10)Kailasahar, (11) Kancharpur, (12) Belonia, (13) Kamalpur, (14) Teliamura, (15) Khowai, (16) Santirbazar, (17) Amarpur, (18) Kumarghat, (19) Panisagar, (20) Sonamura, (21) Karbook, (22) Mohanpur, (23) Sabroom)

Conclusion

Widespread atmospheric warming coupled with changes in long-term regional monsoon dynamics has resulted in limited imminent changes in annual rainfall across the NEI region. Over the past few decades, the NEI region has seen a significant decline in monsoon and winter precipitation amounts, as well as decreased multi-day precipitation extremes. The adverse impact of such gradual climate changes on inhabitants and the existing rural farming economy of the NEI region seemed inevitable in the long run. Such circumstances have paved the way for location-specific climate change vulnerability assessments to strengthen climate change adaptation and disaster preparedness at the sub-district scale in all NEI hill states. The importance of assigning unequal weightage for different exposure, sensitivity, and adaptive capacity indicators was highlighted by an evaluation of intra-district climate change vulnerability that combined climate, agricultural, and socioeconomic datasets. This aided in determining the climate change vulnerability of sub-districts in the Northeast Indian Hills states. The sub-districts that were most vulnerable were identified as Arunachal Pradesh (Doimukh SDO and Koyu EAC), Meghalaya (Shella Bholaganj, Mawsynram, and Amlarem), Manipur (Phungyar and Kasom Khullen), Mizoram (Chawngte and Lawngtlai), Nagaland (Kohima Sadar and Wokha Sadar), Sikkim (Gangtok and Rongli), and Tripura (Jirania, Sadar, Sabroom, and Mohanpur). To ensure the monsoon-dependent local economy and other livelihood options of residents in the identified vulnerable sub-districts are protected from future eventualities, it is imperative to enhance community preparedness at a pilot scale.

Funding Acknowledgement This research work was supported under the National Innovations in Climate Resilient Agriculture (NICRA) project.

Author Statement (Disclaimer) The Authors declare no conflict of interest related to the article.

Data Availability Statement The datasets during and/or analyzed during the current study are available from the corresponding author on reasonable request.

References

Ajay Kumar, Gupta Subrata, Nandy Arun Jyoti, Nath Divya, Mehta Rajiv, Pandey (2024) Spatially Explicit climate change vulnerability assessment of ecological systems along altitudinal gradients in the Indian Himalayan region Environmental and Sustainability Indicators 22. https://doi.org/10.1016/j.indic.2024.100377

Azharuddin, S., & Dogar, M. M. (2016) Decadal climate variability and predictability. CURRENT SCIENCE, 110(8), 1397. https://www.currentscience.ac.in/Volumes/110/08/1397.pdf

Bora SL, Bhuyan K, Hazarika PJ, Gogoi J, Goswami K (2022) Analysis of rainfall trend using non-parametric methods and innovative trend analysis during 1901–2020 in seven states of North East India. Curr Sci 122(7):801–811

Brooks N, Adger WN (2004) Assessing and enhancing adaptive capacity. In: Lim LB (ed) adaptation policy framework. United Nations Development Programme, New York, NY. http://www.undp.org/cc/apf.htm

Chakraborty D, Roy A, Singh NU, Saha S, Das SK, Mridha N, Yumnam A, Paul P, Gowda C, Biam KP, Patra S (2025) Assessing climate change impact on rainfall patterns in northeastern India and its consequences on water resources and rainfed agriculture. Earth 6(1):2

Chakraborty D, Saha S, Singh RK, Sethy BK, Kumar A, Saikia US, Das SK, Makdoh B, Borah TR, Chanu AN, Walling I (2017) Spatio-temporal trends and change point detection in rainfall in different parts of North-eastern Indian states. J Agrometeorol 19(2):160–163

Choudhury BU, Nengzouzam G, Islam A (2022) Evaluation of climate change impact on soil erosion in the integrated farming system based hilly micro-watersheds using revised universal soil loss equation. Catena 214:106306. https://doi.org/10.1016/j.catena.2022.106306

Das A, Ghosh PK, Choudhury BU, Patel DP, Munda GC, Ngachan SV, Chowdhury P (2009) Climate change in Northeast India: recent facts and events—worry for agricultural management. In: ISPRS Archives XXXVIII-8/W3 Workshop Proceedings: Impact of Climate Change on Agriculture, pp 32–37

Das A (2013) Mapping the regional variation in potential vulnerability in Indian agriculture to climate change-an exercise through constructing vulnerability index. Afr J Environ Sci Technol 7(4):112–121

Das A, Datta D, Samajdar T, Idapuganti RG, Islam M, Choudhury BU, Mohapatra KP, Layek J, Babu S, Yadav GS (2021) Livelihood security of small holder farmers in eastern Himalayas, India: pond based integrated farming system a sustainable approach. Curr Res Environ Sustain 3:100076. https://doi.org/10.1016/j.crsust.2021.100076

Deka S, Borah M, Kakaty SC (2009) Distributions of annual maximum rainfall series of North-East India. Eur Water 27(28):3–14

Devi NL, Choudhury BU (2013) Soil fertility status in relation to fallow cycles and land use practices in shifting cultivated areas of Chandel district Manipur, India. IOSR J Agric Vet Sci 4(4):1–9

Dogar M M, Stenchikov G. Osipov S, Wyman B, Zhao M (2017) Sensitivity of the regional climate in the Middle East and North Africa to volcanic perturbations. Journal of Geophysical Research: Atmospheres 122(15) 7922–7948. https://doi.org/10.1002/2017JD026783

Dogar MM, Leon, Hermanson Adam A., Scaife Daniele, Visioni Ming, Zhao Ibrahim, Hoteit Hans-F., Graf Muhammad Ahmad, Dogar Mansour, Almazroui Masatomo, Fujiwara (2023) A Review of El Niño Southern Oscillation Linkage to Strong Volcanic Eruptions and Post-Volcanic Winter Warming. Earth Systems and Environment 7(1) 15–42. https://doi.org/10.1007/s41748-022-00331-z

Dogar MM, Fujiwara M, Zhao M, Ohba M, Kosaka Yu (2024) ENSO and NAO Linkage to Strong Volcanism and Associated Post-Volcanic High-Latitude Winter Warming. Geophysical Research Letters, 51(1), e2023GL106114. https://doi.org/10.1029/2023GL106114

Dogar MM, Fujiwara M, Ohba M, Almazroui M (2025) Investigating post-eruption amplified winter cooling in the Middle East and North Africa—unraveling ENSO and NAO dynamics. Environmental Research Communications, 7(4), 045012. https://doi.org/10.1088/2515-7620/adc675

Dubey SK, Ranjan RK, Misra AK, Wanjari N, Vishwakarma S (2022) Variability of precipitation extremes and drought intensity over the Sikkim state, India, during 1950–2018. Theor Appl Climatol 148:1–14

FAO (2021) The impact of disasters and crises on agriculture and food security: 2021. Tech Rep FAO, Rome. https://doi.org/10.4060/cb3673en

Feroze SM, Aheibam M, Singh R, Ray LI, Rai M, Singh KJ, Singh RJ (2014) Assessment of agricultural vulnerability to climate change in Manipur: a district-level analysis. Indian J Hill Farming 27(1):22–29

Gautam R, Borgohain A, Pathak B, Kundu SS, Aggarwal SP (2024) Investigation of meteorological variables and associated extreme events over north-East India and its adjoining areas using high-resolution IMDAA reanalysis. Nat Hazards 121(5):5149–5184

Gbetibouo GA, Ringler C (2009) Mapping south African farming sector vulnerability to climate change and variability: a subnational assessment. In: Conference on the Human Dimensions of Global Environmental Change 'Earth System Governance: People, Places and the Planet
Geeta, Kumari Yatendra, Sharma Haroon, Sajjad (2023) Assessing livelihood vulnerability of rural communities in Dimapur district of Nagaland state India: policy implications GeoJournal 88(3) 3143-3162 10.1007/s10708-022-10800-7
Guhathakurta P, Revadekar J (2016) Observed variability and long-term trends of rainfall over India. In: Observed climate variability and change over the Indian region. Springer Singapore, Singapore, pp 1–15. https://doi.org/10.1007/978-981-10-2531-0_1
Guhathakurta P, Sreejith OP, Menon PA (2011) Impact of climate change on extreme rainfall events and flood risk in India. J Earth Syst Sci 120(3):359–373
Halder P, Dey RK, Mandal S (2023) Long-period trend analysis of annual and seasonal rainfall in West Bengal, India (1901–2020). Theor Appl Climatol 154(1–2):685–703. https://doi.org/10.1007/s00704-023-04577-z
Hazarika A, Nath AJ, Reang D, Pandey R, Sileshi GW, Das AK (2024) Climate change vulnerability and adaptation among farmers practicing shifting agriculture in the Indian Himalayas. Environ Sustain Indic 23:100430
Hiremath DB, Shiyani RL (2013) Analysis of vulnerability indices in various agro-climatic zones of Gujarat. Indian J Agric Econ 68:122–136
IPCC (2007) Climate change 2007: impacts, adaptation and vulnerability. In: Contribution of Working Group II to the Fourth Assessment Report of the IPCC
IPCC (2022) Summary for policymakers. In: Climate change 2022—impacts, adaptation and vulnerability. Cambridge University Press, pp 3–33. https://doi.org/10.1017/9781009325844.001
IPCC (2023) Climate change 2023: synthesis report. In: Lee H, Romero J (eds) Contribution of Working Groups I, II and III to the Sixth Assessment Report of the Intergovernmental Panel on Climate Change. Core Writing Team. IPCC, Geneva
Jain SK, Kumar V, Saharia M (2012) Analysis of rainfall and temperature trends in Northeast India. Int J Climatol 33(4):968–978. https://doi.org/10.1002/joc.3483
Jakariya M, Rahman A, Sayem SM, Saad S, Alam MS, Sarker SR, Ali MS, Akter D (2020) Development of livelihood vulnerability index for the coastal fishermen communities of Bangladesh using spatial information technique. Groundw Sustain Dev 11:100475. https://doi.org/10.1016/j.gsd.2020.100475
Jhajharia D, Singh VP (2011) Trends in temperature, diurnal temperature range and sunshine duration in Northeast India. Int J Climatol 31(9):1353–1367
Jhajharia D, Shrivastava SK, Sarkar D, Sarkar S (2009) Temporal characteristics of pan evaporation trends under the humid conditions of Northeast India. Agric For Meteorol 149:763–770
Jagmohan, Sharma Sujata, Upgupta Mathangi, Jayaraman Rajiv Kumar, Chaturvedi Govindswamy, Bala N. H., Ravindranath (2017) Vulnerability of Forests in India: A National Scale Assessment Environmental Management 60(3) 544–553. 10.1007/s00267-017-0894-4
Kalita R, Kalita D, Saxena A (2023) Trends in extreme climate indices in Cherrapunji for the period 1979 to 2020. J Earth Syst Sci 132(2):74
Khan A, Mandal S, Satpati L, Dogar MM, Hasanean H, Salah AM (2025) Analyzing drought: the impact of climate change on precipitation patterns in the Shilabati River Basin, West Bengal, India. Earth Syst Environ. https://doi.org/10.1007/s41748-025-00679-y
Krishnan R, Sabin TP, Vellore R, Mujumdar M, Sanjay J, Goswami BN, Hourdin F, Dufresne JL, Terray P (2016) Deciphering the desiccation trend of the South Asian monsoon hydroclimate in a warming world. Clim Dyn 47:1007–1027
Kumar D, Dimri AP (2018) Regional climate projections for Northeast India: an appraisal from CORDEX South Asia experiment. Theor Appl Climatol 134:1065–1081
Kumar P, Wiltshire A, Mathison C, Asharaf S, Ahrens B, Lucas-Picher P, Christensen JH, Gobiet A, Saeed F, Hagemann S, Jacob D (2013) Downscaled climate change projections with uncertainty assessment over India using a high resolution multi-model approach. Sci Total Environ 468-469:S18–S30. https://doi.org/10.1016/j.scitotenv.2013.01.051

Lalthanpuia, Laldinpuii H, Lalmalsawma S, Hrahsel JL (2022) Assessment of district level climate vulnerability of Mizoram, India: water resources approach. J Clim Chan 8(3):21–29

Lobell DB, Gourdji SM (2012) The influence of climate change on global crop productivity. Plant Physiol 160:1686–1697. https://doi.org/10.1104/pp.112.208298

Maiti S, Jha SK, Garai S, Nag A, Bera AK, Paul V, Upadhaya RC, Deb SM (2017) An assessment of social vulnerability to climate change among the districts of Arunachal Pradesh, India. Ecol Indic 77:105–113

Mandal S, Choudhury BU, Satpati LN (2015) Monsoon variability, crop water requirement, and crop planning for kharif rice in Sagar Island, India. Int J Biometeorol 59:1891–1903. https://doi.org/10.1007/s00484

Mandal S, Choudhury BU, (2014) Estimation and prediction of maximum daily rainfall at Sagar Island using best fit probability models Theoretical and Applied Climatology 121(1–2) 87–97 https://doi.org/10.1007/s00704-014-1212-1

Mandal S, Satpati LN, Choudhury BU, Sadhu S (2018) Climate change vulnerability to agrarian ecosystem of small Island: evidence from Sagar Island, India. Theor Appl Climatol 132:451–464. https://doi.org/10.1007/s00704-017-2098-5

Mishra V, Bhatia U, Tiwari AD (2020) Bias-corrected climate projections for South Asia from Coupled Model Intercomparison Project-6. Sci Data 7(1). https://doi.org/10.1038/s41597-020-00681-1

Monterroso A, Conde C, Gay C, Gómez D, López J (2014) Two methods to assess vulnerability to climate change in the Mexican agricultural sector. Mitig Adapt Strateg Glob Chang 19:445–461. https://doi.org/10.1007/s11027-012-9442-y

Newton A, Weichselgartner J (2014) Hotspots of coastal vulnerability: a DPSIR analysis to find societal pathways and responses. Estuar Coast Shelf Sci 140:123–133. https://doi.org/10.1016/j.ecss.2013.10.010

O'Brien K, Leichenko R, Kelkar U, Venema H, Aandahl G, Tompkins H, Javed A, Bhadwal S, Barg S, Nygaard L, West J (2004) Mapping vulnerability to multiple stressors: climate change and globalization in India. Glob Environ Chang 14:303–313. https://doi.org/10.1016/j.gloenvcha.2004.01.001

Pai DS, Sridhar L, Rajeevan M, Sreejith OP, Satbhai NS, Mukhopadhyay B (2014) Development of a new high spatial resolution (0.25° × 0.25°) Long period (1901–2010) daily gridded rainfall data set over India and its comparison with existing data sets over the region. Mausam 65:1–18

Pathak S, Chielie M, Satish Y, Kusre BC (2024) Spatial and temporal trend analysis of rainfall in Nagaland (India) using machine learning techniques. J Earth Syst Sci 133:105. https://doi.org/10.1007/s12040-024-02320-4

Paul AR, Maity R (2023) Future projection of climate extremes across contiguous Northeast India and Bangladesh. Sci Rep 13:15616. https://doi.org/10.1038/s41598-023-42360-2

Philippus, Wester Arabinda, Mishra Aditi, Mukherji Arun Bhakta, Shrestha The Hindu Kush Himalaya Assessment Mountains Climate Change Sustainability and People Springer International Publishing Cham

Rao CR, Raju BMK, Rao AS, Rao KV, Rao VUM, Ramachandran K, Venkateswarlu B, Sikka AK, Rao MS, Maheswari M, Rao CS (2016) A district level assessment of vulnerability of Indian agriculture to climate change. Curr Sci 110(10):1939–1946

Ravindranath NH, Rao S, Sharma N, Nair M, Gopalakrishnan R, Rao AS, Malaviya S, Tiwari R, Sagadevan A, Munsi M, Krishna N, Bala G (2011) Climate change vulnerability profiles for North East India. Curr Sci 101:384–394. https://doi.org/10.2307/24078517

Rehman S, Azhoni A, Chabbi PH (2023) Livelihood vulnerability assessment and climate change perception analysis in Arunachal Pradesh, India. GeoJournal 88(2):1427–1447

Roy N, Debnath A, Nautiyal S, Sarma M (2023) Livelihood vulnerability to climate change in rural Tripura, India: a comparative study of the tribal and non-tribal households. Ecol Environ Conserv 29:46–57

Roy P, Roy A, Joshi P, Kale M, Srivastava V, Srivastava S, Dwevidi R, Joshi C, Behera M, Meiyappan P, Sharma Y (2015) Development of decadal (1985–1995–2005) land use and land cover database for India. Remote Sens 7(3):2401–2430

Saha S, Chakraborty D, Hazarika S, Shakuntala I, Das B, Chhabra A, Sadhu S, Chakraborty D, Mukherjee J, Singson L, Mishra VK (2022) Spatiotemporal variability of weather extremes over eastern India: evidences of ascertained long-term trend persistence and effective global climate controls. Theor Appl Climatol 148:643–659. https://doi.org/10.1007/s00704-022-03949-1

Saha S, Chakraborty D, Paul RK, Samanta S, Singh SB (2017) Disparity in rainfall trend and patterns among different regions: analysis of 158 years' time series of rainfall dataset across India. Theor Appl Climatol 134(1):381–395. https://doi.org/10.1007/s00704-017-2280-9

Saha S, Chakraborty D, Singh SB, Chowdhury S, Syiem EK, Dutta SK, Lungmuana, Choudhury BU, Boopathi T, Singh AR, Ramakrishna Y, Roy A (2016) Analyzing the trend in thermal discomfort and other bioclimatic indices at Kolasib, Mizoram. J Agrometeorol 18(1):57–61

Saha S, Chakraborty D, Yadav GS, Das A, Das B, Chowdhury S, Lalhmachhuana P, Chauduri DD, Choudhury BU, Datta M, Shakuntala I (2021) Rainfall analysis across the North East Indian state of Tripura. J Agrometeorol 23(4):471–475

Saha S, Yadav GS, Daschaudhuri D, Datta M, Chakraborty D, Sadhu S, Das B, Chowdhury S, Dayal V, Das A, Kandpal B (2024) Shifts in wetness pattern and periodicity across Tripura state in North East India. Mausam 75(1):119–130

Saha S, Chakraborty D, Choudhury BU, Singh SB, Chinza N, Lalzarliana C, Dutta SK, Chowdhury S, Boopathi T, Lungmuana AR, Ngachan SV (2015) Spatial variability in temporal trends of precipitation and its impact on the agricultural scenario of Mizoram. Curr Sci 109:2278–2282

Sanayanbi H, Bandyopadhyay A, Bhadra A (2021) Vulnerability assessment of Manipur to floods using unequal weights. In: Pandey A, Mishra S, Kansal M, Singh R, Singh VP (eds) Hydrological extremes. Water science and technology library, vol 97. Springer, Cham. https://doi.org/10.1007/978-3-030-59148-9_3

Schröder LS, Bhalerao AK, Kabir KH, Scheffran J, Schneider UA (2024) Managing uphill cultivation under climate change–an assessment of adaptation decisions among tribal farmers in Nagaland state of India. J Environ Manag 349:119473

Shrestha UB, Gautam S, Bawa KS (2012) Widespread climate change in the Himalayas and associated changes in local ecosystems. PLoS One 7(5):e36741. https://doi.org/10.1371/journal.pone.0036741

Shukla R, Sachdeva K, Joshi PK (2017) Demystifying vulnerability assessment of agriculture communities in the Himalayas: a systematic review. Nat Hazards 91(1):409–429. https://doi.org/10.1007/s11069-017-3120-z

Sultan H, Zhan J, Rashid W, Chu X, Bohnett E (2022) Systematic review of multi-dimensional vulnerabilities in the Himalayas. Int J Environ Res Public Health 19:12177. https://doi.org/10.3390/ijerph191912177

Thong P, Thangjam U, Sahoo UK, Pebam R (2022) Socio-economic vulnerability assessment of shifting cultivators (Jhumias) amidst the changing climate in Mizoram, Northeast India. Appl Geogr 147:102790

Tahir Z, Haseeb M, Amer S, Hassan MNE, Dogar MM, Hassan SZ (2025) From Green to Grey: Assessing the Impact of Urbanization on Land Surface Temperature and Thermal Comfort in Multan City Earth Systems and Environment 9(2) 777–800 https://doi.org/10.1007/s41748-025-00598-y

Wassmann R, Jagadish SVK, Sumfleth K, Pathak H, Howell G, Ismail A, Serraj R, Redona E, Singh RK, Heuer S (2009) Regional vulnerability of climate change impacts on Asian rice production and scope for adaptation. Adv Agron 102:91–133

Wester P, Mishra A, Mukherji A, Shrestha AB (eds) (2019) The Hindu Kush Himalaya assessment—mountains, climate change, sustainability and people, Springer-Nature, Switzerland

Yomcha M, Feroze SM, Rymbai D, Singh R (2019) State-level agricultural vulnerability to climate change in North-Eastern Region of India. Indian J Hill Farming 32(1)

You have entered nothing

Chapter 6
Cyclone Remal and the Hydrological Destabilization of the Hooghly Estuary: Climate-Induced Threats to Rural Livelihoods in the Indian Sundarbans

Rajrupa Ghosh, Sumana Bhattacharyya, Prosenjit Pramanick, and Abhijit Mitra

Abstract The increasing frequency and intensity of tropical cyclones in the Bay of Bengal have amplified the vulnerability of rural estuarine communities in the Indian Sundarbans, whose livelihoods are intricately tied to aquatic ecosystem stability. This study investigates the hydrological vulnerability of the Hooghly estuary during Cyclone Remal (May 2024), linking short-term hydrological disturbances to long-term climate risks. *In-situ* measurements of salinity, pH, and dissolved oxygen (DO) were recorded across twelve stations before (22 May), during (27 May), and after (04 June) the cyclone. Salinity rose sharply from 14.11 PSU to 17.65 PSU, peaking at 30.01 PSU at Sagar South, indicating strong seawater intrusion from tidal surges. Although values declined post-event, they remained above pre-cyclone levels, implying residual saline stress. Surface pH increased moderately from 8.03 to 8.14, stabilizing at 8.07, demonstrating the buffering influence of marine water. Conversely, DO levels dropped from 5.75 ppm to 4.34 ppm, reaching 3.93 ppm at Diamond Harbour, reflecting acute oxygen stress with potential impacts on fishery productivity and benthic stability. These physico-chemical changes mirror socio-ecological vulnerability, as livelihoods based on estuarine fisheries, aquaculture, and freshwater availability are directly affected. Cyclone Remal thus exemplifies how extreme weather events disrupt hydrological balance and threaten rural livelihood security. The study underscores the urgency of continuous hydrological monitoring, mangrove restoration, and adaptive livelihood planning to enhance climate resilience in the fragile deltaic ecosystem of the Indian Sundarbans.

R. Ghosh (✉)
Department of Zoology, Banwarilal Bhalotia College, Asansol, India

S. Bhattacharyya · P. Pramanick · A. Mitra
Department of Oceanography, Techno India University, Kolkata, West Bengal, India

S. Mandal et al. (eds.), *Climate Change Vulnerability*, Climate Change Management, https://doi.org/10.1007/978-3-032-16783-5_6

Introduction

Tropical cyclones represent one of the most destructive meteorological phenomena affecting low-lying coastal zones across the globe (Mohanty et al. 2013; Kumar et al. 2025; Lau et al. 2022; Ortiz et al. 2023). These storms, formed through the interplay of oceanic heat, atmospheric instability, and favourable circulation patterns (Dogar 2020), bring intense winds, storm surges, and heavy precipitation that reshape coastal landscapes, disrupt livelihoods, and alter aquatic biogeochemistry. The impacts of cyclones are particularly pronounced in tropical and subtropical estuarine environments where freshwater–seawater interactions already sustain delicate hydrological equilibria. For these systems, storm-induced saline intrusion, turbidity surges, and oxygen depletion can destabilize ecological functions in ways that persist well beyond the immediate aftermath of landfall (Mitra et al. 2011). The Bay of Bengal (BoB), situated in the northeastern sector of the Indian Ocean, stands as one of the most cyclone-prone regions of the world, accounting for nearly 5–6% of all global tropical cyclones but disproportionately higher mortality and socio-economic damage (Table 6.1).

Historical datasets show that between 1891 and 1970, the Bay of Bengal experienced over 346 cyclonic events, of which nearly 133 qualified as severe, compared with just 98 cyclones (55 severe) in the Arabian Sea during the same interval. This asymmetry underscores the critical vulnerability of the Bay of Bengal and the densely populated deltaic ecosystems along its rim.

The physical geography of the northern Bay of Bengal, characterized by shallow bathymetry, funnel-shaped coastal orientation, and high astronomical tidal ranges, magnifies storm surge impacts. When coupled with socio-demographic realities, including extreme population density, agrarian dependence, and poverty in Bangladesh and eastern India, the risks associated with cyclones are accentuated manifold. The Indian Sundarbans, at the apex of this bay, exemplify such vulnerability. Comprising a mosaic of 102 low-lying islands, intricate networks of tidal creeks, and mangrove-dominated shorelines, the Sundarbans function both as a buffer against cyclonic devastation and as a habitat supporting millions of people. Yet, the very attributes that enable this ecological resilience also render the system acutely sensitive to perturbations in hydrological conditions. Sudden changes in salinity, pH, and dissolved oxygen (DO) due to cyclonic storm surges can destabilize biogeochemical cycles, impair fisheries, precipitate fish mortality, and trigger cascading effects across food webs. Consequently, systematic documentation of estuarine responses to cyclonic disturbances is critical both for advancing scientific understanding and for developing effective disaster management frameworks.

The Hooghly estuary, forming the western frontier of the Indian Sundarbans, has long served as a critical natural laboratory for assessing the vulnerability of estuarine ecosystems to climate-induced hydrological disturbances. Earlier investigations, particularly during Cyclone Aila (May 2009), documented the magnitude of storm-driven alterations in estuarine water quality, where surface salinity across twelve monitoring stations rose from an average of 13.64 ppt to 17.08 ppt, a 25.2%

Table 6.1 List of cyclonic storms in BOB (1990–2024)

Year	Storm	Landfall date	Landfall place	Wind (km/h)
1991	1991 Bangladesh cyclone (BOB 01)	29-Apr-1991	Near Chittagong, Bangladesh	215
1999	1999 Odisha Super Cyclone (05B)	29-Oct-1999	Near Paradip, Odisha (India)	260
2007	Akash	15-Jun-2007	Near Chittagong, Bangladesh	85
2007	Sidr	15-Nov-2007	Bangladesh (Barguna/Baleshwar coast)	215
2008	Nargis	02-May-2008	Ayeyarwady Delta, Myanmar	215
2009	Aila	25-May-2009	SW Khulna, Bangladesh / Indian Sundarbans	110
2010	Laila	20-21 May 2010	Near Bapatla, Andhra Pradesh, India	110
2011	Thane	30-Dec-2011	Cuddalore Puducherry, Tamil Nadu, India	140
2013	Phailin	12-Oct-2013	Near Gopalpur, Odisha, India	215
2013	Helen	22-Nov-2013	Near Machilipatnam, Andhra Pradesh, India	100
2014	Hudhud	12-Oct-2014	Near Visakhapatnam, Andhra Pradesh, India	175
2016	Roanu	21-May-2016	Near Chittagong, Bangladesh	85
2016	Vardah	12-Dec-2016	Near Chennai, Tamil Nadu, India	130
2017	Mora	30-May-2017	Near Cox Bazar, Bangladesh	110
2018	Titli	11-Oct-2018	Near Palasa, Odisha- Andhra border, India	150
2018	Gaja	16-Nov-2018	Near Vedaranyam, Tamil Nadu, India	120
2019	Fani	03-May-2019	Near Puri, Odisha, India	200
2019	Bulbul	09-Nov-2019	Sundarbans (India- Bangladesh)	130
2020	Amphan	20-May-2020	West Bengal (India) & Satkhira (Bangladesh)	185
2020	Nivar	26-Nov-2020	Near Puducherry, Tamil Nadu, India	120
2020	Burevi	02-03 Dec 2020	Sri Lanka (crossed to Tamil Nadu, India)	85
2021	Yaas	26-May-2021	Bahanaga, Odisha, India	130
2021	Gulab	26-Sep-2021	Near Kalingapatnam, Andhra Pradesh, India	90

(continued)

Table 6.1 (continued)

Year	Storm	Landfall date	Landfall place	Wind (km/h)
2022	Sitrang	24-Oct-2022	Bangladesh coast (Barisal/Bhola)	85
2022	Mandous	10-Dec-2022	Near Mamallapuram, Tamil Nadu, India	85
2023	Mocha	14-May-2023	Near Sittwe, Myanmar	215
2023	Hamoon	24-25 Oct 2023	South of Chittagong, Bangladesh	120
2023	Midhili	17-Nov-2023	Near Khepupara, Bangladesh	85
2023	Michaung	05-Dec-2023	Near Bapatla, Andhra Pradesh, India	110
2024	Remal	26-27 May 2024	Between Sagar Island (India) & Bangladesh coast	120

Source: https://www.jagranjosh.com/general-knowledge/list-of-cyclones-in-india- 1591178815-1

increase, while surface water pH showed a marginal rise from 7.99 to 8.01 and dissolved oxygen (DO) declined by approximately 5.5% (5.24 ppm to 4.95 ppm) (Mitra et al. 2011). Although partial recovery was evident ten days after the event, these findings underscored the high sensitivity of the Hooghly estuary to marine water incursions and highlighted how cyclonic surges could destabilize benthic communities, impair fisheries, and trigger mass fish mortalities, thereby threatening the livelihoods of estuarine-dependent rural populations. Cyclone Aila thus provided not only a scientific baseline for storm–estuary interaction but also revealed the broader social and economic vulnerabilities that accompany such ecological shocks.

The landfall of Cyclone Remal on 26 May 2024 presented a renewed opportunity to examine these dynamics within the framework of understanding rural vulnerability to climate change. Classified by the India Meteorological Department (IMD) as a severe cyclonic storm, Remal originated over the central Bay of Bengal and advanced northward with sustained wind speeds of 95–105 km h^{-1}, gusting up to 115 km h^{-1}. Making landfall near Sagar Island, West Bengal, it generated powerful tidal surges that forced extensive saline water intrusion deep into the Hooghly estuary. In contrast to weaker tropical depressions, Remal's intensity, coupled with its broad spatial impact, induced pronounced shifts in estuarine salinity, pH, and DO across multiple monitoring sites. The present study, designed on the methodological framework established during the Aila investigation, monitored three distinct temporal phases—pre-cyclone (22 May 2024), peak cyclone (27 May 2024), and post-cyclone recovery (04 June 2024) across twelve strategically located stations ranging from Raichak in the upstream sector to Sagar South in the marine-dominated zone. These hydrological observations not only quantify the physical transformations induced by Remal but also illustrate how extreme climatic events increasingly exacerbate the ecological fragility of the Sundarbans, reinforcing the urgency for integrated monitoring and adaptive livelihood strategies to safeguard the region's vulnerable rural communities.

Materials and Methods

Twelve monitoring sites were established along the Hooghly estuary to evaluate the hydrological alterations caused by Cyclone Remal (Table 6.2 and Fig. 6.1). The precise positions of these sites were recorded using a Garmin GPS. At each station, three replicate surface water samples were taken during both high and low tides, with each sample separated by approximately 100 m. All parameter values are expressed as the mean of the three replicates together with their standard deviation.

Surface salinity was first measured in the field with a handheld optical refractometer (Atago, Japan) to obtain rapid and reliable estimates. To confirm these readings, laboratory checks were carried out following the classical Mohr–Knudsen argentometric titration procedure (Strickland and Parsons 1972). Standard seawater solutions (chlorinity 19.376 PSU; IAPO, Charlottenlund, Denmark) were used for calibration, and a correction factor was derived by titrating silver nitrate solution

Table 6.2 Local names of sampling stations and respective coordinates

S. No.	Sampling Station	Coordinates
1	Raichak (Stn. 1)	22°12'12.00"N and 88°07'42.09"E
2	Diamond Harbour (Stn. 2)	22°11'04.02"N and 88°10'50.52"E
3	Kulpi (Stn. 3)	22°36'28.86"N and 88°23'28.32"E
4	Balari (Stn. 4)	22°07'02.16"N and 88°11'35.34"E
5	Haldi River mouth (Stn. 5)	22°00'26.07"N and 88°03'29.64"E
6	Nayachar (Stn. 6)	22°00'30.42"N and 88°03'32.52"E
7	Khejuri Reserve Forest (Stn. 7)	21°54'51.66"N and 88°00'56.52"E
8	Ghoramara Island (Stn. 8)	21°56'15.24"N and 88°07'33.06"E
9	Harwood point (Stn. 9)	21°56'15.24"N and 88°07'33.60"E
10	Harinbari (Stn. 10)	21°46'54.12"N and 88°04'02.64"E
11	Chemaguri (Stn. 11)	21°39'49.32"N and 88°09'11.88"E
12	Sagar South (Stn. 12)	21°39'04.68"N and 88°01'47.28"E

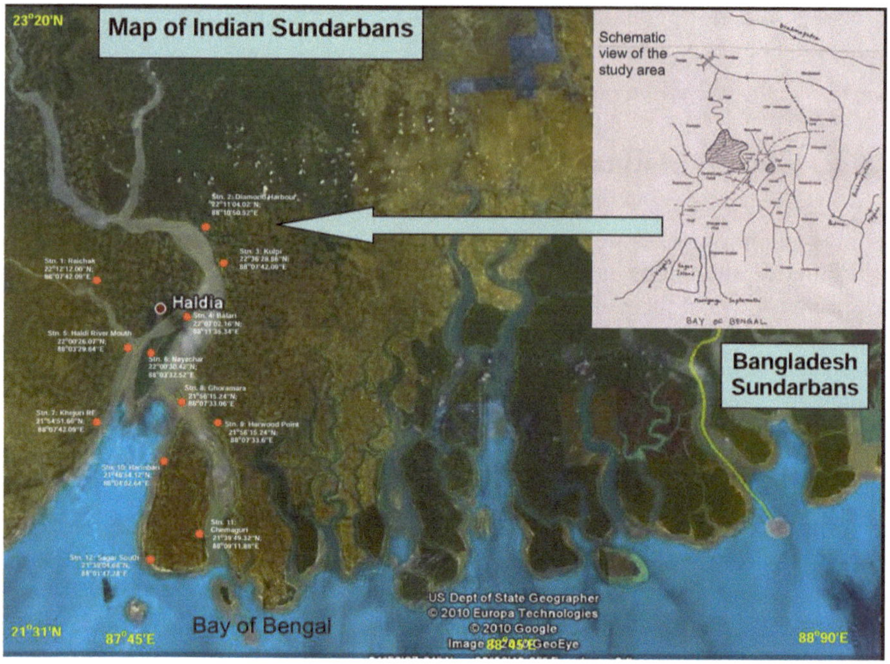

Fig. 6.1 Sampling sites along the Hooghly estuary in Indian Sundarbans

against this reference. Additional quality assurance was achieved using certified standard seawater provided by the National Institute of Oceanography (NIO, India).

The combination of in-situ and laboratory analyses provided a very low variability, with a standard deviation of 0.02 PSU, and the overall precision of salinity estimation was within ±0.30 PSU under the adopted triplicate protocol. For dissolved oxygen assessment, 125 ml glass bottles were filled with water samples and analyzed using the Winkler titration technique (Winkler 1888). Surface water pH was measured using a portable Hanna meter (USA) with an accuracy of ±0.01.

Results

Cyclone Remal generated hydrological disturbances in the Hooghly–Matla estuarine system that exceeded those documented during Aila, particularly in relation to saline intrusion. Mean surface water salinity rose sharply from 14.11 PSU in the pre-cyclone phase to 17.65 PSU at the cyclone peak (Fig. 6.2). The most dramatic spike was observed at Sagar South, where salinity increased from 24.15 PSU to 30.01 PSU. Even after one week of recovery (14.47 PSU on average), salinity levels did not revert fully to pre-cyclone conditions, underscoring the persistence of marine influence in the estuary (Fig. 6.2).

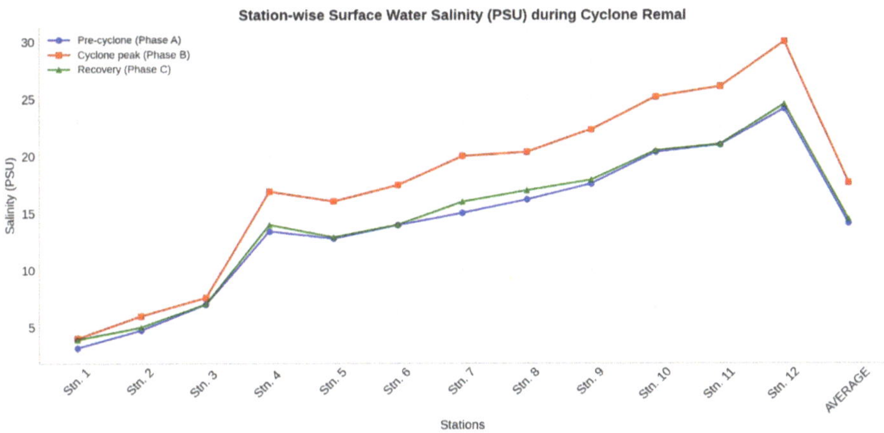

Fig. 6.2 Station-wise Surface Water Salinity (in PSU) level during pre-Remal, Remal, and Post-Remal periods

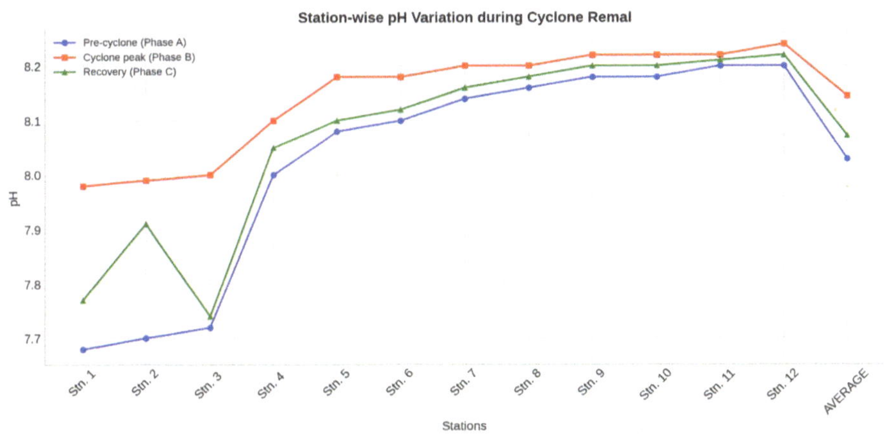

Fig. 6.3 Station-wise pH level during pre-Remal, Remal, and Post-Remal periods

This enhanced seawater incursion was also reflected in pH dynamics. Average values climbed from 8.03 before the cyclone to 8.14 during the storm, before stabilizing at 8.07 in the recovery phase (Fig. 6.3). While the magnitude of change was modest, such alterations are ecologically meaningful, influencing carbonate chemistry, nutrient cycling, and trace-metal solubility, with cascading effects on plankton productivity (Fig. 6.3).

Dissolved oxygen (DO) concentrations showed the most acute disruption. Mean levels dropped from 5.75 ppm pre-cyclone to 4.34 ppm during the storm, before recovering to 5.67 ppm after one week (Fig. 6.4). Localized stress was most evident at Diamond Harbour, where DO sharply declined from 5.60 ppm to 3.93 ppm, highlighting mid-estuary vulnerability to cyclone-driven turbulence, stratification breakdown, and saline intrusion. Although recovery trends were apparent by early June,

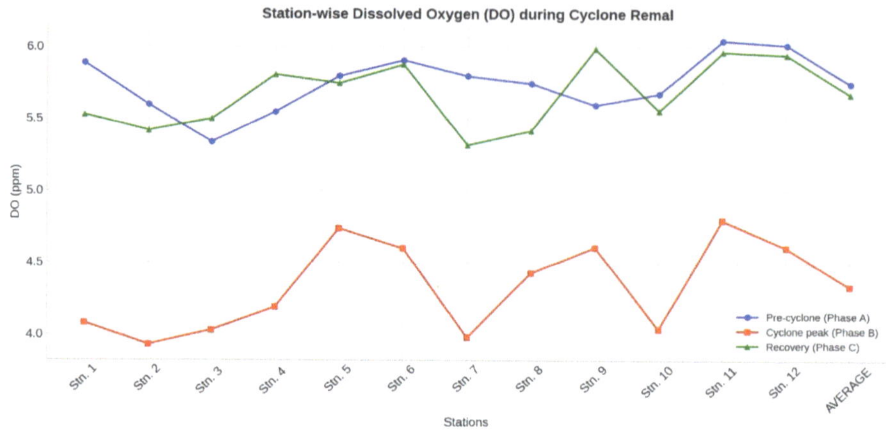

Fig. 6.4 Station-wise Dissolved Oxygen (in ppm) level during pre-Remal, Remal, and Post-Remal periods

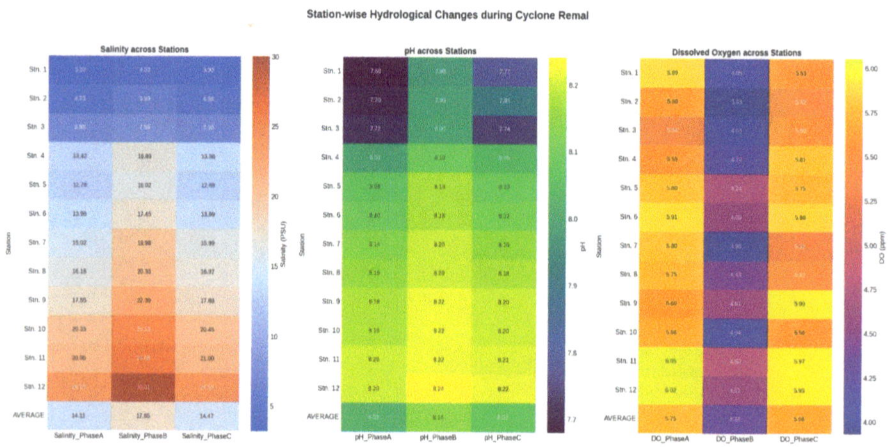

Fig. 6.5 Cyclone Remal - induced Alterations of Salinity, pH, and DO

inter-station differences remained pronounced, revealing spatial heterogeneity in the resilience of estuarine sub-systems (Fig. 6.4).

The station-wise heatmaps presented in Fig. 6.5 provide a detailed overview of hydrological alterations across twelve monitoring stations during Cyclone Remal, spanning from Raichak (Stn. 1) in the upstream region to Sagar South (Stn. 12) in the southernmost estuarine front. The combined heatmaps of the three parameters (salinity, pH and DO) highlight a clear spatial gradient in hydrological disruption during Cyclone Remal. Downstream stations (Stns. 10–12) bore the brunt of saline intrusion but benefited from marine buffering of pH, while upstream sites (Stns. 1–3) faced comparatively lower salinity but greater DO depletion. These results underscore the sector-specific vulnerabilities of the Hooghly estuary, with implications for estuarine biota, particularly fish and benthic organism's sensitive to salinity shocks, pH shifts, and oxygen stress.

The combined responses of salinity, pH, and DO indicate that Remal imposed a stronger and more prolonged perturbation than Aila, with probable implications for osmoregulatory stress in estuarine fish, benthic invertebrates, and the biogeochemical equilibrium of the Sundarbans estuary.

Discussion

The Sundarban delta, located at the apex of the Bay of Bengal, is one of the most cyclone-prone regions in the world due to its unique geomorphology and climatic setting (Mitra 2013). The shallow bathymetry of the bay, its funnel-shaped coastline, and high astronomical tides amplify storm surges when cyclones make landfall. These surges inundate vast areas of the delta, breaching embankments, eroding shorelines, and salinizing agricultural fields. Historically, the Sundarbans has been repeatedly devastated by major cyclones such as Sidr (2007), Aila (2009), Amphan (2020), Yaas (2021), and most recently Remal (2024). Each of these events has reshaped both the physical landscape and the socio-economic fabric of the region.

Ecologically, cyclones disrupt the delicate estuarine hydrology, driving sharp increases in salinity, altering pH, and reducing dissolved oxygen, which collectively stress mangroves, fish populations, and benthic organisms. Socially, millions of inhabitants, largely dependent on agriculture, aquaculture, and forest resources face livelihood insecurity, health hazards, and displacement. The recurring cycle of destruction and recovery underscores the fragility of the Sundarbans, where climate change is projected to intensify cyclone frequency and severity. Building resilience through mangrove restoration, sustainable land-use planning, and integrated disaster management is therefore essential to safeguard both biodiversity and human communities in this vulnerable delta.

Comparison with Aila data provides a valuable perspective. Both cyclones triggered sharp salinity increases and DO declines, yet the magnitude of Remal's effects appears greater in terms of salinity intrusion. Whereas Aila induced a 25.2% rise in salinity, Remal drove localized increases exceeding 30%. The DO depression during Remal (nearly 25% at some stations) also exceeded the 5.5% decline documented during Aila. These differences may reflect variations in cyclone trajectory, tidal amplitude, or freshwater inflow during the observational windows. Regardless, the comparative data confirm that the Hooghly estuary responds dynamically and predictably to cyclonic forcing, with salinity intrusion and oxygen depletion emerging as primary stressors. Such recurring patterns underscore the need for predictive modelling that incorporates cyclone intensity, estuarine morphology, and tidal phase to anticipate ecological outcomes.

From an ecological and socio-economic perspective, the consequences of hydrological alterations triggered by tropical cyclones are both profound and far-reaching, especially in estuarine regions like the Hooghly that sustain dense rural populations.

The surge in salinity following cyclonic intrusion fundamentally alters the ecological equilibrium by displacing freshwater and estuarine fish species and encouraging the dominance of marine-tolerant and low-value trash fishes. Such compositional shifts degrade the overall fishery quality, reducing the abundance of commercially important species such as *Tenualosa ilisha*, *Liza parsia*, and *Mystus gulio* that are vital to the livelihoods of traditional fishers in the Sundarbans. Elevated salinity levels impose osmotic stress on sensitive taxa, impairing growth, and reproductive capacity, while benthic invertebrates that are crucial to estuarine food webs, suffer from reduced diversity and abundance. The ecological implications are therefore deeply intertwined with economic outcomes, as the replacement of high-value species by low-economic-value fauna directly undermines local fish markets and aquaculture profitability (Mitra et al. 2023).

Equally critical is the observed decline in dissolved oxygen (DO) levels during cyclonic events. Oxygen depletion constrains aerobic metabolism in aquatic organisms, heightens physiological stress, and amplifies the risk of mass mortality, particularly among finfish and crustaceans cultured in brackish water ponds. As DO levels plummet, fishes congregate near the surface, becoming vulnerable to disease, predation, and mortality, which in turn disrupts the income of fisherfolk and aquaculture workers who rely on daily catches for sustenance. Such biological stress events, though transient, have cascading impacts on local economies and food security. Moreover, oxygen depletion also slows organic matter decomposition, altering nutrient cycling and primary productivity, which may lead to long-term ecosystem imbalance. The concurrent fluctuations in pH, though seemingly modest, further complicate this dynamic by influencing metal solubility and nutrient bioavailability. Previous studies from the Hooghly estuary (Mitra et al. 2014; Ray Chaudhuri et al. 2014) revealed that even slight alkalinity shifts can remobilize heavy metals like cadmium and lead, increasing the potential for bioaccumulation in edible fish species, posing additional health hazards for dependent communities.

Despite the ecological and livelihood consequences of such hydrological perturbations, comprehensive cyclone-focused monitoring in the Indian Sundarbans remains sparse. Traditional assessments have emphasized visible physical damage like embankment breaches, land erosion, and mangrove destruction—while often overlooking subtler yet equally consequential changes in aquatic chemistry that dictate ecosystem resilience. The present study on Cyclone Remal (2024) bridges this knowledge gap by systematically documenting spatio-temporal variations in salinity, pH, and DO across twelve stations, providing a holistic understanding of both immediate disturbances and the post-cyclone recovery trajectory. These data not only extend the foundational work established during Cyclone Aila (2009) but also refine it through higher-frequency temporal sampling and expanded parameter coverage, thereby enhancing the scientific understanding of estuarine response to climatic shocks.

In the broader context of climate change, such empirical evidence is invaluable. With rising sea surface temperatures, the Bay of Bengal is projected to witness more frequent and intense cyclones, while sea-level rise will further amplify saline water penetration into estuarine systems. These trends heighten ecological fragility and intensify the socio-economic vulnerability of coastal rural populations. The hydrological responses observed during Cyclone Remal serve as critical baselines for developing predictive models that inform climate-resilient fisheries management, adaptive aquaculture practices, and sustainable livelihood strategies. The study thus situates itself at the intersection of hydrology, ecology, and human resilience, underscoring the necessity of integrated monitoring systems that couple meteorological forecasts with estuarine observations. Such interdisciplinary approaches are essential to design proactive adaptation frameworks that protect both ecosystem health and rural livelihoods in one of the world's most climate-sensitive deltaic environments.

Conclusion

The findings of the present study highlight the deep interconnection between climate-induced hydrological disturbances and the vulnerability of rural livelihoods in the Indian Sundarbans. Cyclone Remal (May 2024) produced pronounced and measurable alterations in the hydrological dynamics of the Hooghly estuary, exceeding many of the disruptions recorded during earlier events such as Cyclone Aila (2009). Among all parameters, salinity intrusion emerged as the most persistent and socially consequential change, with elevated values persisting even a week after landfall. This prolonged saline encroachment not only revealed the estuary's hydrological fragility but also translated into freshwater scarcity, soil salinization, and loss of brackish aquaculture productivity, factors that directly threaten the livelihoods of local fishers and farmers. Conversely, dissolved oxygen (DO) exhibited a sharp but short-lived depletion during the cyclone's peak, imposing acute physiological stress on aquatic organisms. Such transient hypoxic conditions frequently result in mass fish mortality and yield reduction in estuarine capture and culture fisheries, intensifying the economic vulnerability of dependent rural households.

While surface pH displayed modest fluctuations, its slight upward drift indicated the buffering influence of seawater and revealed subtle but significant implications for nutrient cycling and metal mobility, processes that govern the productivity and health of aquatic systems. Collectively, these hydrological shifts confirm that the Sundarbans' estuarine ecosystems are highly sensitive to extreme weather events, and that the resulting ecological stress rapidly cascades into socio-economic instability across rural communities. Comparison with Aila further underscores that even cyclones of comparable classification can differ markedly in spatial impact and persistence, depending on tidal phase, storm trajectory, and freshwater inflow. This reinforces the urgency for station-wise, long-term monitoring integrating physical, chemical, and biological parameters. In the broader framework of climate vulnerability assessment, the present study contributes critical empirical evidence for predictive modeling, sustainable fisheries management, and livelihood adaptation planning, thereby offering a scientific foundation for building climate-resilient rural futures in the fragile deltaic landscapes of the Indian Sundarbans.

References

Dogar MM (2020) Study of the Regional Climatic Impacts of Tropical Explosive Volcanism in the Middle East and North Africa Region. Doctoral dissertation, Hokkaido University. https://doi.org/10.14943/doctoral.r7091.

Kumar A, Chakraborty A, Almazroui M et al. (2025) Investigation of the genesis and intensification of simultaneous cyclones in the Arabian sea and Bay of Bengal in November 2020. Theor Appl Climatol 156, 443. https://doi.org/10.1007/s00704-025-05684-9

Lau YY, Yip TL, Dulebenets MA, Tang YM, Kawasaki TA (2022) Review of historical changes of tropical and extra-tropical cyclones: a comparative analysis of the United States, Europe, and Asia. Int J Environ Res Public Health 19:4499. https://doi.org/10.3390/ijerph19084499

Mitra A (2013) Sensitivity of mangrove ecosystem to changing climate. Springer, New Delhi, Heidelberg, New York, Dordrecht, London. https://doi.org/10.1007/978-81-322-1509-7

Mitra A, Halder P, Banerjee K (2011) Changes of selected hydrological parameters in Hooghly estuary in response to a severe tropical cyclone (Aila). Indian J Geo-Marine Sci 40(1):32–36

Mitra A, Zaman S, Jana H, Ray Chaudhuri T, Amin G, De UK, Das S (2014) Status of heavy metal in the lower Gangetic delta: implication of acidification on compartmentation. Environ Anal Chem 1(2):1–4

Mitra A, Zaman S, Pramanick P (2023) Climate resilient innovative livelihoods in Indian Sundarban Delta. Springer. https://doi.org/10.1007/978-3-031-42633-9

Mohanty UC, Mohapatra M, Singh OP, Bandyopadhyay BK, Rathore LS (2013) Monitoring and prediction of tropical cyclones in the Indian Ocean and climate change. Springer. https://doi.org/10.1007/978-94-007-7720-0

Ortiz AMD, Chua PLC, Salvador D Jr, Dyngeland C, Albao JDG Jr, Abesamis RA (2023) Impacts of tropical cyclones on food security, health and biodiversity. Bullet World Health Org 101(2):152–154. https://doi.org/10.2471/BLT.22.288838

Ray Chaudhuri T, Fazli P, Zaman S, Pramanick P, Bose R, Mitra A (2014) Impact of acidification on heavy metals in Hooghly Estuary. J Harmon Res Appl Sci 2(2):91–97

Strickland JDH, Parsons TR (1972) A practical handbook of seawater analysis, vol 157, 2nd edn. Fisheries Research Board of Canada

Winkler LW (1888) Die Bestimmung des in Wasser gelösten Sauerstoffen. Berichte der Deutschen Chemischen Gesellschaft 21:2843–2855

Part III
Climate Resilient Adaptation Strategies for Rural Communities

Chapter 7
Climate Smart Strategies and Technologies for Climate Resilient Hill Agriculture in the Eastern Himalayan Region

R. Krishnappa, Burhan Uddin Choudhury, T. Ramesh, A. Gangarani, Saurav Saha, Sourabh Deori, Vinay Singh, Rumi Narzari, Supriya Debnath, and H. Samarendra

Abstract The Eastern Himalayan Region (EHR) is constantly faced with numerous challenges, many of which are caused by a changing climate, such as the frequent occurrence of high-intensity rainfall, flash floods, and the increase in erratic dry spells and droughts. The region's livelihood is largely dependent on agriculture and its allied sectors, which are exposed to a high degree of climate vulnerabilities. The sustainability of traditional farming and low-input organic agricultural production of major crops in the region is at risk due to climate change. The importance of assessing the effects of climate change and strategizing the strengths and vulnerabilities of hill agriculture cannot be overstated. In recent days, experiments under elevated levels of atmospheric CO_2 concentration and the associated rise in temperature and variation in precipitation patterns have become increasingly significant. Developing climate-resilient technologies, such as tailor-made climate smart crops and customised climate change-ready agro-technologies developed through CO_2 enrichment facilities like CTGC/FATE, is crucial for improving the livelihood security of hill dwellers under a challenging climate. Indigenous superior genetic livestock stocks can be harnessed to ensure livelihood security and climate-smart hill farming. Under the ongoing NICRA (National Innovations on Climate Resilient Agriculture) project, many climate-resilient and climate-smart rice varieties, like NICRA Aerobic Dhan 1, aerobic Dhan 2, and Hill rice have been developed for cultivation in aerobic and upland conditions. The yield they produced was 10–15% greater than the national average. Two drought-stress-toleraNT TomaTo liNes

R. Krishnappa (✉) · T. Ramesh · S. Deori · R. Narzari · S. Debnath · H. Samarendra
ICAR Research Complex for the North Eastern Hill Region, Umiam, Meghalaya, India

B. U. Choudhury · A. Gangarani · V. Singh
ICAR Research Complex for North Eastern Hill Region, Tripura Centre, Lembucherra, Tripura, India

S. Saha
ICAR Research Complex for North Eastern Hill Region, Sikkim Centre, Tadong, Gangtok, India

S. Mandal et al. (eds.), *Climate Change Vulnerability*, Climate Change Management, https://doi.org/10.1007/978-3-032-16783-5_7

(RCM-N-T-1 aNd DVRT-2) with yield potentials of 30.0 t/ha and 32.0 t/ha have been developed under NICRA. Biochar, a product of low oxygen pyrolysis, has been developed as a potential soil ameliorator by utilizing weed, crop biomass, and vegetation inputs to improve the productivity of acid soils. Digital technologies viz., NES+ (North East Soil Plus), android application to assess the status of soil health and to recommend location-specific soil acidity and plant nutrient management for important crops (rice, maize, pulses & different vegetables). Similarly, Hyper Green Indexer (HsGI), a web-based user-friendly online application developed under NICRA, can easily generate a set of 26 essential hyperspectral vegetation indexes from spectral reflectance data. To alleviate cold and heat stress in chickens, animal science recommended a low-cost dietary supplement that contains 2.0% black turmeric, 10% pudhina, and 15% alligator weed. Further, exploring modern and high throughput techniques involving digital phenotyping, artificial intelligence (AI) enabled services like deep learning and internet of things (IoT), role of space applications, leveraging speed breeding and plant-soil-environmental interactions for developing climate smart crops and live stocks are on progress. With enhanced stress adaptability and increased resource use efficiency of fragile agro-ecologies like EHR for improved productivity vis a vis livelihood security of low income and low input agriculture of hill inhabitants under changing climate is the need of the hour.

Introduction

The Eastern Himalayan Region (EHR) of India is endowed with a diverse edapho-climatic situation inhabited by diverse ethnic communities and rich biodiversity in flora and fauna. Agriculture and allied activities, being the main source of livelihood for the people of the region, persistently face multifarious challenges like degradation of natural resources owing to undulating geography, low cropping intensity, and productivity. Increase in the atmospheric CO_2 concentration, the associated rise in temperature, and variation in precipitation patterns will have profound effects on terrestrial plant growth and crop productivity in the region (Kimball et al. 2002). Although the region receives high average annual rainfall ($\approx$2450 mm), the majority (>80%) occurs during a short period (May—October) (Das et al. 2021). The changing climate in the region may strikingly alter total rainfall, the number of rainy days, increase the frequency of high-intensity rainfall-induced flash floods, and manifest prolonged droughts in the hill slopes (Khan et al. 2025; Dogar and Almazroui 2022). Genetic improvement and variety development for challenging climate could be the major thrust objective to increase crop production under warmed and water scarcity conditions of climate change, but actual consequences of temperature rise (above 35° C) and associated increase in atmospheric CO_2 concentration are difficult to predict as these interactive effects are still to be established (Dogar 2019; 2020; Adnan et al. 2021; Dogar et al. 2017a, 2017b). Other collateral and integral environmental variables, including drought stress and nutrient availability, light use efficiency, and canopy photosynthesis, will be prime considerations for holistic and comprehensive studies under a changing climate. Scientific knowledge on the factors that control crop growth and development, assimilate partitioning between organs under an altered environment, is still at a descriptive stage, and many of them are yet to be studied objectively. Hence, controlled research and simulation studies have to go a long way to inculcate and address all these obvious variables and provide authentic data sets to understand the impact of climate change realistically, thereby enhancing the climate resilience of hill agriculture in the North Eastern Region (NER) of India.

The signs of climate change in the NER in terms of rise in mean annual maximum and minimum temperature (Choudhury et al. 2012), changes in rainfall pattern, occurrence of frequent droughts, and high-intensity rainfall during a short period of the year are strikingly evident (Kumar et al. 2026). Being intricately linked to the agricultural sector, these climatic aberrations will have profound effects on terrestrial plant growth and crop productivity. The shift in the climatic scenario, low availability of resources, and lack of appropriate adaptation and mitigation strategies make the challenges for agricultural sustainability in the region further complex. Hence, alarming and unprecedented climate change warrants an urgent need to understand the synergistic and holistic physiological and biochemical mechanisms associated with plant growth and development, and to develop crop genotypes and agronomic practices with potential for climate resilience and enhanced productivity. Currently, levels of greenhouse gas have already increased more than 40% since pre-industrial times and now exceed 400 μmol/mol, which is the highest CO_2 levels that have been since the past 650,000 years and possibly the last 23 million years (Mauna Loa Observatory measurement as of June 5, 2019, Intergovernmental Panel on Climate Change (IPCC 2007). Unfortunately, concentrations are unabated and projected to continue to rise between ≈440 and 550 μmol/mol by 2050 and could reach as high as ≈900 μmol/mol by the end of the century (Meinshausen et al. 2011).

The rapid rise of anthropogenic greenhouse gas levels is leading to unprecedented global warming, with a 1 ° C increase already attained compared to pre-industrial times and up to a 4 °C increase possibly reached by the end of the century under current emission scenarios (Allen et al. 2018; Collins et al. 2013). Escalating temperatures are certain to bring about many direct and other downstream climatic effects, such as the increase in the severity and incidence of droughts, floods, and other extreme weather events, which all have a significant effect on crop production (Allen et al. 2018; Reddy et al. 2010). While small temperature increases may have positive effects in the short term, at least in certain temperate areas like NER, increasingly high temperatures and drought may eventually reduce biomass and seed production, as well as quality (Myers et al. 2014). Indeed, negative effects on crop production arising from a changing climate are already evident in many parts of the world. The exponential increase of CO_2 with the concomitant rise of temperature in the atmosphere is inimical to climate change, which is alarming to the growth and productivity of crop plants. Since CO_2 acts as the primary substrate for photosynthesis of all crop plants, which is critical for crop productivity, current alterations in its atmospheric levels will have profound implications on all agricultural systems (Tausz et al. 2013). Growing evidence across the globe suggests that alterations in [CO_2] can have a substantial impact on various physiological and developmental processes in plants, which could exacerbate consequences on future crop production (Kimball et al. 2002; Long et al. 2004).

Climate Risk and Vulnerability Assessment in Northeast INDIA

An analysis of temperature trends in Northeast India from 1951 to 2015 reveals significant changes in both maximum and minimum temperatures across the eight states (Fig. 7.1). Between 1951 and 1990, the average maximum temperature was 29.0 ° C, rising at a modest rate of +0.04 °C per decade. However, from 1990 to 2015, this rate nearly doubled to +0.086 °C per decade, bringing the average maximum temperature to 29.2 °C. The minimum temperature, initially showing a slight decrease during 1951–1990, shifted to an upward trend of +0.01 °C per decade from 1990 onward, reaching an average of 19.2 °C. Rainfall patterns also demonstrated notable variability; the average annual rainfall increased from 1530 mm in 1901–1930 to 1677 mm in 1961–1990, before decreasing slightly to 1606 mm in 1991–2015. Additionally, rainfall variability has grown, with 47 districts

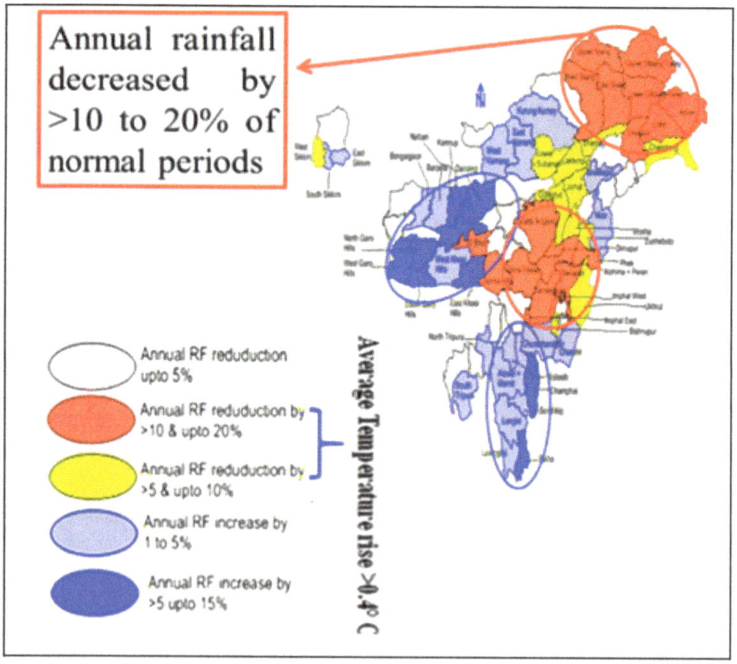

Fig. 7.1 District-wise mapping risk of exposure to abiotic stress

experiencing a 5–20% decline in annual rainfall, particularly in Arunachal Pradesh, Manipur, and Nagaland, while Mizoram, Meghalaya, and the Lower Brahmaputra Valley of Assam observed a 3–15% increase (Fig. 7.1). This shifting climate underscores the need for robust water conservation and management strategies to address the impacts of rising temperatures and inconsistent rainfall patterns. Furthermore, the projected increase in heat wave intensity under RCP 4.5 and RCP 8.6 scenarios highlights the urgency of implementing adaptive measures to safeguard agriculture and water resources in the region.

Projected Rainfall Erosivity and Climate Vulnerability in Eastern India

Climate change poses a significant threat to soil stability and agricultural sustainability in Eastern India, with rainfall erosivity (R factor) emerging as a key driver of soil erosion and land degradation (Choudhury et al. 2022). A study utilising CMIP6 experiments analysed projected changes in rainfall erosivity under four Shared Socioeconomic Pathways (SSPs) using convolution neural network (CNN) downscaling and bias correction through Quantile Data Mapping (QDM) against Indian Meteorological Department (IMD) ground observations. Results indicated that rainfall erosivity is expected to increase significantly under high-emission scenarios, particularly SSP3–7.0 (regional rivalry) and SSP5–8.5 (fossil-fuel-driven development). Among the five meteorological subdivisions of Eastern India, the sequence of mean rainfall erosivity values followed GWB < NMMT < AM < SHWBS < AP, with Arunachal Pradesh (AP) exhibiting the highest projected increase of 192.89 MJ mm ha^{-1} h^{-1} $decade^{-1}$ by 2100. This intensification of rainfall erosivity is expected to accelerate soil erosion due to the increased kinetic energy impact of raindrop splash, thereby threatening agricultural productivity, water retention capacity, and land sustainability. The study also revealed that extreme R factor thresholds demonstrated limited predictability ($R^2 = 0.332$) in CMIP6 experiments compared to IMD observations, underscoring the need for further refinement of climate models. Interestingly, SSP4–6.0 ("high inequality" scenario) showed the least changes in rainfall erosivity, suggesting that societal choices and radiative forcing levels significantly influence future precipitation patterns. These findings highlight the urgent need for adaptive land management strategies to mitigate soil erosion risks and enhance climate resilience in vulnerable regions of Eastern India (Figs. 7.2 and 7.3).

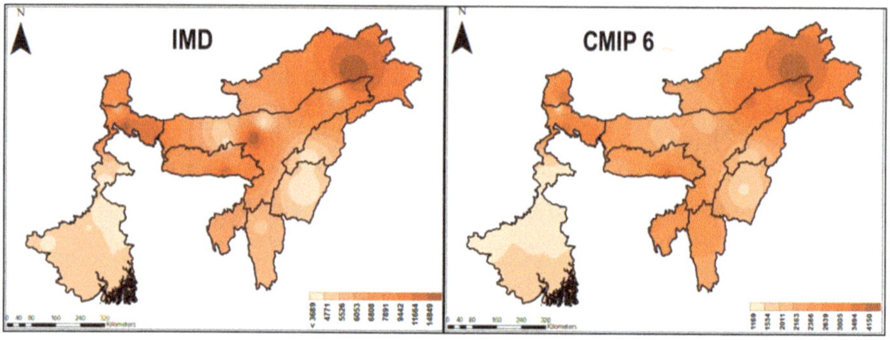

Fig. 7.2 Spatial distribution of observed R factor for IMD historical data and CMIP 6 experiments

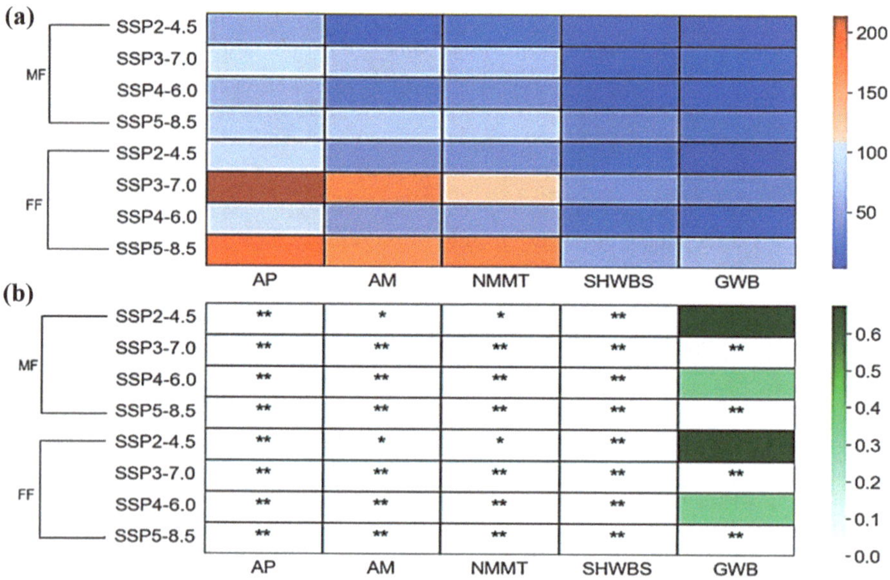

Fig. 7.3 **(a)** Sen slope and **(b)** p-values of R-factor time series of all meteorological subdivisions of Eastern India

Climate Vulnerability in Meghalaya: A Block-Level Perspective

The North East Indian states, particularly Meghalaya, are highly vulnerable to climate change due to their diverse ecosystems, agrarian economy, and proximity to the Himalayas. Rising temperatures, altered precipitation patterns, and glacial melt have led to an increase in extreme weather events such as floods and landslides, disrupting river systems, water availability, and biodiversity. These climatic shifts further threaten agricultural productivity, food security, and public health, particularly by increasing the spread of vector-borne diseases. Socioeconomic factors

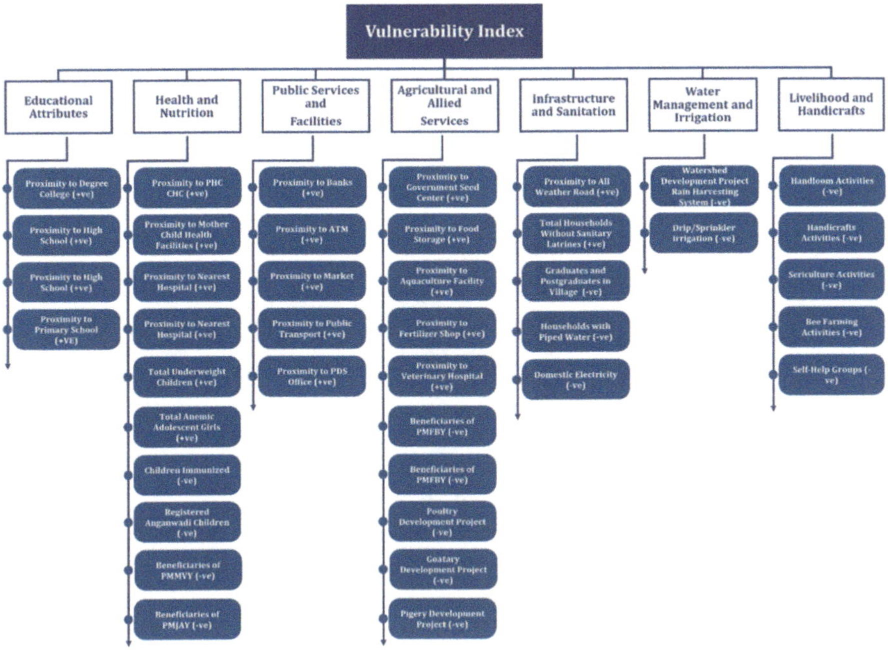

Fig. 7.4 Indicators of vulnerability

exacerbate the vulnerability, as climate-induced disruptions impact rural livelihoods, economic stability, and community resilience. To address these challenges, a district- and block-level vulnerability analysis was conducted in Meghalaya, utilising spatial secondary data from the Ministry of Rural Development (2020). The study employed Exploratory Factor Analysis (EFA) to identify key vulnerability indicators, including sensitivity and adaptive capacity parameters, while Confirmatory Factor Analysis (CFA) could not be fully applied. The sub-indicators under sensitivity and adaptive capacity indicators are presented in Fig. 7.4.

Key Drivers of Vulnerability

In Meghalaya, vulnerability is largely driven by health and nutrition factors, particularly total anaemic pregnant women (11.49%). This highlights a significant concern regarding maternal health. Other key drivers include Self-Help Groups (9.24%), suggesting that economic vulnerability and the role of women's collectives are critical factors, and Drip/Sprinkler Irrigation (9.10%), which indicates sensitivity in agricultural practices due to insufficient irrigation infrastructure. Further, the influence of Registered Anganwadi Children (8.93%) and beneficiaries of Pradhan Mantri Matru Vandana Yojana (PMMVY) (8.90%) underscores the importance of child nutrition and maternal care services as vulnerability drivers.

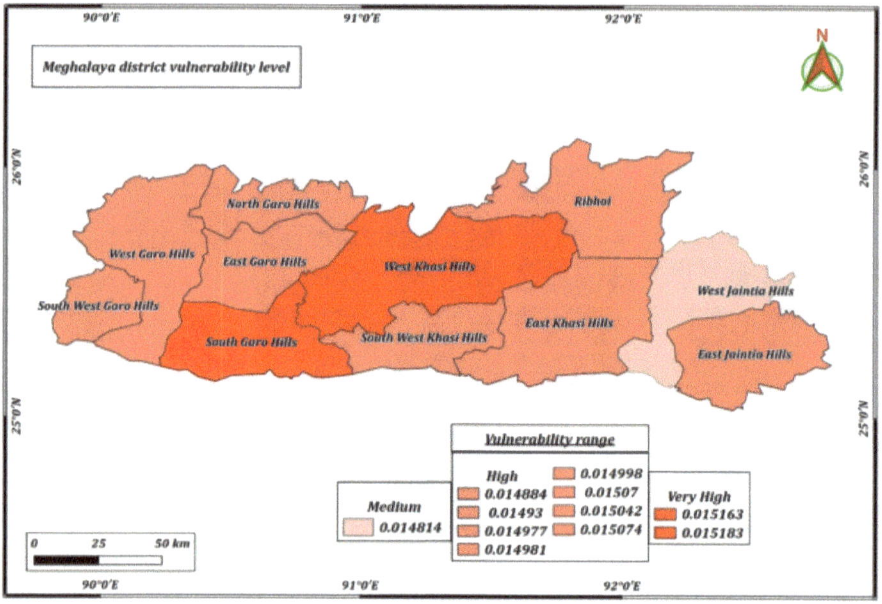

Fig. 7.5 District-level vulnerability in Meghalaya

District-Level Vulnerability in Meghalaya

The vulnerability assessment of districts in Meghalaya presents distinct patterns of vulnerability, classified into five categories: very low, low, medium, high, and very high, based on the vulnerability index. In Meghalaya (11 Districts), 72.73% of the districts (8 districts) exhibit high vulnerability, including East Khasi Hills, South West Garo Hills, East Jaintia Hills, and others. West Jaintia Hills (9.09%) has a medium vulnerability rating. South Garo Hills and West Khasi Hills (18.18%) are classified under very high vulnerability, reflecting severe challenges in health, nutrition, and socio-economic resilience (Fig. 7.5).

Block-Level Vulnerability Analysis in Meghalaya

The block-level vulnerability analysis offers a finer spatial resolution to assess localized disparities within Meghalaya (Fig. 7.6). By classifying blocks into five categories-very low to very high this section identifies critical areas requiring targeted interventions. Such granularity enables precise identification of vulnerability drivers, thereby informing more effective adaptation strategies. In Meghalaya, the majority of the blocks, 69.57% (32 blocks) and 21.74% (10 blocks), were under the high and very high vulnerability category. The remaining 6.52% (Mawryngkneng, Mylliem, and Amlarem block) and 2.17% (Mawpat block) were under medium and low vulnerability, respectively. The most vulnerable blocks were Mawshynrut,

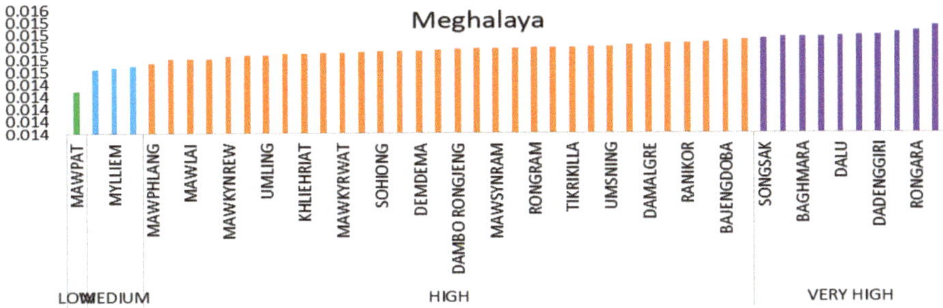

Fig. 7.6 Block-level vulnerability profile of Meghalaya

Rongara, Jirang, Dadenggiri, Nongstoin, Dalu, Gambegre, Baghmara, Chokpot, and Songsak. Among these, Mawshynrut, Dadenggiri, Nongstoin, Dalu, and Gambegre are located in the West Khasi Hills district, while Rongara, Baghmara, and Chokpot are in the South Garo Hills district, making these two districts the most vulnerable in Meghalaya.

Impacts of Climate Change on Crop Plants

Under a climate change scenario, the existing literature demonstrates both positive and negative effects of rising CO_2 on photosynthesis in different groups of higher plants, as it is a major process of carbon sequestration and turnover of the total carbon on the planet (Fig. 7.1). Rising atmospheric CO_2 concentration consistently stimulates the growth and harvestable grain production of C_3 crops (Kimball et al. 2002; Long et al. 2004)· as well as benefiting C_4 crops under stressful conditions. In C_3 plants, mesophyll cells containing ribulose-1, 5-bisphosphate carboxylase-oxygenase (RuBisCO) are in direct contact with the intercellular air space that is connected to the atmosphere via stomatal pores in the epidermis. Hence, in C_3 crops, rising CO_2 increases net photosynthetic CO_2 uptake as RuBisCO is not CO_2-saturated in the present atmosphere, and further CO_2 inhibits the competing oxygenation reaction, leading to decreased photorespiration. At 250, an increase in CO_2 from the present-day value of 390 ppm to that of 550 ppm, projected for the year 2050, would increase C_3 photosynthesis by 38% (Long et al. 2004). In contrast, in C_4 crops such as maize and sorghum, where RuBisCO is localised to bundle sheath cells in which CO_2 is concentrated to three to six times of atmospheric CO_2 will not benefit the crop directly (Von Caemmerer and Furbank 2003). This concentration is sufficient to saturate RuBisCO in these crops, and they may not show a direct response in photosynthetic activity, but an indirect increase in water use efficiency via reduction in stomatal conductance may still increase yield (Long et al. 2004).

The up- and down-regulation of photosynthesis under elevated CO_2 is a complex process. It is regulated by morpho-physiological changes during plant growth and development with particular reference to carbon allocation between source and sink tissues of either annual or perennial plants (Reddy et al. 2010) (Fig. 7.2). As

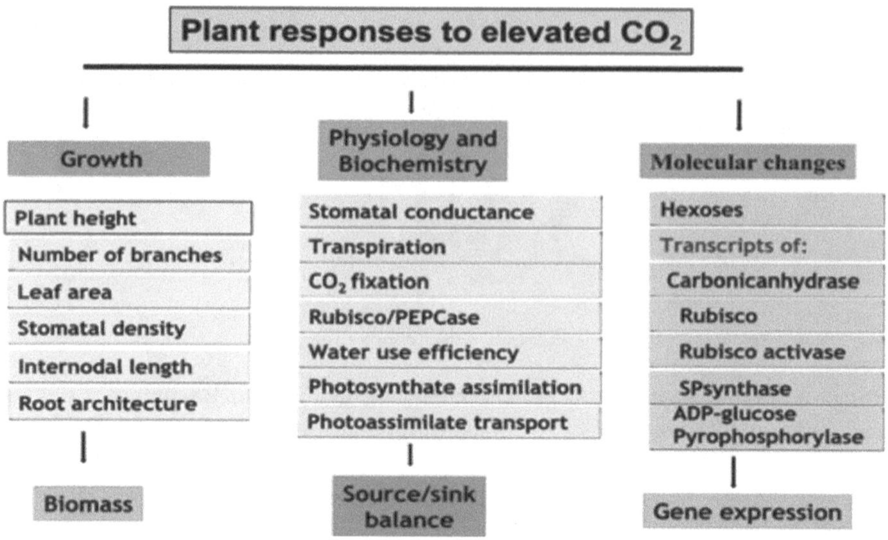

Fig. 7.7 Summary of morphological, physiological, biochemical, and molecular characteristics of plants as affected by rising atmospheric CO_2 concentrations (Reddy et al. 2010)

regulatory events associated with the inter- and intra-specific metabolic plasticity governed by the genetic makeup of different plants and an altered environment are little understood, only some factors affecting the regulation of photosynthesis in annual and perennial plant species at elevated CO_2 have been described (Reddy et al. 2010). However, research knowledge or scientific data on those factors that control crop growth and development, assimilate partitioning under elevated CO_2 and temperature conditions of NER of India, is yet to be studied systematically. Hence, the climate change research initiatives in the region may primarily focus on the evaluation of impacts of climate change on plant genetic, physio-biochemical responses of crop growth and productivity, source–sink balance, as well as its interactive mechanisms with different environmental factors, including elevated temperature, water and nutrient availability as an adaptation and mitigation strategy to climate change in the region (Fig. 7.7).

To date, the findings illustrating the effects of e [CO_2] on the yield and quality of leguminous crops have been inconsistent owing to their complexity, vast number of interacting elements, including genotype/species-specificity, rhizobial strain, soil type, irradiation level, and stress incidence (Singer et al. 2020). Therefore, it remains currently uncertain how future global warming conditions will impact legume yield and quality. But it seems that legumes may exhibit superior yield benefits with reduced protein decreases under e [CO_2] compared to non-N_2-fixing plants (in many instances, they still do not maximize yield gains or completely prevent nitrogen depletion when grown at elevated CO_2 levels).

Approaches/Strategies for Developing Climate-Resilient Crops

Development of Thermo-Tolerance

High temperatures have deleterious effects on several physiological processes like respiration, stomatal conductance, CO_2 uptake, and carbon utilization rates, reducing carbohydrate synthesis, accumulation, and enzyme-modulated functions. Plants tolerate high temperature stress by excess heat dissipation through evaporative cooling, maintaining membrane integrity and differential expression of stress-responsive genes and synthesis of heat shock proteins (HSPs) (Joshi et al. 1997). In crop plants, as temperature increases, it affects canopy growth, enhances anther dehiscence, reduces the duration of grain filling, and increases grain growth rate, etc. Despite observed negative effects of high temperature on leaf photosynthesis and translocation, the optimum temperature for net photosynthesis is likely to increase with elevated levels of atmospheric CO_2. Several studies have concluded that CO_2-induced increases in crop yields are much more probable in warm than in cool environments (Saseendran et al. 2000). However, the actual consequences of the temperature rise (>35 °C), associated with the increase in atmospheric CO_2 concentration, are difficult to predict as these interactive effects are yet to be established in combination with other vital environmental variables. Several studies investigated the phenomenon of acquired thermo-tolerance (Jorgenson et al. 2001) through many techniques. Temperature Induction Response (TIR) is one such efficient technique to screen and identify genotypes for thermo-tolerance (Fig. 7.3) by exposing the seedlings of crop plants to an induction temperature and subsequently to an extremely high temperature (which is otherwise lethal to non-induced seedlings). The surviving seedlings at the end of the recovery period are selected as thermo-tolerant lines (Srikanthbabu et al. 2002). Searching for genetic variability in thermo-tolerance of crop plants through this technique could be the appropriate option for evaluating morphological, physiological, and biochemical mechanisms and investigating the effect of temperature on reproductive growth parameters, viz., pollen viability, stigma receptivity, seed development, and yield attributes.

Induced: Seedlings subjected to an induction temperature (45–50 °C) and transferred to the challenging temperature. Non-induced: temperature maintained at ambient condition (25–30 °C), then exposed to the challenge temperature. Absolute control: temperature maintained at ambient (25–30 °C) throughout the experiment (Fig. 7.8).

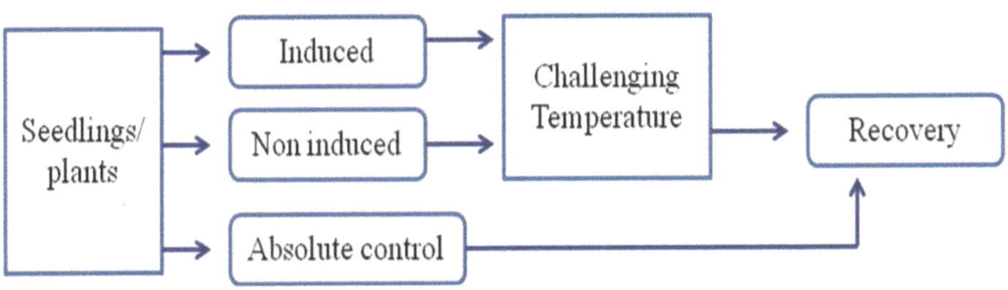

Fig. 7.8 Protocol to study temperature induction response in crop plants at the seedling level (Reddy et al. 2010)

Rice genotypes of Meghalaya mainly favouring a cold climate, and a slight increase in temperature may be detrimental for achieving higher productivity. Thus, breeding rice varieties with higher yield and heat tolerance is necessary to cope with future climate change and to sustain rice production in the future global warming. Information on QTLs/gene(s) responsible for heat tolerance and their mapping would strengthen our breeding programme as well as provide a clue to understand the genetics of rice heat tolerance. Thus, in the present investigation, the responses of rice genotypes to the interaction of elevated CO_2 and temperature stress were studied in the Carbon dioxide Temperature Gradient Chamber (CTGC) to simulate elevated CO_2 concentration and temperature.

Drought Tolerance and its Significance

The impacts of climate change manifest in the region in the form of changes in rainfall patterns, increased frequency high high-intensity rainfall events causing flash floods, prolonged or intermittent droughts. Under a future climate change scenario, with unpredictable/potential increase in the variability of precipitation patterns, drought or moisture stress is going to be one of the major constraints for agricultural productivity in the fragile ecosystems of hill regions. Since rice is the most important staple and semi-aquatic crop grown in the region, it is more likely that rice production as well as productivity is highly affected by the unpredictable drought conditions during its growing season in both lowland and upland conditions. Maize, being a highly water-demanding C_4 crop, minimal changes in rainfall patterns induces drastic changes in its growth and yield. High-intensity rainfall can also lead to submergence or excessive soil moisture in maize and cause a significant decline in its yield. Growth and productivity of *rabi* crops under rice fallows during which very meagre or no rainfall occurs, are also likely to be affected severely due to a lack of sufficient residual soil moisture in case of scanty rainfall in the rainy season.

Plant responses to drought stress are very complex as stress itself involves various climatic, soil, and agronomic factors, frequently and intricately linked to substantial variations in the timing of occurrence, duration, and intensity. Genetic improvement and varietal development for adaptation to drought could be one of the better options as the most drought-tolerant genotypes with the highest drought tolerance efficiency (DTE) (Fischer and Wood 1981), the least drought susceptibility index (DSI) (Fischer and Maurer 1978) to improve drought tolerance. Therefore, future research emphasis should be given to find out the simple and precise technique to detect genotypic differences for drought resistance and to quantify loss in yield by studying (a) the effect of moisture stress on the water relations, root traits, levels of osmo-protectants and antioxidant metabolism, (b) examining the effect of moisture stress on source and sink relationships, reproductive growth and yield attributes of rice and maize under elevated CO_2 and temperature.

Nutrient Uptake and Utilisation Efficiency

Another major constraint for crop production in the NEH region is that even though the soils are rich in organic matter, but they are acidic in nature (medium to strongly acidic in soil reaction), low to medium in available NPK and micronutrients with poor water holding capacity (Table 7.1). In these soils, phosphorus fixation and aluminium toxicity poses bigger challenge to farmers to maintain optimum nutrient balance in the soil. Exchangeable calcium and magnesium being critically low, the soils are poor in fertility and adequate nutrient content for high nutrient demanding crops (deficient in many micro nutrients) (Table. 7.1). To harvest the benefits of climate change with elevated CO_2, appropriate nutrient management strategies in consonance with water management are of paramount importance. Therefore, there is an urgent need to re-orient research activities to evaluate crops and cropping systems with altered climate resilience for soil related constraints under this nutrient stressful environment. Balanced and adequate fertilization, soil organic matter (SOM) improvement and soil acidity amelioration practices need to be continued to improve the sustainability of Hill agriculture in the region. Under the soil acidity-induced nutrient stress environment, carbon and nutrient dynamics, soil microbial activity, harnessing the innate root adaptability in concomitance with plant nutrient uptake and use efficiency, that too under elevated CO_2 and temperature, need to be studied precisely in NER of India (Fig. 7.5).

There is a general agreement that increasing temperature and atmospheric CO_2 will affect both soil and plant productivity, which provides the valuable input to soil organic carbon (SOC) and the rate of SOM decomposition which determines the loss of SOC (Verma et al. 2026). In current climate change scenario, SOC dynamics is governed by microbial activity in the soil and driven by complex interactions. Among physical drivers, soil properties and disturbances like tillage and soil management practices are important (Jorgenson et al. 2001). Deep ploughing and other non-conservative cultivation techniques are known to increase CO_2 emissions from soils (Six et al. 1999), while carbon sequestration could be increased by the adoption of agronomic practices such as diversified cropping systems, reduction of soil

Table. 7.1 Water requirement and critical stages of water requirement for different crops

Crop	Water requirement (mm)	Important moisture-sensitive stage/ critical stage
Rice	900–2500	Panicle initiation and flowering
Maize	500–800	Silking and tasselling
Chilies	600–900	Flowering and fruit set
Potato	500–700	Tuber initiation to tuber maturity
Tomato	600–800	From the commencement of fruit set
Peas	350–500	Flowering and pod formation
Cabbage, cauliflower and broccoli	380–500	Head formation until it becomes firm
Carrot	500–650	Root enlargement
Citrus	900–1200	Fruit setting and enlargement

erosion, application of organic fertilizer, including farm-yard manure and green-manure, improved grassland management and conservation tillage (Freedman et al. 2001). Further, fertilizer demand in Indian agriculture will increase considerably for the sustainability of soil and counteracting adverse impact of climate change on food and nutritional security in future (Manoj Kumar et al. 2011). Innovative investigations are required to study the responses of crop plants to the elevated CO_2 and temperature and nutrients availability, the physiological and biochemical characterization of these responses. Under changing climate, environmental stresses going to be the primary cause of crop losses worldwide, reducing average yields for most major crops by more than 50%. Interactive response of two or more environmental parameters can be successfully and realistically studied in CO_2 enrichment technologies like Carbon dioxide and Temperature Gradient Chamber (CTGC) and free air temperature enhancement (FATE).

Effect of CO_2 Fertilization on Crop Growth through CO_2 Enrichment Technologies

Plant responses to the projected future levels of CO_2 were first characterized since the 1980s by using controlled environment technologies such as phytotrons, closed chambers, cuvettes which are lasting for short-term period (days to weeks). However, technologies like OTC (open top chambers), Free-Air CO_2 Enrichment (FACE), FATE, CTGC experiments and SPAR (Soil Plant Atmosphere Research) with holistic approach have been established and are currently in use for crop response studies (Uprety et al. 2006). The data generated using these facilities are valuable for identifying mechanisms of crop response at the molecular, biochemical and physiological scales for changing environment. FACE and FATE facilities provide the most realistic conditions for evaluating and estimating crop yield responses to elevated CO_2 and temperature. These field CO_2 enrichment technologies facilitate large-scale trials of the major grain crops (Ainsworth, 2008). Therefore, to study the effect of global warming realistic happenings with concomitant rise in CO_2 and altered rainfall patterns on plant growth under field like conditions, CTGC and FATE facilities are established in NEH region to study crop response and to develop resilient crop varieties for sustaining agricultural productivity under elevated temperature and CO_2 conditions. These facilities will be useful for understanding simultaneous changes in CO_2 and temperature for the validation of data sets under global carbon flux analysis. The CO_2 fertilization effects reported in some reviews for the major crops (Long et al. 2004) and estimated the crop performance at a common CO_2 concentration of 550 ppm. Collectively, the fertilization factors averaged across the C_3 crops (rice, wheat, and soybeans) are 24% for yield, 27% for biomass, and 29% for photosynthesis. The responses for maize were lower except for yield, which was reported to increase by 27% but for wheat and soybeans, an even larger yield fertilization factor of 31% was observed (Long et al. 2004).

Although majority of controlled environment and field-based facilities are used to study the crop responses to individual atmospheric factor through 'single component approach', in practice, the future atmosphere may alter in several ways encompassing many factors simultaneously. Therefore, the future climate change impact studies should aim at quantification of many variables with the simultaneous increase in the CO_2 levels and temperature which are major factors predicted to dictate the changed climatic conditions under controlled conditions. Hence, developing germplasm/variety that responds better to elevated CO_2 and temperature or resilient to climate change is a distant goal. Public research and development is needed to extend capacity building beyond the current chamber experiments, which are limited in their geographical distribution, the very narrow range of CO_2 concentrations used in the experiments and the inclusion of important interactions with other climate change factors.

Climate-Resilient Technology Developed and Optimised under NICRA

Maintaining the productivity of mountain soils of Meghalaya is a challenge because of unprecedented land degradation resulting from the periodic nutrient leaching losses upon receipt of intense rainfall, widespread occurrence of low pH soil induced nutrient imbalances (>90% of the region), and prevalence of unscientific cultivation practices. As the region is potentially organic with meager use of agro-chemicals, plenty of crop residue and diverse vegetative biomasses are available (5, 10, 000 t/year), and the region is highly prone to the impacts of apparent climate, long-term carbon sequestration in terms of biochar incorporation is promising and sustainable (Das et al. 2023a). Shift from traditional "slash and burn" culture to "slash and char" might aid in sequestering around 190–213 Tg of carbon from the atmosphere. In addition, under changing the prevailing cropping systems of Hill Agriculture in hill slopes are facing frequent intermittent dry spells cum crop moisture stresses due to less or no rainfall distribution during the Rabi season and low water holding capacity of hill soils.

Under this context, our studies on biochar incorporation into the acid soils demonstrated that it improves the physical, chemical, and biological properties (Das et al. 2023b). Biochar is a solid carbonaceous and porous material produced through pyritization that can be used as a soil amendment to increase water and nutrient holding capacity of acid soil, *a vis-à-vis* serving as a soil carbon sink and to store carbon for a long time. Our studies on biochar that were produced from maize residue, mixed weed biomass (*Lantana camara, Chromolaena odorata, Eupatorium* sp.), and pine wood, and applied at rates of 0, 5, 10, and 20 t/ha. The soil properties after 3 years of incorporation, viz., soil moisture, soil organic carbon (SOC), pH, and available N, P, and K status, were significantly improved under 5 t ha-1 biochar than under no biochar. Among the NMP, 75% RDF + FYM and 100% RDF had higher soil moisture, water infiltration, pH, SOC concentrations, and available N, P, and K status than others. The combined application of 100% RDF and biochar @ 5 t ha-1 produced the highest maize equivalent yield. In another study the biochars displayed distinct physical and chemical properties influencing their performance. Weed biochar had a pH of 9.83, total organic carbon (TOC) of 37.5%, and high potassium (6.14%), while maize biochar showed moderate potassium (1.98%) and a higher TOC (45.7%). Pine Residue biochar (PRB) had the lowest pH (5.84) but the highest TOC (55.9%) and minimal potassium (0.21%). BET surface area, total pore volume, and pore size were higher for weed biochar, enhancing soil physical interactions. Maize Straw Biochar (MSB) is an irregularly shaped material with pore sizes from 206.1 nm to 1.282 µm, resulting from the dehydration and degradation of cellulose, lignin, and hemicelluloses, whereas mixed weed biomass (MWB) indicated variations from quadrangular with a rough surface to elongated tubular with a smooth surface with smaller pores, for capturing nutrients and adsorbents. These variations in surface area and porosity are expected to influence nutrient retention and water-holding capacity in soils. Biochar particles were nearly spherical in the case of maize and weed biochar, with PRB being the exception due to its irregular shape. Particle size range of MSB was 39.6 to 114.9 nm, MWB ranged from 31.2 to 48.6 nm, and PRB had particles ranging from 37.3 to 69.6 nm. All biochar materials (PRB, MSB, and MWB) contain a range of functional groups, including O-H, P-H, N=O, NH_2, C-O, P=O, S=O, C-H, P-OR, Si-OR, and alkyl halides. These functional groups contribute to the chemical composition and reactivity of the biochars, which can have implications for their applications in different contexts. Fe was four times higher in MWB than PRB, exceeding the threshold limit of 4.5 mg kg^{-1}, which is particularly significant in acidic soils. MWB and MSB biochar contained 11 times more Zn than PRB biochar, providing a suitable alternative to manage Zn deficiency in acidic soils.

Maize treated with weed biochar at 20 t/ha exhibited a significant yield increase (4.58 t/ha) and longer cob length (17.7 cm) over the control (4.22 t/ha). During rabi, French bean treated with 20 t/ha weed biochar produced higher pod numbers (15.8) and leaf area index (3.74). In kharif rice, maize biochar (10 t/ha) and weed biochar (20 t/ha) enhanced panicle weight (2.34–2.94 g) and grain yield (3.42–3.53 t/ha), with weed biochar promoting higher filled grains per panicle (143.8). Rapeseed grown after rice recorded higher yields with weed biochar at 20 t/ha (1537.9 kg/ha) compared to maize biochar (1295.4 kg/ha). Maize recorded the highest photosynthetic rate (56.6 µmol/m^2/s) and transpiration (13 mmol/m^2/s) under weed biochar at 20 t/ha, while rice showed notable improvements with weed biochar @20 t/ha (30.1 µmol/m^2/s) and maize biochar (28.3 µmol/m^2/s). Enhanced transpiration and water uptake were linked to altered root architecture as influenced by biochar, and soil moisture retention was also higher under weed biochar at 20 t/ha, supporting better crop performance in moisture-limited conditions of EHR. Moreover, studies conducted under CTGC (Carbon Dioxide Temperature Gradient Chambers) facilities of ICAR RC NEH, Umiam, with elevated temperature up to +3 ° C, around 48% increase in grain yield was recorded under biochar @ 10 t/ha with 100% NPK application, followed by nano-fertilizer application (33%) and 100% NPK application with lime @ 500 kg/ha (30%) as compared to control. Similarly, about a 27% increase in pod yield was recorded under biochar @ 10 t/ha with 100% NPK application, followed by application of 100% NPK application with mulching application (24%) as compared to the control.

Weed and Pine Biochar application @ 5.0–10.0 t/ha and mixed weed biomass @ 20 t/ha is recommended for improved soil physical, biological, and hydrological properties of mountain soils, which in turn improved physiological traits and Yield of Rice-mustard and Maize-French bean cropping systems with improved resource use efficiency of hill slopes. A greater understanding of Carbon neutrality, microbial community, and pesticide degradation is needed. Thus, biochar from crop residues has a lot of scope, especially in EHR for recycling/re-utilisation of locally available bio-resources (Das et al. 2023a, b).

The study investigates the impact of biochar on soil quality, carbon dynamics, and greenhouse gas (GHG) emissions in a maize and upland rice-based cropping system. Various biomass sources, including weed biomass (*Lantana camara, Chromolaena odorata, Eupatorium* sp.), maize residue, and pine wood, were collected for biochar production, applied at rates of 0, 5, 10, and 20 tons per hectare (Fig. 7.9). Biochars produced from these feedstock displayed distinct properties: weed biochar had a pH of 9.83, total organic carbon (TOC) of 37.5%, and high potassium (6.14%); maize biochar showed a pH of 9.31, TOC of 45.7%, and moderate potassium (1.98%); while pine biochar had a lower pH of 5.84 but the highest TOC at 55.9%, though it contained minimal potassium (0.21%).

Technologies for Soil and Crop Health Monitoring

The NICRA played a pivotal role in the development of eco-friendly and innovative technologies aimed at improving soil and crop health monitoring. One such advancement is the non-destructive estimation of foliar nitrogen in tomato plants using hyperspectral remote sensing. This cutting-edge technology enables rapid and accurate monitoring of nitrogen levels, a critical nutrient for plant growth, without the need for conventional destructive sampling methods. In Meghalaya, a study led by B.U. Choudhury, and Krishnappa R. to measure leaf spectral reflectance and correlate it with nitrogen concentration, offering a time-efficient and precise approach to nitrogen management. Apart from nitrogen estimation, NICRA has introduced other digital tools for crop health assessment. The Hyper Green Indexer (HsGI) (Fig. 7.10), a web-based application that allows users to effortlessly generate 26 essential Hyperspectral Vegetation Indices from spectral data. This tool simplifies operations by providing consistent and accurate outputs, improving productivity without requiring advanced technical knowledge. Since its release, HsGI has been used in 236 sessions across seven countries. Another key tool, the Northeast Soil Plus (NES+) (Fig. 7.10), is an Android app developed by the Institute. NES+ assists in assessing soil health and provides location-specific nutrient management

Fig. 7.9 Biomass conversion to biochar and its utilization as soil amendment

Fig. 7.10 Advanced technologies for crop and soil monitoring

recommendations for crops such as rice, maize, pulses, and vegetables. These technologies highlight NICRA's commitment to enhancing sustainable agriculture through innovative solutions.

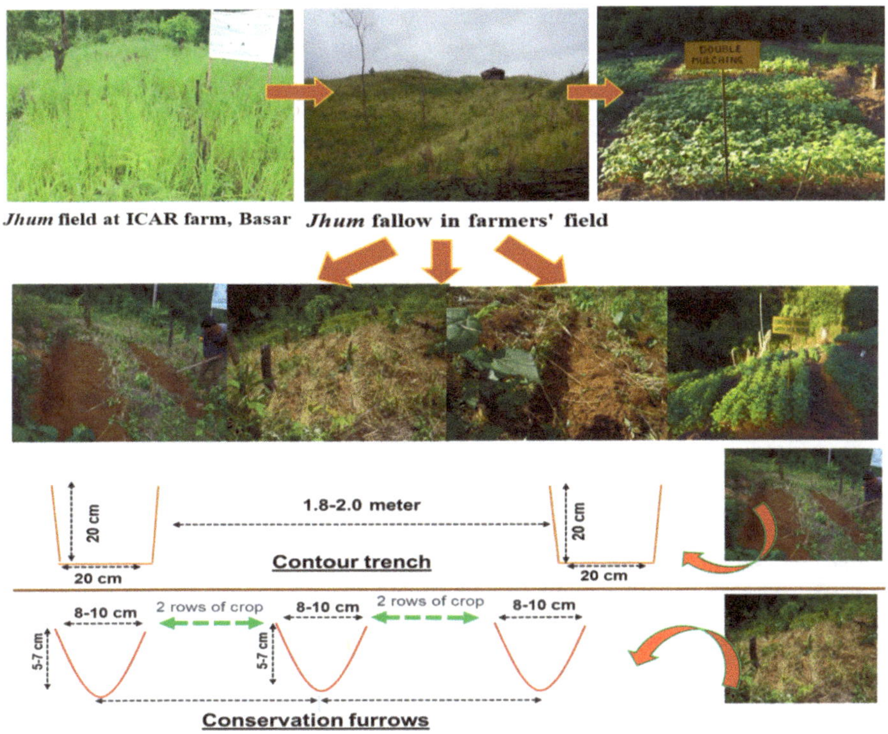

Fig. 7.11 Soil moisture conservation practices developed under NICRA

Soil Moisture Conservation in Jhum Lands

Crop intensification in the *jhum* ecosystem through soil moisture conservation (SMC) practices is crucial for sustainable agriculture in Arunachal Pradesh. A preliminary study was conducted in the upland fallow rice field of Gori Research Farm to assess the feasibility of growing Rabi crops using conserved soil moisture in *jhum* fallows (Fig. 7.11). The experimental layout was established after the harvest of kharif rice, under minimum tillage conditions. Except for the control, all plots retained standing rice straw (~30 cm) to aid moisture conservation, and no irrigation was provided throughout the crop cycle. The study followed with four soil moisture conservation treatments—conservation furrow, trench, mulching, and control (Fig. 7.11). The findings highlight the potential of soil moisture conservation (SMC) practices in improving soil moisture retention and crop productivity in *jhum* lands. Studies in Arunachal Pradesh have shown that SMC methods such as mulching, conservation furrows, and contour trenches increased soil moisture by up to 19% at 0–30 cm depth, with mulching being most effective in surface layers and furrows providing deeper moisture retention, particularly during dry periods. These practices also contributed to improved soil health, with higher organic carbon (up to 1.93%) and available nitrogen (up to 233 kg/ha). Crop yields responded positively to these interventions, with French bean yields increasing by 24–47%, while green gram and buckwheat yields improved by 11–30% and 11–27%, respectively.

Additionally, SMC practices enhanced crop water productivity, chlorophyll content, and pod length, leading to improved plant growth. Mulching and furrows significantly increased root dry weight by 8–22% and leaf area by 20–26% over control plots, with the highest yields observed in treatments involving mulching and double mulching, where French bean yields were 21–45% higher, green gram 7–30% higher, and buckwheat 6–27% higher over two years.

Further investigation was conducted following the harvest of *jhum* paddy to evaluate the performance of French bean and buckwheat under various SMC structures, including mulching, double mulching, conservation furrows, and contour trenches. Among all the treatments, double mulching proved to be the most effective, leading to significant improvements in plant height, relative water content, and economic yield for both crops. Additionally, double mulching accelerated flowering, resulting in earlier flowering compared to other treatments. Data revealed that French bean plants under double mulching reached a height of 51.09 cm, with a relative water content of 87.37%, and produced a green pod yield of 45.62 q/ha, whereas the control plots recorded the lowest values for these parameters.

Similarly, buckwheat plants under double mulching exhibited superior growth, with a plant height of 64.02 cm, a relative water content of 92.38%, and a grain yield of 4.03 q/ha. These results reaffirm the effectiveness of enhanced mulching techniques in improving soil moisture retention and overall crop productivity in the mid-hill regions of Arunachal Pradesh. These findings emphasize the importance of integrating SMC techniques to enable successful double cropping in shifting cultivation areas. Conservation furrows and mulching, particularly double mulching, emerged as the most effective strategies for enhancing soil moisture retention, plant growth, and crop productivity, demonstrating their potential to improve agricultural resilience and sustainability in rainfed, sloping terrains. Implementing these practices in *Jhum* systems can support long-term soil health, sustain productivity, and improve farmers' livelihoods in Arunachal Pradesh. Further studies in actual *jhum* fields will provide deeper insights into their long-term benefits and adaptability.

Carbon Sequestration and Soil Carbon Stock Assessment

A study in Mizoram assessed carbon sequestration across eight land-use systems in the mid-hills, including forests, bamboo, areca nut, oil palm, orange, tree bean, teak, and rubber. Tree bean and forests showed the highest active and passive carbon pools, supporting short- and long-term carbon storage, with forests having the highest biomass and ecosystem carbon stocks, followed by bamboo and teak. Carbon pools, both labile and non-labile, decreased with soil depth. In Sikkim, organic nutrient management in maize-soybean and maize-black gram systems increased soil organic carbon fractions, enhancing active and passive carbon pools for sustainable retention. In Meghalaya, agroforestry systems like Alder + pineapple and Alder + tea exhibited higher carbon content in larger soil aggregates, promoting stable carbon storage. These findings highlight the role of sustainable land-use, organic management, and agroforestry in enhancing carbon sequestration, aiding climate change mitigation, and improving soil health in the Eastern Himalayas.

Jalkund: A Vital Component of Climate Resilient Agriculture in the NEH Region of India

Need for Rainwater Harvesting

The rainwater harvesting, storage, and efficient use of it hold the key to sustainable livelihood in hill agriculture (Balusamy et al. 2024). Uneven distribution of rainfall (spatially and temporally) makes crop cultivation is very difficult in the undulating hill topography of this region. Therefore, rainwater harvesting in the form of Jalkund, adoption of efficient and effective water application techniques at the critical stages of water requirement, and adoption of other soil moisture conservation measures like mulching, cover crops, intercrops, etc., need special attention to improve the crop productivity in the low-productivity, high-risk-prone farming systems of the Northeast (Fig. 7.12). In this sense, Jalkund, an in-situ rainwater harvesting structure well well-suited to the undulating topography of this region, with its use, farmers can be able to give life-saving irrigation to crops at the critical stages. Crop water requirement and critical stages of water requirement are the two important factors that need to be considered while planning (Balusamy et al. 2024) for cultivation of crops (Table. 7.1). Application of water at critical stages/moisture-sensitive stages of the crop will save the crop from partial to complete crop loss. Thus, Jalkund is one of the important water harvesting tools to save crops from complete loss in the undulating hill topography of Northeast India ginger, turmeric, carrot, etc., and animal husbandry and livestock activities (Fig. 7.12). The length, width, and depth of Jalkund vary according to land availability, water requirement for different purposes, etc. (Balusamy et al. 2024).

Jalkund Technology: Impact on Adopting Farmers' Livelihood

Assessment on the impact of Jalkund on the adopting farmers' livelihood revealed that a Jalkund having the capacity to store 90 m^3 (90,000 L) of rainwater could sustain irrigation of about 450 m^2 of cultivated area. With supplementary irrigation water from Jalkund, farmers were able to cultivate high-value vegetables such as cabbage, cauliflower, broccoli, green peas, mustard leaves, tomato, and radish, resulting in the production of 10–13 quintals of vegetables and generating

Fig. 7.12 Multiple beneficial uses of Jalkund water by a Farmer in Lawmai village, East Khashi Hills, Meghalaya

significant gross earnings (Rs. 27,000). Poultry rearing earned them an additional income of Rs. 3625 by selling fully grown poultry birds (33 local). Since piggery is the main activity for the livelihood of the Meghalaya farm households and Jalkund supported it by providing the necessary water during the rainfall shortage period. Farmers received an additional income, on average, of Rs. 10,500 by selling 2 fully grown pigs. In addition to these farm activities, farmers were also able to grow fish in the Jalkund, from which they earned Rs. 5000 annually (Balusamy et al. 2024).

Advancing Climate-Resilient Crops for Stress-Prone Areas of the NEH Region

The NICRA project has been actively engaged in developing climate-resilient agricultural practices in the north-eastern region of India. These efforts address the growing challenges posed by climate change, such as drought and rising temperatures, particularly affecting rice and rabi crops. Through its advanced breeding programs and stress tolerance assessments, NICRA has successfully developed five high-yielding, drought-tolerant rice varieties. Notable achievements include the development of three rice varieties which were officially notified to aid in mitigating drought impacts: ICAR-NEH NICRA Hill Rice, ICAR NEH NICRA-Boro Dhan, and NICRA Aerobic Dhan 2 (Fig. 7.13). Two advanced drought-resilient lines, IET

Fig. 7.13 Notified rice for cultivation in Aerobic conditions

30034/TRC 185-B-B-82-2-11 and IET 30023/TRC184-B-B-76-1-1, have been developed, with the two progressing toward national varietal proposal stages, while two promising rice entries, TRC 2015–12 and TRC BN-1311-B-B-43-11-1, for approved through the State Variety Release Committee. TRC 2015–12, suitable for aerobic conditions and classified under the medium maturity group, demonstrated a high yield potential of 7.698 t/ha over 3 years, with yield advantages of 8.98% over national checks (NC), 11.28% over regional checks (RC), and 15.56% over local checks (LC) in Tripura, along with superior grain quality and biotic stress tolerance. TRC BN-1311-B-B-43-11-1, suited for rainfed low and medium elevations in the Northern Hills, showed remarkable yield stability across diverse environments, with a mean yield of 4533.8 kg/ha, outperforming NC, RC, and LC by 9.7%, 35.96%, and 13.91%, respectively, in coordinated hill rice trials. Additionally, six new drought-tolerant rice lines have been nominated for aerobic cultivation trials, with four advancing to later stages of testing. Research efforts have also targeted the identification of key Quantitative Trait Loci (QTLs) related to drought tolerance and grain yield, leading to the introgression of five QTLs in Tripura, associated with seedling vigor, leaf rolling, panicle length, and yield per plant, which enhance resilience during drought, particularly at the reproductive stage. Through these initiatives, NICRA demonstrates a robust commitment to advancing climate-adaptive agriculture in the Northeast programs. These findings provide valuable insights for breeding high-yielding, drought-tolerant rice varieties to enhance climate resilience and food security in the NEH region. Additionally, 15 promising rice breeding lines were evaluated for yield and yield attributes through multi-location trials (MLT) in Nagaland. Out of 990 rice breeding lines, 14 lowland and 16 upland lines were identified as promising. Among these, seven lowland rice breeding lines were selected for MLT across the state, where they outperformed the check variety, RC Maniphou-7, in terms of yield and early maturity, leading to their submission for AICRP trials. Similarly, four upland rice breeding lines were evaluated, identified as promising, and their seeds were multiplied for further MLT. The evaluation of lowland rice breeding lines was conducted at ICAR, Nagaland Centre, during the kharif season of 2024–25, where NLCJ-308 recorded the highest yield of 77.65 q/ha, followed by NLCJ-395 with 63.43 q/ha, with all breeding lines maturing in

<120 days. In upland rice breeding line trials, eight breeding lines were tested under rainfed conditions at ICAR Nagaland Centre alongside the check variety Bhalum-3. NLCJ-295 achieved the highest yield of 37.22 q/ha, followed by NLCJ-10 with 34.17 q/ha, demonstrating their potential for upland rice cultivation in the region.

Development of Drought-Tolerant Tomato for the NEH Region

Drought stress severely affects tomato growth and yield, highlighting the need for resilient cultivars in the North Eastern Himalayan (NEH) region. This study evaluated four native bacterial strains (RCM-SSR-304, −312, −323, and −326) for their ability to enhance tomato growth under moisture stress. Untreated plants exhibited stunted growth, reduced yield, and increased oxidative damage, while bio-inoculant-treated plants-maintained growth and yield comparable to non-stressed controls, indicating their protective role. Antioxidant enzyme activity, particularly catalase and glutathione reductase, was significantly enhanced in inoculated plants, reducing oxidative stress. Additionally, treated plants retained higher soil moisture, suggesting improved water retention in the rhizosphere. Among the tested strains, RCM-SSR-312 (*Burkholderia* sp.) and RCM-SSR-326 (*Streptomyces canus*) showed the most promising results in mitigating drought effects. Bio-inoculant-treated tomato plants also exhibited higher fruit yield with increased vitamin C and lycopene content, improving both productivity and nutritional value. Complementing microbial interventions, two promising drought-tolerant tomato breeding lines, RCM-N-T-1 and RCM-N-T-2, were developed for the NEH region. These lines completed station trials at the Manipur Centre and are now in seed production for AICRP trials. RCM-N-T-1, with duration of 135–140 days, a fruit weight of 50–70 g, and a yield potential of 30 t/ha, demonstrated strong tolerance to moisture stress at the flowering stage. Similarly, RCM-N-T-2, with a slightly higher fruit weight of 60–70 g and a yield potential of 32 t/ha, exhibited superior drought tolerance during early growth stages. Physiological indicators such as green-to-brown leaf ratio, relative water content, and enzyme activities (ascorbate peroxidase and glutathione reductase) were significantly higher in RCM-N-T-1 compared to the check variety DVRT-2, confirming its superior drought resilience and higher fruit yield under stress conditions. As part of the NICRA Tomato-Subproject, seed multiplication of RCM-N-T-1 was successfully undertaken in 2024, and the line has been proposed for multi-locational trials under AICRP (VC). DNA fingerprinting of RCM-N-T-1 and RC Kamenasenba-1 (Sel-9A) identified SSR marker TGS-192 as a distinguishing marker, with further analysis planned to identify additional polymorphic markers. These findings underscore the potential of microbial and genetic interventions in developing drought-resilient tomato cultivars suited for the NEH region.

Development of Excess Moisture Stress-Tolerant Maize

To address excess moisture stress in maize cultivation across the NEH region, a comprehensive screening of 36 maize lines, including 31 from ICAR, Sikkim, and 5 from ICAR-IIMR, Ludhiana, was conducted under both field and controlled microcosm conditions. These lines were subjected to 15 days of continuous water logging stress at the V5-V6 growth stages, leading to the identification of six promising flood-tolerant genotypes: Mizo-41(A)-2–4, Mizo-21(B)-2–3, UMI-1210 (ICAR-IIMR), Mizo-13(b)-1–3, Mizo-16(B)-2–3, and Mani-3-2-1. Among them, Mani-3-3-2-1 exhibited consistent performance over 3 years, demonstrating strong adaptability under both control and waterlogged conditions. Trials in Sikkim and Tripura revealed delayed tasseling and silking under stress, with similar trends observed across locations, confirming the physiological impact of water logging on maize flowering. Yield assessments further highlighted Mani-3-3-2-1 as a superior genotype, maintaining stable yields with minimal reductions—40.03 g/plant under stress in Rabi 2023–2024 and 44 g/plant under stress in Kharif 2024–2025. A third-year trial at ICAR Tripura Centre in Kharif 2024 evaluated four inbred under water logging stress, confirming Mani-3-3-2-1 and UMI-1210 as the most resilient, with average yields of 43 g/plant and 45.95 g/plant, respectively, and minimal yield reductions of 11.3% and 12.9%. In contrast, GP (H)-1–4-2-1 and Mizo-41(A)-2–4–2-1 showed high sensitivity, with yield losses exceeding 60%. Physiological assessments across 33 inbred lines revealed significant genotypic variation in key stress-related traits, including chlorophyll index, leaf temperature, performance index, and NDVI, with notable treatment and interaction effects. Regardless of genotype, water logging led to increased leaf temperature. These findings provide crucial insights into maize resilience under excess moisture stress and support the identification of superior genotypes for large-scale adoption in the NEH region. Ongoing multi-locational trials will further validate these selections, contributing to the development of climate-resilient maize varieties tailored to the region's agroecological challenges.

Technologies Developed Under the Animal Science Component

A Climate-Resilient, Low-Cost Housing Model

A low-cost, climate-resilient pig housing model validated and disseminated through 34 Krishi Vigyan Kendras, Tribal Development programs, Farmer Field Schools, and Biotech-Krishi networks, supports smallholder farmers in boosting pig productivity under diverse climates. Impact studies are recommended to quantify monetary benefits. A climate-adapted, raised-floor poultry housing model in Meghalaya and Manipur outperforms traditional deep litter systems, with higher growth rates (10.34

vs. 8.76 g/day/bird), lower mortality (9.8% vs. 14.1%), better feed conversion, and enhanced summer resilience. A deep litter piggery system also aids pigs in coping with extreme heat and cold. Validating these models across varied locations will enhance their adaptability and impact.

Developed Low-Cost, Eco-Friendly Heating System for Cold Stress Management in Piglets

The "Briquette Fuel Animal Heating System" (BFAHS) was evaluated as a low-cost, eco-friendly solution to manage cold stress in piglets during winter (January–February, avg. min. temp. 6.6 °C). Thirty piglets were split into three groups: Incandescent bulb (T1), Beehive briquettes (T2), and Control (T3). Over 2 months, piglets in the BFAHS group (T2) showed significantly higher growth (9.45 ± 0.09 kg) compared to T1 (7.28 ± 0.65 kg) and T3 (7.14 ± 0.55 kg). Body condition scores indicated all T2 piglets were healthy, while T1 and T3 had compromised piglets, with mortality rates of 20% and 30%, respectively, versus 0% in T2. The BFAHS, using briquettes made from a 2:1 charcoal-mud mix, compacted and sun-dried for 7 days, effectively reduced cold-stress mortality and enhanced piglet performance.

Dietary Supplementation—To Reduce Cold Stress & Heat Stress

Dietary supplements effectively mitigate heat and cold stress in poultry and pigs, boosting production performance. For heat stress, 0.5% ginger powder enhances body weight gain, feed conversion ratio (FCR), egg production, egg weight, and feed intake in laying hens (exposed to 37 ± 1 °C, 3 h daily for 45 days), while reducing serum corticosterone and cholesterol. Black turmeric (2%) improves egg production, egg mass, and FCR, lowering corticosterone. Pudina powder also alleviates heat stress in poultry. Fermented banana pseudo stem (40%) in pig feed and 3% curry leaf powder in poultry feed reduce thermal stress. For cold stress, 1% Alligator weed supplementation enhances body weight, feed intake, FCR, immune responses, and reduces oxidative stress in poultry. Turmeric and Roselle powders further ameliorate cold stress, enhancing resilience to extreme temperatures.

Climate Resilient Technology for the Active Floodplain

A climate-resilient farming system for active floodplains, developed at Tripura significantly enhances agricultural productivity on a 0.15 ha plot using raised and sunken beds (0.5 m height difference) and drainage to prevent water logging. During the rainy season, raised beds support paddy cultivation, while sunken beds enable fish culture with species like Puntius, Mola, Kanla, and carps. In winter and summer, raised beds are used for vegetables, and a water harvesting structure (10% of the area) with a deep trench supports year-round pisciculture. Edible aquatic plants (Water mimosa, Ipomoea) and vertical gourd farming further boost economic value. The system achieves a productivity of 70.3 kg/ha/day, far surpassing the 20–25 kg/ha/day of traditional methods, with a rice equivalent yield of 14.4 t/ha compared to 2.2 t/ha in mono-cropping. Water use efficiency is 6.78 kg/mm (77% higher than 2 kg/mm), and energy efficiency is 6.55 versus 2.25 in conventional farming. The system records a 300% multiple cropping intensity, 92.5 Cultivated Land Utilization Index, and 6.25 diversity index, compared to 100%, 42, and 1, respectively, in traditional practices. With economic returns of Rs. 71.4/ha/day and a benefit-cost ratio of 2.8, this floodplain farming system offers a sustainable, efficient model for flood-prone areas.

Conclusions

Given the above information, it is understood that the future climate change scenario will have a definite impact on the hill agricultural practices and livelihood security of the NE Region. Since the focus is on important themes like natural resource management, climate resilient crops, and climate resilient livestock, ovrall climate change studies, and climate resilient technologies developed for the region is expected to combat the realistic impacts of elevated CO_2 and temperature conditions in the region. Thus, climate change research initiative cum adhoc programmes specific to the NER of India have to go a long way to address all these variables and provide data sets to comprehend the impact of climate change and thereby hasten the development of climate resilience of important staple crops cultivated under the rain fed production systems.

Acknowledgement Authors are duly grateful and regarded for all the Previous Princiapal Investiagtors(PIs), Co-Principal Investigators and funding Agency (ICAR-CRIDA, Hyderbad) of NICRA project for thier smooth coordination, technical and financial support for holistic implementation of the project in the NE region.

References

Adnan M, Khan F, Rehman N, Ali S, Hassan S.S, Dogar MM,... & Hasson S (2021). Variability and predictability of summer monsoon rainfall over Pakistan. Asia-Pacific Journal of Atmospheric Sciences, 57(1), 89–97. https://doi.org/10.1007/s13143-020-00178-2

Ainsworth EA (2008) Rice production in a changing climate: a meta-analysis of responses to elevated carbon dioxide and elevated ozone concentration. Glob Change Biol 14(7):1642–1650. https://doi.org/10.1111/j.1365-2486.2008.01594.x

Allen MR, Dube OP, Solecki W, Aragón-Durand F, Cramer S, Humphreys S, Zickfeld K (2018) Framing and context. In: Masson-Delmotte V, Zhai P, Pörtner H-O et al (eds) IPCC special report. Retrieved from https://www.ipcc.ch/sr15/download/

Balusamy A, Ramesh T, Moirangthem P, Talang HD, Chanu LJ, Gopalakrishnan B, Islam M, Hazarika S, Kandpal BK, Choudhury BU (2024) Indigenous wisdom for sustainable mountain agriculture in the eastern Himalayas (Meghalaya). Indian J Soil Conserv 52(1):46–57

Choudhury BU, Das A, Ngachan SV, Slong A, Bordoloi LJ, Chowdhury P (2012) Trend analysis of long term weather variables in mid altitude Meghalaya, North-East India. J Agric Phys 12(1):12–22

Choudhury BU, Nengzouzam G, Mandal S, Sethy BK, Hazarika S, Mishra VK (2022) Long-term effect of integrated farming systems on soil erosion in hilly micro-watersheds (Indian eastern Himalayas). Land Degrad Dev 33(14):2554–2566. https://doi.org/10.1002/ldr.4332

Collins M, Knutti R, Arblaster J, Dufresne J-L, Fichefet T, Friedlingstein P, Wehner M (2013) Long-term climate change: projections, commitments and irreversibility. In: Stocker TF, Qin D, Plattner G-K et al (eds) Climate change 2013. Cambridge University Press, Cambridge, UK and New York, NY, pp 1029–1136

Das A, Datta D, Samajdar T, Idapuganti RG, Islam M, Choudhury BU, Mohapatra KP, Layek J, Babu S, Yadav GS (2021) Livelihood security of small holder farmers in eastern Himalayas, India: pond based integrated farming system a sustainable approach. Curr Res Environ Sustain 3:100076. https://doi.org/10.1016/j.crsust.2021.100076

Das SK, Ghosh GK, Avasthe R, Choudhury BU, Mishra VK, Kundu MC, Roy A, Mondal T, Lama A, Dhakre DS (2023a) Organic nutrient sources and biochar technology on microbial biomass carbon and soil enzyme activity in maize-black gram cropping system. Biomass Convers Bioref 13:9277–9287. https://doi.org/10.1007/s13399-021-01625-4

Das SK, Ghosh GK, Mishra VK, Choudhury BU, Dutta SK, Hazarika S, Kalita H, Roy A, Singh NU, Gopi R, Devi EL (2023b) Utilizing dissimilar feedstock's derived biochar amendments to alter soil biological indicators in acidic soil of Northeast India. Biomass Convers Bioref 13:10203–10214. https://doi.org/10.1007/s13399-021-01670-z

Dogar MMA (2019). The Sensitivity of DTR to the increased CO2 over Mid-latitude Semi-dry Regions. JOJ Hortic Arboric; 2(2): 555584. https://juniperpublishers.com/jojha/pdf/JOJHA.MS.ID.555584.pdf

Dogar MMA (2020). Study of the regional climatic impacts of tropical explosive volcanism in the Middle East and North Africa region (Doctoral dissertation, Hokkaido University). https://doi.org/10.14943/doctoral.r7091

Dogar MM, Kucharski F, & Azharuddin S (2017a). Study of the global and regional climatic impacts of ENSO magnitude using SPEEDY AGCM. Journal of Earth System Science, 126(2), 30. https://doi.org/10.1007/s12040-017-0804-4

Dogar MM, Stenchikov G, Osipov S, Wyman B, & Zhao M (2017b). Sensitivity of the regional climate in the Middle East and North Africa to volcanic perturbations. Journal of Geophysical Research: Atmospheres, 122(15), 7922–7948. https://doi.org/10.1002/2017JD026783

Dogar MMA, & Almazroui M (2022). Revisiting the strong and weak ENSO teleconnection impacts using a high-resolution atmospheric model. Atmospheric Environment, 270, 118866. https://doi.org/10.1016/j.atmosenv.2021.118866

Fischer RA, Maurer R (1978) Drought resistance in spring wheat cultivars. I. Grain yield responses. Aust J Agric Res 29(5):897–912. https://doi.org/10.1071/AR9780897

Fischer KS, Wood G (1981) Breeding and selection for drought tolerance in tropical maize. In: Proceedings of the symposium on principles and methods in crop improvement for drought resistance with emphasis on rice. IRRI, Philippines

Freedman JM, Fitzjarrald DR, Moore KE, Sakai RK (2001) Boundary layer clouds and vegetation–atmosphere feedbacks. J Clim 14(2):180–197. https://doi.org/10.1175/1520-0442(2001)013<0180:BLCAVA>2.0.CO;2

Intergovernmental Panel on Climate Change (IPCC) (2007) Climate change 2007: the physical science basis. In: Solomon S, Qin D, Manning M (eds) Contribution of working group I to the fourth annual assessment report of the intergovernmental panel on climate change. Cambridge University Press, Cambridge, UK, p 996

Jorgenson MT, Racine CH, Walters JC, Osterkamp TE (2001) Permafrost degradation and ecological changes associated with a warming climate in Central Alaska. Clim Chang 48(4):551–579. https://doi.org/10.1023/A:1005667424292

Joshi CP, Klueva NY, Morrow KJ, Nguyen HT (1997) Expression of a unique plastid-localized heat-shock protein is genetically linked to acquired thermo-tolerance in wheat. Theor Appl Genet 95(5):834–841. https://doi.org/10.1023/A:1005667424292

Khan, A., Mandal, S., Satpati, L., Dogar, M. M., Hasanean, H., & Salah, A. M. (2025). Analyzing drought: the impact of climate change on precipitation patterns in the Shilabati river Basin, West Bengal, India. Earth Systems and Environment, 9(3), 1585–1614. https://doi.org/10.1007/s41748-025-00679-y

Kimball BA, Kobayashi K, Bindi M (2002) Responses of agricultural crops to free-air CO2 enrichment. Adv Agron 77:293–368. https://doi.org/10.1016/S0065-2113(02)77017-X

Kumar P, Chakraborty A Sagar AK (2026). Investigating Heatwave Features: Creating an Intensity-Duration-Frequency Model for India's Principal Climate Zones. Earth Syst Environ. https://doi.org/10.1007/s41748-025-00979-3

Long SP, Ainsworth EA, Rogers A, Ort DR (2004) Rising atmospheric carbon dioxide: plants FACE the future. Annu Rev Plant Biol 55(1):591–628. https://doi.org/10.1146/annurev.arplant.55.031903.141610

Manoj Kumar MK, Patra AK, Anand Swarup AS (2011) Impact of climate change on fertiliser demand in agriculture: concerns and imperatives for food security in India. Indian J Fert 7(2):48–62

Meinshausen M, Smith SJ, Calvin K, Daniel JS, Kainuma MLT, Lamarque J-F, van Vuuren DPP (2011) The RCP greenhouse gas concentrations and their extensions from 1765–2300. Clim Chang 109(1–2):213–241. https://doi.org/10.1007/s10584-011-0156-z

Myers SS, Zanobetti A, Kloog I, Huybers P, Leakey ADB, Bloom AJ, Usui Y (2014) Increasing CO2 threatens human nutrition. Nature 510(7503):139–142. https://doi.org/10.1038/nature13179

Reddy AR, Rasineni GK, Raghavendra AS (2010) The impact of global elevated CO_2 concentration on photosynthesis and plant productivity. Curr Sci:46–57

Saseendran SA, Singh KK, Rathore LS, Singh SV, Sinha SK (2000) Effects of climate change on rice production in the tropical humid climate of Kerala, India. Clim Change 44(4):495–514. https://doi.org/10.1023/A:1005542414134

Singer SD, Chatterton S, Soolanayakanahally RY, Subedi U, Chen G, Acharya SN (2020) Potential effects of a high CO_2 future on leguminous species. Plant-Environ Interact. https://doi.org/10.1002/pei3.10009

Six J, Elliott ET, Paustian K (1999) Aggregate and soil organic matter dynamics under conventional and no-tillage systems. Soil Sci Soc Am J 63(5):1350–1358. https://doi.org/10.2136/sssaj1999.6351350x

Srikanthbabu V, Krishnaprasad BT, Gopalakrishna R, Savitha M, Udayakumar M (2002) Identification of pea genotypes with enhanced thermotolerance using temperature induction response technique (TIR). J Plant Physiol 159(5):535–545

Tausz M, Tausz-Posch S, Norton RM, Fitzgerald GJ, Nicolas ME, Seneweera S (2013) Understanding crop physiology to select breeding targets and improve crop management under increasing atmospheric CO2 concentrations. Environ Exp Bot 88:71–80. https://doi.org/10.1016/j.envexpbot.2011.12.005

Uprety DC, Garg SC, Bisht BS, Maini HK, Dwivedi N, Paswan G et al (2006) Carbon dioxide enrichment technologies for crop response studies. J Sci Ind Res 65(11):859

Verma P, Kaur J, Arora MS et al (2026). Modeling the Impact of Direct Air Capture on Forest Biomass and Population Dynamics. Earth Syst Environ 10, 407–427. https://doi.org/10.1007/s41748-025-00599-x

Von Caemmerer S, Furbank RT (2003) The C4 pathway: an efficient CO2 pump. Photosynth Res 77(2):191–207. https://doi.org/10.1023/A:1025830019591

Chapter 8
Climate-Smart Agriculture: Bridging Tradition and Technology for Rural Resilience

Himashree Kalita, Bijoy Krishna Deka, Jayanta Barman, and Afifa Kausar

Abstract Climate change has caused a significant threat to rural communities, specifically to those who are solely dependent on agriculture for livelihoods. In the current scenario, the ever-increasing food demand due to population explosion has popularized the concept of climate-smart agriculture (CSA), blending traditional with technology; alongside intensifying resilience, sustainability and productivity in those vulnerable regions. The concept of this chapter is to explain innovative CSA strategies specifically customized for rural development. It also includes blending of advanced agronomic practices, technological innovations, and policy interventions. Main focus is given to the adoption of drought-tolerant crops, precision farming, optimized irrigation techniques and agroforestry. Role of digital tools, such as remote sensing and mobile-based advisory services, and its contribution towards catalyzing the processes of decision making and risk management are also discussed here, including successful and promising case studies from Sub-Saharan Africa (e.g., Kenya and Ethiopia), South Asia (India), and Latin America (Peru), based on previously published works, where CSA has been effectively implemented, demonstrating its potential to upgrade food security, increase in

H. Kalita · B. K. Deka
Institutional Biotech Hub, ADP College, Nagaon, Assam, India

J. Barman
Institutional Biotech Hub, ADP College, Nagaon, Assam, India

Department of Physics, ADP College, Nagaon, Assam, India

A. Kausar (✉)
Institutional Biotech Hub, ADP College, Nagaon, Assam, India

Department of Zoology, ADP College, Nagaon, Assam, India

S. Mandal et al. (eds.), *Climate Change Vulnerability*, Climate Change Management, https://doi.org/10.1007/978-3-032-16783-5_8

farmer incomes, and reduce environmental impacts. By the amalgamation of scientific innovations with traditional knowledge, CSA has emerged as a practicable way toward sustainable rural development and climate-resilient farming communities.

Introduction

Climate change has emerged as one of the most critical challenges of the twenty-first century, affecting global food systems and rural livelihoods (Dogar 2020; Dogar and Sato 2018, 2019). The most vulnerable ones to these changes are the rural communities that are heavily dependent on agriculture, not as a means of income but also as a way of life, and are congenitally linked to ecological patterns and traditional knowledge systems (Wijerathna and Pathirana 2022). Increasing temperature extremes, irregular rainfall patterns, prolonged droughts, and other climate-related stresses have significantly reduced agricultural productivity, posing serious risks to food security, farmer livelihoods, and environmental sustainability (Farooq et al. 2022; Ellahi et al. 2025). This situation is fueled by rapid population inflation at the global level, eventually increasing the demand for agricultural resources, placing immense pressure on natural ecosystems and resource availability (Maja and Ayano 2021).

By seeing such a complicated crisis, the concept of Climate-Smart Agriculture (CSA), an integrative and systematic approach to mitigate the adverse effects of climate change on agriculture and food products, has arisen and is being popularized globally. It does not follow a one-size-fits-all model, unlike other conventional agricultural methods. On the contrary, it blends traditional agricultural practices with modern technological innovations. The concept of CSA was first promoted by Food and Agriculture Organization (FAO) (Alexander 2019; Lipper and Ziberman 2018) having three interlinked objectives: (1) increasing agricultural productivity sustainably and incomes; (2) adapting and building resilience to climate change; and (3) reducing or removing greenhouse gas emissions wherever possible (Belay et al. 2024; Matteoli et al. 2020).

Accumulated over generations through intimate interactions with local ecosystems, the traditional farming systems are already incorporated with sustainable methods such as crop rotation, organic composting, intercropping, agroforestry, and indigenous water conservation techniques (Balusamy et al. 2024). All these indigenous methods not only promote biodiversity and enhance soil health but also build resilience to climatic variability (Kumar et al. 2024; Suryawanshi et al. 2025). However, with the growing unpredictability and severity of climate phenomena, there is a critical need to complement traditional wisdom with scientific innovation (Nwankwo 2025). It is noteworthy that recent advancements in digital agriculture, particularly the use of satellite platforms such as Sentinel-2, Landsat-8/9, and drone alongside Geographic Information Systems (GIS) like ArcGIS and QGIS, have significantly transformed the way farmers monitor weather patterns, crop health, and resource allocation (Choudhury et al. 2019; Narzari et al. 2025). For instance, Sentinel-2 enables high-resolution NDVI-based crop monitoring, while Landsat-8 supports soil moisture analysis and land cover classification. When integrated with mobile advisory tools and precision agriculture platforms, these technologies empower farmers with real-time, spatially-informed decision-making capabilities (Singhal et al. 2024; Karunathilake et al. 2023).

The combination of all these approaches under the broader perspective of CSA creates a transformative prototype for rural development, one that is both resilient and inclusive, alongside sustainable and future-oriented. Additionally, it also encourages participatory practices, ensuring engagement of local stakeholders, especially smallholder farmers, in knowledge-sharing networks and adaptive learning programs. In such a way, it empowers communities to become active participants of change rather than becoming passive recipients of technology (Mert-Cakel and Miele 2022).

Basically, this chapter is all about the rising academic works regarding CSA, emphasizing the importance and application of CSA, particularly in rural and climatically vulnerable regions. Slowly it progresses to how CSA can be implemented in such a way serving as a powerful framework for sustainable rural transformation through some thematic lenses, which can be categorized as- foundational value of conventional agricultural practices, role of modern technologies in optimizing CSA practices, blending of traditional knowledge with scientific research, localized applications and authorized policy implementation, and necessity for a holistic approach bridging awareness gaps and other barriers (Hussain et al. 2022; Khatri-Chhetri et al. 2017). Through all these themes, the promises and also the challenges while implementing CSA as an adaptable model for climate resilience can be pointed out, alongside it can also be emphasized as a suitable adaptation for agricultural sustainability and rural development.

Review of Literature

Global population inflation has caused a threat to food security, which has necessitated the search for modern agriculture techniques blending conventional practices with technology. The concept of CSA is one among them, which can mitigate this problem and can also provide resilience to climate change by reducing greenhouse gas emissions. The main strength of CSA is basically its ability to blend traditional agricultural knowledge with advanced technologies and practices, implementing a holistic approach towards sustainable rural development (Adefila et al. 2024). It discusses key scholarly contributions giving importance to the relevance and implementation of CSA across various agroecological and socio-economic circumstances.

Conventional Agricultural Practices: A Foundation of Climate-Smart Agriculture

Singh and Singh (2017) emphasized the fundamental role of conventional agricultural practices in the evolution of climate-smart strategies. Traditional farming practices by native communities are specific to specific local environments and inherently aligned with the principles of sustainability, with the particular weather conditions over centuries. Different environment-specific agricultural techniques, like crop rotation, intercropping, agroforestry, and organic composting not only enhance biodiversity (Balusamy et al. 2024; Suryawanshi et al. 2024) but also help in building soil health, catalyzing water retention, and can also help in reducing dependence on chemical inputs. These traditional practices hold the ability to offer natural resilience to climatic variability and support carbon-efficient agriculture systems. The overdependence on Green Revolution technologies, mainly dominated by chemical inputs and technologization, triggering soil degradation, biodiversity loss, and increased greenhouse gas emissions (Choudhury et al. 2013; Singh and Singh 2017). That's the reason why rejuvenating traditional agricultural knowledge is considered as a vital step toward achieving CSA goals.

If we have to mention real-life examples, in Rajasthan, India, traditional methods like intercropping of legumes and pearl millet have helped in sustaining soil fertility in arid zones. Likewise, in Zambia, indigenous agroforestry methods involving *Faidherbia albida* trees have helped in enhancing nitrogen levels in soil and have also improved maize yields under drought stress (Yengwe et al. 2018).Alongside, in Andean regions of Peru, methods like organic composting and terrace farming have mitigated soil runoff and erosion problems (Leceta et al. 2024). These practices not only hold the ability to mitigate region-specific climatic issues, but also have enough potential to serve as adaptable CSA solutions, if properly blended with modern techniques.

Blending Traditional Knowledge and Modern Science

Okoronkwo et al. (2024) investigate the reciprocation between conventional practices and modern CSA techniques. The key finding of their study is that in many parts of sub-Saharan Africa, farming is entirely connected with cultural beliefs and local knowledge systems. Conventional practices like organic composting, intercropping, and crop rotation have contributed to soil fertility and ecological balance. But, regarding the intensification of climate threats, dependence on these methods is no longer sufficient. Okoronkwo and his group of researcherstherefore proposed a hybrid approach where traditional methods are strengthened by scientific interventions, implementing drought-tolerant crop varieties, precision agricultural tools, and advanced water management techniques, emphasizing farmer participation in knowledge-sharing and adaptive learning initiatives, strengthening local capacities, and improving resilience.

Technological Advancements in CSA Implementation

In the words of Musa and Lim (2025), CSA works within five interconnected themes: climate-smart farming techniques, efficient irrigation systems, nutrient management, conservation practices, and integrated production systems. Recent innovations in smart farming, like mobile-based advisory services, remote sensing and GIS, and real-time monitoring tools, are individually or collaboratively transforming how farmers access climate information and utilize it to make decisions. In countries like Vietnam, the use of real-time agricultural data has led to considerable improvements in crop productivity and resource use efficiency. Additionally, practices like intercropping (e.g., sorghum-cowpea and maize-pea combinations) have been shown to reduce carbon emissions and enhance system pliability. They also indicated the role of nanotechnology in improving food processing, input efficiency, and post-harvest storage, ultimately contributing to the overall sustainability of farming systems.

Despite these technological advancements, hurdles remain. Acquisition among smallholder farmers is often hampered by a lack of awareness, infrastructural limitations, and financial constraints. There is a need for integrative frameworks that combine technological innovation with context-specific needs, ensuring that advancements are affordable, accessible, and acceptable to local farming communities.

Regional Applications and Policy Support

Okolie et al. (2023) offer a region-focused perspective by analyzing CSA initiatives within Africa. Their bibliometric study reveals a steady increase in CSA-related publications, indicating its growing relevance in both research and policy discussions. Evidence from the continent shows that interventions such as sustainable land management practices, the adoption of hybrid crop varieties, and agroforestry systems have led to notable yield gains, in some cases surpassing 200% even under drought-prone conditions. When these practices are coupled with supportive policies and strong community participation, they have significantly improved food security while reducing vulnerability to climate-induced risks. The authors further argue that CSA is most effective when approached through a multidisciplinary, systems-based framework—one that integrates traditional knowledge with modern science and innovative policies.

Similar efforts appear worldwide beyond Africa. India's National Mission on Sustainable Agriculture promotes resource-conserving technologies and weather-based advisories specialized to monsoon variability; Colombia has embedded CSA into national adaptation work through participatory extension targeting coffee growers; and Central America's Agroecological Corridor advances diversified farming systems to lessen hurricane impacts. Although strategies differ by context, these examples share a common aim-strengthening resilience and productivity through locally adapted innovations and integrated policy support (Ghosh 2019; Eise et al. 2021).

Toward a Holistic and Inclusive CSA Paradigm

The literature recurrently finds a way by blending traditional knowledge systems with modern scientific innovations to build flexible and sustainable farming communities. While traditional practices can provide ecological grounding and cultural relevance, on the other hand, modern technologies contribute towards precision, expandability, and predictive capacity. The success of CSA depends largely on participatory processes that empower farmers, can enhance institutional capacities, and strengthen cross-sectoral collaboration. Challenges such as fragmented research, limited extension services, and socio-economic disparities must be addressed through policy alignment, capacity-building programs, and context-driven research agendas (Gardezi et al. 2022; Nagothu et al. 2016).

Mechanisms of GIS and Other Mobile Advisory Services in the Context of Climate-Smart Agriculture

The integration of smart cum digital tools within traditional agriculture is a revolutionary shift to the concept of CSA. Though many such tools are available, two have emerged as the most prominent ones: Geographic Information Systems (GIS) and mobile-based advisory platforms (Khatri et al. 2024). These two tools not only provide critical support for climate-resilient, resource-efficient, and site-specific decision-making but also enable the transition from reactive to anticipatory farming practices by collecting real-time data acquisition, analysis, and dissemination. Here is a detailed description of the scientific mechanisms through which GIS and mobile advisory services contribute to the implementation of CSA, particularly in data-constrained, climate-vulnerable agricultural landscapes (Gangopadhyay et al. 2019; Haworth et al. 2018).

GIS for Climate-Agriculture Interface

GIS is a geographical decision-support system that collects remotely sensed data and, after combining it with its algorithms of geostatistics and environmental modeling to perform precision agriculture (Ghosh and Kumpatla 2022; Raihan 2024).

Though it has a broader role, in the case of implementation with CSA, its role is primarily focused on three domains-agroecological zoning, risk assessment, and resource optimization. If we observe its methodology, GIS utilizes multi-temporal satellite imaging and georeferenced datasets to depict climate zones, monitor vegetation indices (e.g., NDVI), and analyze land use/land cover changes (Choudhury et al. 2013; Usman et al. 2015). GIS also helps the stakeholders to recognize those areas which are susceptible to extreme climatic conditions, such as droughts or flood-prone areas, which can guide preplanning for any disaster conditions, hence pre-preparedness for adaptation. Other facilities of GIS include GIS-based soil fertility mapping and hydrological modeling, which can previously inform about crop suitability assessments, irrigation scheduling, and nutrient application strategies (Abdel Rahman et al. 2021; Lanki and Onwu 2024). Prior information regarding these can enhance both productivity and input-use efficiency. Also, if these tools are used in conjunction with crop simulation models (e.g., DSSAT, APSIM, CropSyst, etc.), it can further enhance yield dynamics in different climatic conditions by guiding risk-informed planning and early warning systems (Banerjee et al. 2024; Ahmed et al. 2024).

Mobile Advisory Platforms for Smallholder Adaptation in the Context of CSA

Mobile advisory services play a critical role in centralized data systems in the case of farm-level decision-making. These platforms often work via SMS, interactive voice response (IVR), or digital apps. Through these ways, mobile advisory platforms provide climate-informed agronomic advisories customized to specific crops, regions, and temporal conditions (Van Campenhout et al. 2021). Their main mechanism is combining meteorological data, market information, and agronomic models to generate location-specific advisories. For example, predictive analytics for weather forecasting can inform farmers regarding optimal planting windows, crop disease outbreaks, crop-pest outbreaks, even water conservation and irrigation measures (Malik 2023; Stankovic et al. 2022). This information later often goes through a scrutiny process with localized data inputs, like temporal conditions and soil types, which contribute to the enhancement of their reliability and applicability. They also function as two-way communication systems, allowing farmers to report field-level conditions, receive diagnostic feedback, and participate in real-time learning loops directly, which enhances adaptive capacity, particularly among smallholder farmers with limited access to traditional extension systems (Prajapati et al. 2025; Sharma et al. 2020) (Table 8.1).

Table 8.1 Key tools, platforms, and their applications in CSA implementation

Tool/platform	Type	Specific use in CSA	Region of use	Reference
Sentinel-2	Satellite	NDVI crop monitoring; detailed monitoring of crops, water, and soil, which helps in detecting stress conditions, resource use optimization, yield prediction, and land management for climate adaptation and mitigation; optimize irrigation and fertilization; land cover changes	Global	Segarra et al. (2020)
Landsat-8/9	Satellite	Soil moisture, long-term vegetation tracking, water use management, and drought stress impact assessment	Global	Sholihah et al. (2024)
ArcGIS/QGIS	GIS tool	Agroecological zoning, yield modeling, hazard mapping, precision agriculture	Asia, Africa	Haque et al. (2025)
CropSyst/DSSAT/APSIM	Crop simulation models	Dynamic yield predictions under different climate scenarios	Africa, South Asia	Mihret et al. (2024), Boote et al. (2021)
mKRISHI	Mobile advisory	Localized advisories (crop health, pest alerts)	India	Singh (2017)
iCow/DigiFarm	Mobile advisory	Market info, voice-based guidance for rural farmers	Kenya, sub-Saharan Africa	Burns et al. (2022)
SEMA/CCAFS-CSA tool	Decision support tool	CSA benchmarking and prioritization	Latin America, Asia	Adimassu et al. (2024)

Synergistic Integration as a Real-Time Responsive CSA Ecosystem

Although GIS and mobile platforms each strengthen agricultural systems on their own, their integration offers far greater benefits. When used together, they enhance the development of predictive and prescriptive analytics frameworks for climate-smart agriculture. GIS contributes by delivering large-scale spatial assessments and risk mapping, whereas mobile technologies convert these insights into practical, localized guidance. This integration allows complex climate and environmental data to be quickly simplified into farmer-friendly formats, thereby supporting smallholders and informing centralized policy decisions (Kakamoukas et al. 2021; Yu et al. 2024).

Implications for Policy and Scaling

Transformative innovations rarely come without obstacles. In the case of digital tools for CSA, issues such as limited digital literacy, weak infrastructure, and difficulties in data integration often restrict their effectiveness. Addressing these barriers requires targeted policy interventions, stronger collaboration between public and private actors, and sustained efforts in capacity-building. Only through such measures can smallholder farmers fully benefit from GIS and mobile platforms, gaining timely insights into weather and climate that are vital for adaptation. To achieve meaningful scaling of CSA, these technologies must be strategically woven into national agricultural policies, climate strategies, and extension systems. Embedding them in this way fosters inclusive digital ecosystems, enabling farmers to act as informed decision-makers and active contributors to climate resilience (Muriithi et al. 2023; Makate 2019).

Conclusion

In the context of increasing climatic variability, CSA stands out as both a practical strategy for farming systems and a transformative way for sustainability. By blending traditional ecological knowledge with digital innovations such as GIS and mobile-based advisories, CSA strengthens productivity, resilience, and resource-use efficiency. For smallholder farmers, it not only supports adaptive practices but also contributes to building secure livelihoods within enabling policy environments. The convergence of science, tradition, and technology signals a new era in which agriculture can actively address, rather than merely endure, climate challenges. Future efforts should focus on evaluating CSA's long-term impacts across different agro-ecological settings, while developing robust frameworks that align real-time data with local farming calendars. Strengthening institutional linkages among research organizations, extension services, and climate modeling centers will be critical. Equally, expanding participatory and gender-responsive CSA initiatives that engage marginalized communities will help ensure that agricultural transformation is both inclusive and equitable.

References

Abdel Rahman MA, Hegab RH, Yossif TM (2021) Soil fertility assessment for optimal agricultural use using remote sensing and GIS technologies. Appl Geomat 13(4):605–618. https://doi.org/10.1007/s12518-021-00376-1

Adefila AO, Ajayi OO, Toromade AS, Sam-Bulya NJ (2024) Integrating traditional knowledge with modern agricultural practices: a sociocultural framework for sustainable development. J Sustain Agric Dev. https://doi.org/10.30574/wjbphs.2024.20.2.0850

Adimassu Z, Tamene L, Tibebe D, Ebrahim M, Abera W (2024) Identification and prioritization of context-specific climate-smart agricultural (CSA) practices in Ethiopia: a framework approach. Environ Res Commun 6(6):065007. https://doi.org/10.1088/2515-7620/ad54a1

Ahmed M, Ahmad S, Abbas G, Hussain S, Hoogenboom G (2024) Cropping systems modeling under changing climate. Spring. https://doi.org/10.1007/978-981-97-0331-9

Akila, Wijerathna-Yapa Ranjith, Pathirana (2022) Sustainable Agro-Food Systems for Addressing Climate Change and Food Security Agriculture 12(10) 1554-10.3390/agriculture12101554

Alexander S (2019) What climate-smart agriculture means to members of the global Alliance for climate-smart agriculture. Future Food: J Food Agric Soc 7(1):21–30. https://doi.org/10.17170/kobra-2018122073

Arun, Khatri-Chhetri P.K., Aggarwal P.K., Joshi S., Vyas (2017) Farmers' prioritization of climate-smart agriculture (CSA) technologies Agricultural Systems 151184-191 10.1016/j.agsy.2016.10.005

Balusamy A, Ramesh T, Moirangthem P, Talang HD, Chanu LJ, Gopalakrishnan B, Islam M, Hazarika S, Kandpal BK, Choudhury BU (2024) Indigenous wisdom for sustainable mountain agriculture in the eastern Himalayas (Meghalaya). Indian J Soil Conserv 52(1):46–57

Banerjee K, Dutta S, Das S, Sadhukhan R (2024) Crop simulation models as decision tools to enhance agricultural system productivity and sustainability—a critical review. Technol Agron 5(1). https://doi.org/10.48130/tia-0024-0032

Belay A, Mirzabaev A, Recha JW, Oludhe C, Osano PM, Berhane Z, Solomon D (2024) Does climate-smart agriculture improve household income and food security? Evidence from Southern Ethiopia. Environ Dev Sustain 26(7):16711–16738. https://doi.org/10.1007/s10668-023-03307-9

Boote KJ, Adam M, Ahmad I, Ahmad S, Cammarano D, Chattha AA, Wajid SA (2021) Understanding differences in climate sensitivity simulations of APSIM and DSSAT crop models. In: Handbook of climate change and agro-ecosystems: climate change and farming system planning in Africa and South Asia: AgMIP stakeholder-driven research Part 1, vol 5(5), pp 15–46. https://doi.org/10.1142/9781786348791_0002

Burns S, Dittmer KM, Shelton SW, Wollenberg EL (2022) Global digital tool review for agroecological transitions. Cali, Colombia, Alliance of Bioversity & CIAT

Burhan U., Choudhury Richard, Webster Vikas, Sharma Jonali, Goswami Thounaojam T., Meetei Rangappa, Krishnappa Penumetcha L. N., Raju (2019) Frost damage to maize in northeast India: assessment and estimated loss of yield by hyperspectral proximal remote sensing Journal of Applied Remote Sensing 13(04) 1-10.1117/1.JRS.13.044527

Choudhury, B.U., Mohapatra, K.P., Das, A., Das, P.T., Nongkhlaw, L., Fiyaz, A.R., Ngachan, S.V., Rajkhowa, D.J. and Munda, G.C. (2013). Spatial variability in distribution of organic carbon stocks in the soils of North East India. Curr. Sci., 104(5): 604–614.

Dogar, M. M. A. (2020). Study of the regional climatic impacts of tropical explosive volcanism in the Middle East and North Africa region (Doctoral dissertation, Hokkaido University).

Dogar, M. M., & Sato, T. (2018). Analysis of climate trends and leading modes of climate variability for MENA region. Journal of Geophysical Research: Atmospheres, 123(23), 13–074.

Dogar, M. M., & Sato, T. (2019). Regional climate response of Middle Eastern, African, and South Asian monsoon regions to explosive volcanism and ENSO forcing. Journal of Geophysical Research: Atmospheres, 124(14), 7580–7598.

Eise J, Lambert NJ, Wiemer EC (2021) Leveraging communities' network strengths to support climate change adaptation information-sharing: a study with coffee farmers in Risaralda, Colombia. Clim Change 168(1):12

Ellahi A, Nafisah IA, Almazah MMA et al. (2025) Innovative drought monitoring: development and application of the multi-regional aggregated standardized drought index (MRASDI). Earth Sci Inform 18, 365. https://doi.org/10.1007/s12145-025-01877-y

Farooq MS, Uzair M, Raza A, Habib M, Xu Y, Yousuf M, Ramzan Khan M (2022) Uncovering the research gaps to alleviate the negative impacts of climate change on food security: a review. Front Plant Sci 13:927535. https://doi.org/10.3389/fpls.2022.927535

Gangopadhyay PK, Khatri-Chhetri A, Shirsath PB, Aggarwal PK (2019) Spatial targeting of ICT-based weather and agro-advisory services for climate risk management in agriculture. Clim Chang 154(1):241–256. https://doi.org/10.1007/s10584-019-02426-5
Gardezi M, Michael S, Stock R, Vij S, Ogunyiola A, Ishtiaque A (2022) Prioritizing climate smart agriculture: an organizational and temporal review. Wiley Interdiscip Rev Clim Change 13(2):e755. https://doi.org/10.1002/wcc.755
Ghosh M (2019) Climate-smart agriculture, productivity and food security in India. J Dev Policy Pract 4(2):166–187
Ghosh P, Kumpatla SP (2022) GIS applications in agriculture. In: Geographic information systems and applications in coastal studies. Intech Open. https://doi.org/10.5772/intechopen.104786
Haque SJ, Hossain S, Billah MM (2025) Precision agriculture through remote sensing and GIS: advancing sustainable farming and climate resilience. Am J Smart Technol Solut 4(1):30–36. https://doi.org/10.54536/ajsts.v4i1.4418
Haworth BT, Biggs E, Duncan J, Wales N, Boruff B, Bruce E (2018) Geographic information and communication technologies for supporting smallholder agriculture and climate resilience. Climate 6(4):97. https://doi.org/10.3390/cli6040097
Hussain S, Amin A, Mubeen M, Khaliq T, Shahid M, Hammad HM, Nasim W (2022) Climate smart agriculture (CSA) technologies. In: building climate resilience in agriculture: theory, practice and future perspective, pp 319–338. https://doi.org/10.1007/978-3-030-79408-8_20
Jayant, Singhal Ankur, Rajwadi Guljar, Malek Padamnabhi S., Nagar G., Rajashekar C. Sudhakar, Reddy S. K., Srivastav (2024) Characterization of tropical forests at community level: combining spectral phenological structural datasets using random forest algorithm Biodiversity and Conservation 33(12) 3459–3474 10.1007/s10531-024-02835-8
Kakamoukas G, Sarigiannidis P, Maropoulos A, Lagkas T, Zaralis K, Karaiskou C (2021) Towards climate-smart farming—a reference architecture for integrated farming systems. In: Telecom, vol 2, No. 1. MDPI, pp 52–74. https://doi.org/10.3390/telecom2010005
Karunathilake EMBM, Le AT, Heo S, Chung YS, Mansoor S (2023) The path to smart farming: innovations and opportunities in precision agriculture. Agriculture (Basel) 13(8):1593. https://doi.org/10.3390/agriculture13081593
Khatri A, Lallawmkimi MC, Rana P, Panigrahi CK, Minj A, Koushal S, Ali MU (2024) Integration of ICT in agricultural extension services: a review. J Exp Agric Int 46(12):394–410. https://doi.org/10.9734/jeai/2024/v46i123146
Kumar P, Raj A, Kumar VA (2024) Approach to reduce agricultural waste via sustainable agricultural practices. In: Valorization of biomass wastes for environmental sustainability: green practices for the rural circular economy. Springer Nature Switzerland, Cham, pp 21–50. https://doi.org/10.1007/978-3-031-52485-1_2
Lanki AD, Onwu CA (2024) Geographic information system (GIS) application in soil fertility management: a review. J Glob Agric Ecol 16(2):29–40. https://doi.org/10.56557/JOGAE/2024/v16i28592
Leceta F, Binder C, Mader C, Mächtle B, Marsh E, Dietrich L et al (2024) The impact of agriculture on tropical mountain soils in the western Peruvian Andes: a pedo-geoarchaeological study of terrace agricultural systems in the Laramate region (14.5° S). Soil 10(2):727–761. https://doi.org/10.5194/soil-10-727-2024
Lipper L, Zilberman D (2018) A short history of the evolution of the climate smart agriculture approach and its links to climate change and sustainable agriculture debates. Clim Smart Agric: Build Resil Clim Change 52:13–30
Makate C (2019) Effective scaling of climate-smart agriculture innovations in African small holder agriculture: a review of approaches, policy, and institutional strategy needs. Environ Sci Pol 96:37–51. https://doi.org/10.1016/j.envsci.2019.01.014
Malik SA (2023) Linking climate-smart agriculture to farming as a service: mapping an emergent paradigm of datafied dispossession in India. In: Climate change and critical agrarian studies. Routledge, pp 545–567. https://doi.org/10.4324/9781003467960
Matteoli F, Schnetzer J, Jacobs H (2020) Climate-smart agriculture (CSA): an integrated approach for climate change management in the agriculture sector. In: Handbook of climate change

management: research, leadership, transformation, pp 1–29. https://doi.org/10.1007/978-3-030-22759-3_148-1
Mengistu M., Maja Samuel F., Ayano (2021) The Impact of Population Growth on Natural Resources and Farmers' Capacity to Adapt to Climate Change in Low-Income Countries Earth Systems and Environment 5(2) 271-283 10.1007/s41748-021-00209-6
Mihret YC, Ketsela GM, Mintesinot SM (2024) Implementation and application of APSIM for crop modelling in Ethiopia: a comprehensive review. Heliyon 10(10). https://doi.org/10.1016/j.heliyon.2024.e31612
Muriithi C, Mwongera C, Abera W, Chege CG, Ouedraogo I (2023) A scalable approach to improve CSA targeting practices among smallholder farmers. Heliyon 9(10). https://doi.org/10.1016/j.heliyon.2023.e20526
Musa SFPD, Ariff Lim S (2025) Revitalizing agriculture through climate change mitigation: a systematic literature review on smart technologies and sustainable practices. Int J Climate Change Strateg Manag. https://doi.org/10.1108/IJCCSM-05-2024-0071
Nagothu US, Kolberg S, Stirling CM (2016) Climate smart agriculture: is this the new paradigm of agricultural development? In: Climate change and agricultural development. Routledge, pp 1–20
Narzari R, Choudhury BU, Singhal G, Choudhary KK (2025) A critical review of how UAVs can transform precision agriculture in the realm of agroecology. Discover Soil 2(1):28. https://doi.org/10.1007/s44378-025-00055-2
Nwankwo SC (2025) Harnessing indigenous knowledge for climate action: bridging traditional wisdom. In: Empowering resilience: women, environment, and socio-economic challenges in Nigeria, vol 38
Okoronkwo DJ, Ozioko RI, Ugwoke RU, Nwagbo UV, Nwobodo C, Ugwu CH, Mbah EC (2024) Climate smart agriculture? Adaptation strategies of traditional agriculture to climate change in sub-Saharan Africa. Front Clim 6:1272320. https://doi.org/10.3389/fclim.2024.1272320
Okolie, C.C., Danso-Abbeam, G. Groupson-Paul O. and Ogundeji, A. A. (2023) Climate-Smart Agriculture Amidst Climate Change to Enhance Agricultural Production: A Bibliometric Analysis. Land, 12(1), 50. https://doi.org/10.3390/land12010050
Prajapati CS, Priya NK, Bishnoi S, Vishwakarma SK, Buvaneswari K, Shastri S, Jadhav A (2025) The role of participatory approaches in modern agricultural extension: bridging knowledge gaps for sustainable farming practices. J Exp Agric Int 47(2):204–222. https://doi.org/10.9734/jeai/2025/v47i23281
Rahul, Suryawanshi Rashmi, Thakare Hrishikesh, Kamat Onkar, Deshmukh Onkar, Shinde Renuka, Alapure Ashutosh, Kamble (2024) Climate Change Impacts and Risks for Animal Health: Indian Context: A Review Agricultural Reviews (Of) 10.18805/ag.R-2712
Raihan A (2024) A systematic review of geographic information systems (GIS) in agriculture for evidence-based decision making and sustainability. Glob Sustain Res 3(1):1–24. https://doi.org/10.56556/gssr.v3i1.636
Segarra J, Buchaillot ML, Araus JL, Kefauver SC (2020) Remote sensing for precision agriculture: Sentinel-2 improved features and applications. Agronomy 10(5):641. https://doi.org/10.3390/agronomy10050641
Sharma, S., Rana, D. S., Jat, M. L., Biswal, S., Khatri-Chhetri, A., & Pathak, H. (2020). A compendium of technologies, practices, services and policies for scaling climate smart agriculture in Odisha (India)
Sholihah RI, Trisasongko BH, Kusdaryanto S, Karyati NE, Panuju DR, Iman LOS, Shiddiq D (2024, June) Climate smart agriculture strategy for drought-prone areas: the role of land surface temperature data. In: IOP conference series: earth and environmental science (vol. 1359, No. 1, p. 012094). IOP Publishing. https://doi.org/10.1088/1755-1315/1359/1/012094
Singh DK (2017) Power of ICT to connect each farm and each farmer—mKRISHI® CCA: a case study. Agric Under Clim Change: Threats Strategies Policies 1:399
Singh R, Singh GS (2017) Traditional agriculture: a climate-smart approach for sustainable food production. Energy Ecol Environ 2:296–316. https://doi.org/10.1007/s40974-017-0074-7
Stankovic M, Neftenov N, Gupta R (2022) Use of digital tools in fighting climate change: a review of best practices. Available online: https://bit.ly/3Gxodt6 (accessed on 21 October)

Suryawanshi P, Pagar V, Patel N, Kumbhar NM, & Kalasare RS (2025) Modern Approaches to Future Agriculture.
Tezcan, Mert-Cakal Mara, Miele Geoff, Desa Xiangping, Jia Social Innovation and Sustainability Transition 'Workable Utopias' for Social Change Through Inclusion and Empowerment? Community Supported Agriculture (CSA) in Wales as Social Innovation Springer Nature Switzerland Cham 307–326
Usman M, Liedl R, Shahid MA, Abbas A (2015) Land use/land cover classification and its change detection using multi-temporal MODIS NDVI data. J Geogr Sci 25:1479–1506. https://doi.org/10.1007/s11442-015-1247-y
Van Campenhout B, Spielman DJ, Lecoutere E (2021) Information and communication technologies to provide agricultural advice to smallholder farmers: experimental evidence from Uganda. Am J Agric Econ 103(1):317–337. https://doi.org/10.1002/ajae.12089
Yengwe J, Amalia O, Lungu OI, De Neve S (2018) Quantifying nutrient deposition and yield levels of maize (Zea mays) under Faidherbia albida agroforestry system in Zambia. Eur J Agron 99:148–155. https://doi.org/10.1016/j.eja.2018.07.004
Yu L, Du Z, Li X, Zheng J, Zhao Q, Wu H, Huang X (2024) Enhancing global agricultural monitoring system for climate-smart agriculture. Clim Smart Agric:100037. https://doi.org/10.1016/j.csag.2024.100037

Chapter 9
Adaptive Breeding Strategies to Combat Climate Vulnerability in Rural Ecosystems

Rahul Kumar, Bapi Das, and B. U. Choudhury

Abstract Climate change is a serious threat to agricultural activities in all regions of the world, and small-scale agricultural societies are in danger. Climate pressures like drought, extreme heat, floods, and soil salinity are reducing crop yields, and burdening marginalized agricultural societies. Adaptive plant breeding can be identified as an essential approach to improve resilience by developing crops that will be able to sustain in severe environmental conditions. Scientists are making use of genetic resources, biotechnologies, and methods like marker-assisted breeding, genomic prediction, and CRISPR-based gene editing to create stress-tolerant crops. A notable example is the introduction of 109 improved varieties in 61 crops by ICAR, which is a deliberate move to enhance climate resilience. Genetic properties of landraces and wild relatives that adapt to stressful conditions are also used in these developments. This kind of innovation requires strong socioeconomic systems that bring the accessibility of high-quality seeds, financial services, education, and local breeding models. Programs such as the National Innovations in Climate Resilient Agriculture (NICRA) and international partnerships through CGIAR are some of the crucial elements in research, knowledge transfer, and capacity-building. Furthermore, technologies such as climate modelling and artificial intelligence are assisting in determining future breeding interests. Essentially, adaptive breeding is a pillar of climatologically intelligent agro-ecosystems, which are critical in protecting food production and rural livelihoods amidst increases in the rate of climate change. This will necessitate the need to have inclusive policies and cooperative systems of innovation in order to ensure that these benefits are availed to the vulnerable populations.

R. Kumar (✉) · B. Das · B. U. Choudhury
ICAR Research Complex for NEH Region, Lembucherra, Tripura Centre, West Tripura, Tripura, India

S. Mandal et al. (eds.), *Climate Change Vulnerability*, Climate Change Management, https://doi.org/10.1007/978-3-032-16783-5_9

Introduction

Climate change is a major challenge to global agriculture. It is especially affecting the rural regions where agricultural activities are the main source of livelihood and a major determinant of food security. The growing number of stresses related to climate change, such as drought, floods, heat waves, and salinity, is also significantly reducing crop production (Dwivedi et al. 2021). Smallholder farmers in these areas are especially at risk. They often rely on traditional varieties and lack resources to adapt. Plant breeding, in turn, has also become an essential strategy to develop the varieties of crops, which will be able to work in the changing environmental conditions (Kumar et al. 2021). There are genetic variation approaches, biotechnology, and better breeding methods that are used to develop climate-resistant crops. These crops perform well in unfavourable climatic conditions compared to the traditional varieties. Adaptive breeding enhances the genetic ability of crops to survive or resist a number of climatic stresses. This includes the selection of such traits that include heat and drought tolerance, early maturity, salinity tolerance, and resistance to pests and diseases. Traditional breeding is, however, significant, yet it is not quick enough or sufficient to adapt to climatic changes. As a result, recent genomic tools of marker-assisted selection (MAS), genomic selection (GS), and CRISPR gene editing are changing the breeding programs by allowing faster and more precise identification of traits (Kumar et al. 2024). As an example of national efforts, Prime Minister Narendra Modi introduced 109 newly developed seed varieties from the Indian Council of Agricultural Research (ICAR) on August 11, 2024. These high-yielding, climate-insensitive varieties are biofortified and cover 61 crops. They are a crucial step in improving agricultural resilience in India.

The newly developed climate-resilient varieties of rice are distinguished by lower water demand, enhanced resistance to submergence and salinity, as well as other prevalent diseases. This project addresses the negative impacts of climate change on agricultural production. It also aims to enhance the livelihoods of rural farming communities (Kassie et al. 2024). They are created through the cooperation of several institutions and the strong investment in the research of agriculture and technological development. The introduction of crop wild relatives (CWRs) and older landraces is an essential part of adaptive breeding (Dwivedi et al. 2021). These genetic materials contain traits developed over centuries of evolution. These traits are important for developing resilient crop varieties. CWRs have tolerance to stresses not typically found in modern crops. This tolerance is due to genetic bottlenecks during domestication and selective breeding. An example is the successful introduction of drought and heat resistance genes of wild plants of rice, wheat, and legumes into cultivated types to enhance resilience to stress (Dwivedi et al. 2021). Similarly, landraces maintained by the indigenous people provide regional peculiarities, which increase the crop performance when the climatic conditions change. However, science cannot do it on its own. Economic conditions and institutional structures are important to the successful introduction of these resilient varieties. Access to quality seeds, credit facilities, markets, training, and extension support is the determinant of farmers' capacity and their desire to use improved varieties. Income and information access are key to adopting innovations by households. However, marginalized farmers often face barriers to accessing these benefits (Kassie et al. 2024). The only way to fill this gap is through the inclusive policies that will bring equity in access to technology. Some initiatives that can be used to make sure that more people adopt the technology and share the benefits include seed subsidies, development of rural infrastructure, and participatory breeding, where farmers are directly engaged.

Participatory plant breeding (PPB) is important in rural areas where farmers actively participate in the selection and development of new crop varieties (Ceccarelli and Grando 2020). The participatory approach develops cultivars that are highly responsive to the locality and environmental factors. This makes those crop varieties more applicable, approved, and practical in their locality. The development of location-specific varieties can be quicker when it is based on decentralized breeding programs. They also contribute to the maintenance of agrobiodiversity. It requires policy support and institutional involvement in order that adaptive plant breeding can be the norm. National Innovations in Climate Resilient Agriculture (NICRA) is one of the most outstanding examples of an integrated approach. It manages indoor and outdoor agricultural risks in India, focusing on climate change by scientific research, technological development, and communication (ICAR 2024). NICRA aims at breeding resilience to stress with the objectives of ensuring resilient agriculture and building capacity at various levels to strengthen adaptation to varying regions. On the international scale, initiatives, such as those of the Consultative Group on International Agricultural Research (CGIAR), also play a similar role through promoting genome-assisted breeding and germplasm exchange to develop more resilient food systems (Ashby et al. 2009). In the future, climate forecasts and long-term predictions must be integrated into breeding goals. They should not only be considered for present difficulties. Crop modelling, forecasting, and predictive breeding are tools that can identify regions that could experience future climate stress, facilitate proactive varietal development (Varshney et al. 2013). Additionally, the decision-making process can be significantly simplified and contribute to trait selection. It can be reduced with the help of artificial intelligence and big data analytics integrated into breeding software. Adaptive breeding thus offers the foundation for climate-sensitive agriculture. It is one of the solutions to maintaining rural livelihoods under increasing climate pressure (Ceccarelli and Grando 2020). This is achievable through equitable impact, though, by having an inclusive policy, coordinated institutional intervention, and active participation of farming communities. There is a growing risk in climatic conditions, and adaptive breeding investment is essential in the long term. Such investments will boost productivity, security of the people in the rural regions, and food security across the world. In this paper, a discussion has been made on how breeding strategies in plants could be adapted to achieve climate-resistant varieties in a shorter period of time. These are the strategies that combine crop wild relatives, genomic tools, and participatory strategies to enhance food security in the world.

Climate Vulnerability in Rural Ecosystems

There are many climate risks on rural communities, such as long-term droughts, the severe flooding, excessive heat waves, the invasion of saltwater, unpredictable rainfall, and cyclones (Mandal et al. 2022; Azharuddin and Dogar 2016; Aggarwal et al. 2019). These climate issues pose a great threat to the major natural resources, including soil, water, and biodiversity. Rural people are dependent on these resources. Droughts decrease moisture in the soil and decrease harvests. Nutrient depletion, soil erosion, and extreme heat, which is caused by floods, interfere with crop growth. Intrusion of water into the coastal region raises challenges to the planning of crops, and the problem of saltwater intrusion also compounds arable land. According to the Intergovernmental Panel on Climate Change (IPCC) climate change increases the threat to the security of food. These threats have an impact on food availability, access, use, and stability (Mandal et al. 2021; IPCC 2022). The outcome is a lack of agricultural production, malnutrition, insecure food prices, and low living standards.

Agriculture-dependent rural economies are vulnerable. Smallholder farmers and women are disproportionately affected due to limited access to assets and support systems. According to the estimates of the World Bank (2020), over 100 million people may face the risk of poverty under the impact of climate change by 2030. Tangible examples show that droughts in Maharashtra and Rajasthan have led to water scarcity and decreased agricultural productivity in India. In Bangladesh, rising sea levels have caused soil salination, impacting rice plots and aquaculture. Sub-Saharan Africa, including the Sahel and East Africa, is struggling with unpredictable rain and extended droughts (Dogar et al. 2017a, 2017b, 2022, 2023, 2024, 2025a, 2025b; Dogar and Sato 2018, 2019; Dogar and Almazroui 2022; Dogar 2018, 2019, 2020). These droughts have led to food shortages, livestock losses, and population displacement (Funk et al. 2021). These patterns highlight the necessity of climate-resilient agricultural methods and innovative adaptations in order to protect rural livelihoods.

Principles of Adaptive Breeding

Adaptive plant breeding may be considered as a proactive approach that tends to generate the form of crops that is capable of adapting to the changing and unfavourable environmental factors (Razzaq et al. 2021). It is basically about resisting crops to climate change, stabilizing the output, and helping in the sustainable production of agriculture. This would deal with abiotic and biotic stressors like drought, salinity, extreme temperatures, floods, and pest infestations (Ceccarelli and Grando 2020; FAO 2022). As opposed to the traditional breeding that is a common activity wherein the breeding objective tends to be the most productive breeding under optimal conditions, adaptive breeding takes into account future climatic trends, local stressors, and traits that are of interest to farmers. The given methodology is based on three approaches: Conventional Breeding, which is grounded in the hybridization and breeding based on visible traits but might be inaccurate and ineffective; Participatory Plant Breeding (PPB), farmers are directly involved in the varieties selection; it will contribute to the focus of new crops on local needs and increase adoption rates (Weltzien et al. 2003); Genomic-Assisted Breeding (GAB), which is based on the implementation of modern molecular tools viz. marker-assisted selection, genomic selection, and gnome editing (Varshney et al. 2021).

The basis of adaptive breeding is the choice of certain traits that bring greater cultivation resistance. As an example, the drought tolerance breeding includes: deep root system, high initial growth, and efficiency in the use of water, as seen in rice varieties such as Sahbhagi Dhan (ICAR-IRRI 2014). In order to fight heat stress, breeders are interested in preserving pollen during high temperatures and enhancing the synthesis of heat shock proteins that are particularly important in wheat (CIMMYT 2020a, b). In response to floods in South and Southeast Asia, there has been a tremendous increase in the ability of the *Sub1* gene in rice to produce a higher level of submergence tolerance, such that the plants can withstand a period of up to 2 weeks of flooding (Xu et al. 2006). Salt-tolerant crops that are able to control ion uptake and osmotic balance are essential in salty coastal habitats. Also, early-season varieties enable crops to avoid late droughts and heat, and adapt them to more intensive production systems. The pest and disease resistance also increases crop stability and decreases the use of chemicals. Adaptive breeding, combining resilience to climatic conditions with high-tech genomic approaches, provides vulnerable rural areas with climate-resilient varieties that meet their needs (Jarvis et al. 2011; IPCC 2022).

Participatory and Community-Based Breeding Approaches

Participatory and community-based breeding approach aims at linking formal crop improvement work with the individualized requirements of smallholder farmers, especially in underserved rural areas. Such strategies are proactive in involving the farmers during the breeding process by setting up breeding objectives. They also involve testing and selection of the new varieties within the local settings. Participatory Plant Breeding (PPB) makes crop innovations more relevant, adoptable, and successful in the long term by integrating farmers' preferences, indigenous knowledge, and real situation environmental challenges in breeding decisions (Fig. 9.1) (Ceccarelli and Grando 2007). As an example, PPB was used in rice breeding programs in India and Nepal to create drought-resistant varieties with enhanced grain quality. These varieties were faster embraced by farmers than those breeding programs that entirely relied on traditional breeding (Witcombe et al. 2005). The iterative and decentralized character of PPB ensures the maintenance of agrobiodiversity and improves resilience to fluctuating climate conditions. Principles like Community Seed Banks and farmer-led trials are also complementary activities

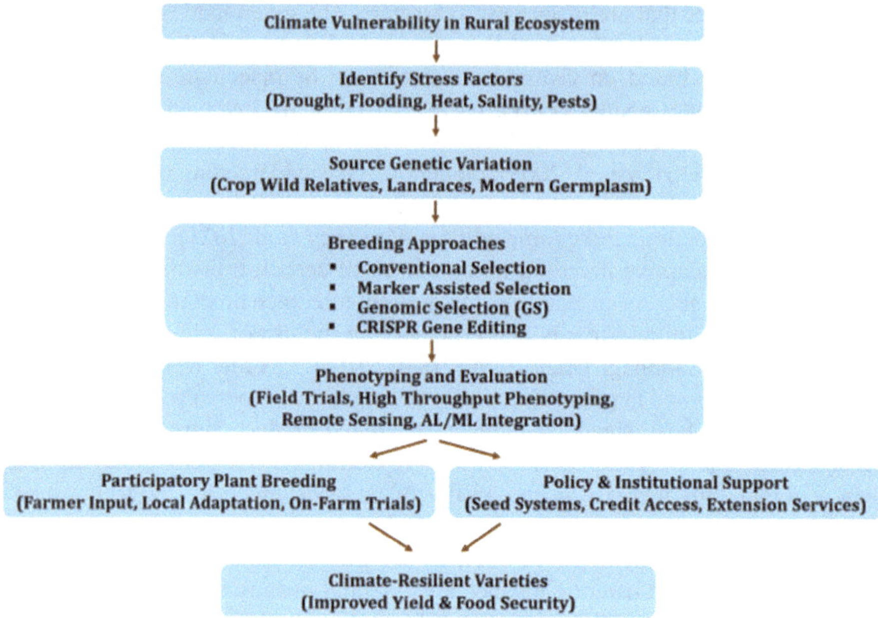

Fig. 9.1 Adaptive breeding strategies for climate smart agriculture

that enhance the conservation of genetic resources and stimulate grassroots innovation.

Community-based strategies are also useful, as testified by effective practices already practiced in Africa and Latin America. Sheep and goats in Ethiopia have been subjected to Community-Based Breeding Programs (CBBPs) where thousands of smallholder households have taken part in the program, leading to genetic gains on traits favoured by the farmers. This has seen household income increase by around 20%. This type of program has become a part of the livestock policy of countries (ICARDA 2023). Similarly, the Demand Led Breeding project in Africa has enabled bean farmers in Uganda and Ethiopia to use the consumer and market-based demand as a part of the official breeding programs. The project has come out with increased adoption of varieties (ACIAR 2023). Farmer Research Committees (CIALs) have enabled participatory breeding in Latin America, whereby communities evaluate, select, and distribute varieties based on their agroecological situations and cultural preferences and strengthen local seed systems and seed sovereignty (Ashby et al. 2000).

Traditional information and local seed management are important for effective breeding outcomes. The farmers usually select the seeds depending on the features of taste, shelf life, pest resistance, and the early maturing traits, which are occasionally neglected by the centralized breeding programs. Integrating such indigenous knowledge will aid in making sure that new varieties have practical and site-specific demands (Sperling et al. 2001). Gender-sensitive breeding is especially important in rural settings where women play a central role. They select seeds, conduct on-farm trials, and ensure household food security. However, the breeding process is usually done in a formal way by disregarding the interests and adversities of women. Gender responsive breeding entails a proactive program where women are not merely taken into consideration during the process of evaluating and selecting varieties, but also breeding varieties that have qualities of significance to them. It includes traits such as short cooking time, processability, or low fuel consumption (Ashby et al. 2009). The adoption of new varieties is improved through the support of the women and marginalized communities that engage in participatory breeding. It also leads to more accommodating, strong, and equitable food systems.

Advances in Breeding Technologies

The increasing climatic uncertainty has been witnessed in rising cases of drought, extreme temperatures, saltiness, floods, and irregular rainfall. Such conditions have been threatening agricultural production and food supply to a great extent. This has been addressed through plant breeding initiatives by the use of more advanced technology. These technologies will increase the accuracy, efficiency, and frequency of producing crops resistant to such stresses. Newer technologies such as marker-assisted selection (MAS), genomic prediction, gene editing technologies, high-throughput phenotyping, and other advanced biotechnological systems are making a substantial and high contribution to the rapid development of adaptive traits into new varieties of crops. The following are the highlights of these methodologies.

Marker-Assisted Selection (MAS)

- MAS employs molecular markers linked to target traits to guide breeding choices with greater accuracy (Kumar et al. 2022).
- It identifies favourable genetic traits at early stages, even at seedlings.
- It is specifically applicable to traits that are regulated by a limited number of genes, like flood and salt tolerance

Examples:

- Rice: Submergence tolerance trait in Swarna and IR64 through incorporation of the *Sub1* gene (Xu et al. 2006).
- Rice: Salinity tolerance using *Saltol* QTL (Babu et al. 2017).
- Wheat: *Lr34* and *Sr2* resistance genes against rust in Avocet S and Thatcher (Vanegas et al. 2008).
- Soybean: MAS for *FAD2-1A* and *FAD2-1B* to increase the oleic acid content (Pham et al. 2010).

- Barley: Heat-stable α-amylase gene transfer for improving malting quality (Xu et al. 2018).

  It allows the pyramiding of traits to bring drought-tolerance, pest resistance and early maturity together.

Genomic Selection (GS) and Gene Editing

- GS analyses the entire genome to estimate the breeding values of individual plants, using statistical models built from combined genotype and phenotype data (Budhlakoti et al. 2022).
- It can be applied for complex traits like drought tolerance or yield under stress.
- Software tools include rrBLUP, solGS, GMStool, BWGS, BGLR, GenSel, GSelection, and lme4GS (Budhlakoti et al. 2022).

Applications:

- Breeding of wheat, rice, and maize breeding under water-scarce and heat-stressed conditions (Heffner et al. 2009).
- Combination of genomic selection with dense genotyping enhanced breeding for drought and heat tolerance in maize. Widely implemented through the Drought Tolerant Maize for Africa (DTMA) project (Cairns et al. 2013).
- Incorporation of salt-tolerance gene *OsHKT1;5* in rice enabled varieties suitable for coastal and saline regions (Ren et al. 2005).
- It has showed prediction accuracies of 0.13–0.78 in rice trials (Cerioli et al. 2022).
- It is efficient in disease resistance and abiotic stress tolerance (Zhang et al. 2022a; Zhang et al. 2022b).
  - Gene editing technologies, particularly CRISPR-Cas9, provide a precise and targeted way to modify specific regions of the plant genome (Errum et al. 2023).

  CRISPR allows for the alteration of native genes without introducing foreign DNA compared to more traditional transgenic techniques, which often makes regulatory approval simpler.

Examples:

- Negative regulators have been turned off, leading to increased drought tolerance in rice (Zhang et al. 2019).
- Editing transcription factors like *TaDREB2* for drought resilience in wheat (Kim et al. 2021).
- *Ppd-1* editing altered spike structure and grain traits in wheat (Errum et al. 2023).
- CRISPR-edited lines that are resistant to wheat blast (Wang et al. 2021).

Phenotyping Platforms

Phenotyping of plant traits has been a major challenge in contemporary breeding of crops, although genotyping is becoming quicker and cheaper. The high throughput phenotyping (HTP) systems combine field equipment and remote sensing equipment to record information about several traits, viz. drought or heat. These phenotyping platforms are discussed herein.

Field-Based Phenotyping Methods

- It includes handheld devices such as chlorophyll meters (SPAD) and infrared thermometers.
- Mounted sensors on tractors (multispectral sensors).
- Ground-based platforms like phenomobiles and gantry systems.
- These tools track real-time characteristics such as; Canopy temperature, Leaf area, Biomass, Plant height, and Stomatal conductance (Tardieu et al. 2017).

Remote Sensing Technologies

- Drones, manned aircraft, and satellites permit non-invasive massive data gathering.
- Imaging methods include: RGB, Multispectral, Hyperspectral and Thermal imaging (Yang et al. 2020).
- Applications:
  - Computation of vegetation indices such as NDVI, PRI, and canopy water content

  Identification of early stress signs (e.g., drought, heat) with drone thermal cameras (Zaman-Allah et al. 2015).

Advanced Phenotyping Systems

- LemnaTec, Field Scanalyzer, and TERRA-REF are the types of platforms that are changing the performance evaluation.
- They increase the knowledge of GxExM interactions (Genetics x Environment x Management).

AI and Machine Learning

Integration of AI and ML algorithms with phenotyping datasets improves trait prediction accuracy and decision-making in breeding programs (Cobb et al. 2013; Araus et al. 2018).

Role of Biotechnology in Fast-Tracking Climate-Resilient Traits

Biotechnology is very crucial towards expediting the process of finding, verifying, and incorporating the trait of resilience to stress into superior varieties of crops. The present-day methods enable researchers to address the problems of heat, drought, salinity, and disease resistance faster and better.

Advanced Biotechnological Techniques

- It includes QTL mapping, RNA sequencing, and gene expression analysis for dissecting complex traits (Kumar et al. 2023b).
- CRISPR/Cas9 genome editing for precise modifications of stress-responsive genes.
- Genome-wide association studies (GWAS) for detecting novel loci linked with abiotic stress tolerance (Kumar et al. 2023a; Kumar et al. 2024).

Discovery of Stress-Responsive Genes

- Transcriptomic and GWAS research identified drought-responsive genes, such as DREB and EREB transcription factor families, in Chickpea and Pearl millet.
- These genes have been successfully introgressed into elite lines for drought resilience (Varshney et al. 2018; Jaganathan et al. 2020).

Stacking Multiple Tolerance Traits

- Breeding of Rice under IRRI's STRASA program.
- Drought, submergence, and salinity resistance genes have been combined into a single cultivar of rice under this program.
- It has been achieved via marker-assisted backcrossing and genetic engineering (Ismail et al. 2013).

Speed Breeding Approaches

- Light, temperature and humidity are controlled in order to speed up crop life cycles.
- Up to six generations per year in crops such as wheat, chickpea, and barley has been taken.
- Institutions such as ICRISAT and CIMMYT using speed breeding to combine multiple tolerance traits (Watson et al. 2018).

Simple traits (e.g., ***Sub1*** in rice submergence tolerance) are best studied using MAS, but GS and CRISPR can be used to study more complex, polygenic traits, such as drought tolerance. Together, they represent a continuum from identifying single major genes to editing or predicting multiple small-effect loci. MAS, GS, and CRISPR can all act at the genetic level, whereas high-throughput phenotyping platforms and remote sensing technologies can give the environmental and performance information required to test genetic predictions in the field. This G × E × M (Genetics × Environment × Management) combination is essential to climate resilience (Tardieu et al. **2017**; Araus et al. **2018**). The use of speed breeding and CRISPR lowers the number of generations between the improvement of specific traits in crops, whereas MAS and GS guarantee the rapid and precise integration of traits of interest. This expedites the entire breeding process, including the process of discovering to cultivar release. These tools have the potential to stack several stress-tolerant traits, drought, salinity, and submergence in individual cultivars, as in the case of rice under the IRRI STRASA program (Ismail et al. **2013**). This shows that rather than isolated contributions, each technology plays a role in building integrated, climate-smart solutions.

National and International Initiatives

The world has achieved significant progress in creating climate-resistant crops of major staples such as rice, wheat, and maize (Table 9.1). The Indian Council of Agricultural Research (ICAR) has been at the forefront of various initiatives in India under the National Innovations in Climate Resilient Agriculture (NICRA) initiative in enhancing crops to be climate resistant (ICAR 2021). In the case of rice, Sahbhagi Dhan was bred to work in the rainfed and drought-infested areas of eastern India, as it was more tolerant to water shortage (Verulkar et al. 2010). Swarna-Sub1 is another major rice and was developed through the aid of marker-assisted backcrossing. It possesses the Sub1A gene, which enables the crop to withstand submergence up to 2 weeks, a desirable quality in flood-prone areas (Xu et al. 2006; Septiningsih et al. 2009). CR Dhan 801 targets salty coastal areas in West Bengal and Odisha, which makes use of the Saltol QTL to enhance salt resistance (Thomson et al. 2010).

Table 9.1 Institutional focus on climate resilient varieties

Crop	Variety	Trait	Developing institute	References
Rice	Sahbhagi Dhan	Drought-tolerant for rainfed uplands	ICAR-NRRI	Verulkar et al. (2010)
	Swarna-Sub1	Submergence-tolerant (Sub1A gene)	ICAR-NRRI	Septiningsih et al. (2009)
	CR Dhan 801	Salt-tolerant for coastal areas	ICAR-NRRI	Thomson et al. (2010)
	IR64-Sub1	Submergence tolerance	IRRI	Ismail et al. (2013)
	IRRI 154, IRRI 156	Drought and heat-tolerant	IRRI	IRRI (2023)
Wheat	HD 2967, HD 3086	High-yielding, rust and heat resistance	ICAR-IARI	Joshi et al. (2014)
	HI 8759 (Pusa Anmol)	Drought-tolerant, Central India	ICAR-IARI	ICAR (2021)
	HTWLs (heat-tolerant wheat lines)	Heat and drought tolerance, rust resistance	CIMMYT	Bukhari et al. (2021)
Maize	Pusa Vivek hybrid 27	Early maturity, drought tolerance	ICAR-IARI	ICAR (2021)
	HQPM-1, HQPM-5	Quality protein maize (QPM), moderate drought tolerance	ICAR-IARI	ICAR (2021)
	ZM 309, ZM 401	Drought tolerance (DTMA project)	CIMMYT	CIMMYT (2020a, b)
Sorghum	CSV 23, CSV 27	Drought-tolerant, dryland adaptation	ICAR-IIMR	ICAR-IIMR (2018)
	ICSV 112, ICSV 745	Early maturity, heat and drought resilience	ICRISAT	ICRISAT (2022)
Pearl millet	HHB 67 improved	Drought and heat tolerance, early maturing	HAU/ICRISAT	Yadav et al. (2012)
	AHB 1200	Dual-purpose, heat and disease resistance	ICAR-AICPMIP/MPKV	AICPMIP (2021)
Chickpea	JG 14, JG 16	Drought-tolerant, fusarium resistance	ICAR-IIPR/JNKVV	IIPR (2020)
	Super chickpea	Drought and heat tolerance	ICRISAT	Varshney et al. (2013)
Soybean	NRC 127, NRC 128	Drought-tolerant, pest resistance	ICAR-IISR	IISR (2021)
	Plenish®, Vistive gold®	High oleic acid, oxidative stability	DuPont Pioneer/Monsanto	Kinney (2009), Wilson and Hildebr (2016)
Groundnut	ICGS 76, TG 37A	Drought-tolerant, suitable for rainfed	ICAR-DGR	Muthaiah et al. (2020)
	ICGV 91114, ICGV 00350	Water-use efficiency, aflatoxin control	ICRISAT	ICRISAT (2022)

(continued)

Table 9.1 (continued)

Crop	Variety	Trait	Developing institute	References
Tomato	Arka Rakshak	Heat and bacterial wilt resistance	ICAR-IIHR	IIHR (2020)
	AVTO1429	Heat and drought tolerance, pest resistance	AVRDC (world vegetable center)	AVRDC (2021)
Brinjal	Bt Brinjal (EE-1 lines)	Fruit and shoot borer resistance	MAHYCO/ BARI	Shelton et al. (2018)
	Arka Anand	Heat-tolerant, high yield	ICAR-IIHR	IIHR (2020)

The International Rice Research Institute (IRRI) operates a huge gene bank with more than 130,000 rice samples worldwide. IRRI includes projects such as Stress-Tolerant Rice in Africa and South Asia (STRASA) that have resulted in the creation of varieties like IR64-Sub1 (flood-tolerant) and IRRI 154 and 156, which have been bred to be resistant to heat and drought (Ismail et al. 2013; IRRI 2023). In wheat, ICAR has published varieties that are popular, such as HD 2967 and HD 3086, which are heat-tolerant and resistant to rust, and HI 8759 (Pusa Anmol), which is specifically designed to tolerate drought in central India (Joshi et al. 2014). On the international level, CIMMYT has the program Climate Resilient Crops, which has been involved in the improvement of wheat by coming up with heat-tolerant lines of wheat in South Africa and Sub-Saharan Africa (Bukhari et al. 2021). In the case of maize, drought-tolerant hybrids and early-maturing hybrids like Pusa Vivek Hybrid 27 and Quality Protein Maize varieties HQPM-1 and HQPM-5 that exhibit moderate drought resistance have been launched by the ICAR (2021). The Drought Tolerant Maize for Africa (DTMA) program by CIMMYT has led to the development of high-tech hybrids such as ZM 309 and ZM 401, utilizing cutting-edge techniques including genomic selection and doubled haploids (CIMMYT 2020a, b).

In addition to staple cereals, substantial progress in the form of the enhancement of climate resilience in the legumes, millets, oilseeds, and horticultural crops has been achieved. However, in the example of chickpea (Cicer arietinum), ICAR, in collaboration with JNKVV and ICAR-IIPR, has developed drought-tolerant hybrid varieties like JG 14 and JG 16 that are suitable in semi-arid rainfed areas (IIPR 2020). ICRISAT has come up with Super Chickpea that has drought and heat resistance by applying genome-wide association studies (GWAS) and molecular markers (Varshney et al. 2013). ICRISAT gene bank also stores over 20,000 accessions of chickpea in Hyderabad. The CSV 23 and CSV 27 varieties of ICAR have proved useful in the dryland farms, and pearl millet PM 1121 has an early maturity and heat tolerance (ICAR-IIMR 2018).

Meanwhile, ICRISAT is also breeding sorghum and pearl millet in India and Sub-Saharan Africa, encouraging free open-source systems of seed that will support the small-scale farmers (ICRISAT 2022). In the case of groundnut (Arachis hypogaea), ICAR-DGR has recommended ICGS 76 and TG 37A, which are resistant to rainfed areas. ICRISAT has also released lines of drought-tolerant, early-maturing types by marker-assisted selection (Muthaiah et al. 2020). The varieties are also useful in combating the foliar diseases as well as reducing aflatoxin contamination, which is a major post-harvest problem. Under the vegetable and horticulture segments, the ICAR NICRA has developed a plant breeding program that has resulted in superior breeding of tomato, brinjal, and onion, based on climatic resilience. In tomato, the Arka Rakshak variety is highly immune to bacterial wilt and heat (IIHR 2020). The World Vegetable Center (AVRDC) also makes a contribution by creating resilient crops like tomato, cowpea, and amaranth, and incorporating lesser-known native species to support climate change adaptation and global nutrition security (AVRDC 2021).

Challenges and Limitations in Advancing Climate-Resilient Agriculture

Although there have been significant achievements in the production of climate-resistant crop species, the wide adoption of such agriculture remains constrained by a number of systemic and institutional challenges. Access to basic technologies, quality seeds, and finances to enable the adoption of the climate-smart farming practice are some of the key challenges faced by smallholder farmers, especially in developing countries. These challenges are compounded by the high cost of improved seeds, lack of awareness, and the poor agricultural extension services.

- **Access and affordability**: Aggarwal et al. (2019) noted that the awareness of stress-resilient crop varieties was low among farmers in the rainfed areas of India due to poor dissemination mechanisms. The World Bank (2015) has recorded in sub-Saharan Africa that less than one in ten smallholder farmers access formal credit, which seriously limits their practice of investing in resilient agricultural technologies. Indicatively, drought-tolerant maize seeds cost in Kenya is almost 30–40% more than their normal counterparts, which restricts their uptake by low-income farmers.
- **Seed system inefficiencies**: Fragmented seed system, insufficient seed production and distribution, poor quality control systems, and insufficient involvement of the private sector all continue to be a big bottleneck. IFPRI (2018) noted that unreliable seed supplies and lack of trained agro-dealers have slowed the diffusion of submergence-tolerant rice in Bangladesh. Adoption remains under 25% of the potential target area. Equally, in Ethiopia, a pilot project on heat-tolerant wheat was a success, but the adoption of the technology was <15% among the target farmers due to a lack of an efficient last-mile delivery system.
- **Policy and regulatory hurdles**: The release of genetically improved cultivars, especially those that have been produced using modern biotechnologies, may be slowed down by long and demanding varietal approval and biosafety processes. Tripp and Louwaars (2007) pointed out that such delays have an unequal effect

on the breeding programs of the public sector. In Nigeria, GM cowpea regulatory approvals took virtually 8 years to halt farmer access following field benefits.

- **Policy misalignment**: Agricultural policies are usually focused on high-input, high-yield crops, which leave out stress-tolerant but lower-yielding varieties. As an example, Indian government rice and wheat procurement policies do not give farmers much incentive to grow millets and other resilient crops, which are proven to be drought and heat-resistant.
- **Biodiversity and genetic erosion risks**: The massive use of homogenous, high-performing cultivars threatens the replacement of traditional landraces, which contain adaptively distinctive characteristics. Jarvis et al. (2016) warned that a decline in genetic diversity would threaten the future breeding developments. There was an introduction of new high-yielding rice varieties in Nepal, and within 20 years the traditional landraces decreased by 75% due to the introduction of modern technology, which decreased the genetic base of future stress-adaptive genes. All in all, the innovation in breeding technology is rapidly increasing, but the prospects are spoiled by economic, institutional, and policy-related obstacles. These constraints will continue to be unevenly distributed and will not be adequate to protect the vulnerable smallholder farming systems unless these constraints are addressed.

Future Perspectives

To speed up the adoption and spread of climate-resilient farming, a proactive policy approach must tackle fundamental weaknesses in seed systems, research funding, farmer involvement, and institutional cooperation. Enhancing seed systems and extension services is particularly vital in rainfed and marginal regions where farmers often lack access to high-quality seeds. Governments should prioritize investments in decentralized seed production, quality control, and digital platforms for climate-smart advisory services. McGuire and Sperling (2016) highlight that formal seed systems currently reach fewer than 20% of farmers in Sub-Saharan Africa, pointing to the importance of integrated, farmer-led seed networks. At the same time, public-private partnerships (PPPs) can help close the gap between innovation and delivery, as demonstrated by CGIAR's STRASA program, which has benefited more than 18 million farmers with stress-tolerant rice varieties (IRRI 2019a, b). Coordinated efforts between public research institutions and private sector entities, supported by favourable policies like tax breaks and streamlined licensing, can substantially enhance the scalability of climate-resilient agricultural technologies.

An actionable roadmap may be envisioned into short, medium, and long-term:

Short-term (1–3 years): Expand farmer-led seed networks and strengthen local seed banks to ensure reliable access to climate-resilient varieties. Research and invest in digital advisory systems that would provide farmers with access to the climate-smart information on their mobile phones (Tripp and Louwaars 1997). Initiate special subsidy and voucher schemes to assist the disadvantaged, such as women and small holders, to obtain better seeds and training. Climate analogs (farmer experiments) can be used to speed up adaptation strategies by conducting preliminary pilot tests with varieties subjected to conditions that mimic future climatic conditions (Ramírez-Villegas et al. 2011).

Medium-term (4–7 years): institutionalize participatory breeding and decentralized on-farm experimentations in National Agricultural Research Systems (NARS) to ensure that innovations are made in line with the preferences of the farmers (Ceccarelli and Grando 2007). Expand PPPs to cover private seed companies, cooperatives, and NGOs to cover bigger distribution networks. Establish agronomy, genomics, and precision breeding research centers in the region to develop research capacity (Fuglie et al. 2020). Enhance synthetic biology foundations of improving crops by engineering bioengineered traits of resilience to stress and nutrient efficiency, with ethical controls and socialization (Parisi et al. 2016).

Long-term (8–15 years): Mainstream climate-resilient agriculture into national food security and rural development policies, ensuring stable funding streams for agricultural R&D. Institutionalize inclusive innovation platforms where farmers, researchers, and private actors co-create technologies and practices (Scoones and Thompson 2009). Integrate gender responsive and equity-based approaches in breeding and extension services to have resilient food systems. Increase transnational cooperation and harmonization of seed policies in regional economic blocs to increase varietal flow and resilience at scale (Spielman and Smale 2017). Invest strategically in frontier areas such as synthetic biology for climate-ready crops and large-scale climate analog trial networks. This will predict the future adaptation requirement and guide long-term varietal pipelines (Halewood et al. 2020).

Research and development (R&D), as well as capacity building, are also important to invest in. Achievement of the future advances in genomic technologies, precision breeding, and climate-smart agriculture relies on the continuity of funding and professional training of the national agricultural research systems (NARS). Fuglie et al. (2020) state that growing the agricultural R&D budgets of developing countries (by 10% a year) would be a valuable step to increase productivity and resilience. In addition, increased innovation by the farmers needs to be encouraged to make sure that new technologies are applicable and extensively adopted. The farmers should be empowered as active collaborators in solution development by ensuring that they are engaged in participatory breeding trials where the trials are done locally and the traditional knowledge is valued. Scoones and Thompson (2009) pointed out the significance of grass-root innovation in the development of equitable and sustainable agricultural systems. Women, smallholders, and marginalized groups can be given specific support, including subsidies, training, and access to innovation platforms to democratize the process of agricultural progress and establish resiliency at the community level.

Conclusion

Climate-resilient agriculture lies at the intersection of ensuring food security, promoting environmental sustainability, and strengthening rural livelihoods amid increasing climate fluctuations. In this chapter, much effort has been highlighted on the development of stress-tolerant varieties of the major crops like rice, wheat, maize, chickpea, sorghum, and groundnut, through collaboration between national institutions (ICAR-IARI, ICAR-NRRI, ICAR-IIPR) and international research institutes (IRRI, CIMMYT, ICRISAT). Nonetheless, these innovations cannot be widely adopted due to issues of poor access to technologies, insufficiency in finances, poor seed systems, regulation, and the risk of genetic diversity loss. Increasing the presence of extension services, encouraging inclusive innovation development, and making the organization of institutional structures correspond to the requirements of farming communities are the essential steps forward. The adaptive breeding, particularly in the case of farmer engagement and advanced genomic technologies, becomes instrumental in creating crop varieties that would be resistant to drought, heat, flooding, and salinity-related stressors (Reynolds et al. 2020). These stressors adversely impact agricultural systems. In the future, studies need to be integrated and systemic in nature to include climate prediction, digital technology, and indigenous knowledge. The solutions to be implemented to prevent any failure in the future will involve scaling of the public-private partnerships, alignment of the regional policies on seeds, and community-based adaptation. Due to the increasing climate challenges worldwide, the future of food security will require effective incorporation of science, policy, and practical innovation in building inclusive resilience in agriculture.

Potential Conflicts of Interest On behalf of all authors, the corresponding author states that there is no conflict of interest.

References

ACIAR (2023) Demand-led plant variety design for emerging markets in Africa. Australian Centre for International Agricultural Research

Aggarwal PK, Bandyopadhyay SK, Pathak H, Kumar A, Vyas S (2019) Climate change impact & adaptation in agriculture: evidence from India. ICAR-NICRA, New Delhi

AICPMIP (2021) Annual progress report. All India Coordinated Research Project on Pearl Millet, ICAR, Jodhpur

Araus JL, Kefauver SC, Zaman-Allah M (2018) Precision phenotyping for drought tolerance in crops. In: Hossain MA et al (eds) Drought stress tolerance in plants, vol 1. Springer, Cham, pp 159–180

Ashby JA, Braun AR, García T, Guerrero MP, Hernández LA, Quirós CA, Roa JI (2000) Investing in farmers as researchers: experience with local agricultural research committees in Latin America. CIAT, Cali

Ashby JA, Polar V, Braun A (2009) Gender equity in agricultural innovation: when does it matter & what works? CGIAR Systemwide Program on Participatory Research & Gender Analysis, Working Document No. 8

AVRDC – The World Vegetable Center (2021) Heat-tolerant & pest-resistant vegetable varieties for Asia. AVRDC, Shanhua

Azharuddin S, & Dogar MM (2016). Decadal climate variability and predictability. CURRENT SCIENCE, 110(8), 1397. http://www.currentscience.ac.in/Volumes/110/08/1397.pdf

Babu NN, Krishnan SG, Vinod KK, Krishnamurthy SL, Singh VK, Singh MP, Singh R, Ellur RK, Rai V, Bollinedi H, Bhowmick PK (2017) Marker aided incorporation of Saltol, a major QTL associated with seedling stage salt tolerance, into Oryza sativa 'Pusa basmati 1121'. Front Plant Sci 8:41

Budhlakoti N, Kushwaha AK, Rai A, Chaturvedi KK, Kumar A, Pradhan AK, Kumar U, Kumar RR, Juliana P, Mishra DC, Kumar S (2022) Genomic selection: a tool for accelerating the efficiency of molecular breeding for development of climate-resilient crops. Front Genet 13:832153

Bukhari A, Dreisigacker S, Sansaloni C, Wenzl P, Amri A, Ogbonnaya FC (2021) Harnessing wheat genetic resources for climate resilience: recent developments at CIMMYT. Crop Sci 61(1):79–89

Cairns JE, Hellin J, Sonder K, Araus JL, MacRobert JF, Thierfelder C, Prasanna BM (2013) Maize production in a changing climate: impacts, adaptation, & mitigation strategies. Adv Agron 114:1–58

Ceccarelli S, Grando S (2007) Decentralized-participatory plant breeding: an example of demand-driven research. Euphytica 155:349–360

Ceccarelli S, Grando S (2020) Participatory plant breeding: who did it, who does it & where? Exp Agric 56(1):1–11

Cerioli T, Hernez CO, Angira B, McCouch SR, Robbins KR, Famoso AN (2022) Development & validation of an optimized marker set for genomic selection in southern US rice breeding programs. Plant Genome 3:e20219

CIMMYT. (2020a). DTMA final report

CIMMYT (2020b) Wheat & climate change: adapting to heat & drought. International Maize & Wheat Improvement Center, Mexico City

Cobb JN, Declerck G, Greenberg A, Clark R, McCouch S (2013) Next-generation phenotyping: requirements & strategies for enhancing our understanding of genotype–phenotype relationships & its relevance to crop improvement. Theor Appl Genet 126(4):867–887

Dogar MM, Kucharski F, Azharuddin S, (2017a). Study of the global and regional climatic impacts of ENSO magnitude using SPEEDY AGCM. J Earth Syst Sci 126, 30. https://doi.org/10.1007/s12040-017-0804-4

Dogar MM, Stenchikov G., Osipov S., Wyman B, & Zhao M (2017b). Sensitivity of the regional climate in the Middle East and North Africa to volcanic perturbations. Journal of Geophysical Research: Atmospheres, 122(15), 7922–7948. https://doi.org/10.1002/2017JD026783

Dogar MM, Kucharski F, Sato T, Mehmood S, Ali S, Gong Z, Arraut J (2019). Towards understanding the global and regional climatic impacts of Modoki magnitude. Global and Planetary Change, 172, 223–241.

Dogar MM, Hermanson L, Scaife AA et al. (2023). A Review of El Niño Southern Oscillation Linkage to Strong Volcanic Eruptions and Post-Volcanic Winter Warming. Earth Syst Environ 7, 15–42. https://doi.org/10.1007/s41748-022-00331-z

Dogar MM, Fujiwara M, Zhao M, Ohba M, & Kosaka Y (2024). ENSO and NAO linkage to strong volcanism and associated post-volcanic high-latitude winter warming. Geophysical Research Letters, 51(1), e2023GL106114.

Dogar MM, Fujiwara M, Ohba M, & Almazroui M (2025a). Investigating post-eruption amplified winter cooling in the Middle East and North Africa—unraveling ENSO and NAO dynamics. Environmental Research Communications, 7(4), 045012. https://doi.org/10.1088/2515-7620/adc675

Dogar MM, Watanabe S, Fujiwara M et al. (2025b) Tropical volcanic impacts on MENA climate via ENSO and NAO dynamics in a high-top model. npj Clim Atmos Sci 8, 330. https://doi.org/10.1038/s41612-025-01201-x
Dogar MM, & Sato T (2018). Analysis of climate trends and leading modes of climate variability for MENA region. Journal of Geophysical Research: Atmospheres, 123(23), 13–074. https://doi.org/10.1029/2018JD029003
Dogar MM, & Sato T (2019). Regional climate response of Middle Eastern, African, and South Asian monsoon regions to explosive volcanism and ENSO forcing. Journal of Geophysical Research: Atmospheres, 124(14), 7580–7598. https://doi.org/10.1029/2019JD030358
Dogar MMA, & Almazroui M (2022). Revisiting the strong and weak ENSO teleconnection impacts using a high-resolution atmospheric model. Atmospheric Environment, 270, 118866. https://doi.org/10.1016/j.atmosenv.2021.118866
Dogar MM (2018). Impact of tropical volcanic eruptions on Hadley circulation using a high-resolution AGCM. Current Science, 114(6), 1284–1294. https://www.jstor.org/stable/26797333
Dogar MMA (2019). The Sensitivity of DTR to the increased CO2 over Mid-latitude Semi-dry Regions. JOJ Hortic Arboric. https://juniperpublishers.com/jojha/pdf/JOJHA.MS.ID.555584.pdf
Dogar MMA (2020). Study of the regional climatic impacts of tropical explosive volcanism in the Middle East and North Africa region (Doctoral dissertation, Hokkaido University). https://doi.org/10.14943/doctoral.r7091
Dogar MMA, Fujiwara M, & Shahid A (2022). Disentangling the volcanic signals from natural climate variability. In AGU fall meeting abstracts (Vol. 2022, pp. A15H-1327).
Dwivedi SL, Britt AB, Tripathi L, Sharma S, Upadhyaya HD, Ortiz R (2021) Harnessing crop wild relatives for climate resilience. Genes 12(5):783
Errum A, Rehman N, Uzair M, Inam S, Ali GM, Khan MR (2023) CRISPR/Cas9 editing of wheat Ppd-1 gene homoeologs alters spike architecture & grain morphometric traits. Funct Integr Genomics 23(1):66
Food and Agriculture Organization (FAO). (2022). The state of Food Security & Nutrition in the world 2022. Rome: FAO
Fuglie K, Gautam M, Goyal A, Maloney WF (2020) Harvesting prosperity: technology & productivity growth in agriculture. World Bank Publications, Washington, DC
Funk C, Husak G, Michaelsen J, Shukla S (2021) Climate hazards & food security in the Sahel: climate variability & food production. Brookings Institution
Halewood M, Lopez Noriega I, Ellis D, Roa C, Rouard M, Sackville Hamilton R (2020) Using genebanks to harness genetic diversity. Ann Rev Plant Biol 71:629–660
Heffner EL, Sorrells ME, Jannink JL (2009) Genomic selection for crop improvement. Crop Sci 49(1):1–12
ICAR (2021) NICRA annual report. Indian Council of Agricultural Research, New Delhi
ICAR (2024) Annual report 2023–24. Indian Council of Agricultural Research, New Delhi
ICARDA (2023) Community-based breeding programs take root. International Center for Agricultural Research in the Dry Areas
ICAR-IIMR (2018) Annual progress report. ICAR-Indian Institute of Millets Research, Hyderabad
ICAR-IISR (Indian Institute of Soybean Research) (2021) Annual report 2020–21. ICAR-IISR, Indore
ICAR-IRRI (2014) Development & dissemination of climate-resilient rice varieties. Indian Council of Agricultural Research—International Rice Research Institute, New Delhi
ICRISAT (2022) Crop improvement strategy report. International Crops Research Institute for the Semi-Arid Tropics, Patancheru
IFPRI (2018) Improving seed systems for smallholder farmers. International Food Policy Research Institute, Washington, DC
IIHR (2020) Annual report. ICAR-Indian Institute of Horticultural Research, Bengaluru
IIPR (2020) Annual report 2019–2020. ICAR-Indian Institute of Pulses Research, Kanpur
Intergovernmental Panel on Climate Change (IPCC) (2022) Climate change 2022: impacts, adaptation, & vulnerability

IRRI (2019a) Annual report: stress-tolerant rice for Africa & South Asia (STRASA). International Rice Research Institute

IRRI (2019b) STRASA project highlights: reaching farmers with climate-smart rice varieties. Los Baños: International Rice Research Institute

IRRI (2023). International rice genebank & STRASA

Ismail AM, Singh US, Singh S, Dar MH, Mackill DJ (2013) Physiological characterization of stress-tolerant rice varieties developed through the STRASA project. Field Crop Res 134:33–44

Jaganathan D, Ramasamy K, Sellamuthu G, Jayabalan S, Venkataraman G (2020) Genomics-assisted breeding for drought tolerance in crops: status & prospects. Front Genet 11:579

Jarvis A, Lane A, Hijmans RJ (2011) The role of crop genetic diversity in adapting to climate change. In: PGRFA: thematic background studies. FAO, Rome

Jarvis DI, Hodgkin T, Sthapit BR, Fadda C, Lopez-Noriega I (2016) A heuristic framework for identifying multiple ways of supporting the conservation & use of traditional crop varieties in agricultural landscapes. Int J Agric Sustain 14(3):283–300

Joshi AK, Mishra B, Chatrath R, Ferrara GO, Singh RP (2014) Heat stress-tolerant wheat for South Asia: technologies & partnerships. J Integr Agric 13(2):267–275

Kassie M, Alemayehu M, Motbainor A, Hirvonen K (2024) Determinants of adoption of climate-resilient crop varieties among smallholder farmers. Cogent Food Agric 10(1):2430404

Kim D, Alptekin B, Budak H (2021) CRISPR/Cas9 genome editing in wheat. Funct Integr Genomics 21(1):31–41

Kinney AJ (2009) Genetic modification of soybean oil for improved oxidative stability & nutrition. In: Halford NG (ed) Genetic modification of plants. Springer, Berlin, pp 131–150

Kumar A, Sandhu N, Singh VP, Singh RK (2021) Advances in genomic tools for climate-resilient crop breeding. Front Plant Sci 12:674136

Kumar R, Nagarajan M, Bollinedi H, Bhowmick PK, Singh AK, Ellur RK, Sathee L, Mehrotra S, Kumar D, Krishnan SG, Vinod KK (2022) Meta-QTLs linked tonitrogen use efficiency are randomly distributed in Indian rice germplasm. Indian J Genet Plant Breed 82(01):7–15

Kumar R, Saini M, Taku M, Debbarma P, Mahto RK, Ramlal A, Sharma D, Rajendran A, Pandey R, Gaikwad K, Lal SK (2023a) Identification of quantitative trait loci (QTLs) & candidate genes for seed shape & 100-seed weight in soybean [Glycine max (L.) Merr.]. Front Plant Sci 13:1074245

Kumar R, Saini M, Taku M, Debbarma P, Sharma D, Pandey R, Gaikwad K, Lal SK, Talukdar A (2023b) Genetic studies & mapping of quantitative trait loci (QTL) for yield and yield-related traits in vegetable and seed soybean (Glycine max L.)-derived populations. Indian J Genet Plant Breed (The) 83(03):301–309

Kumar R, Das SP, Choudhury BU, Kumar A, Prakash NR, Verma R, Chakraborti M, Devi AG, Bhattacharjee B, Das R, Das B (2024) Advances in genomic tools for plant breeding: harnessing DNA molecular markers, genomic selection, & genome editing. Biol Res 57(1):80

McGuire S, Sperling L (2016) Seed systems smallholder farmers use. Food Secur 8(1):179–195

Mandal S, Sadhu S, Rajbanshi J, Patra P, Halder A, Khan A, Satpati LN (2022) Surface Temperature Mapping to Assess Local Climatic Zones: A Study using WUDAPT in Kolkata-The Colonial City in Transition, edited by Sumana Bandyopadhyay. Chapter 11, pp 177–205. https://doi.org/10.4324/9781003205968. ISBN:978-1-032-20596-8 (ebk)

Mandal S, Patra P, Halder A, Satpati L N (2021) Extreme Climatic Events: A review of trends, vulnerabilities, adaptations in the South Asian Region. In South Asia and Climate change, edited by Mousumi Kar, Jayita Mukhopadhayay and Manisha Deb Sarkar. Chapter 2, pp 15–47. https://doi.org/10.4324/9781003045731

Muthaiah AR, Bhatnagar-Mathur P, Thirunavukkarasu N, Mallikarjuna MG, Arun B, Hake A, Varshney RK (2020) Genomic breeding for drought tolerance in groundnut: challenges and opportunities. Agronomy 10(4):539

Parisi C, Tillie P, Rodríguez-Cerezo E (2016) The global pipeline of GM crops out to 2020. Nat Biotechnol 34(1):31–36

Pham AT, Lee JD, Shannon JG, Bilyeu KD (2010) Mutant alleles of FAD2-1A and FAD2-1B combine to produce soybeans with the high oleic acid seed oil trait. BMC Plant Biol 10:1–13

Ramírez-Villegas J, Lau C, Köhler AK, Signer J, Jarvis A, Arnell N et al (2011) Climate analogues: finding tomorrow's agriculture today. In: Working paper, CGIAR Research Program on Climate Change, Agriculture & Food Security (CCAFS)

Razzaq A, Kaur P, Akhter N, Wani SH, Saleem F (2021) Next-generation breeding strategies for climate-ready crops. Front Plant Sci 12:620420

Ren ZH, Gao JP, Li LG, Cai XL, Huang W, Chao DY, Zhu MZ, Wang ZY, Luan S, Lin HX (2005) A rice quantitative trait locus for salt tolerance encodes a sodium transporter. Nat Genet 37(10):1141–1146

Reynolds M, Braun HJ, Atlin G, Xia X, Sonder K, Payne T (2020) Improving global integration of crop research: insights from the CGIAR research program on wheat. Adv Agron 160:1–59

Scoones I, Thompson J (2009) Farmer first revisited: innovation for agricultural research & development. Practical Action Publishing, Rugby

Septiningsih EM, Pamplona AM, Sanchez DL, Neeraja CN, Vergara GV, Heuer S, Ismail AM, Mackill DJ (2009) Development of submergence-tolerant rice cultivars: the Sub1 locus & beyond. Ann Bot 103(2):151–160

Shelton AM, Hossain MJ, Paranjape V, Azad AK, Rahman ML, Hossain MA, Majumder R (2018) Bt Brinjal in Bangladesh: a genetically modified eggplant with benefits for farmers. Front Bioeng Biotechnol 6:1–7

Sperling L, Ashby JA, Smith ME, Weltzien E, McGuire S (2001) A framework for analysing participatory plant breeding approaches & results. Euphytica 122(3):439–450

Spielman DJ, Smale M (2017) Policy options to accelerate variety change among smallholder farmers in South Asia & Africa South of the Sahara. IFPRI discussion paper 01666

Tardieu F, Cabrera-Bosquet L, Pridmore T, Bennett M (2017) Plant phenomics, from sensors to knowledge. Curr Biol 27(15):R770–R783

Thomson MJ, de Ocampo M, Egdane J, Rahman MA, Sajise AG, Adorada DL, Tumimbang-Raiz E, Blumwald E, Ismail AM (2010) Marker-assisted breeding for salt tolerance in rice. Int J Plant Genomics 2010:953195. 1–10

Tripp R, Louwaars N (1997) Seed regulation: choices on the road to reform. Food Policy 22(5):433–446

Tripp R, Louwaars N (2007) Seed regulation: choices on the road to reform. Food Policy 2(3):356–373

Vanegas CDG, Garvin DF, Kolmer JA (2008) Genetics of stem rust resistance in the spring wheat cultivar Thatcher and the enhancement of stem rust resistance by Lr34: stem rust resistance in TcLr34. Euphytica 159:391–401

Varshney RK, Thudi M, Nayak SN, Gaur PM, Kashiwagi J, Krishnamurthy L, Jaganathan D, Koppolu R, Bohra A, Tripathi S, Rathore A (2013) Genomics-assisted breeding for drought tolerance in chickpea. Funct Integr Genomics 13(4):535–547

Varshney RK, Bohra A, Yu J, Graner A, Zhang Q, Sorrells ME (2018) Integrated genomics and breeding approaches for crop improvement. Trends Plant Sci 23(7):611–623

Varshney RK, Bohra A, Yu J, Graner A, Zhang Q, Sorrells ME (2021) Designing future crops: genomics-assisted breeding comes of age. Trends Plant Sci 26(6):631–649

Verulkar SB, Mandal NP, Dwivedi JL, Singh BN, Sinha PK, Mahato RN, Pandey P, Yadav R, Singh ON, Dongre P, Atlin GN (2010) Breeding resilient and productive rice varieties for drought-prone rainfed uplands of India. Field Crop Res 117(2–3):197–206

Wang S, Zeng L, Chen L, Chen Y, Xie X, Liu L, Wang Y, Lin Y, Chen Z (2021) CRISPR/Cas9-mediated genome editing of wheat genes for resistance to wheat blast. J Integr Plant Biol 63(7):1286–1300

Watson A, Ghosh S, Williams MJ, Cuddy WS, Simmonds J, Rey MD et al (2018) Speed breeding is a powerful tool to accelerate crop research and breeding. Nat Plants 4(1):23–29

Weltzien E, Smith ME, Meitzner LS, Sperling L (2003) Technical and institutional issues in participatory plant breeding—from the perspective of formal plant breeding. PPB Monograph. CGIAR

Wilson RF, Hildebr DF (2016) High-oleic soybeans: genetic and agronomic improvements. In: Boerma HR, Specht JE (eds) Soybeans: improvement, production, & uses, 3rd edn. ASA, CSSA, & SSSA, Madison

Witcombe JR, Joshi A, Joshi KD (2005) Participatory plant breeding is better described as highly client-oriented plant breeding. Exp Agric 41(3):299–319

World Bank (2015) Agricultural finance & risk management: a review of country experiences. World Bank, Washington, DC

World Bank (2020) Shock waves: managing the impacts of climate change on poverty. World Bank, Washington, DC

Xu K, Xu X, Fukao T, Canlas P, Maghirang-Rodriguez R, Heuer S, Ismail AM, Bailey-Serres J, Mackill DJ (2006) Sub1A is an ethylene-response-factor-like gene that confers submergence tolerance to rice. Nature 442(7103):705–708

Xu Y, Zhang XQ, Harasymow S, Westcott S, Zhang W, Li C (2018) Molecular marker-assisted backcrossing breeding: an example to transfer a thermostable β-amylase gene from wild barley. Mol Breed 38:1–9

Yadav OP, Rai KN, Rajpurohit BS, Mahala RS, Gupta SK (2012) Improved cultivars and production technologies of pearl millet in India. All India Coordinated Pearl Millet Improvement Project (AICPMIP), Jodhpur

Yang W, Duan L, Chen G, Xiong L, Liu Q (2020) Crop phenomics and high-throughput phenotyping: past decades, current challenges, and future perspectives. Mol Plant 13(2):187–214

Zaman-Allah M, Vergara O, Araus JL, Tarekegne A, Magorokosho C, Zarco-Tejada P, Cairns JE (2015) Remote sensing and phenotyping for maize improvement in sub-Saharan Africa. Field Crop Res 170:166–178

Zhang H, Zhang J, Wei P, Zhang B, Gou F, Feng Z, Mao Y, Yang L, Zhang H, Xu N, Zhu JK (2019) CRISPR/Cas9-mediated gene editing in rice for improved drought tolerance. Plant Biotechnol J 17(1):47–59

Zhang J, Gill HS, Brar NK, Halder J, Ali S, Liu X, Bernardo A, Amand PS, Bai G, Gill US, Turnipseed B (2022a) Genomic prediction of fusarium head blight resistance in early stages using advanced breeding lines in hard winter wheat. Crop J 10(6):1695–1704

Zhang A, Chen S, Cui Z, Liu Y, Guan Y, Yang S, Qu J, Nie J, Dang D, Li C, Dong X (2022b) Genomic prediction of drought tolerance during seedling stage in maize using low-cost molecular markers. Euphytica 218(11):154

Part IV
Community-Based Climate Change Management and Policy Decision

Chapter 10
Managing Rural Sustainable Livelihood and Climate-Resilient Adaptation by Garo Tribal Knowledge System: A Case Study in Assam's Cleanest Village, Rangsapara

Tapas Pal

Abstract This study explores the role of Garo Tribal Knowledge Systems (TKS) in rural sustainable livelihood management in Rangsapara, recognised as the cleanest village in Assam. The study highlights how the Garo community integrates indigenous knowledge with natural resource management to sustain agriculture, health, and the local economy. The findings emphasise the potential of tribal knowledge systems as a grassroots model for environmentally sound and culturally inclusive rural development. This study explores the intersection of indigenous knowledge and climate resilience from the perspective of the Garo tribal community of Rangsapara. Set within the framework of sustainable rural livelihoods, the study explores how traditional ecological knowledge, cultural practices, *jhum* (shifting) farming, ethno-medicine, bamboo crafts, community-led sanitation, and community-led governance contribute to climate-resilient adaptation. The deep-rooted practices of the Garo people, from agro-forestry, organic farming, and water conservation to collective decision-making and sacred landscape management, provide a compelling model for sustainable development. Using a qualitative case study methodology including Participatory Rural Appraisal (PRA), in-depth interviews, and ecological mapping, the study documents how these indigenous practices reduce climate risk while ensuring food security, biodiversity conservation, and socio-economic well-being. The findings emphasize the critical role of indigenous systems in building adaptive capacity at the grassroots level and promoting environmental stewardship. The paper concludes that integrating Garo tribal knowledge with contemporary rural development policies can inform inclusive and sustainable climate adaptation strategies in Northeast India and beyond.

T. Pal (✉)
Department of Geography, Raiganj University, Raiganj, West Bengal, India

S. Mandal et al. (eds.), *Climate Change Vulnerability*, Climate Change Management, https://doi.org/10.1007/978-3-032-16783-5_10

Introduction

Ethnically, the Garo tribe belongs to the *Mongoloid* race and is spread mainly in the northeast states of India, especially in Meghalaya, Assam, and Tripura. *Andrew Playfair* (1998), in his book, '*The Garos*', indicated that the word '*Garo*' is a derivation from the word 'Gara'. The Garo belong to the *Tibetan-Burman Chinese*-speaking family that migrated to northeast India many centuries ago (Sangma 2001). The thought of sustainable development was then dignified as the root for the United Nations Conference on Environment and Development (UNCED 1992) held in Rio de Janeiro (1992) and marked the first international efforts to prepare action plans and strategies to move towards more sustainable patterns of development (UN 1993). Rangsapara village in Goalpara district of Assam bears the outstanding example of purity in India (DNN24 2023). This village nourishes the rural sustainable livelihood and climate-resilient adaptation as an environmental thinking. This study examined the present climate perception of the Garo tribe of Rangsapara village regarding managing rural sustainable livelihood and climate-resilient adaptation by Tribal Knowledge System (TKS) to find out the way and techniques of tribal resource management.

Materials and Methods

Study Area

There is are total of 505 Garo population; among these, 230 males and 275 females in Rangsapara with 92 households (sbmgramin 2023) covers an area of 128.35 hectares (RGCC 2011). Rangsapara is a tourist village about 16 km from the Goalpara district, inhabited by the Christian Garo tribes (Marak 2019). The 88 families residing in this village (Dutta 2019) can bring about systematic changes in their lives for rural sustainable livelihood and climate-resilient adaptation. They truly believe that cleanliness is next to godliness, making the community take care of their households along with the village areas. In 2000, the Rangsapara Management Committee (RMC) was formed to monitor the no-open-defecation policy, peace and unity, good sanitation and hygiene in toilets, fresh drinking water (Ministry of Jal Shakti 2021), and the cleanup of the village every Wednesday by males and on Saturdays by females (Assam Rural Development Society 2001). In 2021, this village achieved ODF (Open Defecation Free) plus village in India (sbmgramin 2023). In 2017, this village was affirmed as the cleanest village of Assam (Fig. 10.1) by the ASCOSAN (Assam Conference on Sanitation 2017). This village stopped alcohol, even though it is prevalent amongst all. It is compulsory to maintain good village health and sanitation and around the surroundings of their household, church and the local community used to clean public places like churches and schools once per month, which is managed by monthly meetings. They decided that a fine of Rs. 5001 be imposed

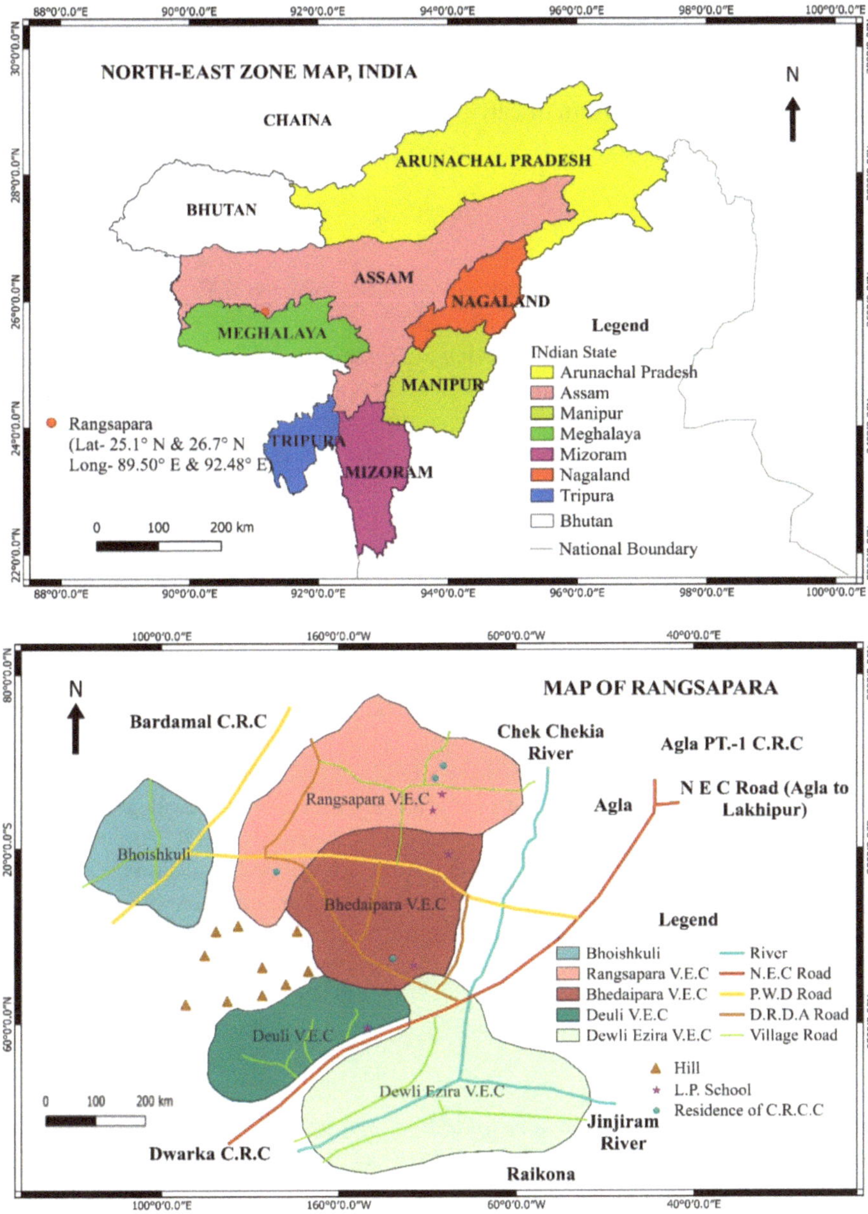

Fig. 10.1 Location of the study area

on the rule breaker (Kumar 2023). Though the community did not defecate, they used the *kuccha* latrine at the initial stage of development. A mission of the committee, headed by the changing tactics and capacity building, continues to maintain dirtiness, and toilets were converted to *pucca* sanitary (SBM Gramin 2023).

M.R.N. Momin took up activities that triggered the demand for sanitary facilities in the Chief Secretary of the National Council of (some organisation) Public Forum (Momin 2023), and finally, the Garo community took an expletive to forbid the exercise of plastics in the form of carry bags for climate protection.

Database and Its Sources

Primary Database

Primary data collected from the field visit observation, open-ended questions based on their indigenous livelihood, and geo-tagged photographs to study the scenario of rural sustainable livelihood, climate-resilient adaptation, and the Garo Tribal Knowledge System. During the field survey, the researcher visited door-to-door houses of the Garo tribe and took lots of evidence of their contributions and practices for natural resource-based rural sustainable livelihoods and bamboo-mud--rubber-bitternut-cattle-based traditional farming for climate-resilient adaptation.

Secondary Database

The secondary information had been taken from Assam Conference on Sanitation (Sentinel Digital Desk 2017), the Public Health Engineering Department of Goalpara (PHEDG 2017), the Rangsapara CRC (Cluster Resource Centre) office, District Primary Education Programme of Goalpara (Gargaon College 2016), the office of Rangsapara L. P. School), the Executing Agency of A.P. (Assam Project) Level, Balijana Development (Assam State Rural Livelihoods Mission 2024), Local Panchayat, Youth Activities Report 2017), the Rangsapara SSA office), and the Goalpara District Water and Sanitation Committee (Sentinel Digital Desk 2019).

Methodological Aspects

The whole research is based on an *explanatory sequential design* of *mixed methods research* (Creswell and Plano Clark 2018) to select the Garo tribe for *semi-structured interviews*, *participatory rural appraisal (PRA)*, and *narrative interviews* (Chambers 1994). *Purposive and snowball sampling* (Palinkas et al. 2015) was used to select elderly persons (especially the elderly males and females who are busy cleaning their courtyard or village path); members of the community centre, *anganwadi*, and the schoolmaster of L.P. School of the village are interviewed. Ninety-two households, 147 males, and 163 females of the total Garo population have been surveyed, and the *Yamane* (1967) method was used to select the sample size. *Direct field observation* (Kawulich 2005) and *geotagged photography* (Pfeifer and Budke 2020) were used to get the field data and relate all natural resource-based rural sustainable livelihoods and farming for climate-resilient adaptation by the Garo tribe in Rangsapara village. So, data was collected in three corners from 92 *households*, 7 *key informants*, and 3 *focus group discussions* (Krueger and Casey 2014). To discuss the Garo Tribal Knowledge System, the *content analysis* (Krippendorff 2018) was used. As per ethical considerations (Beauchamp and Childress 2019), the *informed consent, anonymity and confidentiality* (Bryman 2016), and *community participation* (Cornwall and Jewkes 1995) methods were applied.

Results and Discussion

Managing Rural Sustainable Livelihood

Natural Resource-Based Livelihood

From the house-building materials to the pleasure for all of this, the Garo tribe of Rangsapara village is directly dependent upon nature. Among their various incredible works, their climate-resilient house-making process, architecture, and materials are quite interesting matters to disaster management. There are only a few *pucca* houses in this village. Most of the houses are *kuccha*, which are made of mud, bamboo, mud tile, wood, cane, bait, etc. In some cases, they also use wood for making doors and windows, chairs, etc., which are very clean and beautiful to see. Each house is well decorated with various pure natural materials. Firstly, they cut the bamboo from the forest and then let it dry for a few days. After drying, they cut the bamboo into several pieces as per their requirement. And then arrange the small bamboo pieces into a certain process. They use these bamboos for making walls of their climate-resilient rooms, bamboo walls to surround their house, and making bamboo chairs and tables (Fig. 10.2). These Garo people are quite different from modern society. They choose the natural materials that are easily available instead of modern housing accessories. They maintain their village by using pure natural resource materials, which do not cause any harm to nature. Even when these house-making materials have expired, they try to reuse them for other works. Figure 10.3 shows that all the houses of the village have proper numbering. They covered their

Fig. 10.2 Bamboo is used for different purposes in the Garo tribe's livelihood, like the bamboo fencing, school playground, school furniture, house boundary, and sustainable bridges

Fig. 10.3 Climate-resilient *Kuccha* houses of a farmer and labourer at Rangsapara village

Fig. 10.4 Bamboo, stones, and soils are used for fencing around the forestland and houses

Fig. 10.5 The bamboos are used to make the sustainable natural fences of the houses

drainage sites with bamboo fencing. This bamboo fencing is totally biodegradable. Instead of bricks and cement, the villagers have used stones and soil to make the walls. They dig the soil at the corners of the house to make a natural manure zone. Garo people use agricultural fields as playgrounds and prepare gardens in every house. The sign of ancient Garo culture is still fully present in each and every house. They are very much concerned and aware of nature. The houses are not only clean from inside but also from outside. They have planted several trees around their houses to cope with the harsh climate naturally. Another important aspect is that those dustbins are mostly made of eco-friendly materials like bamboo (Fig. 10.4). The villagers use bamboo and wood to make fences (Fig. 10.5) instead of concrete

Fig. 10.6 Traditional rainwater harvesting by the Garo tribe

Fig. 10.7 Social forestry by the Garo tribe to make unique green layers of village

walls. It is their tribal indigenous knowledge. Instead of concrete slabs, the villagers have used natural ingredients to make the small over-bridge for drainage, which is a fine example of sustainable, clean construction. Almost all the houses are planted with flowers to enrich the beauty of the household. In both the primary and middle village schools, the furniture is made out of bamboo or wood. They are not only utilizing an eco-friendly way of living in their real life but also encouraging their future generation and all the tourists to do so.

Agro-Forestry, Gardening and Rain Water Harvesting

The villagers traditionally harvest the rainwater by digging a small area (Fig. 10.6). They use this water mainly for agricultural purposes. In the village, there are many newly planted *social forestry* (Fig. 10.7) areas to help fight against heat waves. The tribal villagers are not only utilising the forest material effectively and conserving the forest cover, but they are also planting new trees to increase the forest area of the village, adjoining areas, and around the house. It's a model of green layers to protect the village from climate hazards. To manage solid waste, every household in the village must have a dustbin according to the village law, and so far, tourists have not seen a single area without one. Every year, they plant several trees in the village.

Fig. 10.8 Stones around the newly planted saplings to enhance the protection of the model green layers'

Fig. 10.9 Sustainable organic agriculture is practiced to support climate sustainability

Gardening and nurseries are both integral parts of Rangsapara. The inside as well as the outside of the property looks properly arranged and clean. They plant trees on their side of the courtyard and use natural rocks to make a protected boundary (Fig. 10.8). The Garo tribe maintains organic farming (Fig. 10.9) to sustain the climate.

'Gender-Sanitation Responsibility Cycle' (GSRC) and 'Green Livelihood' (GL)

Different approaches have been taken by the Garo tribe as well as by the village committee to make Rangsapara the cleanest village in Assam state. To promote awareness about sanitation among students and adults, various awareness programmes like workshops and road shows have been organised by the Rangsapara L.P. school (e.g., drawing competition during the national sanitation campaign on 13th October, 2015) and different committees (e.g., awareness programmes organised by *Swachh Bharat Abhiyan*) (Fig. 10.10) and organizations (e.g., a sanitation camp organised by ASCOSAN of WSC, Goalpara, in 2017). To promote sanitation among students, hand hand-washing programme had been organised in the local school. Within-village path shows are also part of the awareness drive in the region. The Mason training programme has been organised every year, which is a training programme to train the local people about climate-resilient and green livelihood construction. The village has its own proper tribal waste management system by using non-plastic dustbins at every probable waste-throwing zone. In the village,

Fig. 10.10 A community sanitary complex made by MANRSA in Rangsapa village, which gives the sustainable sanitation of this village

Fig. 10.11 'No smoking' message is placed to reduce the noxious village smoke

throwing garbage haphazardly outside is considered a crime. In each and every place, whatever it is—a playground, house, or community hall, people can find several dumpsters for the segregation of garbage. This village is also restricted from taking any kind of tobacco product, alcohol, or other toxic materials. That is why one can find "No smoking" (Fig. 10.11) and "No alcohol" posters everywhere in the village. There is a fine of Rs. 5001 for taking any kind of alcohol or cigarette, but so far nobody has had to pay because no one has broken the rule. There is another fine of Rs. 501 for throwing garbage out of wastebaskets. According to the villagers, the village is now free from tobacco and alcohol. For an open defecation-free plus zone, *Pucca* sanitary complexes are developed by the community to make the aim possible in real life. In the year 2017, the Rangsapara village was declared a clean village in Assam by the Public Health Engineering Department (PHED) (Fig. 10.12), and the journey towards the goal began way back in the year 1999. It's a unique model of rural cleanness and livelihood for developing resilient villages in the weather. To promote sanitation, villagers followed the handwash programme by the school, the inside-outside cleaning model, and awareness about sanitation among students and adults. Even spitting in the open is restricted here. The villagers clean their houses and campuses every day and follow a unique model of the '*gender-sanitation responsibility cycle*', i.e., even as tribal men clean public places (like roads and community halls) on Wednesday, and tribal women clean public places on Saturday. Residents here raise special flowers around their houses to harmonize the magnificence of their small cottages (Fig. 10.13). Even a 4–5-year-old child (Fig. 10.14) is seen standing with a broom to clean the courtyard.

Fig. 10.12 Road show during National Drinking Water and Sanitation Week. A clean drive by the villagers on the occasion of Gandhi *Jayanti* in 2016

Fig. 10.13 A climate thinker village woman cleaning the courtyards of her house

Fig. 10.14 A Garo child standing with a broom to clean the house

Climate-Resilient Local Adaptation (CRLA)

The Garo tribe fully depends upon nature. Though, as per the 2011 census, the average literacy rate of this village was about 70%, the villagers are very serious about their development in a sustainable way. They follow their indigenous Garo culture. They have planted some trees around the house. Every household in the village must have a bamboo-based litter bin (Fig. 10.15) according to the village law. The villagers not only use the trees to build their houses; they also plant many trees to keep the climate balanced every year and follow several tree plantation programmes (Fig. 10.16) and nature conservation-related campaigns. To enhance the weather of their house, the villagers have planted colourful flowers instead of artificial structures. The park in the village has colourful bamboo chairs and bamboo trash bins of different designs. All the things in the park are made of bamboo and wood. Cleaning

Fig. 10.15 Types of recyclable bamboo-made dustbins in every corner of Rangsapara village

Fig. 10.16 Young saplings are guarded with bamboo rolls, and natural drainage is used for proper rainwater infiltration

of the park is the responsibility of the villagers. Children come here to play and spend time with their friends and families. This is their model village park for nourishing sustainable climate philosophy. The villagers are very aware of environmental and climate degradation. They maintain everything in a proper, sustainable way. They adopted a sustainable developmental approach. Their houses are mainly built with natural ingredients, and the rubbish bins, as well as chairs and furniture, are mostly made of natural materials. They have shown us how we can make our landscape beautiful with naturally built construction and how climate and development can live side by side. They have planted some trees around the houses to restrict heat waves (Fig. 10.17). Only the church is made with *pucca* materials. They collect byproducts of the climate, like bamboo and wood from the jungle, and finally sell them to local buyers. They have a special technique to collect leaves and burn them in a dump area. They are using modernized signboards to identify their house and trees. They also create weapons such as bows and arrows. Electricity and television, mobile and internet service are available everywhere in the village. So, they never burn the waste for safe climate health. Each house in the village has a campus where they have planted different local plants, like coconut, for multiple purposes. The Garo tribe always managed climate issues with the Garo indigenous knowledge system, so they use every inch of the tree. The people of Rangsapara village depend on traditional equipment. Frequently, they bring in money from banana farming,

Fig. 10.17 A local natural resource-based heat wave resilience house with a bamboo-made dustbin is an example of resilient housing and critical infrastructure of the house are coated by mud to restrict heat wave

Fig. 10.18 Sample of rubber plantations by the villagers

Fig. 10.19 *Borang* of the Garo Tribal Knowledge system

rice and rubber cultivation (Fig. 10.18), pig nurturing, agriculture, handloom, betel nuts, wage labour, etc. The villagers are alert about excessive rubber cultivation and its bad effects on disturbing the groundwater reserve and soil quality. So, they do it minimally. *Aborang*' (Fig. 10.19) is a type of house built on a tree with a ladder attached. Earlier, the Garo people used to build these houses where there was danger from wild animals, which is seen at the entrance of the village, mainly to attract tourists. The Garo tribes plant a fruit garden beside their house in connection with the climate status. They use *Jamsireng* (Fig. 10.20), a small traditional hut, to store grains and fruits even during climate hazards. The village offers an experience of

Fig. 10.20 Researcher is in *Jamsireng* of the Garo Tribal Knowledge system

Fig. 10.21 Bamboo-made traditional tribal *dolna* in the village

the true potential of nature. They only have their nature and natural resources to display, which is becoming rare these days. Away from all the traffic and pollution, a clean environment and a simple way of living—that's what this village is all about (Fig. 10.21).

Models of GTKS (Garo Tribal Knowledge System), and GICLAF (Garo Integrated Climate-Livelihood Adaptation Framework)

The Garo Tribal Knowledge System (GTKS) of Rangsapara village presents a striking example of how indigenous ecological knowledge can build climate-resilient adaptation and sustainable rural livelihoods (Das 2017). Built on experiential learning gained over generations, GTKS incorporates a range of practices; such as *jhum*-modified *agroecology*, seed conservation, ethno-forestry, and bamboo-based construction, that not only enhance food and climate security but also provide protection against climate vulnerabilities). The role of the *nakmas* (village head) as traditional environmental stewards ensures the integration of customary governance with natural resource management (Das 2017). Through decentralized governance, seasonal forecasting based on biological cues, and the conservation of sacred landscapes, the Garo community demonstrates a holistic sustainability model aligned

Fig. 10.22 A Garo woman wearing traditional cloth and carrying the future generation of the village on her back to teach climate narratives practically through watching cleanness education

with the United Nations SDG-13 ('*Climate Action*') and SDG-15 ('*Life on Land*'). Policy implications suggest that the coding of GTKS in district-level climate plans can serve as a replicable framework for adaptive governance in other tribal areas of Northeast India. This concept proposes a *Garo-informed Livelihood-Climate Integrated Model* (GLIM), which emphasizes the need for policy mainstreaming of *biocultural* rights recognition, Community-Based Adaptation (CBA), and indigenous resilience strategies. The Garo tribe of Rangsapara village in Assam represents an important indigenous community whose traditional ecological knowledge can be transformed into a decentralized (Fig. 10.22), sustainable, and climate-resilient system. Below is a novel conceptual model called the *Garo Integrated Climate-Livelihood Adaptation Framework* (GICLAF), which (Fig. 10.23) integrates traditional knowledge, biodiversity, adaptive agriculture, and policy inclusion. This model is unique because it is the combo of tribal indigenous practices (e.g., jhum, sacred groves, and oral traditions) plus climate-smart techniques (e.g., rainwater harvesting, climate-connected cultivation) under Nokma governance. This model reflects the importance of customary laws, local institutions, sustainable resource management, and climate-resilient crops. This model will be helpful for integrated state climate plans, balanced traditional-cum-modern practices, and guarantees of tribal governance.

Strategic Policy Recommendations

The agro-ecological concept of promoting Integrated Farming Systems (IFS) was adopted from traditional Garo practices for increasing water use efficiency and crop diversity (Das 2017). Integrate and integrate Garo ethnobotanical knowledge into regional climate adaptation planning through documentation of indigenous knowledge communities and participatory policy frameworks. Rural youth and women should be trained in climate-resilient agriculture using local dialects and tribal teachers for capacity building so that they can better disseminate (Das 2017). Local

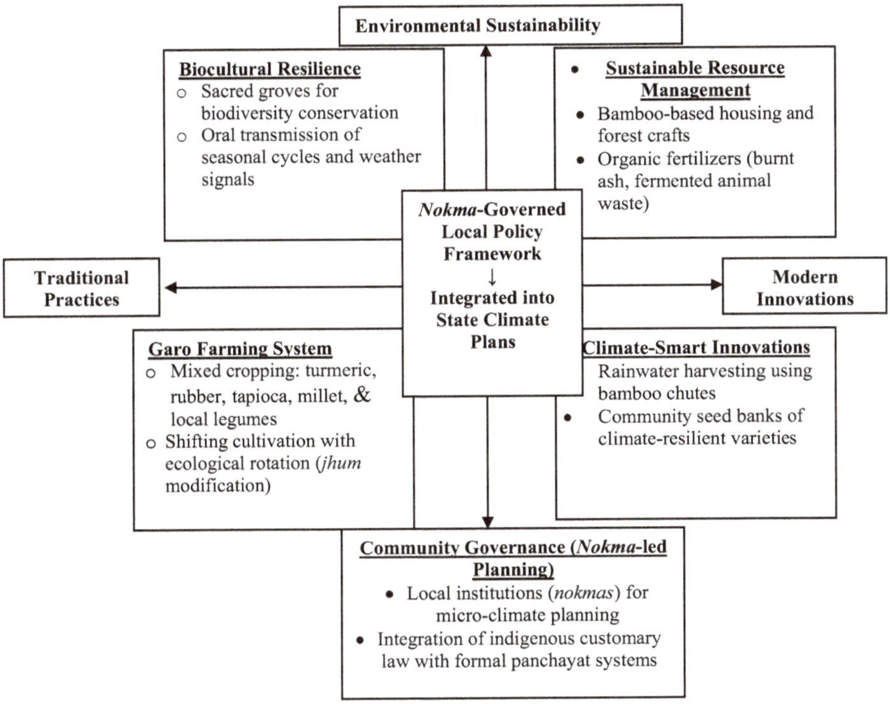

Fig. 10.23 GICLAF Model (Garo Integrated Climate-Livelihood Adaptation Framework)

Fig. 10.24 The welcome board of the Rangsapara village to attract the Garo tribal knowledge system-based climate-resilient village

administrations should encourage non-timber forest products, weaving, and traditional handicrafts in Garo villages through micro-economic policies for livelihood diversification (Fig. 10.24). Again, community participation through institutionalization of Gram Panchayats (*Nokma*) in climate governance through recognition in regional policy charters may be recommended. An accurate community-led climate monitoring system can be developed using oral history and seasonal change indicators. Finally, land tenure reform through legal recognition of customary land uses in the forest-*agroecotone* should be recommended to support sustainable resource management (Table 10.1). *Digi Dhan Mela* is quite a famous *mela* in the village

Table 10.1 Nature of policies, recommendations, strategy cum implementation

Nature of policies	Recommendations	Strategy cum implementation
Agroecology	Promote indigenous organic farming methods (e.g., jhum with soil--restoring cycles)	Integrate with ICAR/NICRA; provide climate-resilient training to Garo youth
Village Green Governance	Legal recognition of Garo-managed forests	Implement the Forest Rights Act with the consultation of *the Nokma*
Water and Sanitation	Pull traditional bamboo pipe irrigation and Swaccha Bharat sanitation scheme	Pilot Garo community watershed programme
Climate-friendly Livelihood Diversification	Revamp forest-based crafts, beekeeping, and Ayurveda	Develop rural cooperatives and disaster management skill development programmes
Sharing Climate Data	Community Climate Station (CCC) linked with the Indian Meteorological Department	Allow NGOs to train the Garo people as "climate fighters"
Cultural Sustainability	Implement Garo cosmology into village-school curriculum and policy framing	Local school-based syllabus formation on indigenous education
Disaster Preparedness	Traditional-smart early warning systems with the oral memory method	Give importance to local language climate resilient adaptation programmes
Traditional Behavioral Maintenance	Recognize the traditional tribal behavior for climate resilience	Apply the Climate philosophy and climate psychology camps

Source: Field Survey, 2024

Fig. 10.25 Non-pitched pavement built under the MANREGA scheme for tourist accessibility

where the Garo tribe sees the public health engineering department has put up a stall to create awareness among people about sanitation. Though the main street of this village is still non-metal, with the help of the government, construction is going on. After winning the award for the cleanest village in the state, the village has opened up its doors for global tourists, and the local government supported contracting the brick road rather than paving it fully with concrete due to the infiltration of water in the soil through the joints of the bricks (Fig. 10.25).

Discussions

In the year 2016–2017, the Rangsapara village was declared the cleanest village (also open defecation-free) in the state of Assam by the Public Health Engineering Department (PHED); the journey towards the goal began way back in the year 1990. In the year 2000, Rangsapara villagers created a 10-person administrative committee for resolutions on peace, harmony, clean, disinfected toilets, and fresh drinking water for all and observance of the village being clean every Wednesday by men and Saturday by women. Now the *Swachh Bharat Abhiyan* has helped the villagers to achieve their goal much faster. Every establishment in the village has dustbins so that people can discard the garbage in a proper way, and it is the responsibility of the villagers to keep their village clean and free of plastics. Every day in the morning, local people clean their houses. This is a daily habit of the villagers. They not only keep the inside of their house clean but also the outside. Every individual in the village is motivated to keep their village clean; as we can see, from a small child to an aged person, everybody is engaged in different ways to keep the village and their house clean. It's very amazing to understand that the way of managing the village is in a proper, sustainable way. Climate stewardship is an idea that exemplifies the accountable planning and management of natural resources. The notions of stewardship can be pragmatic to the environment and nature, economics, health, property, information, theology, etc. Natural resource management of Rangsapara village denotes the management of land, soil, plants, water, and animals, with a specific importance on how supervision marks the superiority of existence for both present and future generations and environmental stewardship. The main occupation of the Garo tribe in this region is agriculture. Being farmers, the people of this region are much closer to the climate. The need for climate resource management was felt by them and involves the conservation and development of natural resources like forests, wetlands, etc. Various approaches have been taken by the Garo tribe for the management of the natural resources in the village. Rangsapara village has good tourism potential. Though the village is small, it has its own cleanliness identity to attract people to itself. Every structure of this village is a scenic beauty. The houses are made of mud and are well decorated, clean, and equipped with beautiful flower gardens. Though the Garo people are very isolated from our modern society, they are very advanced in climate protection. They have been staring at the conservation of nature since ancient times. The Garo people prefer to use natural things instead of artificial or man-made things, and so you can see very few household things used by the Garos.

Conclusions

This longitudinal study examined how indigenous practices can support rural sustainability and climate adaptation. This study is rich in ethnographic detail and has strong potential to be a model of good sanitation and hygiene. The village started its cleanliness journey in the 1990s and finally achieved its position in 2017. It is a single case study rather than a comparative study. This village proved the importance of community effort to success in Swaccha Bharat and developed the role model of ODF (Open Defecation Free) plus status. The whole study described the importance of tribal culture to maintain peace, health, unity, sanitation, and hygienic efforts for sustainable development. Further research could highlight the community-supported solid waste management and behavioural importance study, or how the integrated village knowledge system is fruitful for health and well-being. So, the demographic decision model is important to develop a population study. Digital campaigning and AI-based cleanliness management in any village can be applied in the future for smart village development. The negative effect of development can also be seen. Although there are many dustbins and the village is a no-drinking zone, it is still seeing these bottles being dumped in the open. It is suspected that with increasing tourism, these kinds of incidents might increase, so tourists need to be more responsible and respectful. The study has underlined the potential of the village and its unique climate-resilient activity and the Garo tribal knowledge system. The study will also help the government to formulate the policies and decision-making for facilitating the rural sustainable tribal tourism potential. The villagers are hard workers, and they are trying to give an identity to their village as a major tourist destination in Assam.

Acknowledgements The researcher would like to extend his deepest gratitude to *S. Dutta and D. Mandal* for their support during the field visit.

Funding No funding was received for this research work.

Informed Consent Informed consent was obtained from all individual participants included in the study.

Author's Contribution All authors have contributed equally to this work.

Data availability statement The datasets analysed during the current study are available from the corresponding author on reasonable request.

References

Assam Conference on Sanitation (2017) Cleanest villages of Assam: ASCOSAN awards 2017. Government of Assam. http://www.ascosan.assam.gov.in/awards2017

Assam Rural Development Society (2001) Community-led sanitation and hygiene initiatives in Rangsapara. http://www.ards-assam.org/reports/rangsapara-initiatives

Assam State Rural Livelihoods Mission (2024) Visit to Balijana Development Block, Goalpara. Facebook. https://www.facebook.com/asrlmassam/posts/visit-to-balijana-development-block-goalpara/1234567890

Beauchamp TL, Childress JF (2019) Principles of biomedical ethics, 8th edn. Oxford University Press

Bryman A (2016) Social research methods, 5th edn. Oxford University Press
Chambers R (1994) The origins and practice of participatory rural appraisal. World Dev 22(7):953–969. https://doi.org/10.1016/0305-750X(94)90141-4
Cornwall A, Jewkes R (1995) What is participatory research? Soc Sci Med 41(12):1667–1676. https://doi.org/10.1016/0277-9536(95)00127-S
Creswell JW, Plano Clark VL (2018) Designing and conducting mixed methods research, 3rd edn. SAGE Publications, Thousand Oaks, CA
Das A (2017) Training manual on integrated farming system for sustainable hill agriculture: an option for climate smart agriculture and natural resource management. ResearchGate. https://www.researchgate.net/publication/320923043
DNN24 (2023) Rangsapara: the cleanest village of Assam—know the story of determination. DNN24. https://dnn24.com/rangsapara-the-cleanest-village-of-assam-know-the-story-of-determination
Dutta S (2019) Natural resource management and sustainable development by Garo tribe: study on Assam's cleanest village Rangsapara. Unpublished M.Phil. dissertation, Raiganj University
Gargaon College (2016) A district level study on primary education in Assam. Soc Sci J Gargaon Coll 4:39–53. https://gargaoncollege.ac.in/pdf/publications/1/pub_more/62.pdf
Kawulich BB (2005) Participant observation as a data collection method. Forum Qual Soc Res 6(2):43. https://doi.org/10.17169/fqs-6.2.466
Krippendorff K (2018) Content analysis: an introduction to its methodology, 4th edn. SAGE Publications
Krueger RA, Casey MA (2014) Focus groups: a practical guide for applied research, 5th edn. SAGE Publications
Kumar D (2023) Rangsapara: the cleanest village of Assam, know the story of determination. DNN24. https://dnn24.com/rangsapara-the-cleanest-village-of-assam-know-the-story-of-determination
Marak SA (2019) Rangsapara: the cleanest village of Assam [unpublished manuscript]. Academia. edu. https://www.academia.edu/39613859/Rangsapara_The_cleanest_village_of_Assam
Ministry of Jal Shakti (2021) ODF Plus villages in India: annual report 2021. Government of India. http://www.swachhbharatmission.gov.in/odfplus2021-report
Momin MRN (2023) Community-led sanitation and environmental initiatives in Rangsapara. DNN24. https://dnn24.com/rangsapara-the-cleanest-village-of-assam-know-the-story-of-determination
Office of the Registrar General & Census Commissioner (RGCC), India (2011) Primary census abstract data: Rangsapara Village. Ministry of Home Affairs, Government of India. https://censusindia.gov.in
Palinkas LA et al (2015) Purposeful sampling for qualitative data collection and analysis in mixed method implementation research. Adm Policy Ment Health Ment Health Serv Res 42(5):533–544. https://doi.org/10.1007/s10488-013-0528-y
Pfeifer C, Budke A (2020) Using geotagged photographs for geographic fieldwork and education. J Geogr High Educ 44(1):130–147. https://doi.org/10.1080/03098265.2019.1661366
Playfair AM (1998) The Garos. (Reprint 2016). Spectrum Publications, Guwahati, p 31
Public Health Engineering Department (PHED), Goalpara (2017) Annual report 2017–18. Government of Assam
Sangma MS (2001) History and culture of the Garos. Spectrum Publications, New Delhi
sbmgramin (2023) Rangsapara's journey from the cleanest to an ODF Plus Model village. Swachh Bharat Mission (Grameen). https://sbmgramin.wordpress.com/2023/05/23/rangsaparas-journey-from-the-cleanest-to-an-odf-plus-model-village/
Sentinel Digital Desk (2017) ASCOSAN-2017 in city from Feb 16. The Sentinel Assam. https://www.sentinelassam.com/guwahati-today/ascosan-2017-in-city-from-feb-16
Sentinel Digital Desk (2019) Rangsapara: Assam's cleanest village in the year 2016–17 situated in Goalpara. The Sentinel. https://www.sentinelassam.com/north-east-india-news/assam-news/rangsapara-assams-cleanest-village-in-the-year-2016-17-situated-in-goalpara

Swachh Bharat Mission Gramin (2023) Rangsapara's journey from the cleanest to an ODF Plus Model village. SBM Gramin. https://sbmgramin.wordpress.com/2023/05/23/rangsaparas-journey-from-the-cleanest-to-an-odf-plus-model-village
United Nations (1992) Agenda 21: Programme of action for sustainable development. Rio Declaration on Environment and Development. United Nations conference on environment and development (UNCED), Rio de Janeiro, Brazil, 3–14 June 1992. United Nations, New York
United Nations (1993) Earth summit: agenda 21, the United Nations Programme of action from Rio. United Nations Publications, New York. ISBN: 9789211005097
Yamane T (1967) Statistics: an introductory analysis, 2nd edn. Harper and Row, New York

Chapter 11
Tokbari: A Dual-Purpose Poultry Bird for Enhancing Rural Livelihoods Under a Changing Climate in Northeast India

Vinay Singh, Asit Chakrabarti, Lopamudra Sahoo, B. U. Choudhury, Chandan Debnath, Mahak Singh, and Sourabh Deori

Abstract For rural populations in northeastern India, backyard poultry rearing is a vital source of additional revenue and nutritional stability. Most of the farmers in this region traditionally rear local, native birds that yield minimal outputs in terms of both meat and egg production. Despite an increasing poultry population over the years, low productivity remains a major constraint limiting the economic potential of backyard poultry in the region. The absence of location-specific improved poultry breeds/varieties adapted to the unique agro-climatic conditions of the Northeast has long been recognized as a significant barrier to improving rural poultry productivity. To address this gap, the ICAR Research Complex for NEH Region Tripura Centre at Lembucherra has created the location-specific, promising improved poultry variety "Tokbari" for backyard farming in rural, resource-poor, and delicate ecosystems of Northeast India. This variety has the potential to outperform native breeds while thriving in local conditions. With a genetic mix of 25% Tripura Black, 25% Coloured Broiler, and 50% Dahlem Red, this dual-purpose variety outperforms native birds in terms of performance. Field data confirms Tokbari produces 133% more eggs annually (140 vs. 60), 26% larger eggs (48 g vs. 38 g), and achieves 45–66% higher body weights (males: 2.5 kg vs. 1.5 kg at 40 weeks). Economic analysis reveals an additional profit of Rs. 436 per bird compared to native breeds. Till now, over 1.30 lakh Tokbari germplasms have been distributed across Northeast India, validating its suitability for local agro-climatic conditions while significantly enhancing rural income and nutritional security through improved productivity. The widespread farmer adoption rates and documented success stories further validate Tokbari's potential to uplift rural livelihoods in Northeast India.

V. Singh (✉) · A. Chakrabarti · L. Sahoo · B. U. Choudhury
ICAR Research Complex for NEH Region, Tripura Centre, Lembucherra, Tripura, India

C. Debnath · S. Deori
ICAR Research Complex for NEH Region, Umiam, Meghalaya, India

M. Singh
ICAR Research Complex for NEH Region, Nagaland Centre, Medziphema, Nagaland, India

S. Mandal et al. (eds.), *Climate Change Vulnerability*, Climate Change Management, https://doi.org/10.1007/978-3-032-16783-5_11

Introduction

Backyard poultry farming is vital for the food and nutritional security of rural communities, particularly in delicate ecosystems. It has been acknowledged as a means to alleviate poverty and enhance food and nutritional security among the rural poor. Poultry meat and eggs serve as the most economical and superior sources of high-quality protein, vitamins, minerals, and other essential nutrients. The poultry industry significantly contributes to improving the economic conditions of rural populations in India (Chatterjee and Rajkumar 2015; Rajkumar et al. 2021). There is a strong demand for poultry and its products in the Northeast Region of India (Kumar et al. 2007). Backyard poultry farming can contribute to a sustainable food production system by enhancing productivity with appropriate measures (Singh et al. 2018a; Rajkumar et al. 2021). In India, particularly in the Northeastern states, backyard poultry farming plays a crucial role in the livelihoods of rural communities. Across the entire Northeast Region, poultry farming acts as a vital source of additional income and nutritional stability for rural households.

Most of the farmers in the area maintain small flocks of 5–15 local bird species in the backyard under traditional management types (Malik and Singh 2013). The native birds of the Northeast area are believed to have better disease resistance, better predator escape behavior, and higher consumer acceptance due to their multi-colored plumage and better taste in meat (Haunshi et al. 2009). Due to the native chicken's lower productivity and reproductive performance, rural farmers are now seeking improved breeds with better production performance (Khan et al. 2004). Despite an increasing population of birds over the past several years, poor productivity still imposes a significant challenge to the economic development of backyard poultry in the area. The region's challenging topography, characterized by steep slopes and scattered settlements, limits access to modern poultry inputs and veterinary services. Additionally, the intense monsoon seasons and harsh winter conditions in higher altitudes (Azharuddin and Dogar 2016; Dogar et al. 2024) create stress conditions that traditional breeds struggle to withstand, often resulting in mortality rates exceeding 30% during adverse weather periods.

The enhancement of productivity within the backyard poultry production system can be achieved through the incorporation of superior germplasm or by the execution of enhanced management practices (Singh et al. 2018b; Chaiban et al. 2020). A pertinent intervention in this regard involves the introduction of improved poultry breeds specifically tailored for backyard production in rural and tribal regions. The longstanding absence of poultry breeds that are uniquely adapted to the specific agro-climatic conditions of the Northeast has been widely acknowledged as a substantial impediment to the enhancement of productivity in rural poultry systems.

To address these gaps, ICAR initiated a targeted breeding program combining indigenous resilience with exotic productivity traits. In order to overcome the problems highlighted above, there has long been a need to develop location-specific poultry varieties incorporating desirable indigenous characteristics with suitable exotic breeds with preferable egg and meat production potential. To address the absence of egg layers and dual-purpose birds, the ICAR Research Complex for the NEH Region Tripura Centre, Lembucherra, began developing a specific dual-purpose bird suited to local conditions and capable of significantly outperforming native bird types. The outcome of the experimental work is 'Tokbari', a new dual-type poultry variety that has been developed for use in rural backyard farming in the Northeastern Region of India.

Breeding Plan and Development of Tokbari Chicken

The development of Tokbari is an attempt to combine the desirable qualities of improved and indigenous breeds. The composition is 25% Tripura Black (Local indigenous breed), 25% Coloured Broiler (from ICAR-Central Avian Research Institute, Izatnagar), and 50% Dahlem Red (from ICAR-Directorate of Poultry Research, Hyderabad). As part of developing the Tokbari chicken variety, benchmark surveys were conducted in different parts of the Tripura state to understand the farmers' preferences for poultry breeds/varieties before the development of the Tokbari chicken variety. The surveys identified farmers' preferences for dual-purpose birds with good growth potential and egg-laying ability, larger, brown-shelled eggs, low mortality, good scavengers, and foragers. Based on these survey results, a systematic breeding approach was developed to produce a variety combining the desirable characteristics of indigenous breeds and improved production. As part of the breeding plan, selected native Tripura Black hens were crossed with improved meat-type colored broiler parent stock males, and the offspring (F1) were evaluated. To create the desired cross with a genetic composition of 25% Colored broiler, 25% native Tripura Black, and 50% Dahlem Red inheritance, the chosen first-generation (F1) crossbred males were then mated with improved egg-type Dahlem Red hens (Fig. 11.1). This three-way cross was further refined and stabilized through selective breeding in several generations before substantial testing occurred at the farm and field level, leading to the release of the 'Tokbari' variety. This genetic combination was intentionally planned to capitalize on the local adaptation of Tripura Black, the growth potential of Coloured Broiler, and the egg-laying ability of Dahlem Red.

The 50% Dahlem Red inheritance was specifically chosen to maximize egg production potential, as this breed demonstrates consistent laying performance with annual production exceeding 180 eggs under optimal conditions. The 25% Tripura Black contribution ensures retention of critical local adaptation traits, including disease resistance, heat tolerance, scavenging ability, and superior foraging behavior that enables survival in the region's challenging environmental conditions. The remaining 25% Colored Broiler genetics provides enhanced growth rate and improved feed conversion efficiency, contributing to better meat production characteristics while maintaining the hardiness essential for backyard systems. The term "Tokbari" recognizes its local context and use—"Tok" refers to birds in the Kokborok language (indigenous language of Tripura), while "Bari" refers to

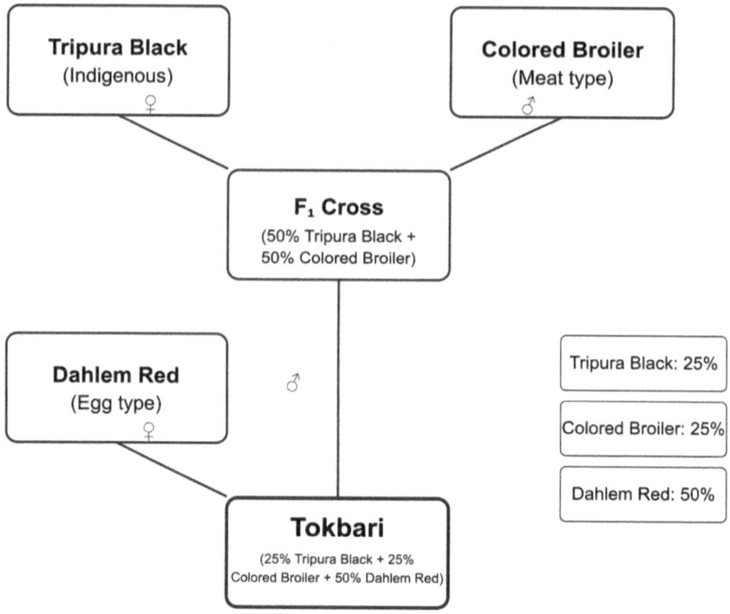

Fig. 11.1 Breeding diagram illustrating the development of Tokbari chicken variety through crossing of Tripura Black (25%), Colored Broiler Parent (25%), and Dahlem Red (50%)

backyard or homestead in Bengali. This connection provides cultural context and stresses that Tokbari is for backyard rearing.

Salient Features of Tokbari Birds

Tokbari's birds possess the following prominent characteristics, which make them unique from others.

- It exhibits an appealing multicolored plumage pattern comprising hues of brown, black, and grey, alongside possessing a moderate body mass. In comparison to indigenous birds, it demonstrates a greater weight accompanied by an enhanced potential for rapid growth.
- Farmer-preferred traits: The multicolored plumage closely mimics native birds, ensuring strong market acceptance and consumer preference, while the brown egg coloration meets traditional buyer expectations in rural markets.
- Tokbari male birds attain body weights of approximately 1.9 kg at 20 weeks and 2.5 kg at 40 weeks, representing improvements of 58% and 66% respectively, over native birds. Similarly, females reach 1.4 kg at 20 weeks and 1.6 kg at 40 weeks, showing improvements of 56% and 45% respectively.

- Most significantly, annual egg production ranges from 120 to 160 eggs per bird, a remarkable 133% increase over the 45–75 eggs typically produced by native breeds.
- Tokbari variety produces eggs of considerable size, exhibiting a brown coloration akin to that of indigenous hens. These eggs are comparatively larger, with a weight range of 45–51 g, in contrast to the 35–43 g characteristic of native avian species, thereby indicating a 26% enhancement in size.
- Tokbari hens attain sexual maturity at approximately 165–175 days, which is roughly 8% earlier than that observed in native birds.
- This breed demonstrates superior scavenging capabilities, coupled with adequate evasive skills, enabling their effective rearing within a backyard, free-range scavenging system. Due to their elongated shank and robust body conformation, they exhibit increased mobility in free-range environments for scavenging activities, while also possessing the ability to safeguard themselves against potential predators.
- As a dual-purpose variety, Tokbari chickens are suitable for both egg production and meat consumption, making them an ideal choice for rural backyard poultry farming.
- They exhibit commendable immunity and are capable of thriving on a low nutritional plane.
- It is well-suited to the prevailing local agro-climatic conditions and has garnered acceptance among farmers as the preferred stock for backyard poultry farming in rural settings.

Physical Characteristics of the Tokbari Bird

Tokbari chickens possess distinctive physical characteristics that make them immediately recognizable and suitable for rural conditions (Fig. 11.2). They feature attractive multi-colored plumage patterns ranging from brown and reddish-brown to blackish and various combinations (Table 11.1). This color diversity not only makes them aesthetically appealing but also provides natural camouflage against predators in free-range conditions. The birds demonstrate standard feather morphology characterized by rapid feather growth, thereby facilitating an expedited development of defense mechanisms against ecological adversities. They are distinguished by singular red combs of diverse dimensions—medium to large in males and small to medium in females. The integumentary coloration is predominantly pinkish-white, whereas the shank pigmentation exhibits a range from yellow and creamy white to assorted patterns. A distinctive feature of the Tokbari breed lies in its anatomical conformation. The birds exhibit relatively elongated shanks and a robust bodily structure that enhances their locomotion in free-range environments for foraging activities and provides improved self-defense capabilities against predatory threats. This physical adaptation makes them particularly suited to the traditional free-range scavenging systems prevalent in rural areas.

Fig. 11.2 Mature Tokbari chickens (right: male; left: female)

Table 11.1 Physical characteristics of Tokbari birds

Characteristic	Male	Female
Plumage and coloration	Mixed colors including brown, reddish brown, blackish, and orange with a greenish tinge. Pattern varies from solid to mottled and barred	Mixed colors of brown, reddish brown, blackish, brown with black, and grey black with an orange tinge. Patterns include solid and barred
Body structure	Medium to large single red comb, large reddish wattles, pinkish white skin, relatively long shanks for enhanced mobility	Small to medium single red comb, small reddish wattles, pinkish white skin, sturdy build suitable for free-range conditions
Feather characteristics	Normal feather morphology with a fast growth rate, enabling rapid development of protective covering	Normal feather morphology with a fast growth rate, providing quick adaptation to environmental conditions
Distinguishing features	Yellow to creamy white shank coloration with black markings, black eyes, red earlobes	Yellow to creamy white or greyish shank coloration, black eyes, red earlobes

Performance of Tokbari Birds at the Farmer's Field

The performance metrics of Tokbari under field conditions demonstrate their superiority over native birds (Fig. 11.3). As shown in Table 11.2, males attain body weights of approximately 1.9 kg at 20 weeks and 2.5 kg at 40 weeks, representing improvements of 58% and 66% respectively, over native birds. Similarly, females reach 1.4 kg at 20 weeks and 1.6 kg at 40 weeks, showing improvements of 56% and 45% respectively. The reproductive traits are equally impressive. Tokbari hens reach sexual maturity around 165–175 days, approximately 8% earlier than native birds. The eggs are larger, weighing 45–51 grams compared to 35–43 g for native birds, representing a 26% improvement. Most significantly, annual egg production

Fig. 11.3 Tokbari chickens thriving in tribal villages of Tripura, demonstrating successful field performance in typical backyard farming environments

Table 11.2 Performance comparison of Tokbari birds with native birds at farmer's field

Parameter	Native birds	Tokbari birds
Body weight at 20 weeks (males) kg	1.2 (1.0–1.5)	1.9 kg (+58%)
Body weight at 20 weeks (females) kg	0.9 (0.7–1.0)	1.4 kg (+56%)
Body weight at 40 weeks (males) kg	1.5 (1.3–1.9)	2.5 kg (+66%)
Body weight at 40 weeks (females) kg	1.1 (1.0–1.3)	1.6 kg (+45%)
Age at sexual maturity (days)	185 (175–205)	171 days (+8% earlier)
Egg weight (g)	38 (35–43)	48 g (+26%)
Annual egg production (number)	60 (45–75)	140 eggs (+133%)

ranges from 120 to 160 eggs per bird, a remarkable 133% increase over the 45–75 eggs typically produced by native breeds.

In terms of mortality, Tokbari demonstrates good survival rates under field conditions. The mortality during the brooding period (0–6 weeks) averages around 5.75%, while during the growing period (7–20 weeks) it is approximately 7.52%. During the laying phase (21–72 weeks), mortality rates are minimal at just 1.55%. These figures reflect the breed's robust health and adaptability to local conditions. These mortality rates are significantly lower than the 10–15% typically observed in native breeds under similar conditions, reflecting the breed's robust health and adaptability to local conditions.

Economic Impact and Livelihood Benefits

Economic analysis comparing Tokbari birds with native birds is presented in Table 11.3. A unit of 20 Tokbari birds (assuming 8 females completed laying and 8 males reached marketable age) generated a total income of ₹20,384 compared to ₹10,624 from native birds. After accounting for all expenses, the net profit from Tokbari birds was ₹18,844 per unit compared to ₹10,124 from native birds, resulting in an additional profit of ₹8720 per unit or ₹436 per bird (Fig. 11.4).

Tokbari birds deliver superior economic returns through their enhanced production parameters and market advantages compared to native breeds. Their accelerated growth, increased market weight, improved egg production, and larger eggs command premium prices in local markets, further enhanced by consumer preference for their colored plumage and brown eggs.

Success stories from farmers across Tripura further validate the economic potential of Tokbari birds (Fig. 11.5). For instance, Mr. Safarmia from Aralia village in West Tripura, who previously engaged in broiler farming, switched to rearing Tokbari birds after receiving 500 chicks from the ICAR Tripura centre. With minimal mortality (less than 3%), he was able to sell cockerels weighing 2000–2500 g at 6 months of age for ₹280/kg, generating an income of ₹90,000 within 6 months. Additionally, his flock produced 160–200 eggs daily during peak production, providing a daily profit of ₹1000 by selling eggs at ₹10–12 each. Over 9 months, he

Table 11.3 Economic assessment of Tokbari chicken compared to native birds (unit of 20 birds)

Particulars	Native birds	Tokbari birds
Income (₹)		
Sale of males	3360	5600
Sale of eggs	4800	11,200
Sale of females	2464	3584
Total income (₹)	10,624	20,384
Expenditure (₹)		
Cost of chicks	500	500
Labor and rearing cost (free range system)	–	–
Supplementary feeding, etc.	–	1040
Net profit (₹)	10,124	18,844
Extra revenue, in contrast to native (₹/unit)	–	8720
Extra revenue, in contrast to native (₹/bird)	–	436

Note: The economic analysis is based on a 20-bird unit for standardization purposes. Scaling to 50 birds increases net profit proportionally, as demonstrated by farmer case studies presented in this chapter

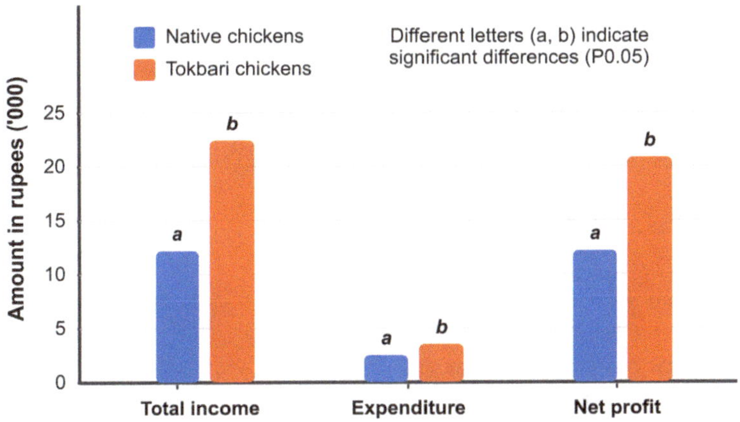

Fig. 11.4 Economic comparison between Tokbari and native chickens showing income, expenditure, and net profit per unit of 20 birds

Fig. 11.5 Successful Tokbari farmers: (left) Shri Safarmia, who earned 1.50 lakhs over 9 months; (right) Shri Tapan Debnath, who achieved 45,600 profit within 3 months

earned a net profit of approximately ₹1.50 lakhs, significantly higher than what he previously earned from broiler chickens.

Similarly, Mr. Tapan Debnath from Rani Bazar village in West Tripura, who was struggling financially with his six-member family, received 350 Tokbari chicks. With minimal mortality (only 10 birds), he sold 3-month-old birds weighing 900–1200 g at ₹290/kg, earning a net profit of ₹45,600 in just 3 months. Such success stories highlight how Tokbari birds can transform rural livelihoods through improved productivity and income generation. The wider acceptance of Tokbari among farmers is evident from the distribution of more than 1.30 lakhs germplasms to farmers across Northeast India till now. This indicates the popularity and growing demand for this variety in the region.

Distribution and Adoption

The widespread acceptance and adoption of Tokbari chickens across the Northeast region reflect their suitability and performance. From 2014 to till now, the Tripura Centre of ICAR has supplied more than 1.30 lakhs Tokbari germplasms to farmers across Northeast India. The distribution of poultry germplasms has been accompanied by extensive extension activities, including training programs, awareness campaigns, demonstration programs, television talks, and health/vaccination camps. These efforts have significantly enhanced farmers' knowledge and skills in poultry management, contributing to the successful adoption and performance of Tokbari birds under field conditions.

Conclusion

Tokbari represents a significant achievement in developing location-specific poultry germplasm tailored to the unique agro-climatic conditions of Northeast India, particularly Tripura. Because of its multicolored plumage, moderate weight, outstanding scavenging habits, and good escape ability, it is perfect for backyard farming in rural areas. Its superior performance metrics—133% higher egg production, 26% larger eggs, and 45–66% higher body weights compared to native birds—translate into substantial economic benefits for rural farmers. The economic analysis demonstrating an additional profit of Rs. 436 per bird over native breeds underscores the potential of Tokbari to transform rural livelihoods. As backyard poultry continues to gain recognition for its role in addressing local requirements for eggs and meat, especially in remote areas, Tokbari offers a sustainable solution that aligns with traditional farming practices while significantly enhancing productivity. Tokbari birds, due to their ability to adapt to local conditions, hardiness, and take-off capacity within low-input systems, are particularly relevant to the socio-economic context of rural Northeast India. The success in developing Tokbari and its subsequent uptake in the region further supports the need for investigative breeding programs that take livestock-based local needs, local preferences, and local conditions into account. Given the spread of this improved variety across the region, the role in improving rural livelihoods, protein malnutrition, and self-reliance in poultry production in Northeast India is likely to grow.

Acknowledgments The senior author appreciates the financial support for the research study by the All India Coordinated Research Project (AICRP) on Poultry Breeding at the ICAR Tripura Centre. Thanks to the directors of the ICAR-Directorate of Poultry Research in Hyderabad, Telangana, India, and the ICAR Research Complex for NEH Region in Umiam, Meghalaya, India, for establishing the facilities used in the study. We also appreciate the farmers' cooperation in taking part in field assessments.

Funding The financial support for the research by the All India Coordinated Research Project (AICRP) on Poultry Breeding.

Potential Conflicts of Interest There is no conflict of interest.

Informed Consent Informed consent was taken from all individual participants included in the study.

Authors Contribution All authors have contributed equally to this work.

References

Azharuddin S, & Dogar MM (2016). Decadal climate variability and predictability, CURRENT SCIENCE 110, no. 8: 1397. http://www.currentscience.ac.in/Volumes/110/08/1397.pdf

Chaiban C, Robinson TP, Fevre EM, Ogola J, Akoko J, Gilbert M, Vanwambeke SO (2020) Early intensification of backyard poultry systems in the tropics: a case study. Animal 14:2387–2396

Chatterjee RN, Rajkumar U (2015) An overview of poultry production in India. Indian J Anim Health 54(2):89–108

Dogar MMA, Shahid A, Fujiwara M, Watanabe S, Ohba M, & Almazroui M (2024). ENSO and NAO Dynamics in the Context of Post-Eruption Winter Cooling Across South Asia and MENA. AGU24.

Haunshi S, Doley S, Shakuntala I (2009) Production performance of the indigenous chicken of the northeastern region and improved varieties developed for backyard farming. Indian J Anim Sci 79:901–905

Khan MKI, Bhuiyan MSA, Khatun MJ, Dey BC (2004) Phenotypic characterization of Hilly chicken of Bangladesh. Prog Agric 15(2):47–52. ISSN 1017-8139

Kumar A, Staal S, Elumalai K, Singh DK (2007) Livestock sector in North-Eastern region of India: an appraisal of performance. Agric Econ Res Rev 20(2):255–272

Malik S, Singh NP (2013) Evaluation of Tripura black native germplasm of poultry at the organised farm conditions. Indian J Anim Sci 83(2):197–200

Rajkumar U, Rama Rao SV, Raju MVLN, Chatterjee RN (2021) Backyard poultry farming for sustained production and enhanced nutritional and livelihood security with special reference to India: a review. Trop Anim Health Prod 53:176

Singh M, Islam R, Avasthe RK (2018a) Factors affecting fertility, hatchability and chick survivability of vanaraja birds under intensive rearing in sub-temperate condition. Indian J Anim Sci 88:331–334

Singh M, Mollier RT, Rajesha G, Ngullie AM, Rajkhowa DJ, Rajkumar U, Paswan C, Chatterjee RN (2018b) Backyard poultry with vanaraja and srinidhi: proven technology for doubling the tribal farmers' income in Nagaland. Indian Farm 68(1):80–82

Chapter 12
Climate Change Impacts and Adaptive Capacity in the Thar Desert: Challenges and Pathways for Indigenous Communities

Arghya Saha and Lakshman Chandra Pal

Abstract Climate change has adverse effects on the natural resource base and poses difficulties for the natural resource-dependent indigenous population of the desert region in Rajasthan. The Thar desert is also known for its ancient culture system, and tradition reflects time-tested techno-scientific knowledge of sustainability. Due to climate change, this fragile ecosystem is facing increasing stress from the interplay of climate and eco-hydrological processes coupled with increasing anthropogenic pressure. These events are leading to an impact on the indigenous agro-pastoral communities and their livelihoods. The purpose of this article is to analyze the outcomes of climate change on communities to quantify the level of environmental threat and adaptive capacity. The study is based upon secondary data collection and scientific analysis with a blend of participatory approach by previous governmental and non-governmental projects. Criterion variables were used to identify the vulnerability hotspots, while component indices were used to depict key characteristic features of the study unit. The aim of the study is to determine the focus areas for interventions and community-based natural resource management to address water stress and ecosystem vulnerability with appropriate technology. The research identifies target areas and methodologies for intervention, and the initiative strives to ensure management and access to resources for the disadvantaged desert communities of Rajasthan.

Introduction

The Thar Desert, also known as the Great Indian Desert, is an expansive arid region situated across northwestern India, primarily in Rajasthan, and extending into southeastern Pakistan. As the world's ninth-largest hot subtropical desert, it is remarkably the most populated desert globally, supporting a population exceeding 16 million residents (Agarwal and Narain 1997). The region is characterized by an

A. Saha (✉) · L. C. Pal
Department of Geography, Bidhan Chandra College, Hooghly, West Bengal, India

S. Mandal et al. (eds.), *Climate Change Vulnerability*, Climate Change Management, https://doi.org/10.1007/978-3-032-16783-5_12

arid climate having extremely high temperatures that can reach between 49 and 51 °C, coupled with low and unpredictable rainfall, averaging between 25 and 500 mm annually, predominantly during the monsoon months from June to September (Borras and Franco 2012). Frequent droughts are the resultant climatic feature of this harsh environment. Despite its challenging climate, the Thar Desert sustains a rich biodiversity, including drought-resistant flora and endemic species, which are critical for the livelihoods of millions of people who depend on its delicate natural resources. Over centuries, the indigenous agro-pastoral communities of the Thar have developed ancient cultural systems and traditions that embodied time-tested techno-scientific knowledge of sustainability, allowing them to adapt and thrive in this arid landscape. However, this fragile ecosystem is experiencing increasing stress due to the complex interplay of climatic and eco-hydrological processes, exacerbated by growing anthropogenic pressure. Climate change has intensified into extreme weather events, while human activities such as overgrazing, deforestation, unsustainable farming practices, and excessive groundwater extraction are the main reasons behind land degradation and desertification. A notable and seemingly contradictory development in the Thar Desert is the observed "greening" phenomenon (Dhaka and Poonia 2016). Over the past two decades, from 2001 to 2023, the desert has become, on average, 38% greener, with a noticeable increase in vegetation visible in satellite images (Down To Earth 2024). This observed greening might be an improvement in the desert's ecological health or an increase in its natural resilience. However, a closer examination of the underlying drivers reveals a more complex picture. This greening is primarily attributed to a substantial 64% increase in monsoon precipitation, a direct consequence of climate variability and secondarily because of the expansion of irrigation infrastructure and increased groundwater pumping (Encyclopædia Britannica n.d.). This phenomenon led to the expansion of agricultural and urban areas and a considerable increase in crop yields, potentially enhancing short-term food security and poses a significant challenge (Tahir et al. 2025). The intensification of agriculture and urbanization is threatened the traditional agro-pastoralism, indigenous biodiversity and the regional energy balance. The reliance on groundwater pumping, particularly for sustaining greening during non-monsoon seasons raising serious concerns. This suggests that the observed greening is not solely a sign of enhanced natural resilience but rather a human-modified landscape, driven by a combination of erratic climate patterns and potentially unsustainable practices and creating new forms of vulnerability for traditional systems. The purpose of this article is to analyze the outcomes of climate change on the indigenous agro-pastoral communities of the Thar Desert, quantify the level of environmental threat they face, and assess their adaptive capacity. The study aims to identify focus areas for interventions and community-based natural resource management (CBNRM) to effectively address water stress and ecosystem vulnerability through appropriate technology. This analysis is based on secondary data collection and scientific analysis, integrating a participatory approach informed by previous governmental and non-governmental projects. Criterion variables and component indices are utilized to identify vulnerability hotspots and depict key characteristic features of the study unit.

Climate Change Impacts and Environmental Degradation in the Thar Desert

The Thar Desert's delicate balance is increasingly disrupted by pronounced climatic shifts and escalating environmental degradation, impacting its natural resources and the communities dependent on them.

Observed Climatic Shifts and Extreme Events

The Thar Desert is characterized by an arid climate, featuring high temperatures, low and unpredictable rainfall, and frequent droughts. Climate change has exacerbated these conditions, leading to an increased frequency and intensity of extreme weather events, including prolonged droughts and severe heat waves (Dogar et al. 2017a 2017b, 2019, 2023, 2024, 2025, Dogar 2020; Dogar and Almazroui 2022; Dogar and Sato 2018, 2019). Rajasthan, the state encompassing a significant portion of the Thar Desert, has the highest probability of drought occurrence in India. Historical records for Western Rajasthan indicate a high frequency of dry spells (Fairhead et al. 2012; Gebrehiwot and van der Veen 2015; Government of Rajasthan 2018). However, draught periods are occasionally interrupted by heavy downpours that result in flash floods (IPCC 2014a). This pattern reveals a critical challenge that is not simply leading to a reduction in rainfall, but rather to more erratic and intense precipitation patterns. Increased rainfall intensity, annual averages do not drastically change, but lead to rapid surface runoff and soil erosion rather than effective groundwater recharge. This paradoxically intensifies water scarcity in the long run by reducing infiltration and increasing surface runoff, making water management more complex and challenging for agro-pastoral communities.

Desertification and Natural Resource Depletion

Desertification in the Thar Desert is a complex and multifaceted problem, driven by a combination of natural factors and human activities.

Natural Factors

Climate change intensifies the region's susceptibility to extreme temperatures, erratic rainfall, and prolonged droughts, affecting the local resource base and disrupting ecological balance. The sandy, loosely bound soil characteristic of the Thar Desert is highly prone to wind erosion, particularly during dry months. This erosion strips away the fertile top layer of soil, depleting vital nutrients and reducing the soil's capacity to support vegetation. Furthermore, natural water scarcity, a consequence of low rainfall and high evaporation rates, is compounded by limited and steadily depleting groundwater levels critically restrict agricultural productivity and foster desertification (Jodha 2012a, b).

Human-Induced Drivers

Human activities have a significant role in accelerating desertification. Overgrazing of livestock leads to the overuse of pastures, destroying essential grasses and ground cover that protect the soil. This leaves the soil exposed to wind and water erosion, diminishing its fertility and regenerative capacity, thereby perpetuating a cycle of degradation. Deforestation resulting from the cutting of trees and shrubs for fuel, building materials, and agricultural expansion removes vital vegetation that stabilizes the soil and helps retain moisture by providing shade. The loss of these plants increases soil erosion and reduces biodiversity, making the land more vulnerable to desertification. Unsustainable farming practices, including monocropping, excessive use of chemical fertilizers and improper irrigation methods, contribute to soil nutrient depletion and increased salinity. The shift towards water-intensive crops further strains the already limited water resources and accelerates soil degradation. Extensive groundwater extraction, often through tube wells, for irrigation purposes has led to a steady decline in groundwater levels and increased soil salinization, rendering vast tracts of land infertile. Notably, nearly 80% of Rajasthan's groundwater resources are experiencing depletion (Morton 2007; Kumar and Singh 2020, 2023a, b). Socio-economic pressures, such as rapid population growth, exacerbate these effects by increasing the demand for land, water and other resources. Poverty and lack of alternative livelihoods often compel communities to rely on agriculture despite the environmental damage and limit the adoption of sustainable practices. This degradation also creates a feedback loop with degraded soils release stored carbon into the atmosphere, increasing greenhouse gas concentrations and contributing to global warming, which in turn exacerbates arid conditions. Simultaneously, the loss of vegetation reduces the region's capacity to sequester carbon, amplifying climate change effects and making desert regions hotter and drier.

The "Greening" Phenomenon: Complexities and Implications

The Thar Desert has experienced a 38% increase in greenness over the past two decades (2001–2023), primarily due to a 64% rise in monsoon precipitation and the expansion of irrigation infrastructure that brings groundwater to the surface (Pathak and Joshi 2019). This greening has facilitated the expansion of agricultural and urban areas and leading to considerable increases in crop yields. This agricultural intensification and urbanization resulting food security and economic improvement but it also introduces new vulnerabilities. The greening is improving the socio-economic scenario and contributing to food security but also poses a significant threat to traditional agro-pastoralism, indigenous biodiversity and the regional energy balance. This resulted a fundamental and potentially disruptive shift in the ecosystem and the established livelihood strategies. Traditional agro-pastoralists have a way of life deeply intertwined with natural pastures and nomadic practices, which are being disrupted by intensive agriculture, losing traditional practices and access to vital resources. The increasing population density in the Thar Desert put pressure on existing resources and infrastructure, creating additional forms of vulnerability that extend beyond the conventional impacts of climate change.

Human Rights Implications of Renewable Energy Development

India's ambitious energy transition agenda is aiming at mitigating the impacts of climate change through the development of renewable energy, has unintentionally led to significant human rights concerns, particularly in the Thar Desert. A prominent case study involves the 'Oran Land' in the Thar Desert, which has been directly affected by these energy transition initiatives. Orans are sacred groves and community forest lands, integral to the Thar Desert region. They are rich in biodiversity, serving as habitats for endangered species and water sources in the arid environment. These lands have been traditionally managed and protected by indigenous tribal communities, who practice ancient customs of biodiversity conservation (IPCC 2014b; Schlosberg and Collins 2014).

However, private renewable power companies (PRPCs) are acquiring these Orans through government allotment schemes without conducting effective impact assessments, ensuring social inclusion or engaging in meaningful stakeholder consultations (Sharma and Gupta 2021a, b). This practice has resulted in severe violations, including restricting indigenous communities from accessing their traditional Orans and the widespread cutting of trees and vegetation to make space for solar panels, power transmission lines, and ancillary infrastructure. These actions have ignited significant conflict and led to legal challenges. Over 40 indigenous villages have engaged in protests against the arbitrary allotment of Oran Lands and are demanding the protection of their community rights. These protests have garnered the attention of the High Court of Rajasthan and led to the cancellation of some land allotments. The Supreme Court of India has also intervened, addressing concerns regarding the rising deaths of the critically endangered Great Indian Bustard due to collisions with power transmission lines from renewable energy projects and has ordered that transmission lines be laid underground where feasible (Singh and Sharma 2020; Singh and Kumar 2021a; Sharma 2022).

This situation shows how climate change mitigation actions can create localized human rights and environmental injustices. The pursuit of national net-zero emission targets is overriding local environmental and social safeguards. This is called as a "green grab," where traditional resource bases are sacrificed for a perceived larger environmental good and highlighting a critical governance gap where the urgency of climate action bypasses due diligence, consultation and fundamental human rights considerations (Singh and Sharma 2022; Singh and Patel 2023; Singh and Mehta 2023). This undermines the local communities who are often most vulnerable to climate change impacts, indigenous land rights, and environmental justice in climate change mitigation strategies that ensure the solutions to global problems by not creating new injustices at the local level.

Vulnerability Assessment of Indigenous Agro-Pastoral Communities

The understanding of the vulnerability of indigenous agro-pastoral communities in the Thar Desert is crucial for developing targeted and effective adaptation strategies.

Conceptual Framework of Vulnerability

Vulnerability is a widely used concept in climate change studies, generally defined as a function of three core components: exposure, sensitivity, and adaptive capacity (Singh and Kumar 2021b). Exposure refers to the nature and degree to which a system is subjected to climatic variations. Sensitivity describes the degree to which a system is affected, either positively or negatively, by climate variability or change. Adaptive capacity denotes the ability of a system to adjust to climate change, including climate variability and extremes, to moderate potential damages, to take advantage of opportunities, or to cope with the consequences. The state of Rajasthan is identified as an area of "greatest climate sensitivity, maximum vulnerability and lowest adaptive capacity" (Singh and Nair 2018a, b; Singh and Joshi 2020). The vulnerability of arid regions like the Thar Desert is further compounded by low levels of socio-economic development, which limit the resources and options available for adaptation.

Quantifying Vulnerability Hotspots

District-level climate vulnerability assessments in Rajasthan provide a granular understanding of where and how communities are most at risk. These assessments typically employ a comprehensive methodology, integrating over 30 indicators, which are segregated into four key components: exposure, sensitivity, adaptive capacity, and crop production loss. These indicators are then combined to develop a composite vulnerability index. The analysis reveals significant inter-district variations across these calculated indices. The detailed breakdown of vulnerability across districts highlights that a uniform, one-size-fits-all adaptation strategy would be

Table 12.1 Key vulnerability indicators and status in Thar Desert districts

District	Degree of exposure index	Degree of sensitivity index	Degree of adaptive capacity index	Crop production loss index	Degree of potential vulnerability index (composite)
Ganganagar	Medium	High	Low	Highest	High
Hanumangarh	Medium	High	Low	Highest	High
Bharatpur	Medium	High	Low	Highest	High
Jaisalmer	High	Maximum	Lowest	High	High
Bundi	Maximum	Medium	Low	High	High
Jodhpur	Medium	Medium	Low	High	High
Bikaner	Medium	Medium	Low	High	High
Chittorgarh	Medium	Medium	Low	High	High
Alwar	Medium	Medium	Low	High	High
Baran	Medium	Medium	Low	High	High
Pratapgarh	High	High	Lowest	High	High
Pali	Least	Low	Medium	Medium	Low

Source: Singh and Nair (2018a, b)

ineffective. The varying degrees of exposure sensitivity and adaptive capacity across different regions necessitate tailored interventions. For instance, areas with low adaptive capacity require focused capacity-building programs, while regions with high exposure might benefit more from early warning systems or protective infrastructure. This calls for a nuanced, spatially differentiated policy response rather than broad-brush approaches to effectively moderate vulnerability (Thornton and Herrero 2010; Supreme Court of India 2024).

As indicated in Table 12.1, districts such as Ganganagar, Hanumangarh, and Bharatpur experience the highest production losses, while Jaisalmer records the maximum sensitivity level, and Pratapgarh, Jaisalmer, and Banswara exhibit the lowest adaptive capacity. Conversely, Sirohi district is identified as the least vulnerable due to lower exposure, sensitivity, crop production loss, and higher adaptive capacity. Overall, Hanumangarh, Jaisalmer, Ganganagar, Bundi, Bharatpur, Jodhpur, Bikaner, Chittorgarh, Alwar, Baran, and Pratapgarh are identified as districts with a high level of vulnerability to climate change.

Impacts on Livelihoods and Well-Being

Climate change mainly affects agriculture-based livelihoods, particularly in semi-arid tropical regions that are already disadvantaged by erratic precipitation patterns, lower soil fertility, and frequent droughts. The primary outcomes of climate change impacts on agro-pastoral livelihoods, as observed in relevant studies, include significant losses (Singh and Gupta 2019). For instance, studies indicate that 97% of surveyed pastoralists and agro-pastoralists reported loss of livestock, 91% faced food shortages, 84% experienced decreased income, 83% suffered from feed scarcity, 80% from water scarcity, and 79% from crop failure (Singh and Singh 2018; Yadav and Soni 2019). These direct impacts inevitably lead to widespread food insecurity, economic instability, and an increased reliance on external support for survival.

Adaptive Capacity and the Role of Traditional Knowledge Systems

The indigenous communities of the Thar Desert possess a rich heritage of adaptive strategies, rooted in ancient cultural systems and traditions that reflect time-tested techno-scientific knowledge of sustainability.

Traditional Adaptive Strategies

Traditional water harvesting systems are central to survival in this arid environment. *Tobas*, small natural ponds have historically served as the only source of surface water for both animals and humans. Pastoralists traditionally move their herds to new locations when these tobas dry up. The *Khadeen system* represents an ancient and ingenious technology that leverages peculiar geohydrologic formations to create temporary lakes (Agarwal and Narain 1997; Singh and Yadav 2010). This practice effectively captures seasonal runoff for agriculture by ensuring impressive crop yields even in water-scarce environments and has been vital in supporting agricultural activities since the sixteenth century. Other traditional water management practices include *tankas*, *bawris*, and *johads* (Singh et al. 2023; La Vena and Ram 2024).

Beyond water management, communities rely on drought-hardy animal breeds, such as Tharparkar cows and Marwari goats, which are well-adapted to the harsh conditions. Traditional farming systems include practices like *Birani badi* for cultivating watermelons and cucurbits with limited water, and the use of traditional structures like *Kinara*, *Kothi*, and *Kiradi* for storing food grains and seeds. Diversified agriculture, often involving mixed cropping in the *kharif* season, and the nomadic migration of pastoralists during periods of fodder and water scarcity are also crucial adaptive strategies. Agroforestry, particularly involving species like *Prosopis cineraria* (Khejri), has traditionally served as an important life support system during droughts.

Cultural and Religious Foundations of Conservation

The Bishnoi community of the Thar Desert provides a compelling example of how deeply rooted religious beliefs and practices can play a pivotal role in ecological conservation and climate change adaptation. Their core philosophy, "jivo and jinedo" (to live and let live), is enshrined in 29 principles (Unatis niyam) that guide their lives, focusing on the conservation of wildlife and natural resources (Singh 2020).

Specific principles directly relate to environmental protection:

- "Jeev daya palani" (compassion for all living beings) extends to both domesticated and wild animals, including endangered species like the blackbuck and Great Indian Bustard.
- "Rukh leela nahi dhave" (do not cut green patches, trees, or plants) encourages the planting and protection of indigenous trees like Khejri and Rohida, vital for stabilizing the desert ecosystem.
- Other principles advocate for common shelters for abandoned animals, non-castration of bulls, and abstinence from meat, all contributing to a holistic conservation ethic.

Their conservation practices are tangible: Bishnois actively protect wild animals by providing water and grains, reserving portions of crops for wildlife, and enforcing customary laws with significant fines for poachers. They conserve trees by using cow dung cakes for fuel instead of wood and burying their dead instead of cremating them. The Khejri tree, revered for its resilience and ecological benefits, is particularly valued. They practice sustainable rain-fed agriculture, using natural fertilizers and avoiding chemical pesticides. The Bishnois possess extensive Traditional Ecological Knowledge (TEK), observing natural elements and animal behavior to predict climatic conditions, which guides their occupational activities and festival celebrations. Environmental stewardship is deeply integrated into the upbringing of younger generations through socialization and rituals, ensuring a lasting culture of environmental stewardship (Gadgil and Berkes 1991). Community organizations like the Akhil Bhartiya Bishnoi Maha Sabha actively work to protect wildlife and promote these values. The historical Khejarli massacre in 1730 A.D., where 363 Bishnois sacrificed their lives to protect Khejri trees, stands as a powerful testament to their commitment and inspired the world-renowned Chipko movement (Guha 2000).

While these traditional knowledge systems, such as the Khadeen system and Bishnoi practices, are powerful foundations for resilience but they are not static solutions. The Khadeen system faces contemporary threats from mining encroachment, labor outmigration, and road development. Moreover, the prevailing circumstances in Rajasthan suggest that "incremental adaptations in water management have only a limited role" and that "transformational adaptation is the need of the day" (Sharma and Gupta 2018). This indicates that while traditional knowledge is an invaluable foundation for resilience, it is often insufficient on its own to cope with the accelerated and novel impacts of climate change and modern anthropogenic pressures. The challenge lies in integrating and adapting traditional knowledge with modern scientific approaches and technologies, rather than solely relying on or discarding it. External pressures threaten the very existence and efficacy of these systems, highlighting the need for legal and policy protection.

Community-Level Coping Mechanisms

Beyond formal traditional systems, communities employ various coping mechanisms to manage climate risks. These include strategic livestock management and leveraging social capital through support from relatives and friends or borrowing. Storing hay and crop residue, destocking animals during severe droughts, and employing water harvesting techniques are also common. Furthermore, communities engage in savings and seek non-farm employment to diversify income sources and enhance resilience.

Challenges to Traditional Adaptation

Despite their inherent strengths, traditional adaptive strategies face significant challenges in the face of modern pressures. External factors such as mining encroachment, labor outmigration, and road development directly threaten the viability of traditional systems like the Khadeen. The expansion of the Thar Desert, driven by both human activities and climate change, continuously threatens the delicate desert ecology. Paradoxically, the "greening" phenomenon, largely fueled by intensive irrigation, can pose a "great threat to the traditional agro-pastoralism" by converting traditional grazing lands and disrupting established nomadic routes and practices (Sharma et al. 2021).

Strategic Interventions for Resilience Building and Sustainable Resource Management

Addressing the complex challenges in the Thar Desert requires a multi-pronged approach that integrates governmental policies, non-governmental initiatives, appropriate technologies, and community-based natural resource management.

Review of Governmental Initiatives

The Government of Rajasthan, in collaboration with organizations like TERI and GIZ, has developed the **Rajasthan Action Plan on Climate Change (RAPCC)**. This comprehensive plan aims to achieve sustainable development by reducing vulnerability to climate change impacts and enhancing the resilience of ecological, economic, and social systems across the state. The RAPCC aligns with national priorities outlined in the National Action Plan on Climate Change (NAPCC) and existing state policies such as the Rajasthan Environment Policy, 2010 (Singh and Kumar 2022). Key strategies within the RAPCC for the water resources sector include comprehensive groundwater management (assessment, artificial recharge, regulation of borewells, and public awareness), enhancing drought preparedness through integrated monitoring and early warning systems, promoting water conservation (e.g., rooftop rainwater harvesting and wastewater recycling), improving water use efficiency (e.g., drip/sprinkler irrigation and water audits), and developing a comprehensive water database for climate change impact assessment. Another significant state program is the **Mukhya Mantri Jal Swavlamban Abhiyan (MJSA)**, launched to achieve water self-reliance through converging activities for water conservation and storage, aiming for transformational adaptation rather than incremental changes (Government of Rajasthan 2012).

Non-Governmental Initiatives

Organizations like the **Gramin Vikas Vigyan Samiti (GRAVIS)** are actively engaged in integrated community development and drought mitigation in the Thar Desert. A notable project implemented by GRAVIS is "Enhancing Women and Girls' Leadership in Drought Mitigation and Climate Change Adaptation" (EWGL). This project's interventions included the formation and strengthening of Intergenerational Learning Groups (ILGs) and Self-Help Groups (SHGs), construction and renovation of rainwater harvesting structures (such as *taankas*, village ponds, and percolation wells), construction of farm bunds (*khadins*), establishment of Arid Horticulture Units (AHUs), pasture land development, and Community Seed Banks (CSBs). The EWGL project targeted 20 villages across four severely drought-impacted districts (Barmer, Bikaner, Jaisalmer, and Jodhpur), reaching approximately 50,000 people, with a specific focus on empowering women and girls. The outcomes were significant: enhanced women's leadership and empowerment, improved water security (with *taankas* providing water for 4–8 months and saving families INR 1500–2000 per month on water expenses), enhanced food and nutrition security (increased crop produce and AHUs saving INR 1000–12,000 per month on food expenses), improved standard of living and health, and increased awareness and technical capacities for climate change adaptation. The success of the EWGL project demonstrates that effective climate change adaptation and natural resource management in the Thar Desert is not merely about technology or policy, but fundamentally about social empowerment and participatory governance. By placing women and girls in leadership roles, the project not only improved water and food security but also addressed deeply rooted gender inequalities, leading to broader community transformation, such as the near elimination of child marriage and increased confidence among women. This highlights that gender-responsive, community-led approaches are not just beneficial but are critical enablers for sustainable and equitable resilience building in vulnerable regions. Another organization, **RAJMERU**, is also engaged in promoting the sustainable use of natural resources, soil and water conservation, and climate change adaptation (RAJMERU 2023) (Table 12.2).

Table 12.2 Key interventions and outcomes from governmental and NGO projects

Project/ initiative name	Type	Key interventions/ strategies	Target areas/ communities	Observed/expected outcomes
Rajasthan Action Plan on Climate Change (RAPCC)	Governmental	Groundwater Management, Drought Preparedness, Water Conservation, Water Use Efficiency, Comprehensive Water Database	State of Rajasthan	Reduced vulnerability, enhanced resilience of ecological, economic, and social systems
Mukhya Mantri Jal Swavlamban Abhiyan (MJSA)	Governmental	Water conservation and storage (under and above ground)	Rural populations across Rajasthan	Water self-reliance, transformational adaptation, and reduced vulnerability of farm production and livelihoods
GRAVIS EWGL Project	NGO (Community-led)	Rainwater harvesting (taankas, ponds, wells), farm bunds (khadins), Arid Horticulture Units, Pasture land development, Community Seed Banks, Women/Girls' Leadership training	20 villages in Barmer, Bikaner, Jaisalmer, Jodhpur (approx. 50,000 people, focus on women/ girls)	Enhanced women's leadership and empowerment, improved water and food security, increased income, improved health and living standards, increased CCA awareness and capacity
Bishnoi Community conservation	Community-led (Traditional)	Wildlife protection, tree/plant conservation (Khejri), sustainable rain-fed agriculture, Traditional Ecological Knowledge (TEK), inter-generational education, community organizations	Bishnoi villages in Thar Desert	Sustained desert ecosystem, biodiversity conservation, cultural preservation, climate change adaptation

Source: Government of Rajasthan (2012)

Addressing Water Stress and Ecosystem Vulnerability

Strategic interventions are crucial to address the pervasive water stress and ecosystem vulnerability in the Thar Desert. Focus areas for intervention, as outlined in the RAPCC, include comprehensive groundwater management through assessment, GIS-based aquifer studies, artificial recharge, regulation of borewells, mandatory metering for bulk consumers, and mass awareness campaigns. Enhancing drought preparedness involves mainstreaming traditional water storage structures, implementing in-situ water and soil moisture conservation practices, and developing integrated drought monitoring and early warning systems. Water conservation efforts must include mass construction of rooftop rainwater harvesting (RWH) structures, revival of traditional structures, and promoting wastewater recycling and reuse across all sectors. Improving water use efficiency is also paramount through promoting drip and sprinkler irrigation, conducting water audits, and rationalizing water pricing.

Appropriate Technologies for Climate Change Adaptation

The strategic deployment of appropriate technologies is vital for climate change adaptation in the Thar Desert. For water management, efficient water management techniques, sustainable groundwater management, conjunctive use of canal water with non-saline groundwater, and the construction of 'diggis' (surface tanks) are recommended. To enhance drought resistance, the development and promotion of drought-resistant crops and short-duration crop varieties are essential. For heat stress adaptation, renewable energy, particularly solar pumping, holds immense potential, given the Thar Desert's suitability for solar energy capture (21–25 MJ/m^2/day with over 300 sunny days annually). Rajasthan has seen a significant increase in its renewable energy installed capacity, with solar pumps already in operation. Additionally, low-cost, protective cultivation in greenhouses is recommended for vegetable crops. While technologies like solar pumping, drip irrigation, and greenhouses are crucial, their implementation requires careful consideration. Similarly, the "greening" driven by extensive irrigation, while increasing agricultural output, also poses a threat to traditional livelihoods. This suggests that "appropriate technology" is not solely about the technology itself, but its integration within a broader framework of sustainable management, robust policy, and effective community governance. Technology must be accompanied by strong regulatory frameworks, capacity building for communities, and a holistic understanding of its socio-ecological impacts to avoid unintended negative consequences and ensure equitable access and benefit (Kumar and Singh 2023a, b).

Community-Based Natural Resource Management (CBNRM)

Community-Based Natural Resource Management (CBNRM) is a cornerstone for building resilience. Methodologies for intervention should prioritize participatory approaches, actively engaging communities in water management and conservation efforts. Strengthening community institutions, such as the formation and empowerment of Inter-generational Learning Groups (ILGs), Self-Help Groups (SHGs), and Village Development Committees (VDCs), is critical. Empowering women and girls to play a central role in implementing initiatives for water and food security can lead to significant leadership in climate action. The integration and mainstreaming of traditional knowledge, including the renovation of traditional water storage structures, is also vital. Furthermore, comprehensive capacity building and training programs for communities on sustainable agriculture, water-efficient practices, and the maintenance of rainwater harvesting structures are essential for long-term success (Sharma and Gupta 2021a, b).

Target Areas and Resource Access

Interventions must strategically prioritize vulnerable areas to address regional imbalances through need- and location-based approaches. This includes improving access to water and pasture, establishing easily accessible market channels, and developing robust early warning systems to assist communities in preparing for and building resilience against climatic shocks.

Recommendations for Policy, Practice, and Future Research

To foster a more resilient and sustainable future for the Thar Desert and its indigenous communities, the following recommendations are put forth:

Integrated Policy Frameworks

1. **Harmonize Development and Climate Action**: Develop integrated policy frameworks that explicitly harmonize climate change adaptation with broader development goals, with a particular focus on water resources, agriculture, biodiversity conservation, and human health.
2. **Data-Driven Planning**: Develop a high-resolution Climate Risk Atlas (CRA) to map critical vulnerabilities at the district level, enabling precise identification, assessment, and projection of chronic and acute risks such as extreme climate events, heat and water stress, crop loss, and biodiversity collapse.

3. **Centralized Coordination**: Establish a centralized climate-risk commission to coordinate environmental de-risking missions across various sectors and governmental departments.
4. **Climate-Resilient Infrastructure**: Integrate climate risk profiling into all infrastructure planning processes to enhance adaptive capacity and ensure long-term resilience.
5. **Transformational Adaptation**: Shift from incremental adaptations to implementing "transformational adaptation" measures that address the root causes of vulnerability and fundamentally alter unsustainable practices, recognizing that existing systems are overwhelmed by current and projected challenges.

Strengthening Community Resilience

1. **Empower Local Governance**: Empower local communities through genuine participatory approaches and strengthen local governance structures, ensuring their active involvement in decision-making processes related to resource management and climate action.
2. **Integrate Traditional Knowledge**: Actively integrate and support traditional knowledge systems and practices, such as the Khadeen system and Bishnoi conservation ethics, while simultaneously addressing modern threats to their efficacy and legal protection.
3. **Diversify Livelihoods**: Foster sustainable livelihoods by promoting income diversification strategies and supporting the adoption of drought-resistant farming practices tailored to the arid environment.
4. **Gender-Responsive Interventions**: Prioritize gender-responsive interventions, explicitly recognizing and leveraging the crucial role of women and girls as leaders and agents of change in climate action and natural resource management.

Scaling Appropriate Technologies

1. **Sustainable Water Technologies**: Promote and scale up efficient water management techniques, including drip irrigation, conjunctive use of water resources, and solar pumping, but critically, implement these with robust governance mechanisms to prevent groundwater overuse and ensure long-term sustainability.
2. **Climate-Resilient Agriculture**: Invest in research and development of new drought-resistant crops and heat-stress adaptation strategies that are specifically suitable for the unique conditions of the Thar Desert.
3. **Rainwater Harvesting**: Support the widespread adoption and maintenance of rainwater harvesting structures at both household and community levels, including traditional systems and modern adaptations.

Addressing Human-Environment Conflicts

1. **Rigorous Impact Assessments**: Implement strict environmental and social impact assessments for all development projects, especially renewable energy initiatives, ensuring free, prior, and informed consent of indigenous communities before project initiation.
2. **Protect Traditional Lands**: Legally protect traditional community lands, such as Orans and sacred groves, from arbitrary acquisition for industrial or developmental purposes.
3. **Effective Grievance Mechanisms**: Establish effective and accessible grievance redressal mechanisms and ensure robust legal recourse for communities whose rights or resources are adversely affected by development projects.

Research and Monitoring Gaps

1. **Localized Climate Modeling**: Conduct more localized and high-resolution climate change modeling for different climate zones within Rajasthan to provide more precise projections for regional planning.
2. **Geospatial Tools**: Utilize advanced remote sensing and geospatial tools to develop optimal and spatially differentiated water resource management strategies across various zones of the Thar Desert.
3. **Greening Phenomenon Analysis**: Continuously monitor the "greening" phenomenon to comprehensively understand its long-term hydrological and ecological implications, critically distinguishing between drivers that contribute to sustainable resilience and those that mask underlying vulnerabilities or create new forms of stress.
4. **Socio-Economic Impact Research**: Conduct further in-depth research into the socio-economic impacts of the "greening" on traditional agro-pastoral communities and indigenous biodiversity, particularly focusing on livelihood displacement and cultural erosion.

Conclusion

The Thar Desert and its indigenous agro-pastoral communities are at the nexus of complex, interconnected challenges stemming from climate change and escalating anthropogenic pressures. While the region exhibits a unique "greening" phenomenon driven by both increased monsoon rainfall and extensive groundwater irrigation. This greening simultaneously threatens traditional agro-pastoralism and indigenous biodiversity, covering underlying vulnerabilities and fundamentally altering the socio-ecological landscape. The analysis underscores that the Thar Desert is a region of high climate sensitivity, maximum vulnerability, and low adaptive capacity in many areas. The impacts manifest as severe water stress, accelerated desertification, and profound disruptions to livelihoods, leading to cascading health and socio-economic crises. However, the region also possesses a rich reservoir of traditional knowledge and community-based adaptive strategies, exemplified by the Bishnoi community's deep-rooted conservation ethics and the historical efficacy of systems like the Khadeen. The path forward necessitates an integrated, multi-scalar, and participatory approach. This involves harmonizing governmental policies, such as the Rajasthan Action Plan on Climate Change, with the on-the-ground efforts of non-governmental organizations like GRAVIS, which have demonstrated the transformative power of gender-responsive and community-led initiatives. The strategic deployment of appropriate technologies, including efficient water management systems and renewable energy solutions, must be carefully governed to prevent unintended consequences like groundwater over-extraction. Crucially, any intervention must respect and integrate the profound traditional knowledge of the desert communities, while simultaneously addressing the contemporary threats that challenge the efficacy and very existence of these time-tested practices. Ensuring environmental justice is paramount to prevent climate mitigation efforts from creating new localized injustices. By blending scientific advancements with indigenous wisdom and by prioritizing the resilience and rights of the most disadvantaged, and sustainable future can be forged for this fragile and vital socio-ecological system.

References

Agarwal A, Narain S (1997) Dying wisdom: rise, fall and potential of India's traditional water harvesting systems. Centre for Science and Environment

Borras SM, Franco JC (2012) Global land grabbing and trajectories of agrarian change: a preliminary analysis. J Agrar Chang 12(1):34–59. https://doi.org/10.1111/j.1471-0366.2011.00339.x

Dhaka BL, Poonia SK (2016) Indigenous livestock breeds and traditional farming systems in arid Rajasthan: adaptations to climate variability. Indian J Anim Sci 86(2):137–144

Dogar MM, Kucharski F, & Azharuddin S (2017a). Study of the global and regional climatic impacts of ENSO magnitude using SPEEDY AGCM. Journal of Earth System Science, 126(2), 30. https://doi.org/10.1007/s12040-017-0804-4

Dogar MM, Stenchikov G, Osipov S, Wyman B, & Zhao M (2017b). Sensitivity of the regional climate in the Middle East and North Africa to volcanic perturbations. Journal of Geophysical Research: Atmospheres, 122(15), 7922–7948. https://doi.org/10.1002/2017JD026783

Dogar MM, Kucharski F, Sato T, Mehmood S, Ali S, Gong Z, ... & Arraut J (2019). Towards understanding the global and regional climatic impacts of Modoki magnitude. Global and Planetary Change, 172, 223–241. https://doi.org/10.1016/j.gloplacha.2018.10.004

Dogar MM, Hermanson L, Scaife AA, Visioni D, Zhao M, Hoteit I, ... & Fujiwara M (2023). A review of El Niño Southern Oscillation linkage to strong volcanic eruptions and post-volcanic winter warming. Earth Systems and Environment, 7(1), 15–42. https://doi.org/10.1007/s41748-022-00331-z

Dogar MM, Fujiwara M, Zhao M, Ohba M, & Kosaka Y (2024). ENSO and NAO linkage to strong volcanism and associated post-volcanic high-latitude winter warming. Geophysical Research Letters, 51(1), e2023GL106114. https://doi.org/10.1029/2023GL106114

Dogar MM, Fujiwara M, Ohba M, & Almazroui M (2025). Investigating post-eruption amplified winter cooling in the Middle East and North Africa—unraveling ENSO and

NAO dynamics. Environmental Research Communications, 7(4), 045012. https://doi.org/10.1088/2515-7620/adc675
Dogar MMA (2020). Study of the regional climatic impacts of tropical explosive volcanism in the Middle East and North Africa region (Doctoral dissertation, Hokkaido University). https://doi.org/10.14943/doctoral.r7091
Dogar MMA, & Almazroui M (2022). Revisiting the strong and weak ENSO teleconnection impacts using a high-resolution atmospheric model. Atmospheric Environment, 270, 118866. https://doi.org/10.1016/j.atmosenv.2021.118866
Dogar MMA, & Sato T (2019). Regional climate response of Middle Eastern, African, and South Asian monsoon regions to explosive volcanism and ENSO forcing. Journal of Geophysical Research: Atmospheres, 124(14), 7580–7598. https://doi.org/10.1029/2019JD030358
Dogar MM, & Sato T (2018). Analysis of climate trends and leading modes of climate variability for MENA region. Journal of Geophysical Research: Atmospheres, 123(23), 13–074. https://doi.org/10.1029/2018JD029003
Down To Earth (2024) Thar Desert is greening thanks to higher rainfall, excessive groundwater pumping. https://www.downtoearth.org.in/climate-change/thar-desert-is-greening-thanks-to-higher-rainfall-excessive-groundwater-pumping
Encyclopædia Britannica (n.d.) Thar Desert. Encyclopædia Britannica. https://www.britannica.com/place/Thar-Desert
Fairhead J, Leach M, Scoones I (2012) Green grabbing: a new appropriation of nature? J Peasant Stud 39(2):237–261. https://doi.org/10.1080/03066150.2012.671770
Gadgil M, Berkes F (1991) Traditional resource management systems. Resour Manage Optim 8(3–4):127–141
Gebrehiwot T, van der Veen A (2015) Farm-level adaptation and vulnerability to climate change: micro-level analysis of cereal farmers in the Rift Valley of Ethiopia. Environ Manage 55(6):1302–1317. https://doi.org/10.1007/s00267-015-0502-5
Government of Rajasthan (2012) Rajasthan state action plan on climate change. Department of Environment, Government of Rajasthan. https://environment.rajasthan.gov.in/content/dam/environment/Env/Pdf_Files/Draft%20of%20State%20Action%20Plan%20on%20Climate%20Change%202022.pdf
Government of Rajasthan (2018) State action plan on climate change (Rajasthan SAPCC). Department of Environment, Government of Rajasthan. https://envfor.nic.in/sites/default/files/State%20Action%20Plan%20on%20Climate%20Change%20Rajasthan.pdf
Guha R (2000) The unquiet woods: ecological change and peasant resistance in the Himalaya. University of California Press
IPCC (2014a) Climate change 2014: impacts, adaptation, and vulnerability. Part A: Global and sectoral aspects. In: Field CB, Barros VR, Dokken DJ et al (eds) Contribution of working group II to the fifth assessment report of the Intergovernmental Panel on Climate Change. Cambridge University Press. https://www.ipcc.ch/report/ar5/wg2/
IPCC (2014b) Climate change 2014: impacts, adaptation, and vulnerability. In: Field CB, Barros VR, Dokken DJ et al (eds) Contribution of working group II to the fifth assessment report of the Intergovernmental Panel on Climate Change. Cambridge University Press. https://www.ipcc.ch/report/ar5/wg2/
Jodha NS (2012a) Desertification in the Thar Desert: environmental concern and people's livelihoods. Int J Glob Agric Environ Sci 1(2):45–56. https://www.woarjournals.org/admin/vol_issue1/upload%20Image/IJGAES091204.pdf
Jodha NS (2012b) Desertification and land degradation in the Thar Desert: impact of human activities and resource management. Int J Environ Stud 69(6):821–834. https://doi.org/10.1080/00207233.2012.703007
Kumar S, Singh R (2020) Climate variability, drought risk, and adaptation strategies in Rajasthan, India. J Arid Environ 179:104199. https://doi.org/10.1016/j.jaridenv.2020.104199
Kumar A, Singh P (2023a) Renewable energy development and human rights: the case of Oran sacred groves in the Thar Desert, India. J Environ Policy Plan 25(2):156–172. https://doi.org/10.1080/1523908X.2023.2176543

Kumar S, Singh P (2023b) Technological interventions and socio-ecological impacts in arid regions: a case study of Rajasthan. J Arid Land Manag:78–92
La Vena P, Ram B (2024) Traditional water management in the Thar Desert: the Khadeen of Rajasthan, India. Blue Pap 3(2):140–151. https://doi.org/10.58981/bluepapers.2024.2.11
Morton JF (2007) The impact of climate change on smallholder and subsistence agriculture. Proc Natl Acad Sci 104(50):19680–19685. https://doi.org/10.1073/pnas.0701855104
Pathak A, Joshi R (2019) Drought patterns and their impact in Western Rajasthan. Int J Environ Stud 76(3):425–438. https://doi.org/10.1080/00207233.2019.1566865
RAJMERU (2023) Annual report 2022–2023. https://rajmeru.org/Annual_Report_Rajmeru_2022-23-new.pdf
Schlosberg D, Collins LB (2014) From environmental to climate justice: climate change and the discourse of environmental justice. Wiley Interdiscip Rev Clim Chang 5(3):359–374. https://doi.org/10.1002/wcc.275
Sharma R (2022) Sacred groves and biodiversity conservation in Rajasthan's Thar Desert. Centre for Indigenous Knowledge Studies. https://www.ciks.org/research/oran-sacred-groves-thar-desert. Accessed 31 May 2025
Sharma P, Gupta A (2018) Water management in arid regions: challenges and pathways for transformational adaptation in Rajasthan. J Arid Environ 152:68–79
Sharma A, Gupta S (2021a) Strengthening community resilience through participatory natural resource management. Rural Development Institute
Sharma R, Gupta A (2021b) Groundwater depletion and its impact on agriculture in Rajasthan's Thar Desert. Environ Sci Policy 120:89–98. https://doi.org/10.1016/j.envsci.2021.02.003
Sharma K, Singh R, Devi P (2021) The paradoxical greening of the Thar Desert: implications for agro-pastoral livelihoods. J Arid Land Stud 15(3):39–52
Singh R (2020) Ecology and spirituality: the Bishnoi way of life. University Press
Singh R, Gupta S (2019) Traditional water harvesting systems and their relevance in the Thar Desert, Rajasthan. J Arid Environ 163:15–26. https://doi.org/10.1016/j.jaridenv.2019.03.007
Singh M, Joshi P (2020) Climate vulnerability and adaptation assessment in arid regions: a case study of Thar Desert districts, Rajasthan. Environ Sustain 3(1):12–25. https://doi.org/10.1007/s42398-020-00085-x
Singh R, Kumar S (2021a) Climate change impacts on drought and extreme rainfall events in Western Rajasthan, India. Environ Monit Assess 193(4):200. https://doi.org/10.1007/s10661-021-08998-2
Singh A, Kumar R (2021b) District-level climate vulnerability assessment and adaptive planning in Rajasthan, India. J Environ Manage 295:113034. https://doi.org/10.1016/j.jenvman.2021.113034
Singh A, Kumar R (2022) Climate change adaptation policies in India: a case study of Rajasthan. J Environ Pol
Singh A, Mehta S (2023) Indigenous resistance and legal actions on renewable energy projects: the case of Oran lands and Great Indian Bustard conservation in Rajasthan, India. J Environ Law Policy 35(2):112–130. https://doi.org/10.1093/jel/ejad034
Singh P, Nair G (2018a) Climate change and its impact on agro-pastoral livelihoods in semi-arid tropical regions of India. Agric Syst 167:70–79. https://doi.org/10.1016/j.agsy.2018.07.004
Singh P, Nair G (2018b) Impacts of climate variability on agro-pastoral communities in semi-arid regions of India. Clim Risk Manag 21:25–34. https://doi.org/10.1016/j.crm.2018.07.004
Singh R, Patel M (2023) Renewable energy development and indigenous rights: challenges faced by Oran sacred groves in the Thar Desert, India. Energy Pol Soc Justice 19(1):45–61. https://doi.org/10.1016/j.enpol.2023.113567
Singh R, Sharma V (2020) Human-induced desertification and groundwater depletion in Rajasthan, India: causes and consequences. Environ Res Commun 2(4):045001. https://doi.org/10.1088/2515-7620/ab9f45
Singh K, Sharma P (2022) Climate change, soil erosion, and water scarcity in arid regions: a case study of the Thar Desert, India. J Arid Land Stud 32(1):45–58. https://doi.org/10.1016/j.jaridenv.2022.104567

Singh M, Singh S (2018) Adaptive strategies of agro-pastoral communities in arid regions of India: an ethnographic study. Agric Hum Values 35(3):705–716. https://doi.org/10.1007/s10460-018-9865-8

Singh RP, Yadav NK (2010) Ecology of Thar Desert. Scientific Publishers

Singh N, Shreya S, Barman B, Ghosh B, Ranjan A, Quader SW, Khan MA, Narain P (2023) Traditional rain water harvesting systems in Rajasthan: water resources conservation and its sustainable management—a review. Sustain Agric Food Environ Res 11. https://www.researchgate.net/publication/383366650_Traditional_rain_water_harvesting_systems_in_Rajasthan_water_resources_conservation_and_its_sustainable_management_-a_review

Supreme Court of India (2024) Judgment on the protection of the Great Indian Bustard and underground transmission line directive. Writ Petition No. 1234/2023. https://supremecourt.gov.in/judgments/2024/gib-conservation. Accessed 31 May 2025

Tahir Z, Haseeb M, Mahmood SA et al. (2025). From Green to Grey: Assessing the Impact of Urbanization on Land Surface Temperature and Thermal Comfort in Multan City. Earth Syst Environ 9, 777–800. https://doi.org/10.1007/s41748-025-00598-y

Thornton PK, Herrero M (2010) The inter-linkages between rapid growth in livestock production, climate change, and the impacts on water resources, land degradation, and biodiversity. Agric Syst 103(9):681–690. https://doi.org/10.1016/j.agsy.2010.05.009

Yadav RK, Soni RK (2019) Traditional agroforestry practices and their role in drought mitigation in the Thar Desert, Rajasthan. J Arid Land Stud 29(1):45–53

Part V
Socioeconomic Implications of Climate Change and Adaptation

Chapter 13
Assessing the Effects of Climatic Hazards on the Rural Livelihood of Coastal Blocks of Purba Medinipur

Swarnali Mukhhopadhyay, Suman Sinha, and Gupinath Bhandari

Abstract Climate change is gradually affecting littoral societies around the world through rising sea levels, severe storms, and erratic precipitation anomalies. In Purba Medinipur, exactly in the zones of Contai 1 and Contai 2, the climate-induced vulnerabilities have been gradually increasing over the past 30–35 years. This study explores how climate disasters distress these littoral regions, focusing on the consequent environmental, societal, and financial difficulties. The results indicated that climate change is deteriorating issues such as coastal destruction, inundation, and saltwater intrusion in the land in Contai 1 and Contai 2. These threats are triggering destruction to homes, infrastructure, and farmland, resulting in the displacement of societies and threatening their livelihoods. So, the aim of the study is to unfold the effect of inundation due to super cyclone, cyclone on the livelihood of the rural people in Contai 1 and 2 Blocks, of Purba Medinipur, West Bengal, India. NDVI data from the years 2004, 2014, and 2024 reveal a decline in the values, ranging from −0.164 to +0.5272 in 2004, −0.163 to +0.52 in 2014, and −0.162 to +0.536 in 2024, showing a gradual decrease in agricultural yield from 2004 to 2024. Notably, the most definite changes in NDVI values occur between the coastline and 20–25 km inland, an area where soil salinity has had a damaging effect on the land. NDVI values have steadily decreased from 2004 to 2024, indicating reduced agricultural productivity. Agricultural land has decreased from 79.40% in 2004 to 69.42% in 2024.

S. Mukhhopadhyay (✉)
Department of Geography, Baruipur College, Kolkata, India

Department of Geography, Amity University, Kolkata, India

S. Sinha
Department of Geography, Amity University, Kolkata, India

G. Bhandari
Department of Civil Engineering, Jadavpur University, Kolkata, India

S. Mandal et al. (eds.), *Climate Change Vulnerability*, Climate Change Management, https://doi.org/10.1007/978-3-032-16783-5_13

Introduction

Climate change, primarily driven by external forcings including anthropogenic greenhouse gas emissions and modulated by internal climate variability (Mahadevia and Vikas 2010; Halder et al. 2023; Khan et al. 2025; Dey et al. 2025a, b; Dogar et al., 2017a, 2017b, 2023, 2024, 2025a, 2025b; Dogar and Sato 2018, 2019) is a major threat and a vital issue in the recent era, with widespread effects on the ecosystem environment, public health, and economic development. Coastal areas (Mandal et al. 2013, 2015, 2018; Mandal and Choudhury 2014) are extremely prone to its consequences and are the most susceptible to its effects. The coastal regions of West Bengal, specifically Purba Medinipur, are also heavily impacted by climate-related stresses. Sited in eastern India along the Bay of Bengal, Purba Medinipur features even topography areas with low elevation areas, making the zone prone to flooding and erosion due to sea level rise and storm surge. The Contai 1 and Contai 2 blocks are specifically susceptible to climate-related impacts. The zone is facing an intensification in the occurrence and concentration of cyclones and storm surges, along with ongoing beach destruction, all of which have extreme consequences for the local inhabitants. The destruction of homes, set-up, and cultivated land, along with public displacement of people and livelihood disruptions, demonstrates the severe penalties of climate change in this area.

Coastal areas of West Bengal are among the most vulnerable areas in India to the hostile effects of climate change, particularly regarding sea saltwater invasion that affects farming (Bastawesy et al. 2012; Halder et al. 2022). The exceptional geographical landscapes of these low seaside sectors, combined with financial challenges and more population density, aggravate their vulnerability to environmental threats such as inundation and saltwater ingress. It is assessed that between 60% and 70% of the residents live in these susceptible parts appearances risk of saline water invasion (Messner and Meyer 2006) of saline water invasion, which impends both yield production and public living (Alexakis et al. 2018). Over the current decades, the speedy inhabitant's development has led to increased urban expansion, loss of farming land (FAO 2021). However, the coastal area is facing an intensified problem of soil and water salinization, especially in coastal cultivated regions (Bastawesy et al. 2012). The use of water that is polluted or of inferior quality for irrigation, combined with ineffective irrigation practices, accelerates salt deposition in soils—a procedure identified as salinization—which harshly reduces crop yield production and causes land degradation and desertification (Bradd et al. 1997; Yasuor et al. 2020). Worldwide, coastal zones are facing increased frequency and intensity of climatic hazards. For example, in the Caribbean, hurricanes have caused widespread devastation to rural livelihoods, leading to long-term socio-economic uncertainty. Equally, Southeast Asia experiences recurrent typhoons and flooding that harshly interrupt agricultural productivity and fishery-based economies. In Africa, coastal peoples in Mozambique and Nigeria are more and more susceptible to cyclones and sea-level rise, which impede food security and livelihoods. Similarly, in the

Philippines, community-driven adaptation strategies have been active in mitigating climate impacts on agriculture and fisheries.

Seawater invasion is a prime concern in these littoral zones, affecting surface water resources such as ponds, rivers, and irrigation canals (Abedin et al. 2014). This saltwater contamination puts at risk the health of over 25 million inhabitants who depend on these water sources for intake and agricultural purposes (Gopalakrishnan and Kumar 2020). Universally, around 830 million hectares of land—around 7% of the surface of the Earth's surface—are at risk of flooding, highlighting the extensive nature of this issue (Yasuor et al. 2020). Moreover, approximately 40% of agricultural lands in Bangladesh are situated in coastal areas, where salinity levels are gradually rising, severely impacting crop yields and farming viability (Fourati et al. 2017; Yasuor et al. 2020). Study from various areas of the world highlights common susceptibilities and adaptive strategies. For example, in Bangladesh, Cyclone Amphan in 2020 caused extensive damage to rural societies, emphasising the significance of robust infrastructure and calamity awareness. Exactly, in the Khulna and Barisal divisions, about 1.05 million hectares out of 2.88 million hectares of cultivable land are impacted by changing grades of soil salinity, leading to significant portions of farmland becoming unproductive (Bradd et al. 1997). Shallow water sources in these areas are also salty, compelling farmers to depend on groundwater for irrigation. However, surface aquifers are gradually polluted by salinity due to the interface between saline and freshwater sources, prompting people to turn to deeper aquifers to access less contaminated water. Saline water invasion needs integrated management approaches that consider both natural processes and human activities. Effective groundwater management, salt-resistant crop variations, and upgraded irrigation practices are vital to mitigate the effects of salinization (Mukhopadhyay et al. 2025). By emphasising perceptions from worldwide knowledge and up-to-date works and literature, this study aims to fill existing research gaps and inform policy interventions adapted to local circumstances.

Objective of the Study

This research aims to explore the effects of weather change-induced hazards on the coastal communities of Contai 1 and Contai 2, Purba Medinipur. The research will examine the impacts of storm surges, sea-level rise, and beach erosion on the local community, including the resulting environmental, societal, and financial consequences. Additionally, the research will also recognize the vulnerabilities and capacities of the local community to adapt to the impacts of climate change.

1. To analyze the variations of the land use and land cover maps of the years 2024 and 2004 by using geospatial technology.
2. To examine the changes in the percentage of land use area and the analysis of soil parameters.

3. Assessment of the impact of climatic hazard (cyclone, super cyclone, extremely severe cyclone) on coastal dwellers.

The prime goal of the present investigation is to analyze the influence of climatic hazards on the livelihood of rural people in Contai 1 and Contai 2 Blocks of Purba Medinipur. In the study data of satellite images from 2024 and 2004 have been used to recognize the modifications of land use patterns.

Study Area

Contai 1 and Contai 2 (Deshopran) (Fig. 13.1) are two blocks of the Purba Medinipur district of West Bengal, India. It is situated on the eastern coast of India, approximately 169 km southwest of Kolkata. These blocks are located on the banks of the Rasulpur River and are part of the Contai subdivision. Contai is situated at 21.78°N latitude and 87.75°E longitude, with an average elevation of 3 m above sea level. Contai is situated on a littoral plain, which is to gently sloped area and is placed adjacent to the coast. The soil in Contai is mainly silty, which is formed by the Rasulpur River and its branches. The blocks are characterised by a low-lying area, with an average elevation of 3 m above sea level, making them susceptible to inundation during high tides and storms. Contai has a humid tropical climate, with more humidity and high temperatures throughout the year. Temperatures can rise to 37 °C in the summer months (April–June), and mild winters are observed with temperatures varying from 15 to 20 °C during the winter months (December–February). Contai receives heavy rainfall throughout the monsoon season (June–September), with an average annual rainfall of 1400 mm. The rich alluvial soil and climate make Contai suitable for agriculture, with crops such as rice, wheat, and jute being commonly cultivated (Mukhopadhyay et al. 2025). These blocks have a substantial

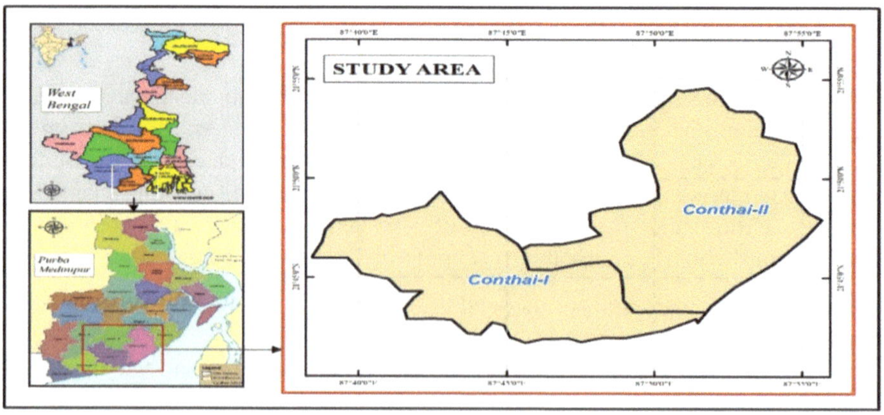

Fig. 13.1 Location map of the study area

fishing industry, with many fishermen operating in the Bay of Bengal. Contai 1 is located in the northern part of the Contai subdivision and includes 12 gram panchayats (village councils). The block has a total population of approximately 200,000 (Census of India, 2011) people and covers an area of 170 square kilometers. Contai 2 (Deshopran) is located in the southern part of the Contai subdivision and contains 11 gram panchayats. The block has a total population of around 180,000 people and covers an area of 160 square kilometres. Both Contai 1 and Contai 2 are mainly rural areas, with cultivation being the main foundation of livelihood for the local population (Statistical handbook 2014). The blocks face several challenges, including poverty, lack of access to basic services, and environmental degradation (Paul and Routray 2010).

Methodology

This research not only studies the influences of climate-induced threats and flooding on the livelihoods of rural coastal inhabitants, but also analyses their coping strategies (Parvin et al. 2016). The study is based on field surveys, with samples collected from several villages that were harshly affected by past flooding due to climatic threats. A total of 97 households were randomly chosen for the household survey from two blocks, with the oldest member of each household serving as the respondent; 75% of these respondents were male, aged between 30 and 65 years. They reflected on their experiences of severe sea inundation over the past 5 years (Parvin et al. 2016) related to climatic hazards.

In addition to the structured questionnaire, members from 97 households from different villages and various professions shared their experiences with climatic hazards and their resulting hardships. This investigation identified several professional groups, including agricultural workers, daily laborers, business owners, farmers, servicemen, and others (Paul and Routray 2010). Additionally, 31 soil samples were collected from various sites in Block Contai 1 and Contai 2. These samples were examined at the Soil Laboratory of the River Research Institute in Haringhata, Kolkata. Exploring the soil properties, features, and quantifiable measurements in this area and the surrounding regions is essential. These soil examinations were conducted on soil samples taken from depths of 1 to 1.5 ft. For plotting, mapping and diagram preparation, software such as ArcGIS 10.8, Google Earth Pro, and Excel 2013 for graphs and charts were applied, while ERDAS Imagine 9.1 was employed for image classification using supervised techniques for improved accuracies (Sinha et al. 2015, 2020). Furthermore, a Landsat 8 image was used for greater clarity to assess changes in land use and land cover in the coastal region. The statistical analysis was conducted using time series analysis, scatter diagrams, and regression analysis to identify trends and patterns of the study (Fig. 13.2; Tables 13.1 and 13.2).

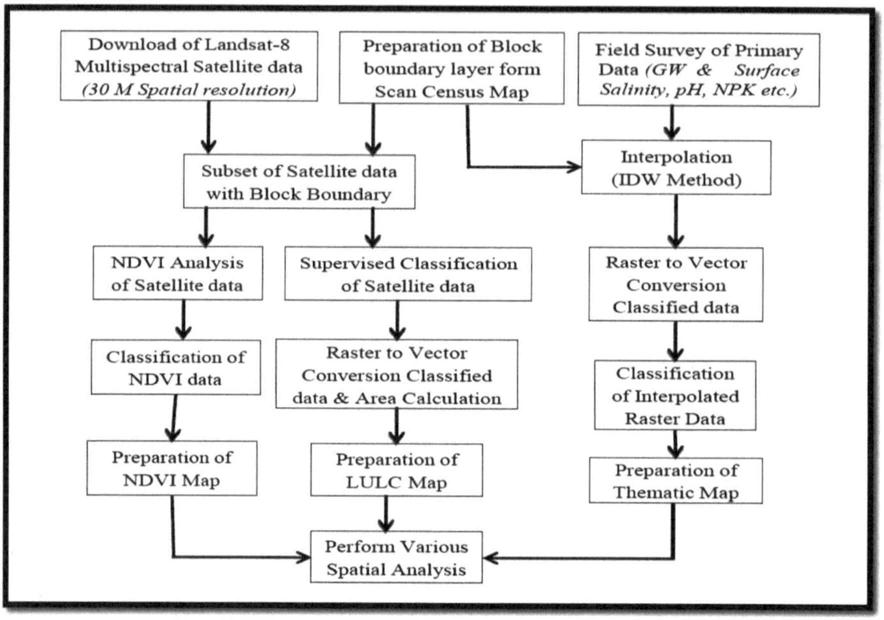

Fig. 13.2 Flowchart of methodology

Table 13.1 Properties of Landsat image

Sensor	Band number	Band name	Wavelength (m)	Resolution (m)
OLI	1	Coastal	0.43–0.45	30
OLI	2	Blue	0.45–0.51	30
OLI	3	Green	0.53–0.59	30
OLI	4	Red	0.63–0.67	30
OLI	5	NIR	0.85–0.88	30
OLI	6	SWIR 1	1.57–1.68	30

Sources: USGS Earth Explorer
OLI operational land imager, *SWIR 1* short-wave, infrared, NIR near infrared

Table 13.2 Distribution of area (%) in 2004, 2024

Land use/land cover class	2004 (% of area)	2024 (% of area)	(% of area change)	Changing pattern
Agricultural land	79.40	69.42	9.98	Negative
Vegetation	9.58	15.11	5.53	Positive
Waterbody	1.28	1.65	.37	Positive
Sand	5.01	2.61	2.40	Negative
Build up area	4.36	11.58	7.22	Positive
Total	100	100		

Results

The varying climate can remarkably distress littoral zones in several ways. These areas are more susceptible to variations in storm occurrence and concentration, increasing sea levels, warmer ocean temperatures, and erratic rainfall. The effects of climate unpredictability are currently crucial issues for coastal societies, leading to destruction, inundation, and saltwater invasion. These difficulties are increasing notably in many parts of this region. Meanwhile 1900, worldwide sea levels had risen by about 6–8 in. (Walsh et al. 2014). This rise is noticeable in several seaside locations, deteriorating the salinization of cultivated land and groundwater sources (Moser et al. 2014). In addition, warmer shallow sea temperatures prime cause of more recurrent and concentrated cyclones, super cyclones, and heavy rainfall in coastal regions. Unpredictable rainfall patterns, caused by higher sea temperatures and climate instability, further increase the vulnerability (Das et al. 2022) of growing populations and developments in coastal areas, like Purba Medinipur (Lavender et al. 2018). Devastating cyclones pose major natural threats to these coastal regions, where 40–50% of rural populations reside. Each year, cyclones disrupt the lives of coastal residents during the monsoon and pre-monsoon seasons, resulting in significant economic losses and increased social vulnerability. Flooding and super cyclones cause damage to yields and vegetation. Major storms include Super Cyclone Aila in 2009, Super Cyclone Fani and Bulbul in 2019, Super Cyclone Amphan in 2020, and Super Cyclone Yash in 2021, along with Cyclone Sitrang in 2022, Remal in 2024, and Dana in 2024. All of these storms destroyed southern West Bengal and Bangladesh, leading to major financial losses and displacing communities (Halder et al. 2021). To evaluate the global impression of climatic hazards such as environmental changes and flooding, Remote sensing technologies and Geographic Information Systems (GIS) are essential tools. The capacity to evaluate environmental degradation or climatic disasters caused by atmospheric threats is significantly enhanced by satellite data and geospatial technologies. This study aims to investigate recent climatic hazards linked to cyclones and severe super cyclones, as well as the impacts of coastal flooding in West Bengal's coastal regions. Therefore, ongoing research on meteorological hazards is crucial for improving future adaptation strategies, awareness, preparedness, disaster response, and community resilience across the entire coastal region (Halder et al. 2021). This study (Fig. 13.3a–c) illustrates the changing patterns of land cover and land use (LULC) in the years 2004 and 2024 for Blocks Contai 1 and 2. The LULC data were generated using standard classification techniques on satellite imagery (Sinha et al. 2013). In 2004, agricultural land consisted of 79.40% of the area, but this decreased to 69.42% by 2024. In contrast, vegetation increased from 9.58% in 2004 to 15.11% in 2024. The area of water bodies grew slightly from about 1.28% in 2004 to 1.65% in 2024. The settlement area extended from 4.36% in 2004 to 11.58% in 2024. These variations examine a substantial shift in land use and land cover patterns in the last two decades, mainly due to an increase in vegetation and settlement areas; cultivated land has decreased, which indicates negative changes. The rise in water bodies may result from storm surges and prolonged waterlogging (Fig. 13.4).

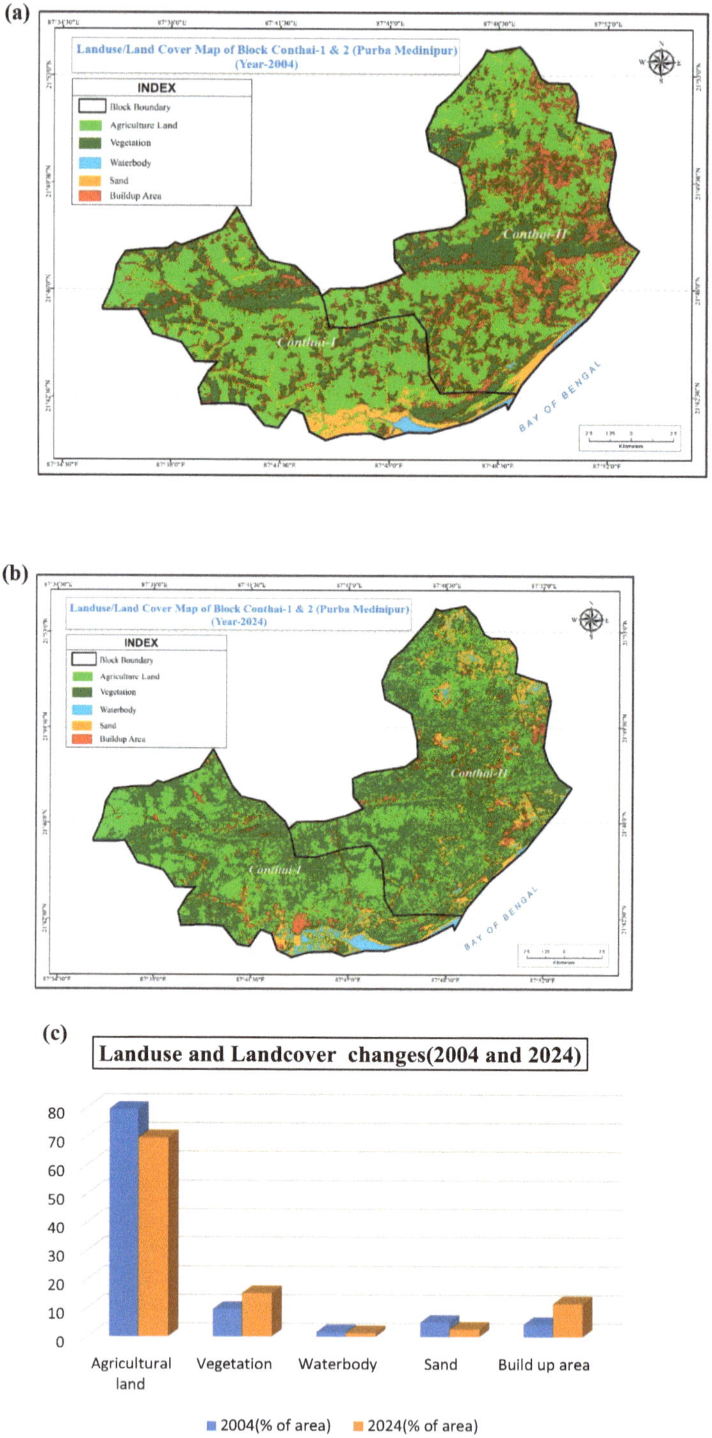

Fig. 13.3 (**a**) Landuse and Landcover Map (LULC) of 2004. (**b**) Landuse and Landcover Map (LULC) of 2024. (**c**) Landuse and Landcover changes (2004 and 2024)

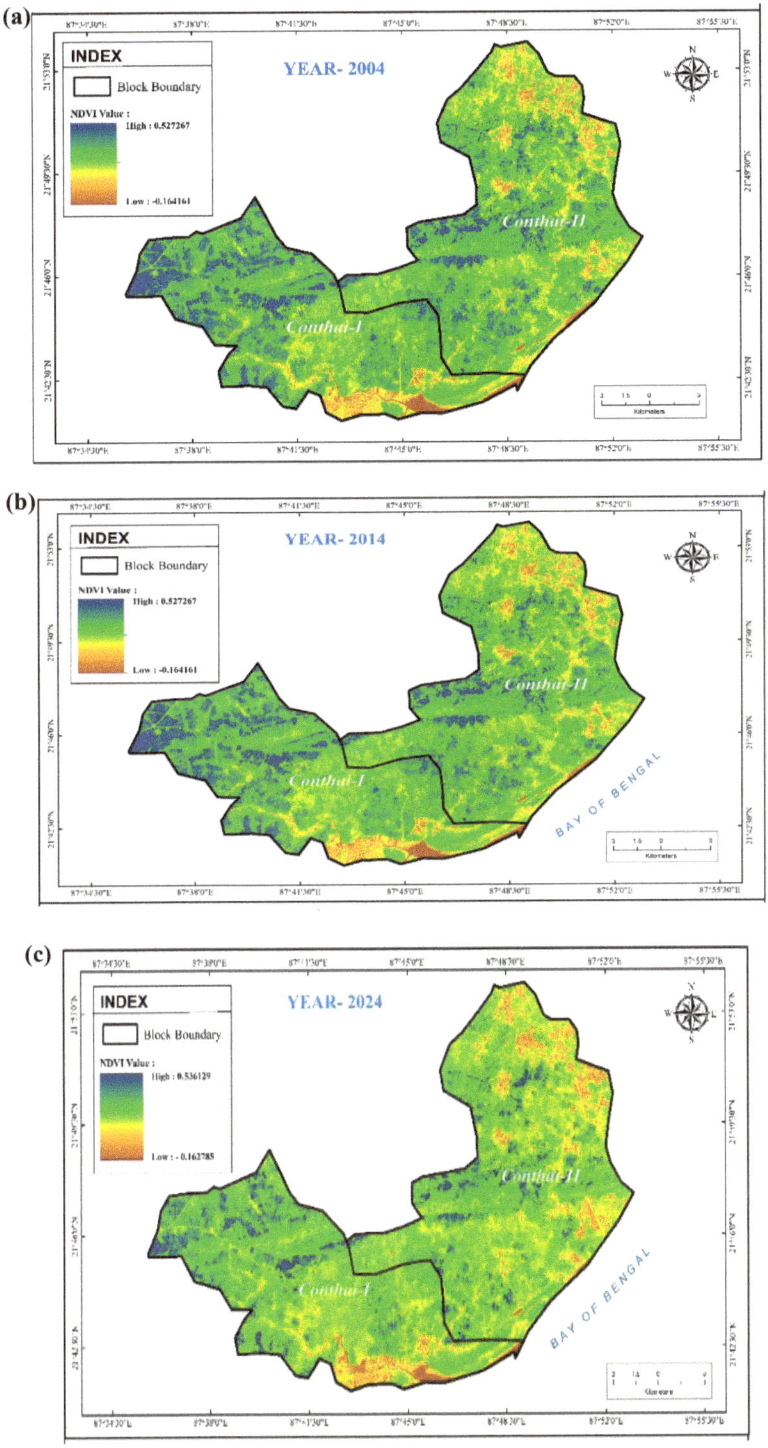

Fig. 13.4 (**a**) Normalized Difference Vegetation Index (NDVI) in 2004. (**b**) Normalized Difference Vegetation Index (NDVI) in 2014. (**c**) Normalized Difference Vegetation Index (NDVI) in 2024

Satellite imagery analysis confirms that significant changes in land use and land cover have taken place over the last 20 years, mainly driven by climate change, environmental factors, afforestation, and frequent weather-related hazards. The impact of these weather events has greatly influenced cultivated land, likely changing land use patterns (Hussein et al. 2019). The size, shape, and form of the Indian Subcontinent accelerate cyclones from nearby seas to affect the coastal areas. An alarming phenomenon is global warming, characterised by the gradual rise in temperature, which may cause severe weather phenomena like cyclones, super cyclones, and storms (Nandargi and Mahto 2019). This research work observes the occurrence and intensity of tropical disturbances affecting the coastal areas from 1970 to till present. The origin of these disturbances over the seas is closely connected to heavy rainfall and storm surges that impact low-lying coastal areas. According to the India Meteorological Department (IMD), various types of cyclones, ranging from severe to extremely severe, have occurred each decade since 1970, leading to significant fatalities and damage. The frequency of these climatic events has steadily increased, with one extremely severe cyclone occurring from 1970 to 1980, four cyclones from 1981 to 1990, and another four from 1990 to 2000. Additionally, five and eight severe climatic disturbances occurred between 2000–2010 and 2010–2024, respectively. The recurring cyclones have represented a deadly natural hazard to the densely populated coastal areas of West Bengal in recent years. Based on IMD data and information from the River Research Institute (Table 13.3), the wind speeds of recent cyclones—Bulbul (2019), Fani (2019), Amphan (2020), Yash (2021), Sitrang (2022), Remal (2024), and Dana (2024)—ranged from 48 to 180 km/h. Such wind speeds contribute to storm surge flooding, exacerbated by high tides and increased water levels (5.915 m during Cyclone Fani, 3.70 m during Cyclone Bulbul, 4.945 m during Cyclone Amphan, 7.2 m during Cyclone Yash, 5.5 m during Cyclone Remal, and 4.8 m during Cyclone Dana, Source: RRI), which inundated low-lying coastal areas (Mukhopadhyay et al. 2025). Storm surge flooding is a recurring event in this region, leading to extensive loss of life and property, with lasting impacts. Field surveys reveal that tidal and storm surges contaminate soil and water with salinity (Mohsenipour 2017), rendering ponds and tube wells unusable due to saline flooding. Coastal residents are significantly affected, as these sources of water are unsuitable for household use. Villagers report that agricultural land becomes saline following storm surge flooding during climatic events, rendering it incompetent for cultivation for the next 3–5 years. Social surveys indicate that between 50% and 60% of cultivated land has turned saline due to the impacts of Cyclones Yash, Amphan, and other climatic hazards.

The increasing sea surface temperature in the Bay of Bengal makes it more prone to cyclones compared to the Arabian Sea. Key factors such as sea surface temperature,

Table 13.3 Number of cyclones that occurred during 1970–2024

Year	Nos. of cyclones
(1970–1980)	1
(1981–1990)	4
(1991–2000)	4
(2021–2010)	5
(2011–2020)	8
(2021–2024)	5

humidity, and kinetic energy contribute significantly to cyclone formation (Nandargi and Mahto 2019). The Bay of Bengal experiences high average rainfall, increasing the likelihood of cyclones due to abundant moisture in the mid-atmosphere (Walsh et al. 2014). The peninsula shape of India also facilitates cyclone development. In the Bay of Bengal, Cyclones are often linked to the low pressure created by the warm ocean waters (Werner et al. 2013). The trough-like configuration of the bay allows storms to gain energy, while the lack of substantial landmass encourages cyclonic winds to move into coastal territories. In the past decade, the eastern coastal region has endured at least eight severe super cyclones, with the most frequent landfalls occurring in West Bengal, Odisha, and Bangladesh, and Andhra Pradesh (Rahman et al. 2014). Figure 13.5d illustrates the cyclonic tracks formed in the central Bay of Bengal, impacting the southern Indian coastline in the last 5 years, with Amphan, Fani, Bulbul, and Yash being among the most devastating super cyclones (Rhein et al. 2013) (Table 13.4).

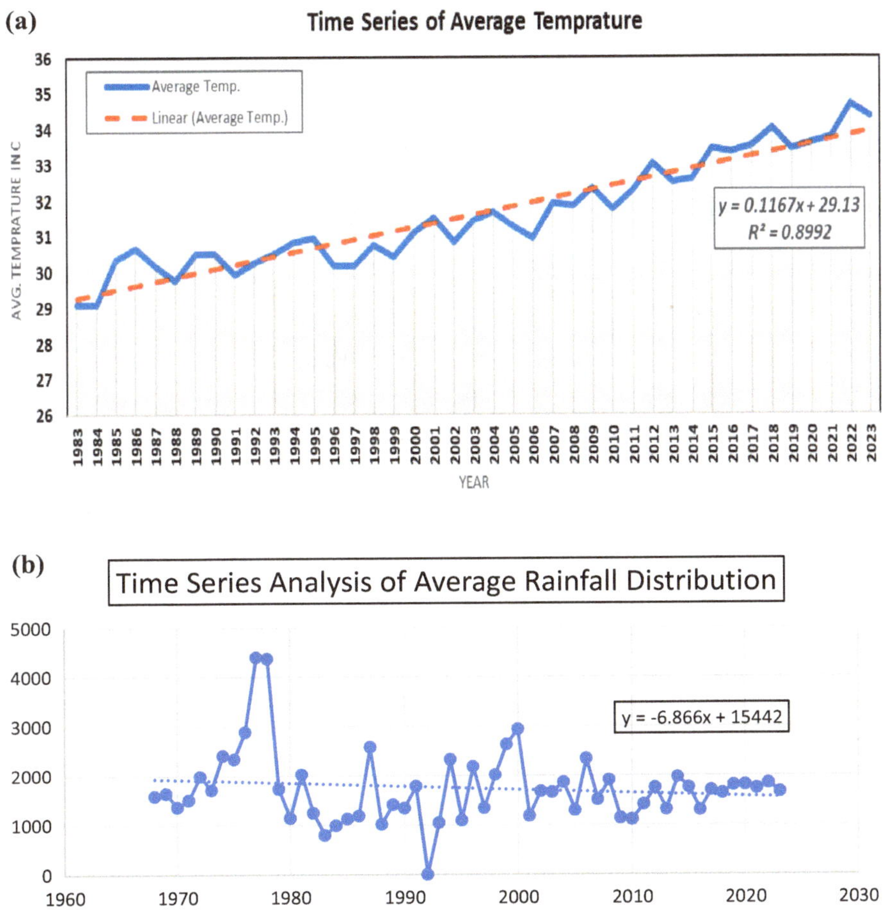

Fig. 13.5 (**a**) Time series analysis of the average Temperature of Purba Medinipore (1983–2023). (**b**) Time series analysis of average rainfall of Purba Medinipore (1960–2024). (**c**) Time series analysis of intensification of cyclone of Purba Medinipore (1970–2024). (**d**) Severe Cyclone track on the Bay of Bengal (last 5 years). (Source: IMD)

(c)

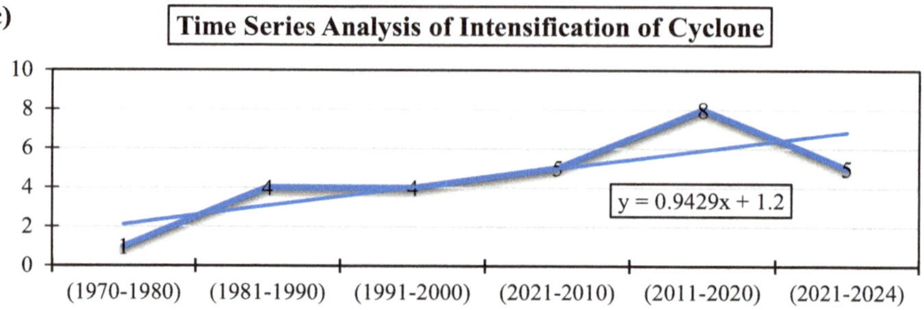

(d)

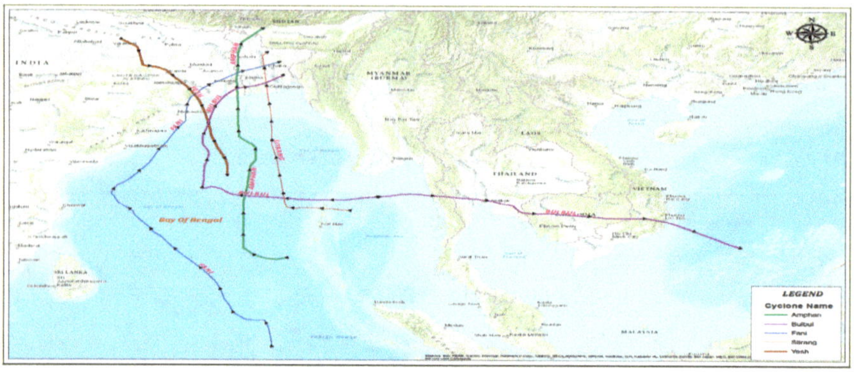

Fig. 13.5 (continued)

Table 13.4 List of cyclones on the Bay of Bengal (BOB) during 1970–2024

Year	SL. No.	Year/month/date	Name	Type of cyclone	Weather conditions	Damages
(1970–1980)	1	1970 WB Border, Bangladesh	Bhola	ESSC	Deadliest cyclone	3–5 lakh death
(1981–1990)	1	1981 WB, Orissa, Bangladesh	BOB02	SSC	Deadliest storm rainfall	200 people died
	2	1984 Orissa	BOB03	Sc	Heavy rainfall, strong storm	Unknown less damage
	3	1988 WB, Sundarban	BOB04	ESSP	Heavy rainfall, strong wind	220–500 death
	4	1990 Andhra Pradesh	BOB01	SC	Worst storm	967 deaths

(continued)

Table 13.4 (continued)

Year	SL. No.	Year/month/date	Name	Type of cyclone	Weather conditions	Damages
(1991–2000)	1	1995 WB	BOB05	CS	Rainfall	Less damage
	2	1997 Orissa, WB	BOB07	SC	Rainfall	Less damage
	3	1998 Orissa, WB	BOB06	SC	Rainfall	Less damage
	4	1999 Orissa	BOB05	SSC	Very strong wind	9887 people lost their lives
(1991–2000)	1	2002 West Bengal	BOB03	SC	Very strong wind, heavy rainfall	173 death
	2	2005 AP, Orissa	Pyarr	CS		Less damage
	3	2008 WB BRODER, BANGHALADAS	Rashmi	SC	Very strong wind	20 died
	4	2009 WB, ORISA	Bijli	CS	Heavy rainfall	Unknown
	5	2009 WB	Aila	ESSC	Worst weather and heavy rainfall	Worst damage, 20 died
(2011–2020)	1	2012 AP	Nilam	CS	Worst weather heavy rainfall	75 people died
	2	2013 ORISHA	Phailin	ESSC	Stormy weather	Worst affected: 45 died
	3	2014 AP	Hudhud	ESSC	Strong wind heavy rain fall	124 people died, the worst damaged
	4	2018 ORISSA	Daya	CS	Heavy rainfall	Less damage
	5	2018 ORISSA, AP	Titli	SSC	Worst weather	77 people died + worst damages
	6	2019 ORISSA	Fani	ESSC	Worst weather	40 people died + worst damaged
	7	2019 WB, ORISSA	Bulbul	SSC	Worst weather	Less damage
	8	2020 WB, ORISSA	Amphan	ESSC	Worst weather	45 lakh people homeless
(2021–till date)	1	2021 WB, ORISSA 2021, WB, ORISSA 2022, WB, ORISSA 2024, WB, ORISSA 2024, WB, ORISSA	Yash Jawad Sitrang Remal Dana	ESSC C C C C	Worst weather	Died 20 people, worst affected total WB and Odisha

Source: IMD

ESSC extremely severe super cyclone, *SSC* severe super cyclone, *SC* super cyclone, *C* cyclone, *BOB* Bay of Bengal

Discussions

Coastal communities predominantly rely on the earth's resources used for their livelihoods, with fishing and agriculture serving as the primary sources of income. However, these vital sectors have been severely compromised due to the increasing occurrence and intensity of cyclones and other severe climatic phenomena. In the previous decade, the agricultural lands in the southern coastal regions experienced significant degradation, largely attributable to saline water accumulation resulting from seawater intrusion during cyclonic storms. This environmental disruption has not only diminished land productivity but has also triggered notable shifts in land use patterns, especially in the study areas of Contai 1 and Contai 2 blocks. As a result, cultivators are affected by a severe threat to the loss of cultivated land, agricultural productivity, with financial stability. Soil quality gradually declined due to high salinity, alkalinity, and deficiencies of nutrients and effects, leading to reduced yield, and farmers are bound to rethink their traditional jobs (Gray 2011).

By examining the soil health in these areas, the study provides a detailed analysis through Fig. 13.6a–h. These figures illustrate important soil factors such as salinity, pH levels, nitrogen, phosphorus, potassium, and groundwater quality. The soil salinity maps show that areas within about 3 km of the coastline are heavily impacted, with salinity levels over 3 dS/m, indicating severe salinity stress. Areas within 3–6 km from the coastline face salinity levels 2–3 ds/m, while areas between 6 and 10 km and beyond 10 km have comparatively lower salinity. According to agricultural department data, about 60–65 square kilometers in each of these blocks are adversely affected by salinity due to recurrent climatic disturbances.

Soil pH, an essential factor influencing soil fertility, also exhibits notable variations across the region. As per Bruckner (2012), soils with a pH below 7 tend to be more acidic, resulting from elevated hydrogen ion concentrations, which can negatively affect nutrient availability and electrical conductivity. The pH maps indicate most zones have pH values under 7 or between 7 and 8, suggesting a tendency toward acidity and alkalinity extremes that are detrimental to crop growth. Soils outside the optimal pH range of 7–8 tend to inhibit plant health, with high salinity and acidity impairing nutrient uptake and overall soil fertility. Nitrogen levels, essential for plant growth, are notably lower in highly saline zones—often falling below 20 kg/hectare—reflecting compromised nitrogen fixation due to saline and alkaline conditions. Conversely, regions with lower salinity maintain nitrogen levels above 24 kg/hectare, indicating better soil fertility. Other soil nutrients, such as phosphorus and potassium, also display troubling patterns: phosphorus levels below 26 kg/hectare delay nutrient absorption, and excessive potassium (above 53–55 kg/hectare) can interfere with micronutrient uptake, further reducing crop yields.

The aftermath of recent super cyclones, including Cyclone Amphan, Cyclone Yash, Remal, and Dana, has intensified soil salinization, leading to a marked decline in agricultural productivity. Field surveys reveal that farmers are actively attempting to mitigate salinity through various means, such as applying lime, tamarind, and decomposed mango leaves. Additionally, there is a shift toward cultivating salt-tolerant crops like brinjal (eggplant), chili, cauliflower, cabbage, rapeseed, and mustard greens (Wang et al. 2010; Kim et al. 2016). To understand the broader agricultural landscape over

(a)

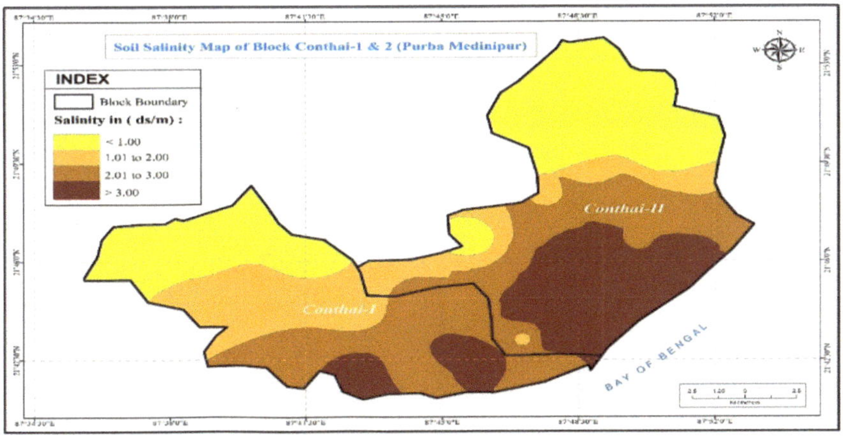

(b)

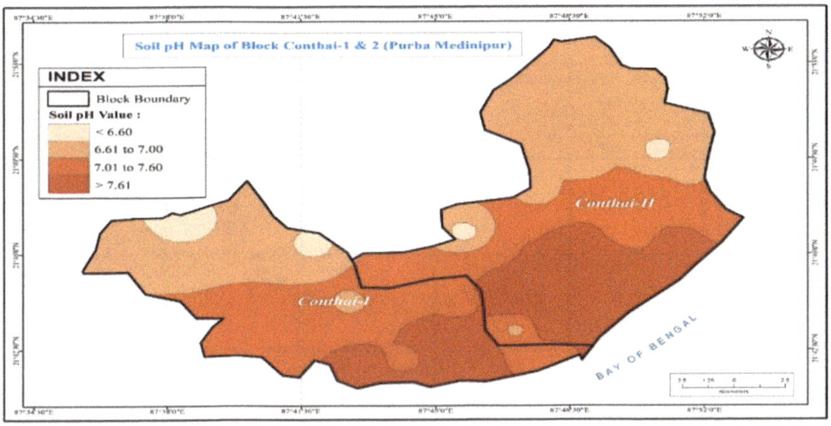

Fig. 13.6 (**a**) Soil salinity distribution. (**b**) Soil pH distribution. (**c**) Relationship between soil salinity and soil pH. (**d**) Nitrogen distribution. (**e**) Phosphorus distribution. (**f**) Potassium distribution. (**g**) Groundwater salinity distribution. (**h**) Groundwater pH distribution. (**i**) Scatter diagram. (**j**) Scatter diagram

two decades, remote sensing and GIS techniques will be employed, particularly the Normalized Difference Vegetation Index (NDVI), a vital tool for assessing vegetation health and crop yield. NDVI, with values ranging from −0.1 to 1, helps identify the density and health of vegetation. Negative values often correspond to water bodies or clouds, while values near zero indicate bare soil. Dense, healthy vegetation typically registers between 0.6 and 1.0. Analyses of NDVI maps from 2004, 2014, and 2024 reveal a gradual decline in vegetation over time, particularly within 15–20 km of the coastline—areas severely impacted by soil salinity. For instance, in 2004, NDVI values ranged from −0.164 to 0.5272; in 2014, from −0.164 to 0.52; and in 2024, from −0.162 to 0.536. The decreasing trend indicates a decline in crop health and productivity, underscoring the long-term adverse effects of climatic hazards on agriculture.

Beyond environmental impacts, climatic events have profound social and economic repercussions for coastal populations. Causes of property destruction, food

(c)

(d)

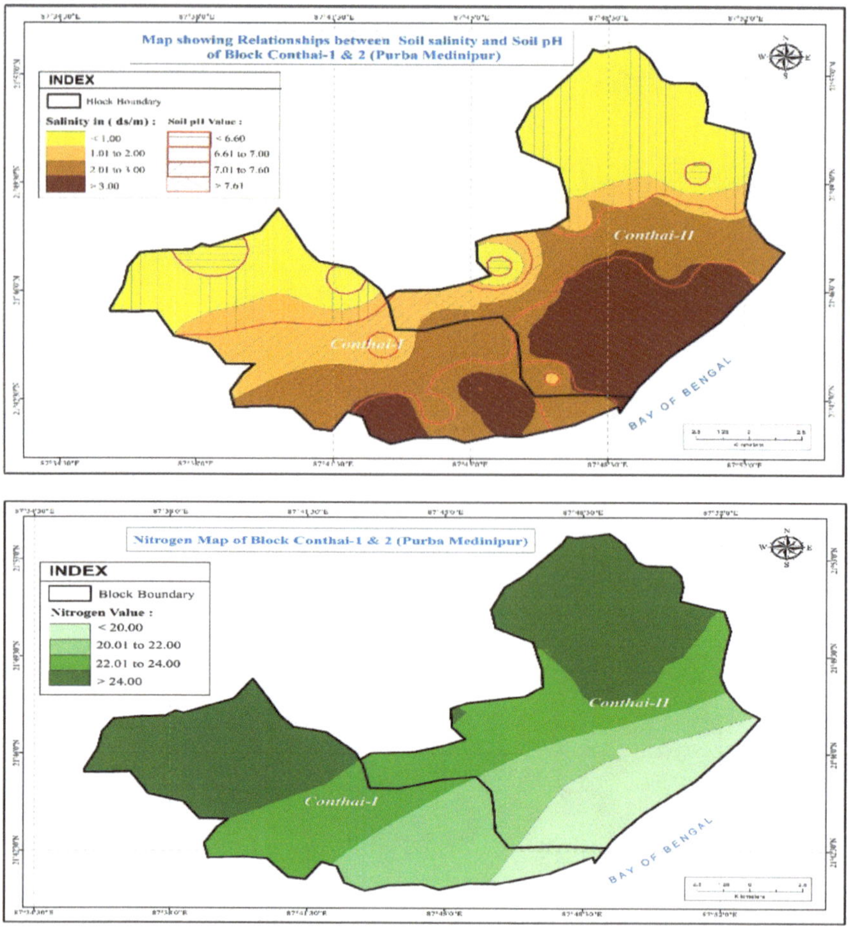

Fig. 13.6 (continued)

shortages, and water scarcity disproportionately affect vulnerable, impoverished communities lacking resources to cope effectively (Mallick and Rahaman 2011). The long-term effects extend beyond immediate damage, disrupting livelihoods and causing poverty. Many affected individuals are forced to seek alternative livelihoods due to the deterioration of traditional income sources like farming and fishing. This occupational shift is often driven by necessity, as primary sources become unviable. Data from social surveys (see Fig. 13.7a–c) illustrate this occupational transition vividly. Before climatic hazards, 56% of households engaged in agriculture, with smaller proportions involved in service sectors (9.2%), business (14.3%), daily wage labour (12.53%), and other activities (7.6%). Post-disaster, the occupational landscape changes significantly; only 31% remain engaged in agriculture, with many households turning to food processing (31%), hotel work (18.2%), transportation (11.6%), shell craft making (14%), and other informal sectors (14%). Notably, about 12% of households become unemployed, highlighting the economic distress inflicted

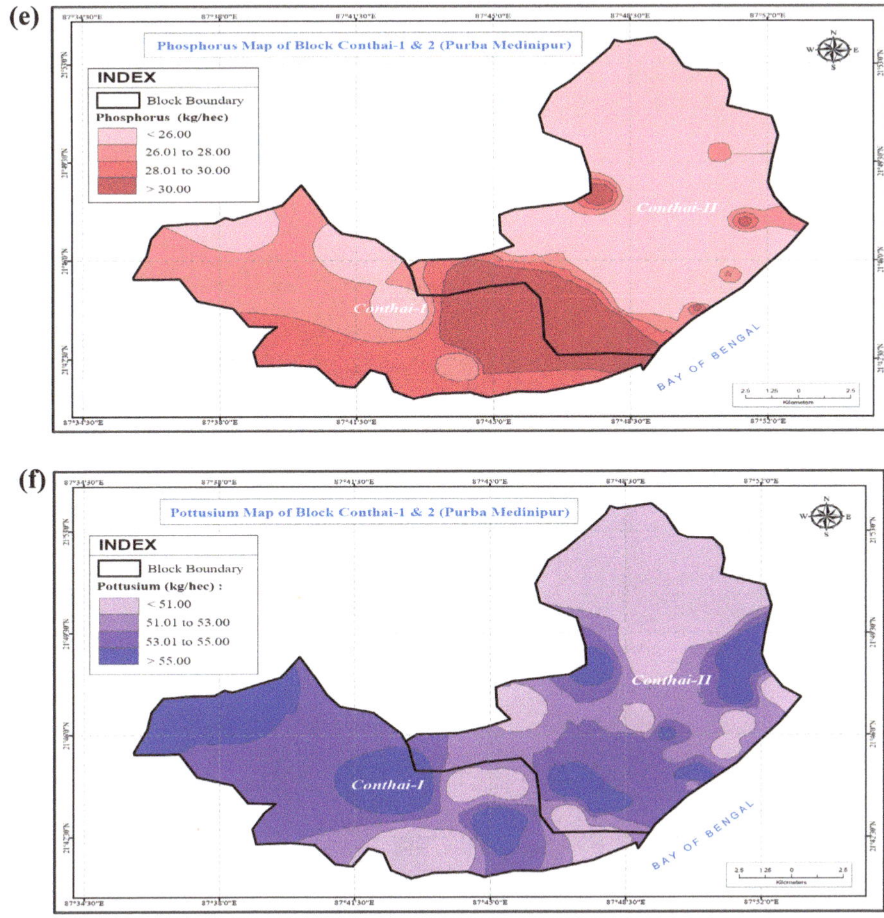

Fig. 13.6 (continued)

by recurrent climatic shocks. The government's relief efforts are insufficient to meet the increasing needs, leaving many households in financial crisis. Further analysis of household income reveals that prior to climatic disturbances, only a small percentage of households earned above 25,000 rupees per month, with most earning between 10 and 25,000. Post-disaster, income levels have dropped further, with a significant proportion earning less than 10,000 rupees monthly. The COVID-19 pandemic has compounded these hardships, further reduced income, and exacerbated poverty (Gray 2011). The data underscores that socio-economic conditions for residents of Contai 1 and 2 are deteriorating, driven by natural calamities and pandemic-related disruptions. The study's analysis illustrates a clear negative correlation between agricultural productivity and soil salinity. Regression analysis (Fig. 13.6i–j) shows that as soil salinity increases, crop yields decline sharply, following the equations:

$Y = -2.5975X + 18.157$ (where Y is productivity and X is soil salinity)
$Y = -0.0694X + 8.2397$ (relationship between pH and crop yield)

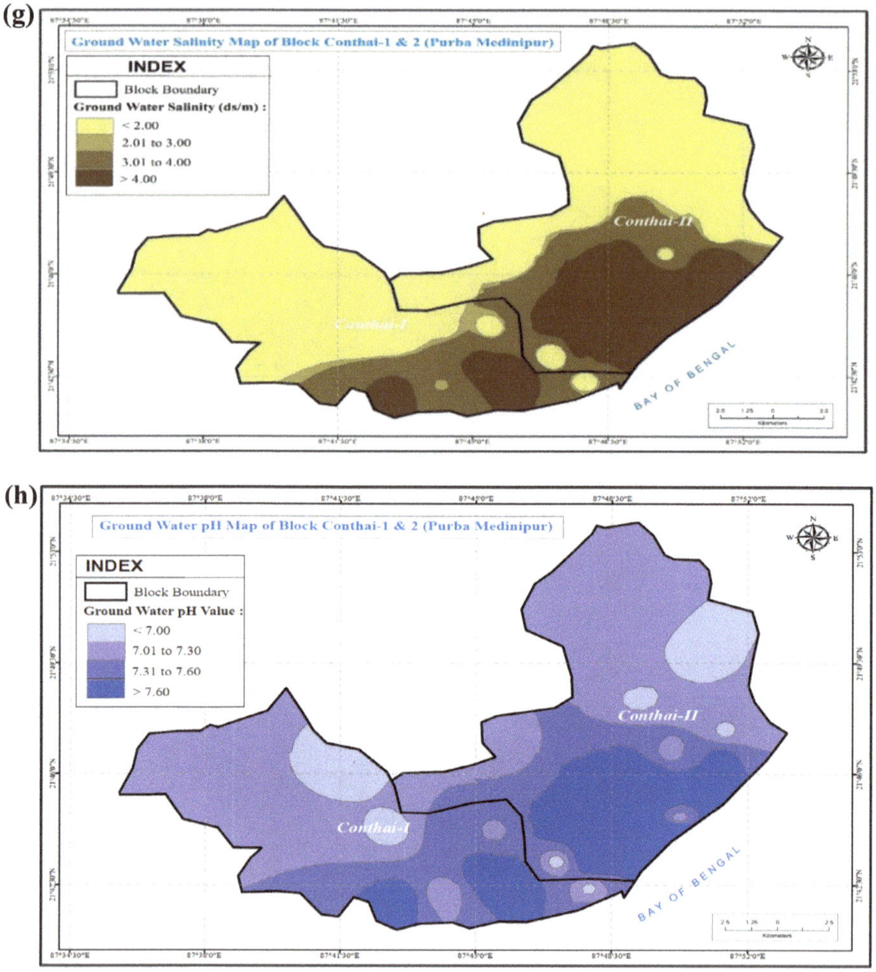

Fig. 13.6 (continued)

These relationships highlight that elevated sodium chloride levels and high alkalinity induce osmotic stress and ion toxicity in plants, impairing root development, seed germination, and nutrient uptake. The negative slopes confirm that managing soil conditions is critical for restoring agricultural productivity (Kang and Li 2020). This thorough examination highlights the diverse effects of climatic hazards on coastal communities, particularly in areas such as Contai 1 and 2. Frequent cyclones and related saline intrusion have resulted in extensive land degradation, diminished agricultural productivity, and socio-economic instability. The deterioration of soil fertility indicators—salinity, pH imbalance, and nutrient deficiencies—has directly led to decreasing crop yields and rising unemployment. Occupational shifts are a common occurrence in coastal communities seeking alternative livelihoods, yet these often lack stability and sustainability. The ongoing impacts necessitate a unified strategy that focuses on climate adaptation and enhancing resilience. Agricultural

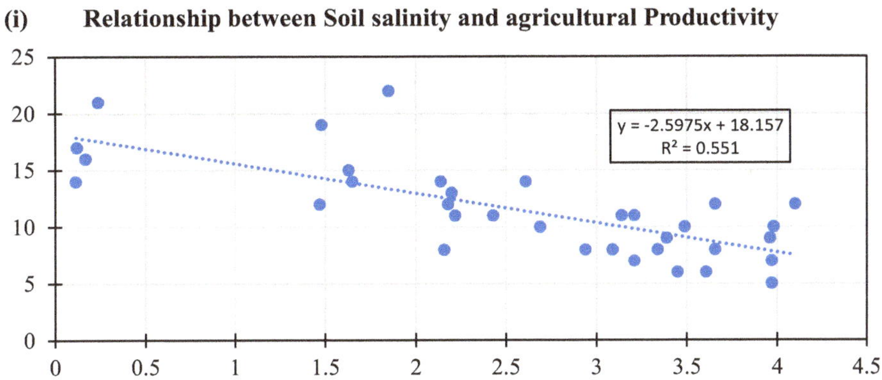

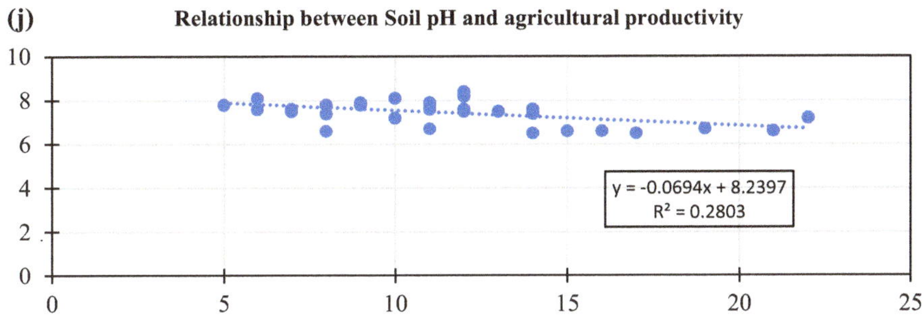

Fig. 13.6 (continued)

communities are striving to address this situation by cultivating salt-tolerant crops and implementing soil modification techniques, improving infrastructure for proper drainage systems, and utilizing remote sensing technologies like NDVI for real-time analysis of plant health. Employing community-driven participatory approaches will also ensure that policies align with local needs and have a greater likelihood of success. In summary, the challenges faced by coastal communities in this region underscore the pressing need for comprehensive strategies that integrate environmental management, social support, and community empowerment. Only through these collaborative efforts can we mitigate the adverse effects of climate change, securing sustainable livelihoods and resilient ecosystems for future generations. Thorough evaluations of current strategies—such as the adoption of salt-tolerant crop varieties or the management of sustainable fisheries—are crucial to determine their scalability and long-term viability. Assessing the effectiveness of policies aimed at climate adaptation and land use regulation can guide the creation of more effective governance frameworks. Similar patterns have been observed in various coastal areas worldwide, including those in India. For instance, indicated that the increasing occurrence of cyclones correlates with a decline in agricultural output and a rise in poverty rates in the Bay of Bengal region. Similarly, Das et al. (2022) found that populations in West Bengal are experiencing frequent displacements and are adapting their livelihoods as a reaction to climate-induced calamities (Table 13.5).

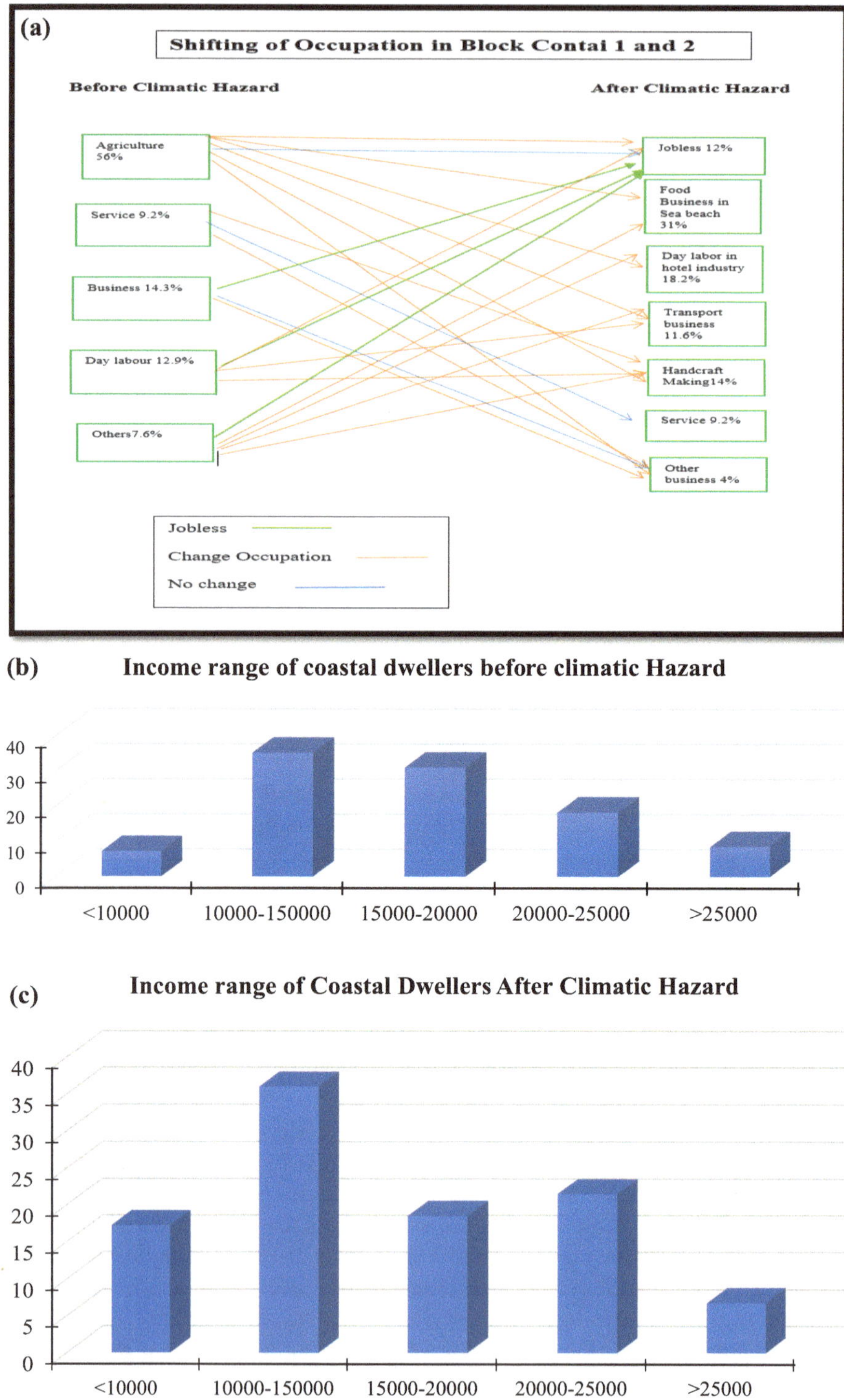

Fig. 13.7 (**a**) Occupational shifting of rural coastal dwellers. (**b**) Income range before climatic hazard. (**c**) Income range before climatic hazard

Table 13.5 Household-level income in pre- and post-cyclonic events

Income range before climatic hazard (in Rupees)	% of Household	Income range after climatic hazard (in Rupees)	% of Household
<10,000	7.1	<10,000	17.2
10,000–150,000	35.2	10,000–150,000	36
15,000–20,000	31.0	15,000–20,000	18.5
20,000–25,000	18.2	20,000–25,000	21.5
>25,000	8.5	>25,000	6.8

Source: Household survey

Conclusion

The study emphasizes on impacts of climate-related phenomena, mainly super cyclones and cyclones, which are becoming more vulnerable because of climate change. Changes in the pattern of land use and variations in land cover by coastal communities increase the risks of land damage, loss of biodiversity, and fewer ecosystem services. The active policies and management are required to reduce negative impacts and figure out the resilience strategies. The findings of this study reveal significant challenges faced by vulnerable coastal communities, particularly in areas like Purba Medinipur, West Bengal. Climatic hazards related phenomena like cyclones, storm surges are extremely impacted on agriculture, local livelihood, income, and jobs (Subhani et al. 2019). The obliteration instigated by these weather events distresses not just individual households but also the extensive local economy, particularly in Contai 1 and 2 blocks, which are very vulnerable to storm surges. These natural disasters lead to extensive property damage, disrupt transportation systems, and create serious risks to public health and safety. As a result, many residents must move to find safer, more stable living conditions (Karl et al. 2009). Rising sea levels make these challenges worse by raising baseline conditions for storm surges, increasing the frequency and severity of coastal flooding (NRC 2010). Recent super cyclones, including Amphan, Yash, Remal, and Dana, have inflicted widespread destruction, destroying homes, damaging infrastructure, and demolishing livelihoods. Many residents were unable to protect their valuable possessions or essential documents amid the chaos of flooding. High tides and storm surges inundated agricultural fields, vegetable gardens, and residential areas, especially when breaches in river embankments allowed saline seawater to flood inland. The destruction of crops and loss of livestock, which are primary sources of income for many families, have plunged communities into economic distress. In the aftermath, many families depend heavily on humanitarian aid, such as food relief provided by NGOs and social organizations, to survive—a clear indication of the urgent need for improved disaster preparedness, early warning systems, and effective relief mechanisms (Subhani et al. 2019). Climate extremes are causing substantial socioeconomic variations in susceptible coastal populations in Purba Medinipore. Gradual increasing of frequent and intense cyclones, storm surge, soil salinization has intermittent the livelihood that those who are depended on agriculture. Deterioration of soil quality due to salinization and alkalinization have decreased the crop productivity. As a result, many agricultural farmers, labours are looking for alternative sources of income. This shift often turns to unstable and less sustainable livelihoods, weakening community stability and eroding the local knowledge and practices that once supported these communities. The proper planning and policies are required from Government administration and relief agencies to mitigate these problems. Sometimes financial benefits are not sufficient to the amount of damage due to climatic hazards. Mainly disaster preparedness, planning of livelihood strategies of affected coastal dwellers, planning of sustainable agricultural practices are required from government organizations to build resilience. Particular strategies commendations are required to encouraging sustainable agricultural systems such as implanting salt-resistant crops in coastal zones, promoting in aquaculture, encouraging to do small scale and tourism industries for the livelihood pf coastal communities. This study evidently ties its objectives and results, screening how climatic extremes have seriously affected the lifestyles of deprived coastal dwellers in Purba Medinipur. The research examined the major challenges of coastal communities to reduce vulnerability. The results of this research indicate that climatic threats, including cyclones, flooding, and salinity intrusion, considerably disrupt rural livelihoods in the coastal areas of Purba Medinipur. These studies support our findings regarding the susceptibility of the agricultural and fisheries sectors, highlighting the necessity for resilient livelihood strategies. Nevertheless, our research builds upon prior studies by specifically examining land use changes over a 10-year period and connecting them to climate impacts, uncovering significant alterations in land cover categories that heighten vulnerability. In contrast to some studies that concentrate exclusively on economic losses, our research encompasses socio-economic and environmental aspects, offering a comprehensive perspective. Despite providing valuable insights, this research has certain limitations. The study mainly depends on data from the past decade, which may not reflect long-term climate variability or historical trends that occurred before this timeframe. The remote sensing data employed for land use analysis possess limited resolution, which may overlook micro-scale land cover changes that are essential for localized decision-making.

Addressing the constraints and broadening future research will enhance the comprehension of climate effects on rural coastal livelihoods and aid in the formulation of modified, sustainable adaptation strategies. Comparative insights derived from global and regional studies emphasize the necessity of integrated approaches that merge environmental management, socio-economic resilience, and policy interventions. Practical implementations may encompass the adoption of climate-smart agricultural techniques, enhancement of early warning systems, and the encouragement of community-driven disaster risk reduction initiatives.

Acknowledgement I want to thank everyone who inspired me to write this chapter and offered helpful feedback and valuable suggestions that improved the quality of this document. Moreover, I spread my honest gratefulness to the office staffs, particularly those from the Irrigation Department in Digha, the Digha Sankarpur Development Authority, and the River Research Institute in Haringhata, West Bengal, as well as Jadavpur University and Amity University in Kolkata, for their assistance in collecting the secondary data.

Some snaps were taken during the field survey

1. Agricultural land damaged due to saline seawater inundation	2. Soil Sample collection from an unproductive agricultural field	3. Soil samples collected from salt salt-tolerant (Chilli) crop field

4. Chilli cultivation field	5. Affected Villagers (social survey)	6. Affected Villagers (household survey)

7. Broken state of concrete structure near the sea beach by Yash	8. Tara pump (Source of non-saline drinking water for villagers)	9. Market Survey after Yash (Affected Villagers)

10. Broken state of concrete structure near the sea beach by Yash	11. Cultivation stopped due to the salinization of agricultural land after inundation	12. Broken state of concrete structure near the sea beach by Yash

13. Homeless Villagers affected by Amphan and Yash	14. House submerged in sea tide after Amphan, Yash	15. Waterlogged Agricultural Land (inundated by saline seawater)

16. Fishery inundated by saline seawater	17. Fallow land due to salinity	18. Broken state of concrete structure near the sea beach by Yash

References

Abedin M, Habiba U, Shaw R (2014) Community perception and adaptation to safe drinking water scarcity: salinity, arsenic, and drought risks in coastal Bangladesh. Int J Disaster Risk Sci 5:110–124

Alexakis DD, Daliakopoulos IN, Panagea IS, Tsanis IK (2018) Assessing soil salinity using WorldView-2 multispectral images in Timpaki, Crete, Greece. Geocarto Int 33(4):321–338

Bastawesy ME, Ali RR, Deocampo DM, Al Baroudi MS (2012) Detection and assessment of the waterlogging in the dryland drainage basins using remote sensing and GIS techniques. IEEE J Sel Top Appl Earth Obs Remote Sens 5:1564–1571

Bradd JM, Milnehome WA, Gates G (1997) Overview of factors leading to dryland salinity and its potential hazard in New South Wales, Australia. Hydrogeol J 5:51–67

Bruckner MZ (2012) Water and soil characterization- pH and electrical conductivity. Microbial Life Educational Resources. Montana State University Bozeman

Dogar MM, Kucharski F & Azharuddin S (2017a). Study of the global and regional climatic impacts of ENSO magnitude using SPEEDY AGCM. J Earth Syst Sci 126, 30. https://doi.org/10.1007/s12040-017-0804-4

Dogar MM, Stenchikov G, Osipov S, Wyman B, and Zhao M (2017b). Sensitivity of the regional climate in the Middle East and North Africa to volcanic perturbations, J. Geophys. Res. Atmos., 122, 7922–7948. https://doi.org/10.1002/2017JD026783.

Dogar MM, Hermanson L, Scaife AA et al. (2023). A Review of El Niño Southern Oscillation Linkage to Strong Volcanic Eruptions and Post-Volcanic Winter Warming. Earth Syst Environ 7, 15–42. https://doi.org/10.1007/s41748-022-00331-z

Dogar MM, Fujiwara M, Zhao M, Ohba M, & Kosaka Y (2024). ENSO and NAO linkage to strong volcanism and associated post-volcanic high-latitude winter warming. Geophysical Research Letters, 51, e2023GL106114. https://doi.org/10.1029/2023GL106114

Dogar MM, Fujiwara M, Ohba M, & Almazroui M (2025a). Investigating post-eruption amplified winter cooling in the Middle East and North Africa—unraveling ENSO and NAO dynamics. Environmental Research Communications, 7(4), 045012. https://doi.org/10.1088/2515-7620/adc675

Dogar MM, Watanabe S, Fujiwara M et al. (2025b) Tropical volcanic impacts on MENA climate via ENSO and NAO dynamics in a high-top model. npj Clim Atmos Sci 8, 330. https://doi.org/10.1038/s41612-025-01201-x

Dogar MM, & Sato T (2018). Analysis of climate trends and leading modes of climate variability for MENA region. Journal of Geophysical Research: Atmospheres, 123, 13,074–13,091. https://doi.org/10.1029/2018JD029003

Dogar MM, & Sato T (2019). Regional climate response of Middle Eastern, African, and South Asian monsoon regions to explosive volcanism and ENSO forcing. Journal of Geophysical Research: Atmospheres, 124, 7580–7598. https://doi.org/10.1029/2019JD030358

Dey RK, Halder P, Nandy S, Ansary R (2025a) Identifying groundwater potential zones using multi-criteria approach in agriculture-dominated landscapes of Malda district, India. Groundwater Sustain Dev 2025:101497. https://doi.org/10.1016/j.gsd.2025.101497

Dey RK, Ansary R, Mandal S (2025b) Impact of climate variability on mango cultivation: a case study in Malda District of West Bengal, India. In: Livelihoods and well-being in the era of climate change: risk to resilience across India. Springer Nature Switzerland, Cham, pp 217–232. https://doi.org/10.1007/978-3-031-81132-6

FAO (2021) World Soil Day: FAO highlights the threat of soil salinization to global food security. https://www.fao.org/newsroom/detail/world-soil-day-fao-highlights-threat-of-soil-salinization-to-food-security-031221/en

Fourati HT, Bouaziz M, Benzina M, Bouaziz S (2017) Detection of terrain indices related to soil salinity and mapping salt-affected soils using remote sensing and geostatistical techniques. Environ Monit Assess 189:177

Gopalakrishnan T, Kumar L (2020) Modeling and mapping of soil salinity and its impact on paddy lands in Jaffna Peninsula, Sri Lanka. Sustainability 12:8317

Gray CL (2011) Soil quality and human migration in Kenya and Uganda. Glob Environ Chang 21:421–430

Halder B, Das S, Bandyopadhyay J, Banik P (2021) The deadliest tropical cyclone 'Amphan': investigate the natural flood inundation over South 24 Parganas using Google Earth Engine. SN Appl Sci 5:1–12. https://doi.org/10.1007/s42797-021-00035-z

Halder P, Satpati L, Dey RK (2022) Water security in the Indian Sundarban under the changing climate. In: Sundarbans: society, environment & development. Raidighi College, pp 94–123. ISBN 978-8-19-511040-7

Halder P, Dey RK, Mandal S (2023) Long-period trend analysis of annual and seasonal rainfall in West Bengal, India (1901–2020). Theor Appl Climatol 154(1–2):685–703. https://doi.org/10.1007/s00704-023-04577-z

Kang Y, Li X (2020) Agricultural utilization and vegetation establishment on saline-sodic soils using a water-salt regulation method for scheduled drip irrigation. Agric Water Manage 231:105995

Karl TR, Melillo JM, Peterson TC (2009) Global climate change impacts in the United States. U.S. Global Change Research Program

Khan A, Mandal S, Satpati L, Dogar MM, Hasanean H, Salah AM (2025) Analyzing drought: the impact of climate change on precipitation patterns in the Shilabati River Basin, West Bengal, India. Earth Syst Environ. https://doi.org/10.1007/s41748-025-00679-y

Lavender SL, Hoeke RK, Abbs DJ (2018) The influence of sea surface temperature on the intensity and associated storm surge of tropical cyclone Yasi: a sensitivity study. Nat Hazards Earth Syst Sci 18(3):795–805. https://doi.org/10.5194/nhess-18-795-2018

Mahadevia K, Vikas M (2010) Climate change–impact on the Sundarbans: a case study. Int Sci J:7–15. http://environment.scientific-journal.com
Mallick B, Rahaman KR (2011) Coastal livelihood and physical infrastructure in Bangladesh after Cyclone Aila. Mitig Adapt Strateg Glob Chang 16:629–648
Mandal S, Choudhury BU (2014) Estimation and prediction of maximum daily rainfall at Sagar Island using best fit probability models. Theor Appl Climatol 117:87–97
Mandal S, Choudhury BU, Mondal M, Bej S (2013) Trend analysis of weather variables in Sagar Island, West Bengal, India: a long-term perspective (1982–2010). Curr Sci 105:947–953
Mandal S, Choudhury BU, Satpati LN (2015) Monsoon variability, crop water requirement, and crop planning for kharif rice in Sagar Island, India. Int J Biometeorol 59:1891–1903. https://doi.org/10.1007/s00484-0150995-9
Mandal S, Satpati LN, Choudhury BU, Sadhu S (2018) Climate change vulnerability to the agrarian ecosystem of the small island: evidence from Sagar Island, India. Theor Appl Climatol 132:451–464. https://doi.org/10.1007/s00704-017-2098-5
Messner F, Meyer V (2006) Flood damage, vulnerability and risk perception—challenges for flood damage research. In: Flood risk management: research and case studies. Springer, pp 149–167
Mohsenipour M (2017) The impact of Cyclone Aila on the Sundarban forest ecosystem. Int J Ecol Dev 32:88–97
Moser SC, Davidson MA, Kirshen P, Mulvaney P, Murley JF, Neumann JE, Petes L, Reed D (2014) Coastal zone development and ecosystems. In: Melillo JM, Richmond T, Yohe GW (eds) Climate change impacts in the United States: the third national climate assessment. U.S. Global Change Research Program, pp 579–618
Mukhopadhyay S, Bhandari G, Sinha S (2025) Impact of climate change induced hazards on the livelihood of marginalized coastal dwellers of West Bengal, India. Turkish Journal of Remote sensing, 7(1): 107–124. https://doi.org/10.51489/1636620
Nandargi SS, Mahto SS (2019) Frequency and intensity of tropical disturbances over the Indian region and its neighboring seas with associated rainfall during the monsoon season. Int J Eng Technol 7(3):17–18. https://doi.org/10.1002/eng2.12069
National Research Council (2010) Adapting to the impacts of climate change. The National Academies Press
Parvin GA, Shimi A, Chanda S, Shaw R, Biswas C (2016) Flood in a changing climate: the impact on livelihood and how the rural poor cope in Bangladesh. Climate 4(2):60. https://doi.org/10.3390/cli4040060
Paul SK, Routray JK (2010) Flood proneness and coping strategies: the experiences of two villages in Bangladesh. Disasters 34(3):489–508
Rahman S, Rahman H, Shahid S, Khan RU, Jahan N, Ahmed ZU, Khanum R, Ahmed MF, Ahsan R, Kellett, J, Karuppannan S (2014) Responses to drought and desertification in the Moroccan Drâa Valley Region: Resilience at the expense of sustainability? The International Journal of Climate Change: Impacts and Responses, 5(2):17
Rhein M, Rintoul SR, Aoki S, Campos E, Chambers D, Feely RA, Gulev S, Johnson GC, Josey SA, Kostianoy A, Mauritzen C, Roemmich D, Talley LD, Wang F (2013) Observations: ocean. In: Climate change 2013: the physical science basis. Cambridge University Press
Sinha S, Sharma LK, Nathawat MS (2013) Integrated geospatial techniques for land-use/land-cover and forest mapping of deciduous Munger forests (India). Universal Journal of Environmental Research & Technology, 3(2):190–198.
Sinha S, Sharma LK, Nathawat MS (2015). Improved Land-use/Land-cover classification of semi-arid deciduous forest landscape using thermal remote sensing. The Egyptian Journal of Remote Sensing and Space Science, 18(2):217–233. https://doi.org/10.1016/j.ejrs.2015.09.005
Sinha S, Santra A, Mitra SS (2020) Semi-automated impervious feature extraction using built-up indices developed from space-borne optical and SAR remotely sensed sensors. Advances in Space Research, 66(6):1372–1385. https://doi.org/10.1016/j.asr.2020.05.040
Subhani R, Morshed M, Ahmad (2019) Socio-economic impacts of Cyclone Aila on migrant and non-migrant households in the southwestern coastal areas of Bangladesh. Geosciences (Basel) 9(1):13

Walsh J, Wuebbles D, Hayhoe K, Kossin J, Kunkel K, Stephens G, Thorne P, Vose R, Wehner M, Willis J, Anderson D (2014) Our changing climate (Ch. 2). In: Climate change impacts in the United States: the third national climate assessment. U.S. Global Change Research Program
Wang QB, Xu W, Xue QZ, Su WA (2010) Transgenic Brassica chinensis plants expressing a bacterial codA gene exhibit enhanced tolerance to extreme temperature and high salinity. Journal of Zhejiang University Science B, 11(11):851–861. https://doi.org/10.1631/jzus.B1000137
Werner AD, Bakker M, Post VEA, Vandenbohede A, Lu C, Ataie-Ashtiani B, Simmons CT, Barry DA (2013) Seawater intrusion processes, investigation and management: recent advances and future challenges. Adv Water Resour 51:3–26
Yasuor H, Yermiyahu U, Ben-Gal A (2020) Consequences of irrigation and fertigation of vegetable crops with variable quality water: Israel as a case study. Agric Water Manag 242:106362

Chapter 14
Influence of Salinity Regimes on Mangrove Carbon Dynamics and Associated Livelihood Risks in the Indian Sundarbans

Sufia Zaman, Rupa Banerjee, Prosenjit Pramanick, and Abhijit Mitra

Abstract Salinity is one of the most influential abiotic stressors shaping the ecological and socioeconomic fabric of the Indian Sundarbans. This study investigates how salinity regimes govern mangrove biomass accumulation, carbon sequestration, and associated livelihood dynamics across two contrasting sites, namely Chemaguri (low salinity) and Jharkhali (high salinity). Over 20 years (2004–2024), significant differences in Above Ground Stem Biomass (AGSB) and Carbon (AGSC) were observed, with Chemaguri exhibiting substantially higher sequestration rates (up to 3.11 t C ha^{-1} $year^{-1}$) than Jharkhali (1.32 t C ha^{-1} $year^{-1}$). Hypersalinity at Jharkhali was linked to stunted mangrove growth, reduced carbon storage, and degradation of fisheries, leading to livelihood insecurity and progressive outmigration. Conversely, the mesohaline regime of Chemaguri sustained higher biodiversity, carbon stock, and economic stability. The findings underscore how climate-driven salinity intrusion acts as both an ecological and socioeconomic stressor, altering resource availability and triggering displacement. Integrating hydrological restoration, halophyte bioremediation, and mangrove afforestation emerges as a strategic adaptation pathway to curb hypersalinity, enhance blue carbon potential, and strengthen coastal rural livelihood resilience.

Introduction

Salinity stands as one of the most defining abiotic drivers regulating the structure, function, and long-term resilience of mangrove ecosystems, particularly under the accelerating influence of climate change (Dogar et al. 2020). Variations in freshwater inflow, sea-level rise, and tidal intrusion have intensified salinity gradients, reshaping the ecological composition of mangrove habitats and their associated livelihood services. In the Indian Sundarbans Biosphere Reserve (ISBR), salinity governs not only the spatial distribution of mangrove flora (Table 14.1) but also the

S. Zaman · R. Banerjee · P. Pramanick (✉) · A. Mitra
Department of Oceanography, Techno India University, Kolkata, West Bengal, India

S. Mandal et al. (eds.), *Climate Change Vulnerability*, Climate Change Management, https://doi.org/10.1007/978-3-032-16783-5_14

Table 14.1 Population density of true mangrove species (qualitative) across sectors of the ISBR

Heatmap of 34 True Mangrove Species across ISBR Sectors

Population Density (4=High, 3=Moderate, 2=Low, 1=Not documented)

Species	Western Sector	Central Sector	Eastern Sector
Acanthus ilicifolius	4	4	4
Acanthus volubilis	4	4	4
Aegiceros corniculatum	4	4	4
Aegialitis rotundifolia	4	2	4
Amoora cucullata	1	2	3
Avicennia alba	3	4	3
Avicennia marina	3	4	4
Avicennia officinalis	3	4	4
Brownlowia tersa	3	2	4
Bruguiera cylindrica	3	2	3
Bruguiera gymnorrhiza	4	3	3
Bruguiera parviflora	4	2	3
Bruguiera sexangula	4	2	4
Ceriops decandra	2	4	4
Ceriops tagal	2	3	4
Cynometra ramiflora	1	1	1
Derris trifoliata	4	2	4
Derris umbrellatum	1	1	1
Excoecaria agallocha	4	4	4
Excoecaria bicolor	4	4	3
Heritiera fomes	2	1	3
Hibiscus tiliaceus	1	1	1
Kandelia candel	1	1	1
Lumnitzera racemosa	1	1	1
Nypa fruticans	3	2	4
Phoenix paludosa	1	4	4
Rhizophora apiculata	1	3	3
Rhizophora mucronata	1	3	3
Sonneratia apetala	4	2	4
Sonneratia caseolaris	2	2	4
Tamarix dioica	1	1	1
Tamarix gallica	1	1	1
Xylocarpus granatum	3	3	3
Xylocarpus mekongensis	1	1	1

productivity and sustainability of local resource-dependent communities. Numerous studies have documented its profound control over mangrove growth, regeneration, and survival (Banerjee et al. 2010, 2013, 2017; Mitra et al. 2010; Bhattacharjee et al. 2013; Zaman et al. 2013; Jana et al. 2014; Pal et al. 2017; Agarwal and Mitra

2018; Dutta et al. 2020a). During a recent field survey (January–March 2025) using 10 m × 10 m quadrats across 30 representative plots per sector, 34 true mangrove species were recorded. The results revealed clear sectoral differentiation across the western, central, and eastern Sundarbans, mirroring the region's hydro-ecological and salinity regimes. Broadly tolerant species such as *Avicennia marina*, *Excoecaria agallocha*, and *Sonneratia apetala* exhibited wide distribution and higher population density, whereas stenohaline taxa like *Heritiera fomes* and *Ceriops decandra* were confined to low-salinity pockets. Several species appeared sparsely or were locally absent, indicating ecological stress under rising salinity. These observations establish an essential baseline for tracking biodiversity responses to climate-driven salinity intrusion and for designing adaptive restoration strategies that sustain both mangrove resilience and rural livelihoods.

The ecological implications of this salinity-driven zonation extend far beyond the distribution of species. Differential tolerance to salinity directly influences biomass accumulation and carbon sequestration potential, thereby modulating the blue carbon economy of the delta (Verma et al. 2026). Areas with moderate salinity support vigorous mangrove growth and higher carbon storage, reinforcing coastal protection and enhancing the livelihood base of local inhabitants through fisheries, apiculture, and timber yield. Conversely, hypersaline zones experience stunted vegetation, reduced carbon uptake, and loss of ecosystem services, which together erode income security and accelerate outmigration. Hence, understanding the ecological and socioeconomic linkages of salinity becomes critical for developing climate-resilient livelihood strategies and policy frameworks tailored to the dynamic hydro-ecological landscape of the Sundarbans. This salinity-mediated transformation exemplifies a broader regional challenge where climatic drivers such as erratic rainfall, reduced freshwater inflow, and sea-level rise interact to redefine the livelihood security of coastal populations. The Sundarbans, as a microcosm of climate vulnerability in South Asia, reveals how environmental degradation and human adaptation are intertwined. The gradual intensification of salinity not only threatens biodiversity, natural resource base, and carbon reserves but also disrupts traditional occupations, from artisanal fishing to small-scale farming. Understanding these intertwined ecological and socioeconomic shifts is essential to frame region-specific adaptation measures that promote resilience and sustainable rural livelihoods under a changing climate.

The salinity tolerance spectrum of major true mangroves in the ISBR shows clear ecological partitioning. Low-salinity specialists such as *Sonneratia caseolaris* and *Heritiera fomes* are confined to oligohaline zones (≤9 PSU). Species like *Sonneratia apetala* and *Nypa fruticans* extend into mesohaline waters (up to 15 PSU), reflecting moderate tolerance. In contrast, *Avicennia marina*, *Avicennia officinalis*, and *Avicennia alba* withstand a much wider range (up to 30 PSU), enabling them to dominate hypersaline stretches. *Excoecaria agallocha* demonstrates remarkable elasticity, spanning nearly the entire gradient (5–29 PSU), which explains its ubiquity across the biosphere reserve (Fig. 14.1).

Salinity fluctuation, particularly the trend toward hypersalinity, exerts profound stress on the mangrove ecosystems of the Sundarbans (Mitra and Zaman 2015).

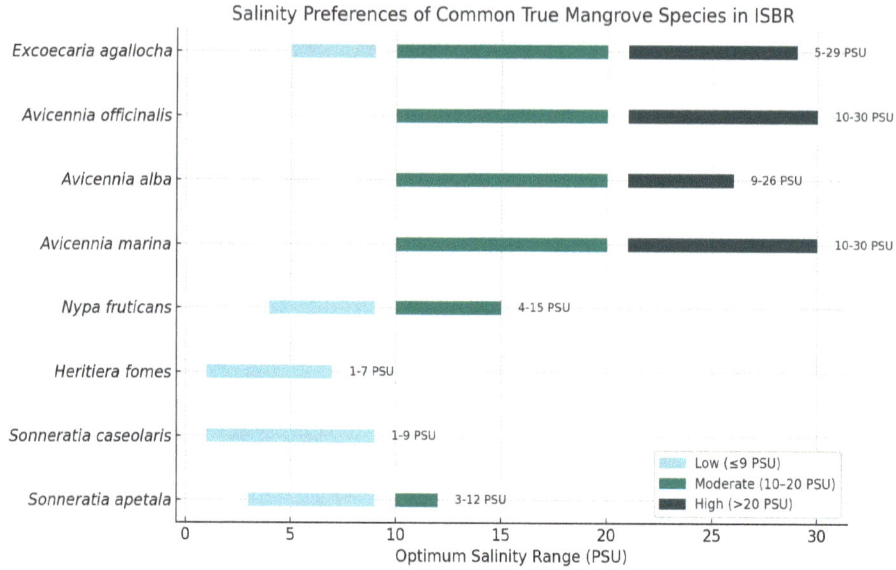

Fig. 14.1 Species with broad ranges (e.g., *Excoecaria agallocha*) show multi-coloured bars, illustrating that they tolerate low, moderate, and high salinity zones simultaneously; Narrow-ranged species (e.g., *Heritiera fomes*) appear in a single colour that matches their specific niche

Elevated salinity disrupts physiological processes such as seed germination, nutrient uptake, and growth, leading to stunted vegetation, higher seedling mortality, and a decline in sensitive species like *Heritiera fomes* and *Sonneratia caseolaris* (Mitra 2013). This ecological imbalance reduces the provisioning services of mangroves, including timber, fuelwood, honey, and medicinal products, while also undermining their regulating functions such as shoreline protection and carbon sequestration. For the island dwellers, whose livelihoods are intimately tied to mangrove resources and estuarine fisheries, hypersalinity triggers dwindling fish catches with more catch volume of trash fishes, scarcity of fodder, and declining apiculture productivity (Fig. 14.2). The cumulative effect leads to livelihood insecurity, forcing households into seasonal migration or risky resource-extraction practices even from within the Reserve Forest (RF), and ultimately amplifying socio-economic vulnerability in one of the world's most climate-sensitive deltas.

Salinity exerts a significant influence on mangrove biomass, with hypersaline conditions often leading to stunted growth, as documented by numerous studies (Mitra et al. 2011, 2012; Banerjee et al. 2013; Bhattacharjee et al. 2013; Bhattacharyya et al. 2015, 2025; Banerjee and Mitra 2015; Roy Chowdhury 2015; Saha et al. 2019; Dutta et al. 2020a). The present research investigates how salinity governs above-ground stem biomass (AGSB) and above-ground stem carbon (AGSC) in two ecologically distinct mangrove zones of the Indian Sundarbans. Chemaguri, located in the western sector along the Hugli estuary, where salinity is comparatively low, and Jharkhali, situated in the central sector along the Matla estuary, characterised by higher salinity levels, have been selected in the present study.

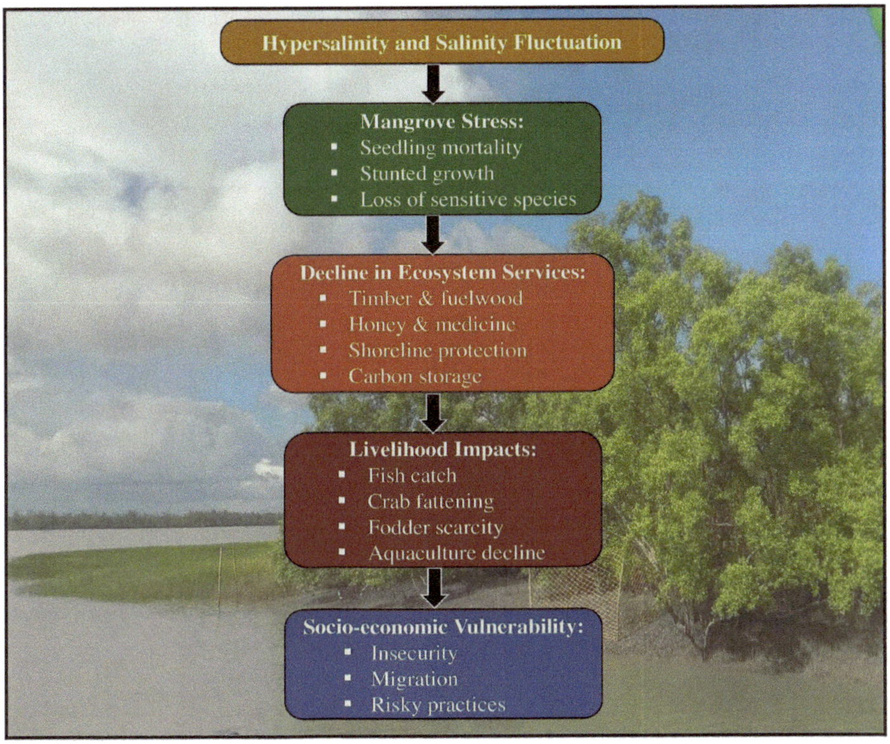

Fig. 14.2 Hypersalinity impacts on mangrove services and livelihoods

This comparative assessment can highlight the contrasting growth responses of mangrove species in low-salinity and high-salinity regimes, providing insight into the broader role of salinity as a key environmental driver shaping biomass and carbon storage capacity in the delta complex that is closely linked to the livelihoods of the islanders.

Materials and Methods

Spatio-Temporal Salinity Variability

This investigation is based on nearly two decades of salinity data (2004–2024), collected through a ground zero survey, and collated from a variety of secondary literature sources (Mitra et al. 2009, 2023, 2024; Banerjee et al. 2013; Mitra 2013; Zaman et al. 2013; Trivedi et al. 2016; Mitra and Zaman 2021). Over this period, numerous ecological and hydrological studies on the Sundarbans have generated long-term datasets and outlined methodological approaches that formed the foundation of the present analysis. To capture spatial heterogeneity across the delta, two

Table 14.2 Geographical coordinates of the selected study sites in the Indian Sundarbans

Sites	Latitude-longitude
Chemaguri (Western sector)	21°39′42.88″N; 88°08′49.01″E
Jharkhali (Central sector)	22°55′58.00″N; 88°41′32.42″E

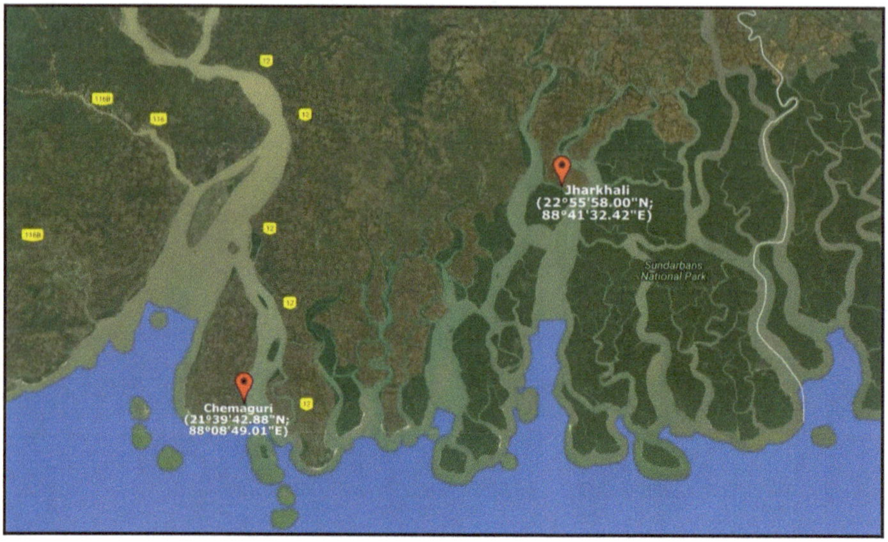

Fig. 14.3 Map of the study sites in the Indian Sundarbans with geographical coordinates

representative stations were selected in the western and central sectors of the Indian Sundarbans (Table 14.2; Fig. 14.3).

At each station, surface water samples were obtained during high tide, with at least three replicates collected within a 250 m radius. The mean value of these replicates was used as the station salinity for a particular sampling event. Seasonal differences were incorporated by sampling across three hydro-climatic phases, namely premonsoon, monsoon, and postmonsoon. Field measurements were conducted with an Atago optical refractometer, providing rapid on-site estimates. These readings were validated through laboratory analysis using the Mohr–Knudsen argentometric titration method (Strickland and Parsons 1972). Additional verification was performed with reference seawater supplied by the National Institute of Oceanography (India), following the protocols described by Mitra et al. (2009). The dual field–laboratory approach ensured high analytical reliability, with a precision of ±0.02 PSU and an overall accuracy of ±0.30 PSU when accounting for triplicate sampling. Importantly, the same protocols and instruments have been maintained consistently since 2004, allowing robust temporal comparison across the dataset.

Selection of Species

The present investigation focuses on five dominant mangrove species, namely *Sonneratia apetala*, *Avicennia marina*, *Avicennia alba*, *Avicennia officinalis*, and *Excoecaria agallocha*, that were examined at two ecologically contrasting sites, Chemaguri and Jharkhali, each representing distinct salinity regimes.

Assessment of Above-Ground Stem Biomass (AGSB), Carbon (AGSC), and Carbon Sequestration Potential (AGSCS)

Field investigations were first carried out in 2004, when the mangrove stands were approximately 11 years old, and re-investigated in 2024, corresponding to a stand age of 31 years. Above-ground stem Biomass (AGSB) of true mangrove species was estimated within 10 m × 10 m quadrats using a non-destructive approach. Tree diameter at breast height (DBH) was recorded using a calliper or measuring tape, while tree height was determined with the assistance of a Phantom-3 Professional drone equipped with GPS, a gimbal, and a 4K high-resolution camera. Stem biomass (B_S) was then computed as $B_S = GV$. The expression for V is $\Pi r^2 HF$, where F is the form factor, r is the radius of the tree derived from its DBH (Diameter at Breast height) and H is the height of the main stem of the target tree. The biomass per species was calculated by multiplying the mean biomass per tree by the number of individuals in all quadrats, and expressed as t ha^{-1}.

To maintain representativeness, the analysis focused only on dominant taxa ($\geq$70% abundance in quadrats), excluding rare and scattered species. Only stems were considered, as they constitute the most persistent carbon reservoir in mangroves. Short-lived components such as branches and foliage, which are frequently lost during tropical cyclones, were omitted to avoid inflating long-term carbon stock estimates.

Carbon content of AGSB, hereafter referred to as Above Ground Stem Carbon (AGSC), was measured directly using a CHN analyser. Stem samples from dominant species were oven-dried at 70 °C, pulverized, and homogenized before being passed through a 0.5 mm mesh. The carbon fraction (%) of each species was then determined using a Vario MACRO Elementar CHN analyser.

A schematic workflow (Fig. 14.4) illustrates the sequential process applied in this study, starting from field-based measurements to laboratory analysis of carbon concentration and final derivation of species-specific AGSC.

Carbon sequestration (t ha^{-1} $year^{-1}$) by the mangrove stem was calculated by dividing the increment in Above Ground Stem Carbon stock between 2004 (11-year stand age) and 2024 (31-year stand age) by the elapsed time (20 years) as per the expression:

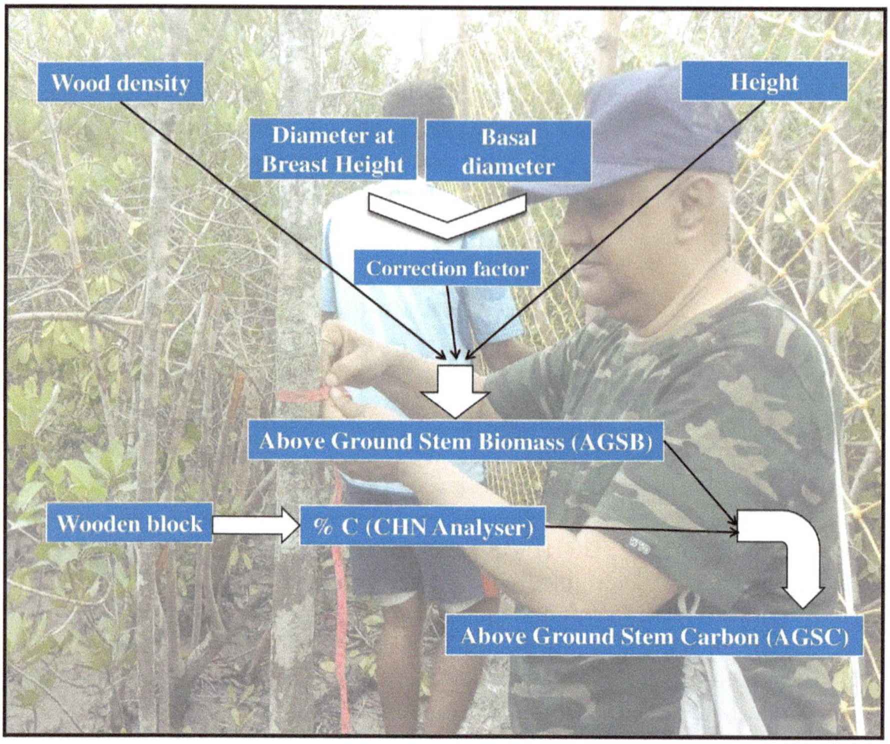

Fig. 14.4 Workflow for estimating AGSB and AGSC

$$\text{Carbon sequestration rate}\left(\text{t ha}^{-1}\,\text{year}^{-1}\right) = \frac{AGSC_{2024\,(31\,\text{years})} - AGSC_{2004\,(11\,\text{years})}}{20}$$

Assessment of Fishery Spectrum Under Shifting Salinity Regimes

Surveys (qualitative) on fish catch composition were carried out to understand ecological responses to salinity stress in the fishery sector of Sundarbans. They provide site-specific evidence of species decline or dominance, helping to track biodiversity shifts over time. These insights are vital for sustainable fisheries management and safeguarding the livelihoods of local communities.

The survey was conducted in December 2024 at selected fish landing units of Chemaguri and Jharkhali to document the qualitative abundance of fish species under prevailing hypersaline conditions. Systematic field visits were made during peak landing hours (early morning and late afternoon), and observations were carried out in collaboration with local fishers and auctioneers. Landed catches were carefully inspected species-wise, with emphasis on identifying both commercially valuable taxa (e.g., *Tenualosa ilisha*, *Liza* spp.) and less valued species (e.g., *Harpadon nehereus*). Species identification was performed using regional fishery manuals and cross-checked with standard taxonomic keys. For each landing unit, repeated observations across multiple days ensured consistency, and fish counts were expressed in terms of relative qualitative abundance (dominant, occasional, rare, and absent) rather than absolute biomass. Informal interviews with fishers supplemented field data, providing insights into temporal shifts in catch composition and perceived changes in abundance. This combined observational and participatory approach ensured a robust qualitative assessment of species assemblages at the two sites.

Results

Salinity Alteration in the Sundarbans: Ecological and Socio-Economic Dimensions

The contrasting salinity regimes have direct ecological and socio-economic implications.

Chemaguri in the western Sundarbans exhibited moderate but highly contrasting seasonal salinity, with sharp monsoon dilution and recent premonsoon spikes, while Jharkhali in the central sector showed persistently high salinity across all seasons. This contrast highlights stronger freshwater influence in Chemaguri versus tidal dominance and hypersalinity stress in Jharkhali.

Over the past two decades, the western sector has shown a moderate salinity regime with strong seasonal fluctuations. Premonsoon values averaged around 23.3 PSU, peaking at 31.5 PSU in extreme years, while monsoon levels declined sharply to as low as 5.9 PSU, reflecting significant freshwater influx (Fig. 14.5). Postmonsoon salinity stabilized between 11.5 and 22.7 PSU. This oscillating pattern underscores the seasonal balance between freshwater discharge and tidal intrusion. However, in the latter years of the record, there is a tendency toward higher premonsoon salinity spikes, suggesting a gradual intensification of dry-season hypersalinity, which could influence seedling recruitment and species composition in the region.

In contrast, the central sector experienced consistently elevated salinity levels, with premonsoon values averaging above 31 PSU and reaching as high as 35 PSU. Monsoon levels remained comparatively high (23–28 PSU), showing limited dilution despite rainfall, while postmonsoon salinity frequently stabilized near 30 PSU (Fig. 14.6). This persistently high salinity reflects the reduced freshwater inflows into the central estuarine stretches, coupled with tidal dominance. Over the 20 years, the central sector exhibited a gradual upward trend in all three seasons, indicating an expanding salinity envelope. Such a regime exerts strong selective pressure on mangrove vegetation, favoring halophytic species like *Avicennia marina*

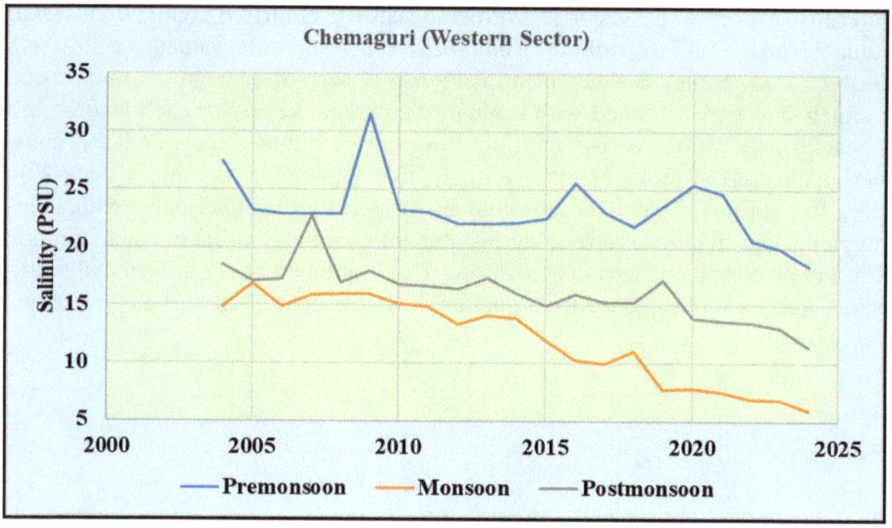

Fig. 14.5 Alteration of salinity level (in PSU) across three seasons in Chemaguri

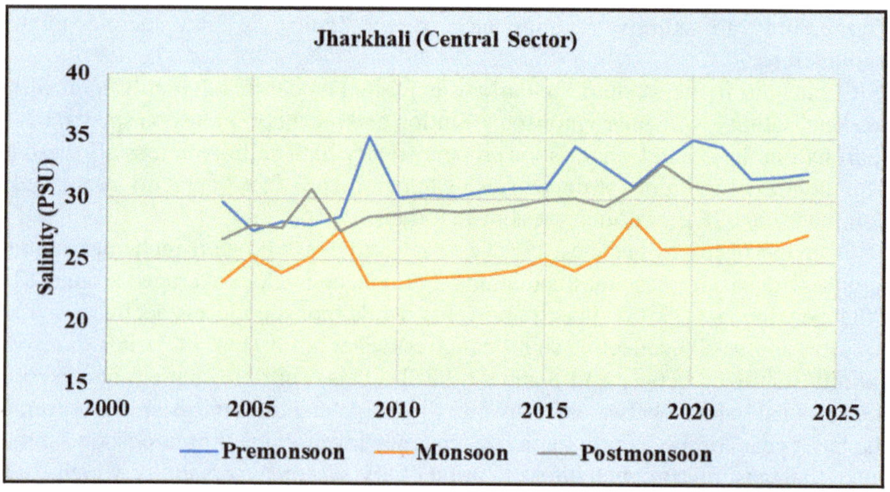

Fig. 14.6 Alteration of salinity level (in PSU) across three seasons in Jharkhali

and limiting the distribution of freshwater-loving taxa, with cascading impacts on fisheries and local livelihoods.

The contrasting salinity regimes trigger the gradual displacement of local people through ecosystem degradation. In Chemaguri, seasonal dilution still sustains fisheries and agriculture, reducing immediate migration pressure. Jharkhali's chronic hypersalinity, however, diminishes freshwater-dependent crops, reduces fish diversity, and shrinks income opportunities. Salinity-driven soil infertility and declining ecosystem services force families to seek alternative livelihoods. Over time, this ecological stress translates into climate-induced displacement from the central Sundarbans.

Influence of Salinity Regimes on Biomass Accumulation and Carbon Storage in Mangrove Ecosystems

Long-term monitoring from 2004 to 2024 revealed marked contrasts in carbon dynamics between Chemaguri and Jharkhali. At Chemaguri, *Sonneratia apetala* showed remarkable growth, with AGSB rising from 41.64 to 165.64 t ha^{-1} and AGSC from 20.11 to 82.32 t ha^{-1}, reflecting a strong sequestration rate of 3.11 t C ha^{-1} $year^{-1}$. *Avicennia marina* also recorded steady gains, with AGSB increasing from 25.76 to 87.76 t ha^{-1} and AGSC from 12.26 to 43.79 t ha^{-1}, yielding 1.58 t C ha^{-1} $year^{-1}$. Comparable upward trends were observed for *A. alba* and *A. officinalis*, with respective sequestration rates of 1.81 and 1.98 t C ha^{-1} $year^{-1}$. In contrast, *Excoecaria agallocha* accumulated far less, with AGSB rising from 12.19 to 40.19 t ha^{-1} and AGSC from 5.02 to 17.52 t ha^{-1}, averaging 0.63 t C ha^{-1} $year^{-1}$.

At Jharkhali, elevated salinity constrained biomass productivity. *S. apetala* recorded lower increments, with AGSB increasing from 29.09 to 81.09 t ha^{-1} and AGSC from 13.79 to 40.14 t ha^{-1}, corresponding to 1.32 t C ha^{-1} $year^{-1}$. Although *A. marina*, *A. alba*, and *A. officinalis* displayed relatively better sequestration (2.56, 2.25, and 2.48 t C ha^{-1} $year^{-1}$, respectively), their values were still below those at Chemaguri. *E. agallocha* again performed modestly, with AGSC increasing from 5.63 to 20.14 t ha^{-1} (0.73 t C ha^{-1} $year^{-1}$). Overall, Chemaguri consistently outperformed Jharkhali across all species, especially *S. apetala*, underscoring the advantage of low-salinity conditions (Table 14.3).

The findings highlighted in Table 14.3 clearly show that reduced salinity enhances mangrove carbon sequestration, while high salinity imposes osmotic stress and ion toxicity, suppressing biomass accumulation. Ecologically, stunted mangrove growth under saline stress weakens shoreline protection, heightens erosion and cyclone vulnerability, and diminishes vital ecosystem services that sustain local livelihoods. Within the broader South Asian context, such salinity-driven transformations exemplify how regional climate variability directly reshapes coastal livelihood security. The Indian Sundarbans, like many deltaic systems across South Asia, thus stands as a critical case study illustrating the intersection of climate-induced hydrological change, carbon cycle disruption, and socio-economic fragility. This linkage underscores the urgent need for adaptive management strategies that integrate ecological resilience with community-based livelihood restoration.

Shifting Salinity Regimes and Their Impact on Fishery Spectrum

Our 2024 seasonal survey, based on systematic observations at site-specific fish landing units across the central Sundarbans, revealed a clear shift in fish species composition under hypersaline conditions. Traditionally important and commercially valuable species such as *Tenualosa ilisha*, *Liza tade*, and *Liza parsia* showed a marked decline in catch volumes at Jharkhali, reflecting their sensitivity to

Table 14.3 AGSB and AGSC of 11- and 31-year-old dominant mangrove tree species in the selected stations; the figures within brackets represent the percentage of carbon in the dominant mangrove tree biomass

Species	Stations	Population density (No./100 m^2)	Age-wise AGSB (t/ha)		Age-wise AGSC (t/ha)		Carbon sequestration = ΔAGSC/Δt (tonnes/ha/year)
			2004 9- year	2024 29-year	2004 9-year	2024 29-year	
Sonneratia apetala	Chemaguri	9	41.64	165.64	20.11 (48.3%)	82.32 (49.7%)	3.11
	Jharkhali	3	29.09	81.09	13.79 (47.4%)	40.14 (49.5%)	1.32
Avicennia marina	Chemaguri	7	25.76	87.76	12.26 (47.6%)	43.79 (49.9%)	1.58
	Jharkhali	7	35.85	139.85	17.46 (48.7%)	68.67 (49.1%)	2.56
Avicennia alba	Chemaguri	5	27.52	99.52	13.35 (48.5%)	49.56 (49.8%)	1.81
	Jharkhali	9	37.84	129.84	18.58 (49.1%)	63.62 (49.0%)	2.25
Avicennia officinalis	Chemaguri	7	27.92	105.92	13.74 (49.2%)	53.38 (50.4%)	1.98
	Jharkhali	7	38.13	134.13	18.76 (49.2%)	68.41 (51.0%)	2.48
Excoecaria agallocha	Chemaguri	4	12.19	40.19	5.02 (41.2%)	17.52 (43.6%)	0.63
	Jharkhali	3	13.62	47.62	5.63 (41.3%)	20.14 (42.3%)	0.73

elevated salinity levels (Table 14.4). In contrast, these declines were accompanied by a noticeable rise in landings of low-value species such as *Harpadon nehereus* (Bombay duck), which are more tolerant of saline stress (Table 14.4). This transition underscores how hypersalinity is reshaping fishery resources, with significant ecological and socio-economic implications (Mitra 2013).

Table 14.4 Seasonal variation in dominant fish species across Western (Chemaguri) and Central (Jharkhali) sundarbans under contrasting salinity regimes

Dominant fish species documented during the survey period	Chemaguri (western sector)			Jharkhali (central sector)		
	Pre	Mon	PoM	Pre	Mon	PoM
 *Tenualosa ilisha* (Family: Clupeidae)	Ab	D	O	Ab	R	Ab
 *Pama pama* (Family: Sciaenidae)	O	O	O	R	Ab	Ab
 *Pampus* spp. (Family: Stromateidae)	O	R	O	R	Ab	Ab
 *Ilisha elongata* (Family: Pristigasteridae)	Ab	D	O	Ab	Ab	Ab
 *Lates calcarifer* (Family: Centropomidae)	D	O	O	Ab	R	Ab

(continued)

Table 14.4 (continued)

Dominant fish species documented during the survey period	Chemaguri (western sector)			Jharkhali (central sector)		
	Pre	Mon	PoM	Pre	Mon	PoM
 *Pangasius pangasius* (Family: Pangasiidae)	O	O	O	Ab	Ab	Ab
 *Liza parsia* (Family: Mugilidae)	D	D	D	Ab	O	R
 *Liza tade* (Family: Mugilidae)	D	D	D	R	O	R
 *Tenualosa toli* (Family: Clupeidae)	Ab	D	Ab	Ab	Ab	Ab
 *Polynemus paradiseus* (Family: Polynemidae)	R	O	O	Ab	R	Ab
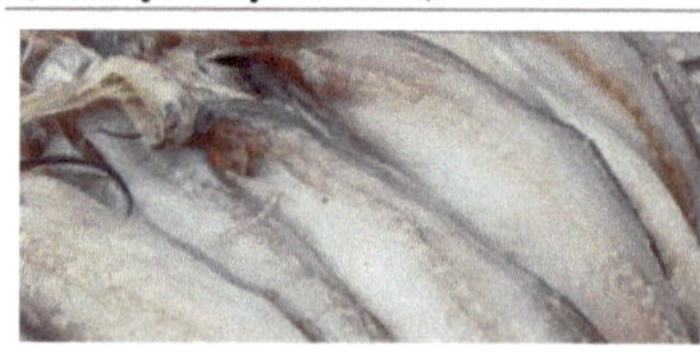 *Harpadon nehereus* (Family: Synodontidae)	R	Ab	R	D	D	D

Index: Pre, Premonsoon; Mon, Monsoon; PoM, Postmonsoon; D, Dominant; O, Occasional; R, Rare; Ab, Absent

Discussion

Contrasting Salinity Regimes and Their Socio-Ecological Consequences in the Western and Central Indian Sundarbans

The Sundarbans estuary demonstrates sharp contrasts in salinity regimes between its western and central sectors, shaped by hydrological and geomorphological processes. Chemaguri, located in the western Sundarbans, benefits from the active discharge of the Hugli River, which introduces large volumes of freshwater during the monsoon and continues to exert influence even in the postmonsoon period. This discharge moderates salinity levels and generates pronounced seasonal oscillations. In premonsoon months, salinity can rise to higher values, but with the onset of monsoon rainfall and fluvial influx, levels decline steeply, often approaching single digits. Postmonsoon conditions stabilize at intermediate ranges, balancing tidal inflow with residual freshwater. This seasonal rhythm has long supported fisheries and small-scale agriculture, providing a measure of livelihood stability for local communities (Mitra et al. 2023).

Jharkhali in the central Sundarbans presents a different picture. Here, heavy siltation in tidal creeks and channels has significantly reduced the penetration of freshwater into the estuarine body. As a result, even during the monsoon, dilution is limited, and salinity remains elevated. Premonsoon values frequently exceed 31 PSU, while monsoon levels rarely fall below 23 PSU, and postmonsoon salinity stabilizes close to marine conditions. Over the past two decades, this pattern has intensified, with a gradual upward trend evident across all seasons, indicating a regime of persistent hypersalinity. Such conditions favour salt-tolerant mangroves like *Avicennia marina* but restrict species requiring brackish or freshwater habitats. The shift in vegetation, combined with reduced fish diversity and abundance, undermines the ecological base that once supported traditional livelihoods.

For residents, these changes carry profound socio-economic consequences. In Chemaguri, seasonal freshwater pulses still sustain crops and estuarine fisheries, delaying large-scale livelihood crises. In Jharkhali, however, chronic hypersalinity leads to soil infertility, crop failure, and the collapse of freshwater-dependent agriculture. Artisanal fishers face declining catches of commercially important species such as *Tenualosa ilisha* and mullets, replaced by lower-value, salt-tolerant species. Household incomes shrink, compelling families to seek alternatives beyond their islands. This ecological stress translates into climate-induced displacement, with many residents migrating westward to areas around Chemaguri where conditions are relatively more favourable, or moving further to the city of Kolkata in search of wage labour and urban livelihoods. Thus, the contrasting salinity profiles are not merely hydrological phenomena but catalysts of social restructuring in the deltaic landscape.

Salinity Stress and Its Cascading Impacts on Mangrove Ecosystem Services and Human Well-Being

Long-term monitoring of mangrove stands in the Indian Sundarbans reveals a clear relationship between salinity regimes, biomass accumulation, and carbon storage. The data from Chemaguri and Jharkhali demonstrate that reduced salinity favours vigorous growth and higher carbon sequestration, whereas persistent hypersalinity imposes physiological stress, constraining mangrove productivity. These differences are not merely ecological in nature but also translate directly into variations in the quality of ecosystem services and, consequently, the socio-economic well-being of island communities.

At Chemaguri, where the Hugli River discharges moderate salinity, mangroves such as *Sonneratia apetala* and *Avicennia* spp. recorded substantial increases in above-ground stem biomass (AGSB) and carbon stock (AGSC) between 2004 and 2024. *S. apetala* sequestered carbon at a rate exceeding 3 t C ha^{-1} $year^{-1}$, while the three *Avicennia* species ranged between 1.5 and 2 t C ha^{-1} $year^{-1}$. Such robust growth strengthens the ecological services these forests provide: shoreline stabilization, attenuation of storm surges, and the maintenance of productive estuarine habitats for fisheries. For local dwellers, this translates into better protection from cyclones and erosion, more reliable fish catches, and access to forest products that supplement household income.

Jharkhali, in contrast, is dominated by hypersaline conditions due to reduced freshwater inflow and siltation in tidal channels. Here, the same species exhibit diminished growth. *S. apetala* shows less than half the sequestration rate recorded at Chemaguri, while *Excoecaria agallocha*—already a poor accumulator—remains stunted. Similar results were supported by several earlier researchers (Mitra et al. 2011, 2012; Sengupta et al. 2013; Agarwal et al. 2017; Ahmed et al. 2021; Dutta et al. 2022; Bhattacharyya et al. 2025). Even though *Avicennia marina* and its congeners show slightly higher sequestration values than at Chemaguri, their productivity remains insufficient to counterbalance the broader ecological stress imposed by persistent salinity. The weakened mangrove canopy provides less effective shoreline defence, increasing the vulnerability of islanders to cyclonic inundation and saline water intrusion into agricultural fields.

The implications for livelihoods are severe. Agricultural productivity in hypersaline zones is declining as soils lose fertility and crops fail, forcing households to abandon agricultural practice. Fisheries, once supported by the nursery role of diverse mangrove habitats, are shrinking in both diversity and volume, particularly in high-value species such as *Tenualosa ilisha*. As fish landings collapse, incomes dwindle, leaving island dwellers with fewer economic options. Women and children, often dependent on fuelwood and honey collection, face heightened hardships as forest yields fall. Over time, these ecological pressures create a socio-economic push toward out-migration, with families moving westward to less saline areas or relocating entirely to urban centres such as Kolkata in search of wage labour.

In essence, hypersalinity not only restricts mangrove growth and carbon sequestration but also undermines the very services that sustain human resilience in the delta. The contrast between Chemaguri and Jharkhali demonstrates how salinity stress cascades from ecological limitation to socio-economic displacement, highlighting the urgent need for freshwater restoration and adaptive livelihood planning in the central Sundarbans.

Salinity-Driven Shifts in Fish Species Composition and Livelihood Risks in the Sundarbans

The seasonal analysis of fish landing stations across the Indian Sundarbans presents a vivid picture of how salinity gradients, intensified by climate-driven hydrological alterations, are reshaping aquatic biodiversity and coastal livelihoods. A stark divergence emerges between the western and central sectors—two regions that, despite their geographic proximity, now represent contrasting ecological and socioeconomic realities. In the western sector, particularly around Chemaguri, the estuarine ecosystem continues to support a diverse and commercially valuable fishery, largely sustained by moderate salinity regimes and periodic freshwater influx from the Hugli River. In contrast, Jharkhali, situated within the central Sundarbans, is increasingly characterized by hypersalinity, reduced freshwater inflow, and diminished ecological productivity, culminating in an alarming contraction of high-value fish species. At Chemaguri, seasonal assessments revealed a relatively stable fish assemblage dominated by estuarine-dependent species that have historically underpinned local livelihoods. Among them, *Liza parsia* and *Liza tade* maintained dominance throughout premonsoon, monsoon, and postmonsoon periods, reflecting the continued ecological suitability of this sector for mullet populations. The iconic *Tenualosa ilisha*—the hilsa shad—remained absent during the dry months but reappeared dominantly in the monsoon, consistent with its well-known migratory behaviour synchronised with freshwater discharge for spawning (Ahsan et al. 2014; Dutta et al. 2020b). Similarly, *Lates calcarifer* (Asian sea bass) was found dominant before the monsoon and occasionally during the monsoon, suggesting that the premonsoon period continues to offer a favourable recruitment window. Even *Pangasius pangasius*, though never abundant, persisted as an occasional species across seasons, underscoring that Chemaguri still retains critical nursery functions for commercially important fishes. Collectively, these patterns signify that the western Sundarbans—though not immune to climatic stress—continues to function as a resilient estuarine zone capable of sustaining biodiversity, fishery productivity, and rural livelihoods. In sharp contrast, Jharkhali's fishery scenario mirrors the ecological degradation associated with chronic hypersalinity. The long-term elevation of salinity, driven by declining freshwater inflows, tidal dominance, and siltation of distributary channels, has precipitated a severe reduction in species richness and commercial value. The once-abundant *Tenualosa ilisha* has nearly disappeared from catches, surfacing only sporadically during brief monsoonal dilution events, an unmistakable indicator of the species' sensitivity to sustained salinity stress. Similarly, mullets such as *Liza parsia* and *Liza tade*, dominant in Chemaguri, have been reduced to rare or occasional occurrences, while other estuarine species like *Lates calcarifer*, *Pama pama*, and *Pampus* spp. have become nearly absent year-round. This decline underscores the loss of ecological connectivity within the central estuary, which has ceased to function effectively as a migratory corridor or nursery ground for brackish water fishes. Replacing these sensitive, high-value taxa are salt-tolerant and low-commercial species such as *Harpadon nehereus* (Bombay duck), which displayed an increased presence during the 2024 survey. Although hardy and adaptable, *H. nehereus* cannot compensate for the economic and nutritional losses associated with the decline of premium species. Its dominance represents a shift from biodiversity-rich, high-value fisheries to a simplified, low-value regime symptomatic of environmental stress. For local fishing households, this transition translates into reduced income, shrinking market opportunities, and heightened livelihood insecurity. The economic repercussions are particularly severe for small-scale and artisanal fishers, whose income depends on daily landings and whose capacity to adapt to ecological change remains limited.

The ecological simplification observed at Jharkhali thus mirrors a broader socio-economic transformation. As high-value fish resources dwindle, households face declining earnings, increased debt, and growing dependence on alternative or supplementary livelihoods such as daily wage labour or migration to urban centers like Kolkata. The interplay between ecological degradation and livelihood erosion exemplifies how climate-induced salinity intrusion extends beyond environmental parameters to reshape the social fabric of deltaic communities. The comparative assessment between Chemaguri and Jharkhali across premonsoon, monsoon, and postmonsoon seasons vividly demonstrates that salinity has become the critical ecological filter in the Sundarbans. Where moderate salinity allows seasonal recruitment pulses and estuarine connectivity, fish diversity, ecosystem function, and income stability are sustained. Conversely, in hypersaline zones, the collapse of ecological complexity is accompanied by the proliferation of opportunistic, low-value taxa—eroding both biodiversity and economic resilience. Without targeted interventions, such as restoring freshwater connectivity, dredging silted creeks, and conserving mangrove buffers, the central Sundarbans faces the risk of irreversible transition into a degraded, salt-tolerant fishery regime.

Thus, it can be stated that salinity is not merely an environmental variable but a socio-ecological determinant of human well-being. Its progressive intensification under changing climate conditions demands an integrated management approach that simultaneously addresses hydrological restoration, habitat conservation, and livelihood diversification. Only through such adaptive interventions can the Sundarbans maintain its ecological integrity, sustain its fisheries, and secure the livelihoods of the millions who depend upon its fragile yet invaluable estuarine resources.

Conclusion

The present investigation reaffirms salinity as a central ecological regulator governing mangrove productivity, carbon sequestration potential, fishery dynamics, and the broader socio-economic resilience of the Indian Sundarbans. The comparative assessment between Chemaguri in the western sector and Jharkhali in the central sector reveals how divergent salinity regimes—shaped by freshwater inflow patterns and tidal dominance—create two contrasting trajectories of ecological and livelihood outcomes. Chemaguri, benefiting from periodic freshwater discharge from the Hugli River, sustains vigorous mangrove growth, higher above-ground biomass, and enhanced carbon storage. These attributes collectively reinforce coastal protection, support biodiversity, and sustain the livelihoods of local communities through productive fisheries, apiculture, and forest-based resources.

Conversely, Jharkhali's chronic hypersalinity, intensified by reduced freshwater inflows and channel siltation, constrains mangrove productivity, depresses carbon sequestration rates, and weakens vital ecosystem services. The physiological stress of salinity manifests in stunted vegetation and reduced ecological functionality, translating directly into the socio-economic sphere. High-value estuarine fish species such as *Tenualosa ilisha*, *Liza parsia*, and *Liza tade* have declined sharply, replaced by salt-tolerant but low-value species like *Harpadon nehereus*, resulting in decreased fishery income and increased livelihood vulnerability. Concurrently, soil salinization undermines agriculture, while the degradation of mangrove buffers heightens exposure to cyclones and erosion—creating a web of interlinked pressures that amplify human insecurity.

The cascading impacts of hypersalinity thus extend beyond ecological degradation to drive demographic and social transformations. As livelihood opportunities shrink, families from the central Sundarbans increasingly migrate toward relatively favorable environments such as Chemaguri or urban centers like Kolkata, marking a growing pattern of climate-induced displacement. In this sense, salinity emerges not simply as a hydrological condition but as a profound agent of socio-ecological restructuring within the deltaic landscape.

The findings underscore the urgent need for integrated and adaptive management interventions that address salinity regulation as a cornerstone of resilience-building. Restoring freshwater connectivity through controlled riverine discharge, promoting halophyte-based bioremediation, and expanding mangrove restoration programs can simultaneously mitigate hypersalinity, enhance blue carbon sequestration, and restore ecosystem services essential for community well-being. Such nature-based interventions, when embedded within participatory livelihood planning, hold the potential to stabilize both ecological processes and rural economies.

Ultimately, this study highlights that managing salinity is not merely an environmental imperative but a socio-economic necessity for sustaining the delicate balance between ecosystems and human livelihoods in one of the world's most climate-sensitive coastal frontiers. By integrating hydrological restoration with climate-resilient livelihood frameworks, the Sundarbans can evolve from a zone of vulnerability into a living model of adaptive sustainability for deltaic systems across South Asia and beyond.

References

Agarwal SK, Mitra A (2018) Salinity: a primary growth driver of mangrove flora. Curr Trend For Res 2:1–9

Agarwal S, Pal N, Zaman S, Mitra A (2017) Role of mangroves in carbon sequestration: a case study from Prentice Island of Indian Sundarbans. Int J Basic Appl Res 7(12):35–42

Ahmed S, Zaman S, Mitra A (2021) Carbon sequestration in a common true mangrove flora, Excoecaria agallocha of Indian Sundarbans. Chapter 7. In: Mitra A, Calma MM, Chakrabarty SP (eds) Natural resources and their ecosystem services, vol II. HSRA Publication, pp 138–143. ISBN 978-93-5506-064-8

Ahsan DA, Naser MN, Bhaumik U, Hazra S, Bhattacharya SB (2014) Migration, spawning patterns and conservation of Hilsa Shad (*Tenualosa ilisha*) in Bangladesh and India: ecosystems for life: a Bangladesh-India initiative. International Union for Conservation of Nature and Natural Resources, p 97

Banerjee K, Mitra A (2015) Mangrove biomass and salinity: is the match made in heaven? Int J Biol Sci Eng:45–54

Banerjee K, Vyas P, Chowdhury R, Mallik A, Mitra A (2010) The affects of salinity on the mangrove growth in the lower Gangetic delta. J Ind Ocean Stud 18(3):389–397

Banerjee K, Sengupta K, Raha AK, Mitra A (2013) Salinity based allometric equations for biomass estimation of Sundarban mangroves. Biomass Bioener 56:382–391

Banerjee K, Gatti RC, Mitra A (2017) Climate change-induced salinity variation impacts on a stenoecious mangrove species in the Indian Sundarbans. Ambio 46:492–499

Bhattacharjee AK, Zaman S, Raha AK, Mitra A (2013) Impact of salinity on above ground biomass and stored carbon in a common mangrove *Excoecaria agallocha* of Indian Sundarbans. Am J Biopharm Biochem Life Sci 2(2):1–11

Bhattacharyya S, Mitra A, Raha AK (2015) Stored carbon in above ground biomass of dominant mangrove floral species in Sagar Island of Indian Sundarbans. J Chem Biol Phys Sci 5(4):4664–4672

Bhattacharyya S, Pramanick P, Zaman S, Mitra A (2025) Salinity-driven model development for mangrove biomass in the Indian Sundarbans. Aegaeum 13(1):61–72

Dogar MM, Sato T & Liu F (2020). Ocean Sensitivity to Periodic and Constant Volcanism. Sci Rep 10, 293. https://doi.org/10.1038/s41598-019-57027-0

Dutta J, Thakur TK, Zaman S, Mitra A (2020a) Regulatory effects of salinity on the mangrove biomass: a brief commentary from Indian Sundarbans. eJ Appl For Ecol 8(1):20–27

Dutta J, Sarker MN, Bepari SP, Biswas P, Mitra A (2020b) The tale of anadromous fish *Tenualosa ilisha* (Hamilton, 1822), the monsoon delicacy of Bengal. In: Mitra A, Calma MM, Chakrabarty SP (eds) Proceedings: natural resources and their ecosystem services. HSRA Publication. ISBN 978-81-947216-7-3, 87-94

Dutta J, Mitra A, Zaman S, Mitra A (2022) Stored carbon in the mangrove vegetation of Lothian wild life sanctuary of Indian Sundarbans, the designated world heritage site. Ind For 148(7):733–741. https://doi.org/10.36808/if/2022/v148i7/151602

Jana HK, Mitra A, Zaman S, Bose R, Raha AK (2014) Will *Avicennia alba* thrive in climate change induced salinity rise? Int J Sci Res 3(2):459–461

Mitra A (2013) Sensitivity of mangrove ecosystem to changing climate. Springer New Delhi. ISBN-10: 8132215087; ISBN-13: 978-8132215080. ISBN 978-81-322-1509-7 (eBook)

Mitra A, Zaman S (2015) Blue carbon reservoir of the blue planet, vol XII. Springer India, p 299. https://doi.org/10.1007/978-81-322-2107-4. e-Book ISBN 978-81-322-2107-4

Mitra A, Zaman S (2021) Estuarine acidification: exploring the situation of mangrove dominated Indian Sundarban estuaries. Springer, Cham. https://doi.org/10.1007/978-3-030-84792-0

Mitra A, Gangopadhyay A, Dube A, Schmidt ACK, Banerjee K (2009) Observed changes in water mass properties in the Indian Sundarbans (Northwestern Bay of Bengal) during 1980–2007. Curr Sci 97(10):1445–1452

Mitra A, Chowdhury R, Sengupta K, Banerjee K (2010) Impact of salinity on mangroves of Indian Sundarbans. J Coast Environ 1(1):71–82

Mitra A, Sengupta K, Banerjee K (2011) Standing biomass and carbon storage of above-ground structures in dominant mangrove trees in the Sundarbans. For Ecol Manage 261(7):1325–1335. https://doi.org/10.1016/j.foreco.2011.01.012

Mitra A, Sengupta K, Banerjee K (2012) Spatial and temporal trends in biomass and carbon sequestration potential of *Sonneratia apetala* Buch-Ham in Indian Sundarbans. Proc Natl Acad Sci India Sect B Biol Sci 82(2):317–323. https://doi.org/10.1007/s40011-012-0021-5

Mitra A, Zaman S, Pramanick P (2023) Climate resilient innovative livelihoods in Indian Sundarban Delta, vol XII. Springer, p 297. https://doi.org/10.1007/978-3-031-42633-9. ISBN 978-3-031-42632-2, eBook ISBN 978-3-031-42633-9

Mitra A, Zaman S, Pramanick P, Ahmed S (2024) Mangrove carbon trading in the lower Gangetic Delta: drawbacks and the potential, vol XIII. Springer, p 282. https://doi.org/10.1007/978-3-031-76316-8. Hardcover ISBN 978-3-031-76315-1, eBook ISBN 978-3-031-76316-8

Pal N, Zaman S, Mitra A (2017) Variation of survival rate of mangrove flora due to salinity in Indian Sundarbans. J Sci Eng Health Manage 1(4):51–54

Roy Chowdhury M (2015) Effect of salinity on the above-ground biomass and growth of dominant mangrove plants in Indian Sundarbans. Ann Biol Res 6(5):20–25

Saha A, Gobato R, Zaman S, Mitra A (2019) Biomass study of mangroves in Indian Sundarbans: a case study from Satjelia Island. Par J Sci Edu 5(2):1–5

Sengupta K, Roy Chowdhury M, Bhattacharyya SB, Raha AK, Zaman S, Mitra A (2013) Spatial variation of stored carbon in *Avicennia alba* of Indian Sundarbans. Discov Nat 3(8):19–24

Strickland JDH, Parsons TR (1972) A practical hand book of seawater analysis, vol 157, 2nd edn. Fisheries Research Board of Canada Bulletin, 310 p

Trivedi S, Zaman S, Ray Chaudhuri T, Pramanick P, Fazli P, Amin G, Mitra A (2016) Inter-annual variation of salinity in Indian Sundarbans. Indian J Geo Mar Sci 45(3):410–415

Verma P, Kaur J, Arora MS et al. (2026). Modeling the Impact of Direct Air Capture on Forest Biomass and Population Dynamics. Earth Syst Environ 10, 407–427. https://doi.org/10.1007/s41748-025-00599-x

Zaman S, Bhattacharyya SB, Pramanick P, Raha AK, Chakraborty S, Mitra A (2013) Rising water salinity: a threat to mangroves of Indian Sundarbans. In: Abedin MA, Habiba U, Shaw R (eds) Water insecurity: a social dilemma. Emerald Group Publishing Limited. ISBN: 9781781908822

Chapter 15
Socioeconomic Implications of Climate Change and Adaptation

Muhammad Tayyab, Sohaib Muhammad, Muhammad Umer Farooq Awan, Tahira Khalid, Zaheer-ud-din Khan, and Muhammad Zahid

Abstract Climate change is now an existing reality rather than a remote threat, and it has significant socioeconomic impacts. The chapter examines the multidimensional effects of climate change on world and regional economies, livelihoods, health, and food security, especially in poor communities. Climate change poses serious threats to the economies of developing countries, increasing hunger, floods risk, and food and energy insecurity over time. It disrupts the hydrological cycles by altering the groundwater levels, freshwater, and the quantity and timing of stream flow. This chapter explores the detrimental consequences of climate change on different sectors and emphasized the adaptation policies to strive for resilience-building and vulnerability mitigation. Long term adaptation strategies including Agroforestry, sustainable agricultural practices, and water conservation practices can be adopted to handle the harsh impacts of climate change. This chapter calls for inclusive, context-sensitive, and forward-looking adaptation strategies that not only reduce the risks of climate change but also drive sustainable and equitable socioeconomic development.

Introduction

This chapter attempts to highlight the socio-economic impacts of climate change on economic growth, health, agricultural productivity, food security, poverty, the energy sector, and water resources, and discusses how pragmatic adaptation and mitigation measures can be implemented to effectively address climate change impacts. The social impacts of extreme and erratic weather events, as well as other consequences of climatic extremes, are discussed in detail. Some sustainable practices for addressing mitigation and adaptation challenges at a broad scale are

M. Tayyab · S. Muhammad (✉) · M. U. F. Awan · T. Khalid · Zaheer-ud-din Khan · M. Zahid
Dendrochronology Lab., Department of Botany, Government College University, Lahore, Pakistan
e-mail: dr.sohaibmuhammad@gcu.edu.pk

S. Mandal et al. (eds.), *Climate Change Vulnerability*, Climate Change Management, https://doi.org/10.1007/978-3-032-16783-5_15

described in this chapter, with particular emphasis on their social, environmental, and economic aspects to minimize socio-economic impacts. It also identifies practical measures to mitigate these impacts. Addressing climate change remains critical, particularly for developing countries, as its impacts are expected to intensify in the future.

Scientific studies indicate an increasing frequency and intensity of climate extremes, including temperature extremes and changes in precipitation patterns (Dogar et al. 2017a, 2017b, 2019; Dogar and Sato 2018, 2019; Dogar and Almazroui 2022; Valenzuela and Anderson 2011). According to latest IPCC report, the mean temperature of earth has increased by 1.09 °C during 2011–2020 from the levels of 1850–1900 (IPCC 2022a, b). The 99% range of estimates made by the IPCC indicates rise of the CO_2 and other greenhouse gases emissions resulted in increase in temperature, with range of 0.8 and 1.5 °C. This report indicates that extreme weather events, which occurred once in every 10 years between 1850 and 1900, now happen 2.8 times per decade and could increase to 4.1 times with 1.5 °C exceeding temperature limit and such climate-induced events can be responsible for the devastation from extreme floods and extreme droughts (IPCC 2022a, b).

Sudden changes in extreme weather affect nature now more and more frequently (Dogar et al. 2020; 2023, 2024, 2025; Dogar 2019, 2020). For instance, it decreases biodiversity; decline in population and local eradication has been shown (De la Fuente and Williams 2022). Integrating biodiversity, climate dynamics, sustainable land management, ecosystem health, and societal well-being, the IPCC (2022) highlights that the consequences of climate changes for society are far-reaching. Research, including that of Batten (2018) and the IPCC (2014), has found that there is overarching economic, health and water security impacts resulting from climate change. Constant climate fluctuations are with extreme impact on food security and other agriculture systems (Abeysekara et al. 2023; Chandio et al. 2023; Miron et al. 2023; von Braun et al. 2023), workforce efficiency (Valenzuela and Anderson 2011), economic development (Adom and Amoani 2021; Arndt and Thurlow 2015; Dell et al. 2012; Duan et al. 2022; Meattle et al. 2022), public health, household income, poverty, global energy economy (Tahir and Al-Ghamdi 2023) and water systems regulation (Hertel et al. 2010; Bibi and Tekesa 2023; Han et al. 2022). Global economies, regions, communities and sectors are affected by climate change and its effect is not uniform (Signe and Mbaye 2022). This problem is equally giving challenge to the developed and under developed nations of the world. More specifically this phenomenon of climate changes is posing serious threats to the economics of developing nations (Tol 2018; Stern 2007).

The developing regions, due to the relatively low adaptive capacity they have, are more exposed to higher impacts compared to the developed regions (Bowen et al. 2012; Cline 2007; Stern 2007; Ludwig et al. 2007; Tol 2018). In the developing economies, climate change posing the risk of pushing millions into poverty with poor infrastructural systems, inadequate food security, clean water and healthcare systems. The change in climate jeopardizes the objectives of United Nations Sustainable Development with reversing the past development achievements in these economies (Ludwig et al. 2007).

Developing nations cannot use single universal strategy to overcome diverse and extreme changes in climate (Nelson et al. 2010). Moreover, there is limited agreement about the extent and distribution of outcomes across sectoral domains and socio-economic outcomes (Arndt and Thurlow 2015). This uncertainty complicates policy design and undermines efforts to address changes in climate globally (Baarsch et al. 2020). This study provides an overview of current understanding and impacts of climate change across economic, social and political realms. On the basis of geostrategic position and income level, countries are more or less susceptible to climate change. Countries with greater economic resources can insulate themselves from climate change to a greater extent than poorer countries, and the poor tend to bear an inequitable burden of climate-related disasters (Islam and Winkel 2017). Climate Risk Index (CRI) range of Asian region illustrated it as the most climatically affected as compared to global findings, shown in Fig. 15.1 (Adil et al. 2025).

Impacts

Impact of Climate Change on Growth and Income

The genesis of the relationship between the environment and economy is well-researched, but the connection is still unclear due to the fact that contextual background (i.e., climate change variability), technological dynamism and the adaptive capacity of countries, among others, constitute important mediating factors. Using literature values, guesswork and extrapolation of Cline (1992), Fankhauser (1992, 1995) and Tol (1995), these studies evaluated the substantial costs of climate change from the doubling of atmospheric carbon dioxide emissions. Generally, climate

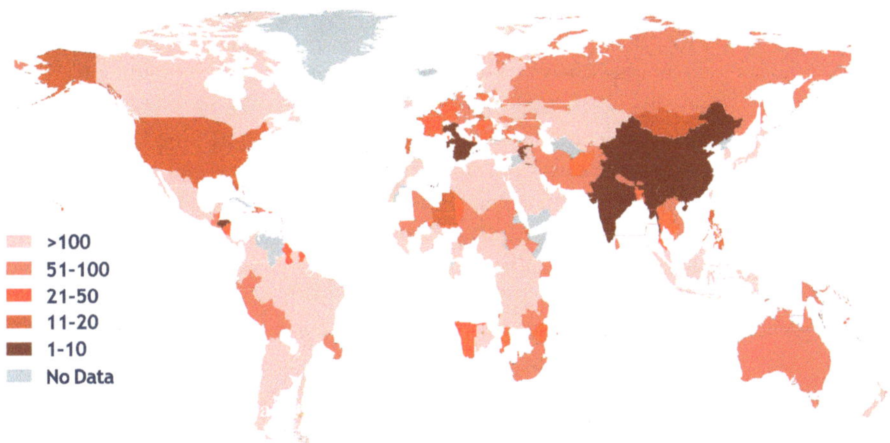

Fig. 15.1 Climate risk index across the globe: overall ranking (1993–2022)

change has negative effects on economic outputs, although the estimated impact varied. Today, climate change impacts on economic outcomes are still uncertain either they are continuous or non-continuous patters, complicates effective in policy development.

Climate change has a significant (negative) effect on economic growth, in particular for agriculture (Lamperti et al. 2021). With rising temperatures and changing rain fall patterns, food production is becoming increasingly difficult in many places around the world. Gleditsch (2021), Alhassan (2021) studied impacts on crop selection, yield, and food security. Many countries, particularly developing ones, are susceptible to climate extremes. i.e. floods, heat waves, storms, and droughts that destroy local economies and threaten lives. Climate change may decrease the worldwide economic output up to 23 trillion dollars in one year by 2050, and developing countries may be most affected by this. The richer countries, which emit most of the greenhouse gases, have benefited more from industrialization, while the poorer ones are bearing a disproportionate share of the damage. This is proof that the world urgently needs fair and effective climate policies.

Today effects of climate change are not uniform across sectors and economies. However, there is strong agreement that some countries will not be affected by its extremes. Studies that have empirically examined the saturating non-linear climatic impacts on economic growth have proposed that further warming will induce growth in temperate regions but hinder the prospect of growth in temporal places. Findings by IPCC (2014), Stern (2006) have widely spread the notion that climate change affects economic production unevenly across global economies.

Impact of Climate Change in Agriculture and Its Productivity

Climate change has prolonged impacts on cultivated land and labor performance, and on food security (Valenzuela and Anderson 2011). This may reduce the food production by (1) direct impacts on agroecological conditions; (2) indirect effects due to changes in consumption of agricultural goods, disparities in income inequalities, and fluctuations in national income growth; (3) constraining arable land availability (Schmidhuber and Tubiello 2007). Industries are responsible for 20% of all greenhouse gases having impacts on agriculture. Floods, droughts and forest fires greatly influence the environment and other agriculture related business outputs. (Chivangulula et al. 2023; Fig. 15.2). Moreover, agricultural system is mainly dependent on finite resources, making it vulnerable and prone to collapse. The declination of agriculture threats the quality of farmer's life (Li 2023) and climate change is dramatically influencing the food and water supplies (Zhang et al. 2023). Farming systems are vital to the economic well-being of nations particularly underdeveloped countries and to address their financial problems and livelihoods of each family unit (Chikafa et al. 2023). According to the IPCC (2022), greenhouse gas concentrations in the atmosphere have increased to levels extraordinary in the previous few thousand years. GHGs are composed of CO_2, N_2O and CH_4. Climate change

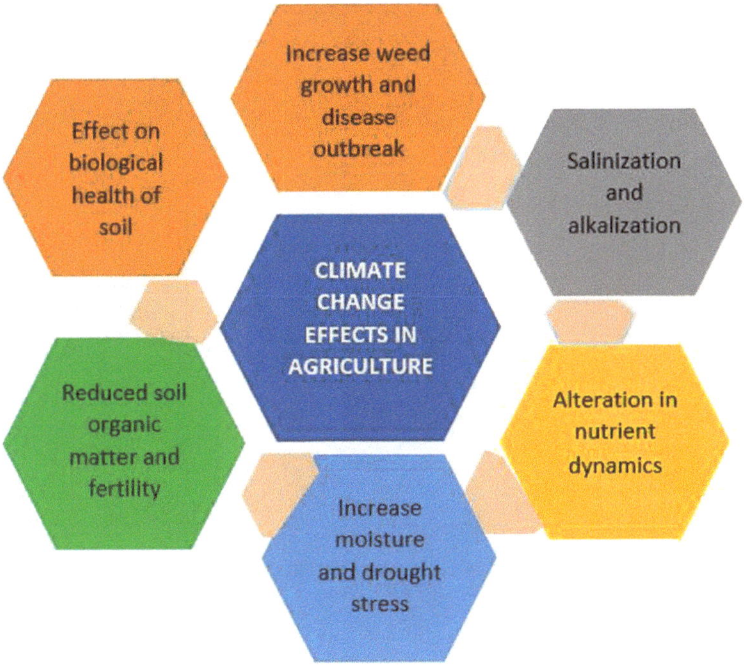

Fig. 15.2 Climate change influence on agro-ecological system

is mainly caused by both natural and anthropogenic factors affect the climate change (Monteleone et al. 2023). It is an estimate that the global mean temperature may rise by a range of 1.1–3.7 degrees at the end of the twenty-first century (Urban et al. 2023). Elevated temperature will have adverse impacts on growth and development of crops, potentially rendering global crop production extremely vulnerable to such global temperature changing trends (Ratnayake et al. 2023).

Literature shows that climate change has far-reaching impacts on biodiversity and has become one of the most pressing environmental issues. A comprehensive and multi-faceted approach is required to address the climate change impacts on biodiversity. It is crucial to mitigate Greenhouse gas emissions from fossil fuels combustion as these are the primary sources of climate change (IPCC 2018). A transition to renewable energy resources is required to reduce reliance on fossil fuels and lower greenhouse gas emissions (Rogelj et al. 2018). One of the main aspects of addressing impacts of climate change on biodiversity is improving energy efficiency. Enhancing energy efficiency lowers energy demand and reduces greenhouse gas emissions, which ultimately reduces pressure on biodiversity (EPA 2022). Sustainable land management practices (such as conservation tillage and agroforestry) can also be adopted to improve the resilience of ecosystems to climate change (Foley et al. 2005).

Climate change has major impacts on abiotic factors (rainfall, temperature, solar radiation, and CO_2 concentration) which are ultimately decreasing crop production

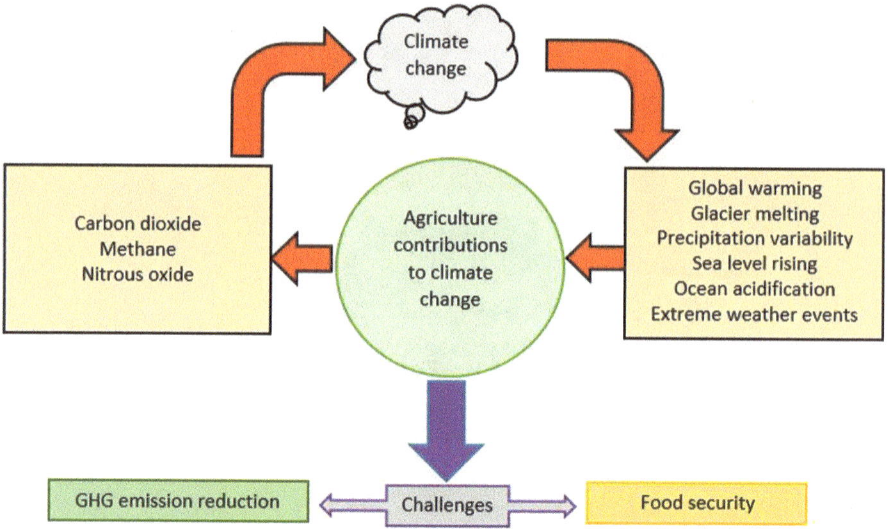

Fig. 15.3 Climate change and food insecurity in agriculture

(Fig. 15.3). High temperature was reported to adversely affect wheat and rice production, reduce growing season length, and interfere with flowering and pollination (Lobell and Field 2007; Rezvi et al. 2023; Mehmood et al. 2023). Studies proved that photosynthesis and reproductive success are inhibited by heat stress, especially in both tropical and subtropical areas (Farhad et al. 2023; Issa et al. 2023). Betel leaf, lemon and pineapple crops in South Asia, particularly East India and Bangladesh were also affected by the weather extremes like the cold, fog and heatwaves during past two decades (2000–2020). These challenges underscore the pressing need for adaptive mechanisms, including crop diversity and climatically strong farming practices, in order to mitigate long-term agricultural losses (Vernooy 2022).

Food Security

UN sustainable target 2.1 and 2.2 objective is to provide safe and nutritious food to people throughout year and eliminate hunger. World food security was relatively stable before COVID-19 and this pandemic brought exceptionally rapid increase in food insecurity. Such increases in food insecurity are exacerbated by the COVID-19 pandemic and mostly affected Latin America, Africa and Asia (FAO et al. 2022). Climate change exacerbates the problem by posing an ever-greater risk to world food supplies, particularly in poor countries (Fig. 15.4). National Academy of

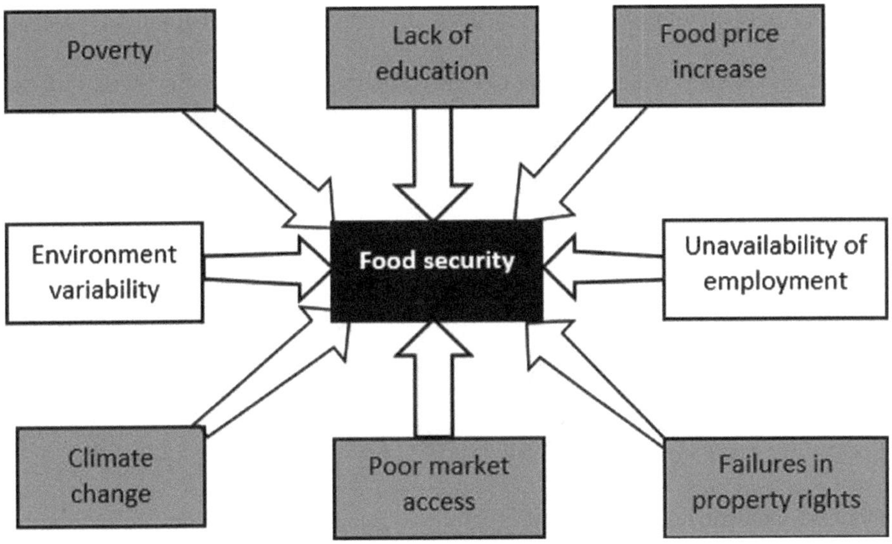

Fig. 15.4 Socioeconomic impact of climate change on food security

Sciences, 2019 have found that Africa and Asia are particularly affected by climate change while the worldwide agricultural system is facing negative effects. By 2080, Africa's agricultural GDP could drop by as much as 9%, as the prices of food, particularly grains and sugar, are set to increase by 39–43% by 2050. This premise makes food less accessible, and may lead to hunger mostly in the poor areas (Adom 2024).

Farmland

A related transmission mechanism of climate change in the agricultural industry is through land use and land value. Farmland is reducing with increasing population which will limit the supply and demand of food in future. Pastor et al. (2019) in a global analysis of the food–trade–water nexus demonstrated that food production will need to be doubled by 2050 under projected food demand and to overcome this situation, 100 million hectares of additional land would be required in order to achieve such increases in production. Many researches have emerged to understand and the linkage between land use, land value and climate change, given their significance to the global food system (Seo and Mendelsohn 2008). Overall, the evidence base is more well form to developed economies where good-quality data is available (Mendelsohn et al. 1994; Mishra et al. 2015; Abidoye et al. 2017; Hossain et al.

2019). From developing economy aspect, the literature is still limited. However, there seems to be some sort of agreement that the value of farmland is negatively affected by climatic changes, although the effect is bound to differ over time and across space.

Although there are temporal and spatial variations, the overall evidence appears to be decisive. Farmland value decreases with increasing temperature, and increases with growing precipitation. Small-scale farmland values are more sensitive to climate change than large-scale farmland values. With respect to timelines, the effect on farmland prices will be greater for longer horizons than for shorter.

Impact of Climate Change on Biodiversity

There is significant loss in biodiversity due to climatic changes, mass species extinction and destabilization of ecosystems. Changes in climate patterns and abundance are modifying the behavior of marine, freshwater, and terrestrial species at levels that frequently surpass their adaptive potential (Manes et al. 2021; Allan et al. 2021). Taxa with little tolerance towards change in environmental conditions and low capacity for dispersal are at particular risk, especially in fragmented habitats where movement to more favorable microclimates is limited (Nehe et al. 2023; Hankin et al. 2023). For example, mangroves are expanding further due to warmer seas, altering sequestration dynamics (Buhaug et al. 2023), and the range shift of carbon tropical fish has led to decline of kelp forests that cannot tolerate tropical conditions (Donham et al. 2023). The consequence of the extinction of such keystone species results in destabilization of whole ecosystems, and hence services like CO_2 capturing and storage, nutrient biodiversity and ecological cycling can be threatened (Kim et al. 2023; Baldrian et al. 2023).

Long-term projections show that climatic changes will cause more increasing number of species to go extinct in the twenty-first century. The American pika is one of many highland species that have already seen population growth and range limitations due to rising temperatures (Palita 2016; Jones et al. 2022) decreases brought on by a lack of long-term ecological data. Additionally, these losses are made worse by the need to apprehend and forecast climatic changes on species and ecosystems in general (Manes et al. 2021). The decline of biodiversity is caused by additional stressors including severe droughts, invading species, or shifting plankton communities (Bedford et al. 2020; Nwosu et al. 2023). These changes in aquatic producers, such diatoms, can alter the carbon cycle and possibly offer an explanation for past 2021. Overall, climate change is caused by variations in atmospheric CO_2 (Lee et al. 2021), which are key factors in the decline of biodiversity and emphasize the necessity of focused conservation and adaptation measures (Fig. 15.5).

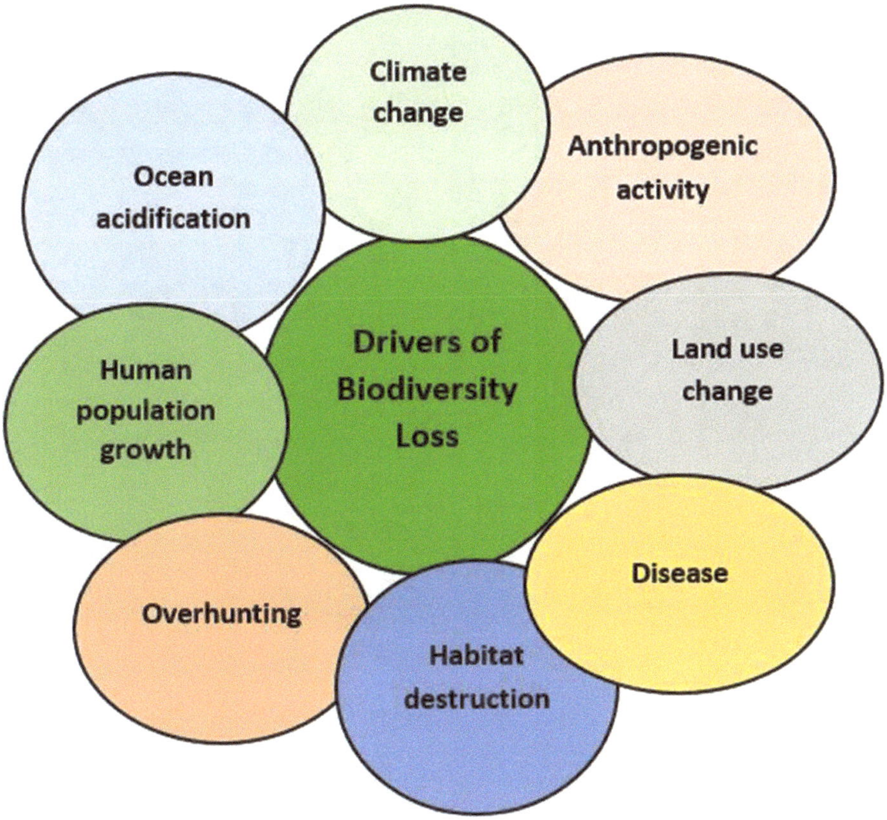

Fig. 15.5 Relationship of human interventions in natural ecosystem (biodiversity loss)

Forestry

Forests are playing climate determining role on a global scale and regulate carbon and nitrogen cycle with many other functions (Tang et al. 2023). Climate has strong impact on the forest's health including increasing temperature, drought, fire, pest outbreak and wildfires (Barati et al. 2023). All these factors influence trees growth, particularly in temperate zones. Approximately 1.6 million forests diminished for food, fuel and other resources, and billions of people globally depend on them in Africa and Asia (Husain et al. 2018; Fonjong and Gyapong 2021). Many communities are still unaware to climate change effects and not well-informed to the reason of their changed forests (Ngoukwa et al. 2023). Smallholders in the Philippines are experiencing reduced crop yields and delayed fruiting in reaction to a greater number of pests (Jawo et al. 2023). In the Himalayas (sensitive to warming) and Bangladesh, increasing temperature caused several insect-borne diseases such as malaria and dengue (Rodway 2023; Farhana and Raihan 2022). Insects, having rapid prolific and dispersal abilities, are outpacing trees in their response to

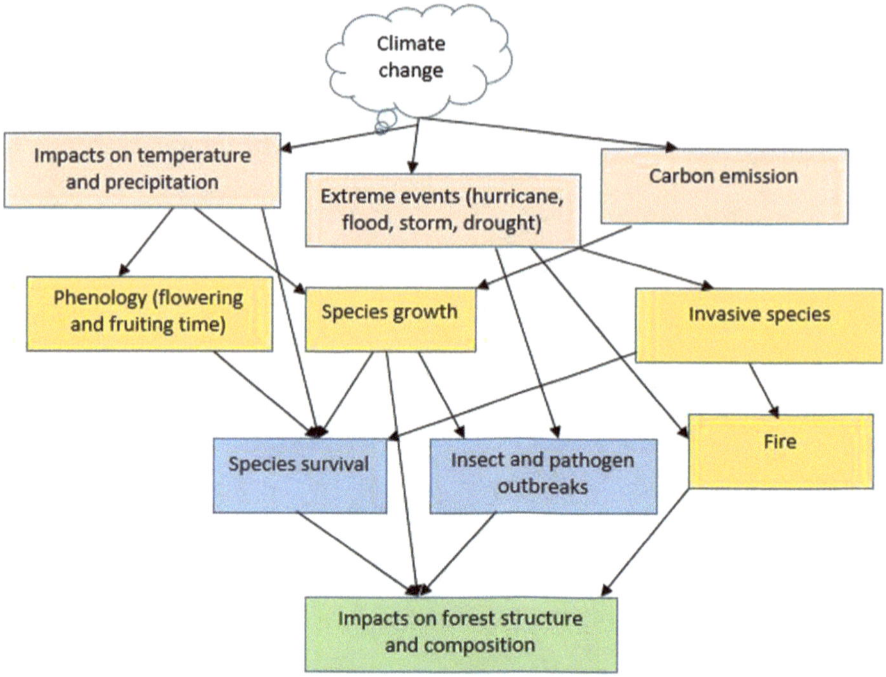

Fig. 15.6 Greenhouse gases impacting forest ecosystem health

warming, making forests more prone to pest outbreaks (Shaw 2020; Subedi et al. 2023). Forest's natural defense systems against insects are weakened with changed global temperature (Fig. 15.6). With changing climate, it becomes crucial to protect forests ecosystems and people living in and around them. There is need to develop management and adaptation strategies for forest ecosystems in response to climate change, and the choice of the specific strategies depends on forest type, species, management goals and the conditions of the climate. Silvicultural measures are applied to enhance the resilience of forests by selecting drought- and fire-resistant species and through sustainable harvesting (Wang et al. 2024). Ecological approaches are adopted to restore degraded ecosystems and maintain the genetic diversity, helping forests withstand the harsh climatic conditions.

Starvation, Food Shortages, and Poverty

A major adverse consequence of climate extreme on the level of agriculture productivity is the widespread increase in extreme hunger, poverty and malnutrition especially in developing countries which depend on agriculture. Climate induced scarcity will ultimately impact the economics of any effected country and people's life as their main source is agriculture (Fischer et al. 2005). Schmidhuber and Tubiello (2007) predicted a 5–26% increase in the number of people affected by hunger situation by 2080 compared to a non-changing climate. On the basis of range of possible scenarios, more than 5–170 million people could be forced to cope with extreme food hunger around the world. This discussion highlights that climate has strong impact on people nourishment, incomes and overall, on their life. These risks are more evident among the people of developing countries as compare to developed countries. Temporally, the impact becomes more significant with time showing evident effects emerging over time rather than immediately.

Global Warming and Water Resources

Water utilized in different sectors of the society, economy and nature (Arnell 1999). This implied that water resources management is paramount to sustainable development. There is large number of the population experiencing extreme scarcity around the world. In the world, approximately 3.6 billion people lack sufficient access to water during at least 1 month of the year. It is projected that by 2050, this figure will increase to 5 billion (WMO, the 2022). As surface and ground water supplies are depleted through time, climate change effects on hydrological cycles are expected to demand more water (IPCC 1996). At a global scale, hydrological cycles are changing to become drier days, severe flooding and more erratic rainfall and melting of glaciers at an increased rate (WMO 2022; Fig. 15.7). In 2021, more parts of the globe experienced drier conditions relative to the 30-year hydrological reference period (WMO 2022). Rainfall and surface runoff, as well as groundwater, will decline with changing climate patterns. In the extreme climate change situation, it is consensus that this could change the water availability with reducing their resources. However, it is feared that it may arise severe water scarcity even in moderate scenarios of maximum resources in the absence of proper water adaptation measures.

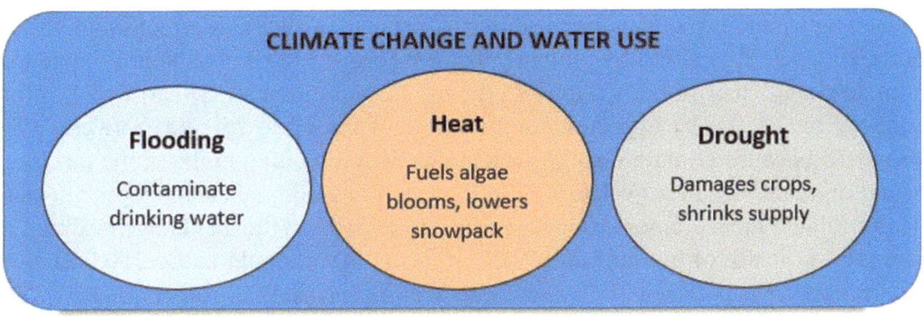

Fig. 15.7 Global warming and water resources

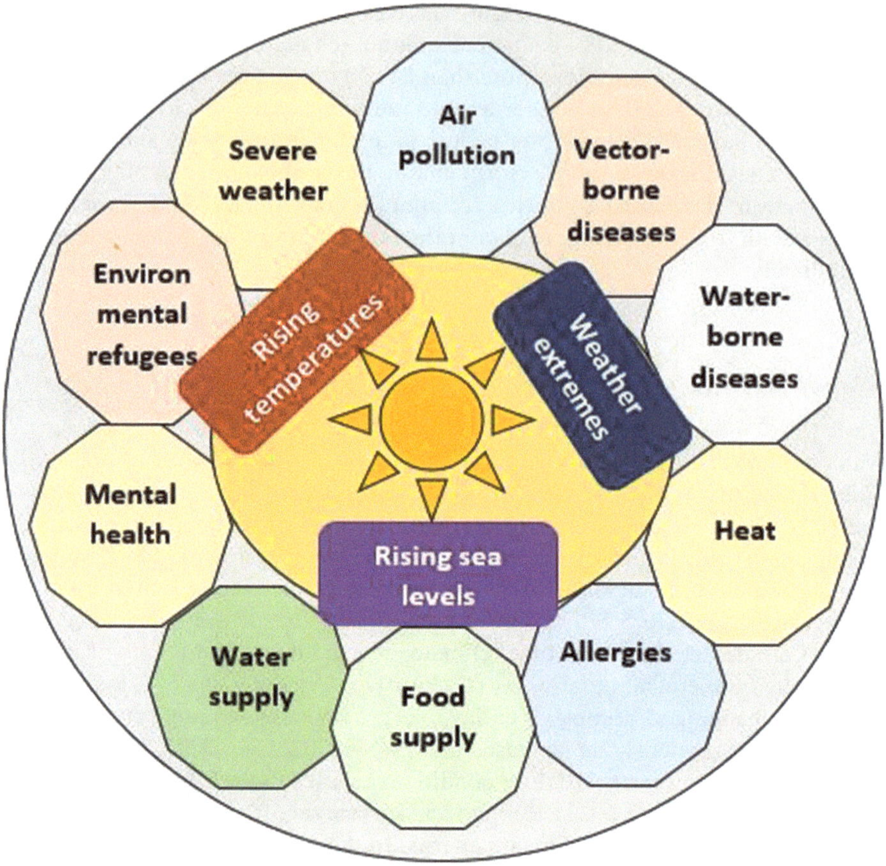

Fig. 15.8 Climate change impacts on human health

Effects of Climate Change on Livelihood and Public Health

Health problems such as respiratory, cardiovascular disease and stroke, illnesses caused by water and food-borne diseases, adverse mental health effects, and vector-borne diseases are caused by climatic changes (Fig. 15.8). It is an approximate that climate change has globally damaged the public health from US$2 to US$4 billion per year by 2030. The expected number of deaths is up to 250,000 between 2030 and 2050 because climate-change related diseases including malaria, malnutrition, heat stress, and diarrhea (WHO 2021). The changing climate needs implications of health-related measurements due to its diverse impacts. Climate change is already a dire risk to health of humans and WHO projects that it could cause 250,000 more deaths each year between 2030 and 2050 (Watts et al. 2015; Sasai et al. 2023). The rise in global temperature is expanding the vector-borne diseases and rebirth of hitherto rare or sleeping diseases (Okoro et al. 2023; Subasinghe et al. 2023). Climate warming influenced the environment such as changed ecosystems and increased levels of CO_2 which ultimately risks the public health. These environmental stressors increase the prevalence of respiratory conditions (asthma), mental health diseases, and the exposure to pathogens. Communities repeatedly facing the climate disasters, like flooding or hurricanes, may experience post-traumatic stress and long-term psychological distress, as evidenced in post-Hurricane Katrina (Ogden 2018). At the same time, antimicrobial resistance (AMR) is emerging as an imminent and significant global health threat that could set back more than a century of medical progress (Adebisi and Ogunkola 2023; Costanzo and Roviello 2023). The overuse and misuse of antibiotics are driving resistance in harmful microbes, complicating common infections treatment, and threatening the safety of major medical systems such as cancer chemotherapy and organ transplantation (Anjani et al. 2023; Anderson et al. 2020). Resistant bacteria spread rapidly across the border, making stronger, the dependence on stronger, expensive, and more toxic antibiotics such as second- and third-generation cephalosporins, becomes increasingly unsustainable (Tiwari et al. 2022; Harris et al. 2023). This overlapping risk highlights the need of integrated response on health and climate to shield the global population.

Climate Change and Energy Security

UN SDG7 is about ensuring the access to resilient, economical, modern energy and sustainable for all by 2030. This target was set in an effort to alleviate energy poverty that is prevalent in low income nations. Although some benefits have been achieved, climate change may negate the initial benefits as well as undermine the long-term progress, as it affects both energy consumption patterns and energy availability (IPCC 2014). Climate change has various impacts on energy system including hurricanes, extreme weather, wildfires, heat waves, rising temperature and precipitation (Zamuda et al. 2018). Effects of climate change on energy system are location specific as climate change phenomena differ across geographies. Energy systems of developing nations are more exposed and affected due to the climate change as compared to developed nations due to their limited and fragile power infrastructure. Yalew et al. (2020) reviewed more than 200 scientific literature studies to show the climate change impacts on the functioning of energy systems of regional and global economies. The simulations predict the decline in thermal power plants capacity and tropocalorific hydropower, rising in cooling and reduction in heating demand (Fig. 15.9). While the results are ambiguous and mixed at the regional level, the energy systems of Latin America and South Asia are expected to be among the most effected by climate extremes. Cronin et al. (2018) noted in the same literature review on the power supply that while there is a general agreement over the anticipated effect of climate change on some energy sources remain inconsistent for hydro and bio-energy sources. Table 15.1 presents the summary of projections across different sectors.

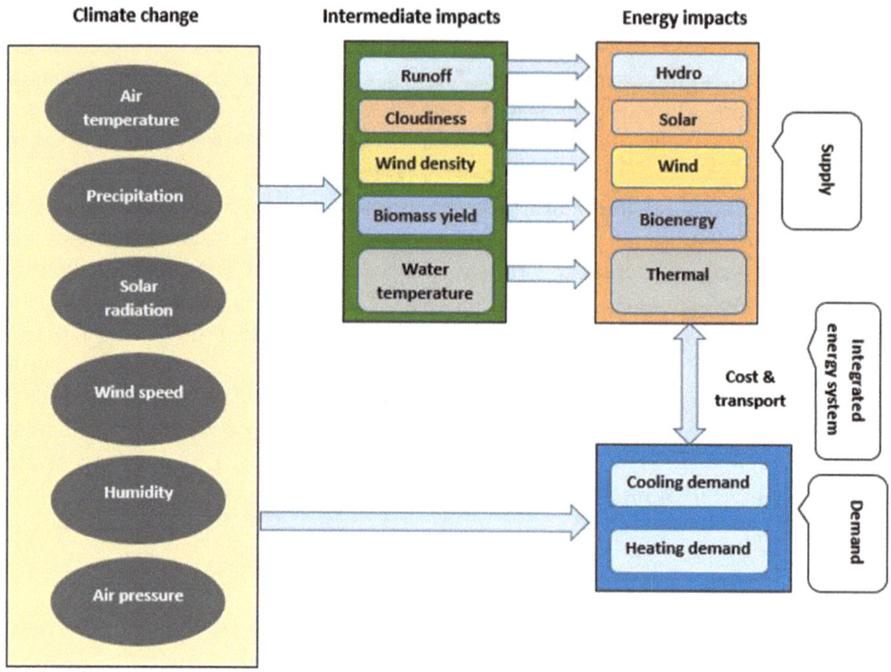

Fig. 15.9 Integrating relationship of energy and climate

Table 15.1 Summary of quantitative projections across sectors

Sector	Projection
Agriculture	By 2080, Africa's agricultural GDP is projected to decline 9% leading to hunger and food insecurity
Undernourishment	The undernourished people number will be increased by 5–26% compared to regions not affected by climate extremes by 2080
Water resources	In current situation, almost 3.6 billion people in the world have no access to water at least a month per year, and this number is expected to be increased to 5 billion in 2050
Health	By 2030, the global economic costs of health impacts are expected to range between $2–4 billion due to extreme change in climate globally
Energy security	Rising temperatures and changing hydrological patterns are anticipated to increase cooling demand due to change in climate, and threaten energy security

Climate Change Impacts on Tourism

Tourism is a main driver of the economy and provides valuable opportunities for both local communities and international stakeholders (Scott 2021). Changes in climate are becoming a challenge to tourist potential of many destinations, as shifting and alteration in weather patterns alter when and where people like to travel

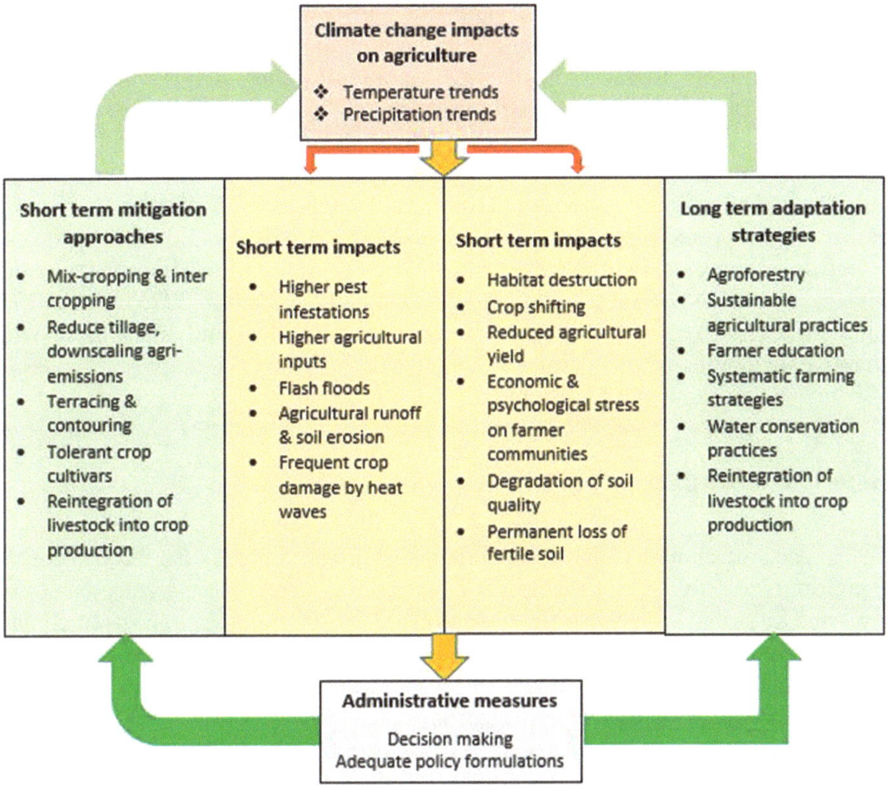

Fig. 15.10 Long/short term impacts, mitigation and adaptation strategies

(Zhao and Liu 2023). Skiing seasons are shortening due to increasing temperatures (IPCC 2022) and rising sea levels threaten to submerge small islands and coastlines. Swimming and fishing could be curtailed because of harmful algae blooms, and wildfires can deteriorate the air quality in affected area, making them less appealing to visit. In certain regions, deforestation and wildlife loss also make the land less appealing. As a consequence, tourism demands robust planning and support to address climate-related threats. Locations with insufficient resources or substandard facilities are relatively vulnerable (Kyriakopoulos and Sebos 2023; Peng et al. 2023).

Strategies for Adaptation and Mitigation

Adaptation and mitigation strategies are required to deal with climate change (Pescaroli et al. 2023). Unlike climate change mitigation, climate change adaptation has an immediate impact on extreme weather, including flooding (Fig. 15.10). From both an economic and an environmental perspective, GHG abatement requires immediate attention, as it dampens emissions (Prokopenko et al. 2023). Scientists are really concerned about the implementation of strategies to mitigate negative climatic impacts in diverse economic and geographic contexts. Agriculture, transportation, industry, forestry and land use sectors need to adapt and mitigate such policies (Waheed et al. 2021). Adaptation and mitigation strategies demand the attention at both international and national levels. Climate change has emerged as a significant problem in recent decades, and adapting to its impacts has become a prerequisite for people of the world for sustainable social and economic developments. International policies and regulations are essential for countries to reduce and adapt the climate changes effects (Pathak 2023). Table 15.2 summarizes the regional impacts, vulnerabilities in Asia and links them with adaptive capacities.

Asia Case Studies

The impacts of climate changes on certain regions or countries has been recorded significantly negative as crop yields reductions in specific region are greater than those projected by global-scale research. Khan et al. (2020) analyzed the impacts of climate change on economics and agriculture system of Pakistan and forecasted that climate caused reductions in wheat and rice yields would erode the country's real GDP by $19.5 billion by 2050. This study further indicates the 14.7% and 20.5% decrease in wheat and rice yield respectively relative to a scenario without climate change.

Chalise et al. (2017) studied the financial impact of climate caused agricultural yield losses in Nepal, and observed substantial adverse impacts on the economy, especially for rural family's dependent on self-sufficient farming. The study models suggested a 10.03% loss of actual GDP in the highest impact scenario, with 6.56% and 2.49% in medium and lowest scenario respectively by 2050. Climate change is projected to decrease the baseline level of agricultural production by between 6.7% and 7.6%. Yields of particular crops are also expected to decrease under various scenarios: 0.8–7.22% for rice, 1.28–5.45 for wheat and 2.43–8.43% for cereal grains.

Do Prado Tanure et al. (2020) assessed how this changing climate may change economy of agriculture land in the Legal Amazon region for the period of 2030 and 2049. This study found that climate-induced changes in agricultural production and employment would lead to 1.18% reduction in economy's inflation-adjusted output by 2049. Certain crops would be affected the most: Rice by 7.55%, soybeans by 7.87%, corn by 7.9%, and sugarcane by 11.34%. Srivastava et al. (2021) assessed the influence of climate changes on production of maize in eastern India. In irrigated zones, maize yields are forecasted to reduce 10.58–23.39% and 15.20–26.83% during 2021–2050 and 2051–2080 respectively. For rainfed maize, products are less pronounced as ranges between −10.55% and +9.20% in 2021–2050 and +4.31% to +10.63% in 2051–2080. Irrigated maize therefore incurs higher losses than rainfed maize, as increased rainfall conditions in future may reduce the effectiveness of irrigation.

Table 15.2 Regional impacts and vulnerabilities to climate change in Asia (Christensen et al. 2007; Cruz et al. 2007)

Impacts	Sectoral vulnerabilities	Adaptive capacity
Temperature • Central Asia, Tibetan plateau, Northern, Eastern and Southern Asia Warming Global mean Figurative • Southeast Asia Warming similar to the global mean • Less number of extreme cold days over East Asia and South Asia	• Change in freshwater availability in Central, South, East and Southeast Asia (including major river basins) could lead to more than half a billion people exposed to water stress • Short-term increase in glacial-melt flooding frequency and strength; slope instability and decline in river flows due to glacier loss of mass	• Difference in Adaptive capacity across countries due to structural and cultural differences that's because resilience is the product of economic capacity, geography, and degree of environmental degradation • Climate research still addresses social processes too little. Although early warning systems in Bangladesh and the Philippines are showing progress, capacity remains limited due to inadequate resources, income inequality, weak institutions and technology gaps
Precipitation, snow and ice • Drier in much of Asia with more precipitation • Reduced summer rainfall in Central Asia • Increased extreme rainfall across portions of South and East Asia • Decreasing snow and ice on glaciers in the Himalaya and Tibetan Plateau	**Agriculture and food security** • Negative effects on crop production in many regions, putting hundreds of millions at risk of hunger • Soil moisture (SM) concentration is reduced and evapotranspiration (ET) evoked, resulting in possibly higher risks of land degradation and desertification • Potential loss of farmland in the Northern high-latitudes	
Droughts and extreme events • Summer droughts will become more frequent and severe • Strengthening of El Niño events • Heavy rain and wind from tropical cyclones in East, South, and Southeast Asia • Rain-induced landslides and floods So as frequency, duration, and intensity of heat waves/hot spells (especially in Eastern Asia)	**Health** • Heat stress and latitudinal re-distribution of vectors for the vector borne diseases • Increase of diarrhoeal disease related morbidity and mortality in south and Southeast Asia • Increase in the burden or severity of cholera on the Indian sub-continent	
Land ecosystems • Increasing extinction risk due to climate change in combination with fragmentation • Northward migration of boreal forests in Asia, limited by increased frequency/intensity of forest fires	**Terrestrial ecosystems** More frequent and intense burning and fire suppression prevents forest advance; biodiversity loss risk increasing	
	Coastal zones • Dozens of millions at risk from sea level rise in low lying coastal areas of South and Southeast Asia • Increased frequency and intensity of tropical cyclones • The impacts of coastal inundation on aquaculture, industry and mega deltas • Rise in danger to wetlands, mangroves and coral reefs	

Jiang et al. (2019) analysed the effects of climate change on yield of rice fields in Vietnam's Mekong Delta. They estimated that yields of rainfed rice are likely to decrease by 35% (2020–2029), 16% (2030–2039) and 21% (2040–2050) as a result of less rainfall in winter. In contrast, irrigated rice yields would increase by 11% in the 2020s, and then to decline by 0.5% in 2030s and by 23% in the 2040s. Mild moisture limiting effects across the summer were expected in the 2020s (49% reduction in rainfed rice yields), 2030s (−56%) and 2040s (−40%). On the other hand, irrigated rice yields are projected to reduce 5%, rise 2% and 5% every decade from 2020s onward, driven by anticipated precipitation. Sinnarong et al. (2019) and Ansari et al. (2021) reported that rice production is decreased due to the negative climate change impacts. Sinnarong et al. (2019) reported that production of rice in Thailand will decline and mean decrease estimated at 9.37% in 2030 and 33.77% in 2090 and the most affected will be the north of the country.

In Indonesia, rice productivity is reduced with the changing pattern of climate. According to the Ansari et al. (2021), implementation of the RCP8.5 scenario could lead to 12% reduction in rice yield by the 2050s. RCPs are constructed by quantifying potential trajectories of greenhouse gas emissions, land-use dynamics, and atmospheric concentrations, which are then expressed as radiative forcing (measured in W/m^2) by the year 2100. These findings emphasize the spatiotemporal heterogeneity of climate change effects on yields. Climate extremes will have an adverse effect on several crops like sugarcane, potato and wheat. Kumar et al. (2014) predicted that wheat yields in India were directed to decrease by 6–23% (2050) and 15–25% in 2080 as compared with a no changing climate scenario. These impacts magnitude differs according to emission intensities, sowing date, and site location, with more effects being observed in the late sowing and in the central and south-central warmer regions. If carbon dioxide fertilization is a factor, there will likely be some benefit, but, in areas with very high seasonal temperature, yields will still reduce. The adaptation strategies (e.g. adoption of productive cultivars and changes in the sowing date) would allow reducing the potential yield losses to 9% by 2050 and to 13% by 2080.

Scott et al. (2019) applied multiple economic, climate, crop, and water models to project potato production in India. They found that potato crop is affected differently under favorable, intermediate and harmful climatic conditions. Economy growth is also affected in these scenarios which ranged from optimistic with strong, to the intermediate and pessimistic with slow economy development. Projections suggested increment in potato production in India over 2010–2030 with 37.6 million tonnes in optimistic and 23.9 million tonnes in the unfavorable scenarios. Pipitpukdee et al. (2020) discovered the impacts of climatic variables on sugarcane yields. In Thailand, sugarcane production has been primarily reduced due to increasing population and shifting climate pattern. Yield reductions reflect the negative consequences of harsh climatic conditions, especially in the eastern and southern part of the central parts of Thailand. This could disrupt the sugarcane farmer's livelihoods and influence sugar prices in worldwide trade.

Studies in Asia highlight the negative impacts of climate change on crop yield, which are moderated by agroecosystem characteristics, cultivated crop varieties, and adaptation options (Table 15.3). Although individual crops may be more

Table 15.3 Summary table of Asian case studies of climate change impact on economic output and crop yield

			Crop (% change in yield)					Forecast	% Change
Author's name	Publication year	Country/region	**Wht	**Ric	**Crn/maiz	**Soybn	**Sgr Can	year	in GDP
Khan et al.	2020	Pak*	−14.70	−20.50				2050	−19.5
Sinnarong et al.	2019	Thai*		−9.37 to −33.77				2090	−33.77
Chalise et al.	2017	Nep*	−1.28 to −5.45	−0.80 to −7.22	−2.43 to −8.43			2050	−10.03
Do Prado Tannure et al.	2020	Legal Amz*		−7.55	−7.90	−7.87	−11.34	2049	−1.18
Pipitpukdee et al.	2020	Ind*					−23.95 to −33.26	2046 to 2050	
Jiang et al.	2019	Viet Me Del*		−35.00				2050	−21
Ansari et al.	2021	Indo*		−12				2050	−12

*Pak = Pakistan, Thai = Thailand, Nep = Nepal, Legal Amz = Legal Amazon, Ind India, Viet Me Del = Vietnam's Mekong Delta, Indo = Indonesia, **Wht = Wheat, **Ric = Rice, **Crn/maiz = Corn/maize, **Soybn = Soybean, **SgrCan = Sugarcane impervious to climate change effects even under intensified conditions, research resilience towards climate change is surprisingly scarce. These dynamics are important for the development of effective adaptation policies to assist vulnerable populations and reduce levels of food insecurity. Hence, more studies on the adaptive dynamics of crops amid changing climate conditions on crop are needed to inform appropriate adaptation policies and agricultural interventions.

Policy Implications

This chapter highlights the urgent need for policy responses to address the rising impacts of climate change on different sectors. An individual's mental health exists within a network of societal, economic, agricultural and physical systems that will all experience disruption as a result of changes in climate. Resilience in both human and natural realms is affected by climate variation, as well as other human and natural pressures. Climate change leads to food insecurity, which could result in poor food quality, higher prices and inefficient delivery systems. Storms, flash floods, drought and heavy rainfall are only a few of the climatic challenges facing forests at a global scale. Their demise, on the other hand is a blessing for mankind. Clearly, susceptibility across the globe is not homogenous; however, appropriate alleviation and other adaptation policies would enable decision makers to develop cost-effective policies targeting its effects (Fig. 15.11).

Adjusting to large scale change like this is vitally important, because modern human society on our world has evolved to thrive under relatively constant climate conditions. As climate extremes intensify, survival and adaptation is becoming difficult. Our primary challenge is to prevent climate change from reaching a scale where its impacts exceed our capacity to manage them. Policies implementation is urgently required to alleviate the damage of vulnerable sectors due to extremes in climate. Extended growing season could increase yields in regions susceptible to frost. Climate change has adverse impacts on agriculture, and effective policy implementation is necessary to protect crops, enhance food security and reduce hunger. Certain season crops, such as barley, cereal crops, wheat, and many vegetables, might benefit from divided growing season with a shortened unplanted summer period if extreme temperature leads to extended high temperatures in warmer months, above critical thresholds. In tropical and subtropical regions, lengthening the planting period could be difficult due to reliance on altered precipitation where yield is curtailed by rainfall happen during the late part of the year. And for some harvests, genetics is extensive, but for others, it's kiwi-style. Genetic mapping and editing can also enable a broader range of traits. The difficulty to ensure features that express during the plant cycle worries about client satisfaction, economics and regulation barriers have partially limited the integration of genetically modified cultivars, particularly in the forecasts, addressed during early forecasts. Plants do not absorb excess fertilizer, which then leaches into the soil, contaminates water, or be emitted into the atmosphere as nitrous oxide from soil. Nitrogen is a major

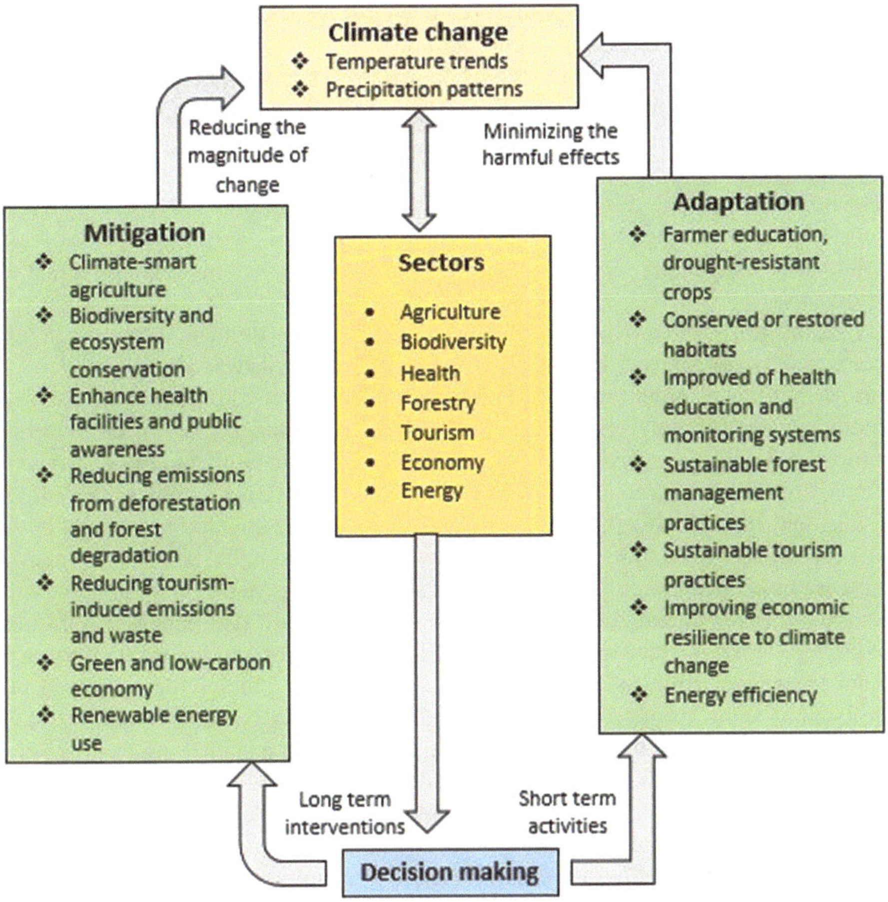

Fig. 15.11 Sector-specific strategies for sustainable ecofriendly system

contaminant in groundwater sources, which is associated with marine ecological destruction and long-term human illness. To reduce impacts of climate-change on agriculture sector, technology adaptation and social/economic adaptation are required.

In consequence of the causal relationship, the production of biofuel is one of the channels through which fluctuations of oil economics away from international macroeconomic problems has policy intimations. The connection between the cost of food and that of oil is because bio fuelling production has barely begun in many of the countries in the scheme, while some there are still significant demand for feedstock to supply the expansion of the CE expansion is based on China and the US. In other words, oil-exporting nations can provide economic incentives to produce food. Credit, seeds, fertilizer and farm implements could all come in handy. Oil-producing countries could ill afford to subsidize the import of food in the short term anyway as oil's worldwide price has fallen and thus, so too must their oil export revenue. Therefore, exports of these countries from the agricultural sectors can improve. Research and development and processing of food products could allow countries to alleviate issues such as exchange rates not in line and/connectivity inefficient trade. For the nations themselves, the drastically volatile nature of global oil markets makes it quite clear that depending on oil exports alone cannot be allowed to remain a viable economic strategy. This makes it feasible in countries exporting oil and other natural resources for non-food (biofuels) renewable energy to be boosted as a percentage of the economy in comparison to hydro, the sun, biofuels, wind, wave and tidal power.) This wouldn't even make a dent in global food and oil reserves.

Actions toward loss in carbon in future of energy will stimulate economic activities and job creation (by dominating damage in the fossil fuels) and enhance welfare, where a full policy menu is available. Many counties, especially those with underdeveloped supply chains and heavy fossil fuels dependence, need systemic reforms to utilize the transformation of the energy system. Governments continue to support fossil fuel extraction with significant policy assistance—in the shape of tax breaks and finance, direct framework spending and circumventions of regulation. The top oil and gas exporting countries have all but unanimously said they plan to increase outcomes. Some regions are trying to cut their coal production, while others are trying to maintain or even raise their level of output. The plans and ambitions of large producing countries have no doubt been diminished by the fact that some of them are now starting to consider and even to execute policies for a just and fair adjustment away from fossil fuel production. Spanning production also needs a steady, reliable source of data on fossil fuels production, and backing from governments and firms. Governments can promote greater transparency by including their goals into their climate commitments under the Agreement of Paris. Without a global shift toward renewable energy, few believe the Paris Agreement targets are likely to be met. Spending on renewable energy technologies is heavily shaped by policy instruments. It is the right policy tool, even though its effects are most apparent in developed renewable energy markets. Generating renewable energy remains more expensive than other, more conventional methods. Government support for R&D in renewable energy can also induce cost savings.

By linking up their separate renewable energy groups, these countries could possibly amplify exports of their technologies and share expertise on policy. The overarching purpose of any of our measures is to decrease costs of production and increase the share of renewable in the power system of the country. Yet, they can benefit from renewable energy technology implementation it its energy sector if they sign long term contracts with renewable energy suppliers, give energy technology a governmental support and control and develop a long-term goal.

Acknowledgements We are thankful to Department of Botany, Government College University, Lahore (Pakistan).

Funding The research was not supported by any funding.

Potential Conflicts of Interest On behalf of all authors, the corresponding author states that there is no conflict of interest.

Informed Consent Informed consent was obtained from all individual participants included in the study.

Authors Contribution All authors have contributed equally to this work.

Data Availability Statement The datasets during and/or analyzed during the current study available from the corresponding author on reasonable request.

References

Abeysekara WCSM, Siriwardana M, Merg S (2023) Economic consequences of climate change impacts on the agricultural sector of South Asia: a case study of Sri Lanka. Econ Anal Policy 77:435–450

Abidoye BO, Mendelsohn R, Ahmed S, Amanullah S, Chasidpon C, Baker L, Dobias R, Ghosh B, Gunaratne LHP, Hedeyetullah MM, Mungatana E, Ortiz C, Simoes M, Kurukulasuriya P, Perera C, Sooriyaarachchi A, Supnithadnapor A, Truong T (2017) South-east Asian Ricardian studies: Bangladesh, Sri Lanka, Thailand, and Vietnam. Clim Change Econ 8(3):1740002. https://doi.org/10.1142/S2010007817400048

Adebisi YA, Ogunkola IO (2023) The global antimicrobial resistance response effort must not exclude marginalised populations. Trop Med Health 51(1):33

Adil L, Eckstein D, Kuenzel V, Schaefer L (2025) Climate Risk Index 2025—who suffers most from extreme weather events. Germanwatch, February. https://www.germanwatch.org/sites/default/files/202502/Climate%20Risk%20Index%202025.pdf

Adom PK (2024) The socioeconomic impact of climate change in developing countries over the next decades: a literature survey. Heliyon 10

Adom PK, Amoani S (2021) Role of climate adaptation readiness in economic growth and climate change relationship: an analysis of output/income and productivity/institution channels. J Environ Manage 293:112923

Alhassan H (2021) The effect of agricultural total factor productivity on environmental degradation in sub-Saharan Africa. Sci Afr 12:e00740

Allan JD, Castillo MM, Capps KA (2021) Stream ecology: structure and function of running waters. Springer Nature

Anderson M, Cecchini M, Mossialos E (2020) Challenges to tackling antimicrobial resistance: economic and policy responses. Cambridge University Press

Anjani QK, Sabri AHB, Hutton AJ, Cárcamo-Martínez Á, Wardoyo LAH, Mansoor Z, Donnelly RF (2023) Microarray patches for managing infections at a global scale. J Control Release 359:97–115

Ansari A, Lin YP, Lur HS (2021) Evaluating and adapting climate change impacts on rice production in Indonesia: a case study of the Keduang subwatershed, Central Java. Environments 8(11):117

Arndt C, Thurlow J (2015) Climate uncertainty and economic development: evaluating the case of Mozambique to 2050. Clim Chang 130:63–75

Arnell N (1999) Climate change and global water resources. Glob Environ Chang 9:S31–S49

Baarsch F, Granadillos JR, Hare W, Knaus M, Krapp M, Schaeffer M, Lotze-Campen H (2020) The impact of climate change on incomes and convergence in Africa. World Dev 126:104699

Baldrian P, López-Mondéjar R, Kohout P (2023) Forest microbiome and global change. Nat Rev Microbiol 21:487–501

Barati AA, Zhoolideh M, Azadi H, Lee JH, Scheffran J (2023) Interactions of land-use cover and climate change at global level: how to mitigate the environmental risks and warming effects. Ecol Indic 146:109829

Batten S (2018) Climate change and the macro-economy: a critical review (Working Paper No. 706). Bank of England. https://doi.org/10.2139/ssrn.3104554

Bedford J, Ostle C, Johns DG, Atkinson A, Best M, Bresnan E et al (2020) Lifeform indicators reveal large-scale shifts in plankton across the North-West European shelf. Glob Chang Biol 26(6):3482–3497. https://doi.org/10.1111/gcb.15066

Bibi TS, Tekesa NW (2023) Impacts of climate change on IDF curves for urban stormwater management systems design: the case of Dodola Town, Ethiopia. Environ Monit Assess 195:170. https://doi.org/10.1007/s10661-022-10781-7

Bowen A, Cochrane S, Fankhauser S (2012) Climate change, adaptation and economic growth. Clim Chang 113(2):95–106

Buhaug H, Benjaminsen TA, Gilmore EA, Hendrix CS (2023) Climate-driven risks to peace over the 21st century. Clim Risk Manag 39:100471

Chalise S, Naranpanawa A, Bandara JS, Sarker T (2017) A general equilibrium assessment of climate change–induced loss of agricultural productivity in Nepal. Econ Model 62:43–50

Chandio AA, Jiang Y, Amin A, Ahmad M, Akram W, Ahmad F (2023) Climate change and food security in South Asia: fresh evidence from a policy perspective using novel empirical analysis. J Environ Plan Manag 66(1):169–190

Chikafa M, Nejadhashemi AP, Moller K, Razavi H, Bizimana JC (2023) Multidimensional evaluation of the impacts of agricultural interventions to achieve food security in Malawi. Food Energy Secur 12:e486

Chivangulula FM, Amraoui M, Pereira MG (2023) The drought regime in Southern Africa: a systematic review. Climate 11(7):147

Christensen JH, Hewitson B, Busuioc A, Chen A, Gao X, Held I, Jones R, Kolli RK, Kwon W-T, Laprise R, Magaña Rueda V, Mearns L, Menéndez CG, Räisänen J, Rinke A, Sarr A, Whetton P (2007) Regional climate projections. In: Solomon S, Qin D, Manning M, Chen Z, Marquis M, Averyt KB, Tignor M, Miller HL (eds) Climate change 2007: the physical science basis. Contribution of Working Group I to the Fourth Assessment Report of the Intergovernmental Panel on Climate Change. Cambridge University Press, pp 847–940. https://www.ipcc.ch/report/ar4/wg1

Cline WR (1992) Economics of global warming. Peterson Institute

Cline WR (2007) Global warming and agriculture: impact estimates by country. Peterson Institute

Costanzo V, Roviello GN (2023) The potential role of vaccines in preventing antimicrobial resistance (AMR): an update and future perspectives. Vaccine 11(2):333

Cronin J, Anandarajah G, Dessens O (2018) Climate change impacts on the energy system: a review of trends and gaps. Clim Chang 151:79–93

Cruz RV, Harasawa H, Lal M, Wu S, Anokhin Y, Punsalmaa B, Honda Y, Jafari M, Li C, Huu Ninh N (2007) Asia. In: Parry ML, Canziani OF, Palutikof JP, van der Linden PJ, Hanson CE (eds) Climate change 2007: impacts, adaptation and vulnerability. Contribution of Working Group II to the Fourth Assessment Report of the Intergovernmental Panel on Climate Change. Cambridge University Press, pp 469–506. https://www.ipcc.ch/report/ar4/wg2

De la Fuente A, Williams SE (2022) Climate change threatens the future of rain forest ringtail possums by 2050. Divers Distrib 29(1):173–183

Dell M, Jones BF, Olken BA (2012) Temperature shocks and economic growth: evidence from the last half century. Am Econ J Macroecon 4(3):66–95

do Prado Tanure TM, Miyajima DN, Magalhães AS, Domingues EP, Carvalho TS (2020) The impacts of climate change on agricultural production, land use and economy of the Legal Amazon region between 2030 and 2049. Economia 21(1):73–90

Donham EM, Flores I, Hooper A, O'Brien E, Vylet K, Takeshita Y et al (2023) Population-specific vulnerability to ocean change in a multistressor environment. Sci Adv 9(3):eade2365

Duan H, Yuan D, Cai Z, Wang S (2022) Valuing the impact of climate change on China's economic growth. Econ Anal Pol 74:155–174

Dogar MM, Kucharski F, Azharuddin S (2017a) Study of the global and regional climatic impacts of ENSO magnitude using SPEEDY AGCM. J Earth Syst Sci 126(2):30
Dogar MM, Stenchikov G, Osipov S, Wyman B, Zhao M (2017b) Sensitivity of the regional climate in the Middle East and North Africa to volcanic perturbations. J Geophys Res Atmos 122(15):7922–7948
Dogar MM, Sato T (2018) Analysis of climate trends and leading modes of climate variability for MENA region. J Geophys Res Atmos 123(23):13–074
Dogar MMA (2019) The sensitivity of DTR to the increased CO_2 over mid-latitude semi-dry regions. JOJ Hortic Arboric 2(2):555584
Dogar MM, Sato T (2019) Regional climate response of Middle Eastern, African, and South Asian monsoon regions to explosive volcanism and ENSO forcing. J Geophys Res Atmos 124(14):7580–7598
Dogar MM, Kucharski F, Sato T, Mehmood S, Ali S, Gong Z et al (2019) Towards understanding the global and regional climatic impacts of Modoki magnitude. Glob Planet Change 172:223–241
Dogar MM, Sato T, Liu F (2020) Ocean sensitivity to periodic and constant volcanism. Sci Rep 10(1):293
Dogar MMA (2020) Study of the regional climatic impacts of tropical explosive volcanism in the Middle East and North Africa region. Doctoral dissertation, Hokkaido University. https://doi.org/10.14943/doctoral.r7091
Dogar MMA, Almazroui M (2022) Revisiting the strong and weak ENSO teleconnection impacts using a high-resolution atmospheric model. Atmos Environ 270:118866. https://doi.org/10.1016/j.atmosenv.2021.118866
Dogar MM, Hermanson L, Scaife AA et al (2023) A review of El Niño southern oscillation linkage to strong volcanic eruptions and post-volcanic winter warming. Earth Syst Environ 7:15–42. https://doi.org/10.1007/s41748-022-00331-z
Dogar MM, Fujiwara M, Zhao M, Ohba M, Kosaka Y (2024) ENSO and NAO linkage to strong volcanism and associated post-volcanic high-latitude winter warming. Geophys Res Lett 51(1):e2023GL106114
Dogar MM, Fujiwara M, Ohba M, Almazroui M (2025) Investigating post-eruption amplified winter cooling in the Middle East and North Africa—unraveling ENSO and NAO dynamics. Environ Res Commun 7(4):045012
EPA (2022) EPA releases 2021 data collected under greenhouse gas reporting program. United States Environmental Protection Agency
Fankhauser S (1992) Economic costs of global warming: some monetary estimates conference presentation. In: International workshop on costs, impacts, and possible benefits of CO_2 mitigation. International Institute for Applied Systems Analysis, Laxenburg, Austria
Fankhauser S (1995) Protection versus retreat: the economic costs of sea-level rise. Environ Plan A 27(2):299–319. https://doi.org/10.1068/a27029
FAO, IFAD, UNICEF, WFP, & WHO (2022) The state of food security and nutrition in the world 2022: repurposing food and agricultural policies to make healthy diets more affordable. FAO. https://doi.org/10.4060/cc0639en
Farhad M, Kumar U, Tomar V, Hossain A (2023) Heat stress in wheat: a global challenge to feed billions in the current era of the changing climate. Front Sustain Food Syst 7:1203721
Farhana S, Raihan A (2022) Comparative study of the prevalence and risk factors associated with diarrhea among the people of rural areas and urban slums in Chittagong, Bangladesh. Int J Adv Res Innov Ideas Educ 8:1389–1398
Fischer G, Shah M, Tubiello FN, van Velthuizen H (2005) Socio-economic and climate change impacts on agriculture: an integrated assessment, 1990–2080. Philos Trans R Soc Lond B Biol Sci 360(1463):2067–2083
Foley JA, DeFries R, Asner GP, Barford C, Bonan G, Carpenter SR, Chapin FS, Coe MT, Daily GC, Gibbs HK, Helkowski JH, Holloway T, Howard EA, Kucharik CJ, Monfreda C, Patz JA, Prentice C, Ramankutty N, Snyder PK (2005) Global consequences of land use. Science 309(5734):570–574. https://doi.org/10.1126/science.1111772

Fonjong LN, Gyapong AY (2021) Plantations, women, and food security in Africa: interrogating the investment pathway towards zero hunger in Cameroon and Ghana. World Dev 138:105293
Gleditsch NP (2021) This time is different! Or is it? NeoMalthusians and environmental optimists in the age of climate change. J Peace Res 58(1):177–185
Han X, Hua E, Engel BA, Guan J, Yin J, Wu N, Sun S, Wang Y (2022) Understanding the implications of climate change and socioeconomic development for the water–energy–food nexus: a meta-regression analysis. Agric Water Manag 269:107693
Hankin LE, Leger EA, Bisbing SM (2023) Reforestation of high elevation pines: direct seeding success depends on seed source and sowing environment. Ecol Appl 33:e2897
Harris M, Fasolino T, Ivankovic D, Davis NJ, Brownlee N (2023) Genetic factors that contribute to antibiotic resistance through intrinsic and acquired bacterial genes in urinary tract infections. Microorganisms 11(6):1407
Hertel TW, Burke MB, Lobell DB (2010) The poverty implications of climate-induced crop yield changes by 2030. Glob Environ Chang 20(4):577–585
Hossain MS, Arshad M, Qian L, Zhao M, Mehmood Y, Kächele H (2019) Economic impact of climate change on crop farming in Bangladesh: an application of Ricardian model. Ecol Econ 164:106354. https://doi.org/10.1016/j.ecolecon.2019.106354
Husain M, Vishwakarma DK, Rathore JP, Rasool A, Parrey AA, Mahendar K (2018) Local people strategies in biodiversity conservation and sustainable development. Pharma Innov J 7(1):444–450
IPCC (1996) Climate change 1995: the science of climate change. In: Houghton JT, Meira Filho LG, Callander BA, Harris N, Kattenberg A, Maskell K (eds) Contribution of working group I to the second assessment report of the Intergovernmental Panel on Climate Change. Cambridge University Press
IPCC (2014) Climate change 2014: impacts, adaptation, and vulnerability. Part A: Global and sectoral aspects. In: Field CB et al (eds) Contribution of working group II to the fifth assessment report of the Intergovernmental Panel on Climate Change. Cambridge University Press
IPCC (2018) Global warming of 1.5°C. An IPCC Special Report on the impacts of global warming of 1.5°C above pre-industrial levels and related global greenhouse gas emission pathways, in the context of strengthening the global response to the threat of climate change, sustainable development, and efforts to eradicate poverty. Masson-Delmotte V, Zhai P, Pörtner H-O, Roberts D, Skea J, Shukla PR, Pirani A, Moufouma-Okia W, Péan C, Pidcock R, Connors S, Matthews JBR, Chen Y, Zhou X, Gomis MI, Lonnoy E, Maycock T, Tignor M, Waterfield T (eds)
IPCC (2022a) Climate change 2022: impacts, adaptation, and vulnerability. In: Pörtner H-O et al (eds) Contribution of working group II to the sixth assessment report of the Intergovernmental Panel on Climate Change. Cambridge University Press. https://doi.org/10.1017/9781009325844
IPCC. (2022b). Climate change 2022: mitigation of climate change. In IPCC (Ed.), Working group III contribution to the sixth assessment report of the Intergovernmental Panel on Climate Change. Cambridge University Press
Islam N, Winkel J (2017) Climate change and social inequality. United Nations Department of Economic and Social Affairs (DESA). https://www.un.org/esa/desa/papers/2017/wp152_2017.pdf
Issa R, van Daalen KR, Faddoul A, Collias L, James R, Chaudhry UA et al (2023) Human migration on a heating planet: a scoping review. PLoS Clim 2(5):e0000214
Jawo TO, Kyereh D, Lojka B (2023) The impact of climate change on coffee production of small farmers and their adaptation strategies: a review. Clim Dev 15(2):93–109
Jiang Z, Raghavan SV, Hur J, Sun Y, Liong SY, Nguyen VQ, Van Pham Dang T (2019) Future changes in rice yields over the Mekong River Delta due to climate change—alarming or alerting? Theor Appl Climatol 137:545–555. https://doi.org/10.1007/s00704-018-2614-2
Jones IL, Timoshenko A, Zuban I, Zhadan K, Cusack JJ, Duthie AB et al (2022) Achieving international biodiversity targets: learning from local norms, values and actions regarding migratory waterfowl management in Kazakhstan. J Appl Ecol 59(7):1911–1924. https://doi.org/10.1111/1365-2664.14178

Khan MA, Tahir A, Khurshid N, Husnain MIU, Ahmed M, Boughanmi H (2020) Economic effects of climate change–induced loss of agricultural production by 2050: a case study of Pakistan. Sustainability 12(3):1216

Kim H, Franco AC, Sumaila UR (2023) A selected review of impacts of ocean deoxygenation on fish and fisheries. Fishes 8(6):316

Kumar SN, Aggarwal PK, Rani DS, Saxena R, Chauhan N, Jain S (2014) Vulnerability of wheat production to climate change in India. Clim Res 59(3):173–187

Kyriakopoulos GL, Sebos I (2023) Enhancing climate neutrality and resilience through coordinated climate action: review of the synergies between mitigation and adaptation actions. Climate 11(5):105

Lamperti F, Bosetti V, Roventini A, Tavoni M, Treibich T (2021) Three green financial policies to address climate risks. J Financ Stab 54:100875

Lee KE, Clemens SC, Kubota Y, Timmermann A, Holbourn A, Yeh SW et al (2021) Roles of insolation forcing and CO_2 forcing on late Pleistocene seasonal sea surface temperatures. Nat Commun 12:5742. https://doi.org/10.1038/s41467-021-25979-6

Li B (2023) Research on the path of promoting the common prosperity of farmers and countryside in the new era. Acad J Humanit Soc Sci 6(13):105–111

Lobell DB, Field CB (2007) Global scale climate–crop yield relationships and the impacts of recent warming. Environ Res Lett 2(1):014002

Ludwig F, Terwisscha van Scheltinga C, Verhagen J, Kruijt B, van Ierland E, Dellink R, de Bruin K, Kabat P (2007) Climate change impacts on developing countries—EU accountability (IP/A/ENVI/ST/2007-04). European Parliament

Manes S, Costello MJ, Beckett H, Debnath A, Devenish-Nelson E, Grey KA et al (2021) Endemism increases species' climate change risk in areas of global biodiversity importance. Biol Conserv 257:109070

Meattle C, Padmanabhi R, de Aragão Fernandes P, Balm A, Wakaba E, Chiriac D, Tonkonogy B (2022) Landscape of climate finance in Africa. Climate Policy Initiative. https://www.climatepolicyinitiative.org/wp-content/uploads/2022/09/Landscape-of-ClimateFinance-in-Africa.pdf

Mehmood M, Qamar R, Joyia FA (2023) Effect of high temperature stress on pollen grains in sunflower (Helianthus annuus L.) inbred lines. Braz Arch Biol Technol 66:e23220927

Mendelsohn R, Nordhaus W, Shaw D (1994) The impact of global warming on agriculture: a Ricardian analysis. Am Econ Rev 84(4):753–771. http://www.jstor.org/stable/2118029

Miron IJ, Linares C, Diaz J (2023) Influence of climate change on food production and food safety. Environ Res 216(3):Article 114674

Mishra D, Sahu NC, Sahoo D (2015) Impact of climate change on agricultural production of Odisha (India): a Ricardian analysis. Reg Environ Change 16:575–584. https://doi.org/10.1007/s10113-015-0774-5

Monteleone B, Borzí I, Bonaccorso B, Martina M (2023) Quantifying crop vulnerability to weather-related extreme events and climate change through vulnerability curves. Nat Hazards 116(3):2761–2796

Nehe A, Martinsson UD, Johansson E, Chawade A (2023) Genotype and environment interaction study shows fungal diseases and heat stress are detrimental to spring wheat production in Sweden. PLoS One 18(5):e0285565

Nelson GC, Rosegrant M, Palazzo A, Gray I, Ingersoll C, Robertson R, Tokgoz S, Zhu T, Sulser TB, Ringler C, Msangi S, You L (2010) Food security, farming, and climate change to 2050: scenarios, results, and policy options. International Food Policy Research Institute. https://www.ifpri.org/publication/food-security-farming-and-climate-change-2050

Ngoukwa G, Chimi CD, Bakonck LM, Zekeng JC, Yonkeu AN, Mboda RBT et al (2023) Perception and adaptation strategies of forest dwellers to climate variability in the tropical rainforest in eastern Cameroon: the case of the inhabitants of the Belabo-Diang Communal Forest. Heliyon 9(4):e14515

Nwosu EC, Brauer A, Monchamp ME, Pinkerneil S, Bartholomäus A, Theuerkauf M et al (2023) Early human impact on lake cyanobacteria revealed by a Holocene record of sedimentary ancient DNA. Commun Biol 6(1):72. https://doi.org/10.1038/s42003-023-04573-0

Ogden LE (2018) Climate change, pathogens, and people: the challenges of monitoring a moving target. Bioscience 68(10):733–739

Okoro OJ, Deme GG, Okoye CO, Eze SC, Odii EC, Gbadegesin JT et al (2023) Understanding key vectors and vector-borne diseases associated with freshwater ecosystem across Africa: implications for public health. Sci Total Environ 862:160732

Palita SK (2016) Climate change and its impact on biodiversity. In: Climate change—India and the world, the future course for cooler planet. SN College, Rajkanika, Kendrapara, Odisha

Pastor AV, Palazzo A, Havlik P, Biemans H, Wada Y, Obersteiner M, Kabat P, Ludwig F (2019) The global nexus of food–trade–water sustaining environmental flows by 2050. Nat Sustain 2(6):499–507

Pathak H (2023) Impact, adaptation, and mitigation of climate change in Indian agriculture. Environ Monit Assess 195(1):52

Peng Y, Welden N, Renaud FG (2023) A framework for integrating ecosystem services indicators into vulnerability and risk assessments of deltaic social-ecological systems. J Environ Manag 326:116682

Pescaroli G, Guida K, Reynolds J, Pulwarty RS, Linkov I, Alexander DE (2023) Managing systemic risk in emergency management, organizational resilience and climate change adaptation. Disaster Prev Manage 32(1):234–251

Pipitpukdee S, Attavanich W, Bejranonda S (2020) Climate change impacts on sugarcane production in Thailand. Atmosphere 11(4):408

Prokopenko O, Prokopenko M, Chechel A, Marhasova V, Omelyanenko V, Orozonova A (2023) Ecological and economic assessment of the possibilities of public-private partnerships at the national and local levels to reduce greenhouse gas emissions. Econ Aff 68:133–142

Ratnayake SS, Reid M, Larder N, Kadupitiya HK, Hunter D, Dharmasena PB et al (2023) Impact of climate change on paddy farming in the village tank cascade systems of Sri Lanka. Sustainability 15(12):9271

Rezvi HUA, Tahjib-Ul-Arif M, Azim MA, Tumpa TA, Tipu MMH, Najnine F et al (2023) Rice and food security: climate change implications and the future prospects for nutritional security. Food Energy Secur 12(1):e430

Rodway GW (2023) Climate change in and around the high ranges of Asia: consequences for human health. Wilderness Environ Med 34(1):1–2

Rogelj J, Popp A, Calvin KV, Luderer G, Emmerling J, Gernaat D, Fujimori S, Strefler J, Hasegawa T, Marangoni G, Krey V, Kriegler E, Riahi K, Vuuren DP, Doelman J, Drouet L, Edmonds J, Fricko O, Harmsen M, Havlík P, Humpenöder F, Stehfest E, Tavoni M (2018) Scenarios towards limiting global mean temperature increase below 1.5 °C. Nat Climate Change 8(4):325–332. https://doi.org/10.1038/s41558-018-0091-3

Sasai F, Roncal-Jimenez C, Rogers K, Sato Y, Brown JM, Glaser J et al (2023) Climate change and nephrology. Nephrol Dial Transplant 38(1):41–48

Schmidhuber J, Tubiello FN (2007) Global food security under climate change. Proc Natl Acad Sci 104(50):19703–19708

Scott D (2021) Sustainable tourism and the grand challenge of climate change. Sustainability 13(4):1966

Scott GJ, Petsakos A, Juarez H (2019) Climate change, food security, and future scenarios for potato production in India to 2030. Food Secur 11:43–56

Seo SN, Mendelsohn R (2008) Ricardian analysis of the impact of climate change on South American farms. Chilean J Agric Res 68:69–79

Shaw AK (2020) Causes and consequences of individual variation in animal movement. Movement Ecol 8(1):12

Signe L, Mbaye AA (2022) Renewing global climate change action for fragile and developing countries (Working Paper No. 179). Brookings Institution. https://www.brookings.edu/wp-content/uploads/2022/11/NOV-2022-Signe_Mbaye_FINAL-1.pdf

Sinnarong N, Chen CC, McCarl B, Tran BL (2019) Estimating the potential effects of climate change on rice production in Thailand. Paddy Water Environ 17:761–769

Srivastava RK, Panda RK, Chakraborty A (2021) Assessment of climate change impact on maize yield and yield attributes under different climate change scenarios in eastern India. Ecol Indic 120:106881. https://doi.org/10.1016/j.ecolind.2020.106881

Stern N (2006) Stern review on the economics of climate change. Part II: The impacts of climate change on growth and development. Cambridge University Press. https://webarchive.nationalarchives.gov.uk/ukgwa/20100407172811/https://www.hm-treasury.gov.uk/stern_review_report.htm

Stern N (2007) The economics of climate change. Cambridge University Press. https://doi.org/10.1017/CBO9780511817434

Subasinghe R, Alday-Sanz V, Bondad-Reantaso MG, Jie H, Shinn AP, Sorgeloos P (2023) Biosecurity: reducing the burden of disease. J World Aquacult Soc 54(2):397–426

Subedi A, Marchand P, Bergeron Y, Morin H, Girona MM (2023) Climatic conditions modulate the effect of spruce budworm outbreaks on black spruce growth. Agric For Meteorol 339:109548

Tahir F, Al-Ghamdi SG (2023) Climatic change impacts the energy requirement for the built environment sector. Energy Rep 9:670–676

Tang T, Hu P, Zhang W, Xiao D, Tang L, Xiao J et al (2023) The role of bedrock geochemistry and climate in soil organic matter stability in subtropical karst forests of Southwest China. Forests 14(7):1467

Tiwari A, Kurittu P, Al-Mustapha AI, Heljanko V, Johansson V, Thakali O et al (2022) Wastewater surveillance of antibiotic-resistant bacterial pathogens: a systematic review. Front Microbiol 13:977106

Tol RSJ (1995) Damage costs of climate change toward more comprehensive calculations. Environ Resour Econ 5:353–374

Tol RSJ (2018) Economic impacts of climate change. Rev Environ Econ Policy 12(1):4–25. https://doi.org/10.1093/reep/rex027

Urban MC, Nadeau CP, Giery ST (2023) Using mechanistic insights to predict the climate-induced expansion of a key aquatic predator. Ecol Monogr 93:e1575

Valenzuela E, Anderson K (2011) Climate change and food security to 2050: a global economy-wide perspective (Presentation No. 100531). In: 55th Australian agricultural and resource economics society conference, Melbourne, Australia. https://doi.org/10.22004/ag.econ.100531

Vernooy R (2022) Does crop diversification lead to climate-related resilience? Improving the theory through insights on practice. Agroecol Sustain Food Syst 46(6):877–901

von Braun J, Afsana K, Fresco LO, Hassan MHA (2023) Science and innovations for food system transformation. Springer, Cham

Waheed A, Fischer B, Khan MI (2021) Climate change policy coherence across policies, plans, and strategies in Pakistan—implications for the China–Pakistan Economic Corridor plan. Environ Manage 67:793–810

Watts N, Adger WN, Agnolucci P, Blackstock J, Byass P, Cai W et al (2015) Health and climate change: policy responses to protect public health. Lancet 386(10006):1861–1914

Wang GG, Lu D, Gao T, Zhang J, Sun Y, Teng D, Yu F, Zhu J (2024) Climate-smart forestry: an AI-enabled sustainable forest management solution for climate change adaptation and mitigation. J For Res 36(1):7–21. https://doi.org/10.1007/s11676-023-01773-5

World Health Organization (2021) Climate change and health [Fact sheet]. World Health Organization. https://www.who.int/news-room/fact-sheets/detail/climate-change-and-health

World Meteorological Organization (2022) State of global water resources 2021 (WMO-No. 13-08). World Meteorological Organization. https://library.wmo.int/idurl/4/58262

Yalew SG, van Vliet MTH, Gernaat DEHJ, Ludwig F, Miara A, Park C et al (2020) Impacts of climate change on energy systems in global and regional scenarios. Nat Energy 5:794–802

Zamuda C, Bilello DE, Conzelmann G, Mecray E, Satsangi A, Tidwell V, Walker BJ (2018) Energy supply, delivery, and demand. In: Reidmiller DR, Avery CW, Easterling DR, Kunkel KE, Lewis KLM, Maycock TK, Stewart BC (eds) Impacts, risks, and adaptation in the United States: Fourth National Climate Assessment, vol 2. U.S. Global Change Research Program, pp 174–201. https://doi.org/10.7930/NCA4.2018.CH4

Zhang Y, Liu H, Qi J, Feng P, Zhang X, Liu DL et al (2023) Assessing impacts of global climate change on water and food security in the black soil region of Northeast China using an improved SWAT-CO_2 model. Sci Total Environ 857:159482

Zhao Y, Liu S (2023) Effects of climate change on economic growth: a perspective of the heterogeneous climate regions in Africa. Sustainability 15(9):7136

Part VI
Establishing Pathways for Resilient and Sustainable Development

Chapter 16
Climate-Resilient and Adaptive Strategies for Hill Farming in Mizoram

Lungmuana, Samik Choudhury, Ramchhanliana Hauchhum, Lalhruaipuii, P. L. Lalrinsanga, Blessa Sailo, Vanlalruati, Jeetendra Kumar Soni, Evelyn Lalparmawii, and Sunil Doley

Abstract The monsoon rainfall in north eastern hills (NEH) like Mizoram is a crucial factor for hill farming, making it vulnerable to certain climate fluctuations. Small and marginal hill farmers' livelihoods are threatened by adverse climate conditions, such as rising temperatures and erratic rainfall patterns, which can negatively impact agricultural productivity. Evidence suggests that in the recent past, the state experienced temperature variability across the state, with reduced rainfall during the pre-monsoon season and increased rainfall during the post-monsoon season. Thus, strategies that have the capacity to adapt and enhance resilience against climate variability are of utmost importance in the fragile ecosystem. Traditional agricultural practices and existing land uses need to be improved through systematic scientific management to enhance soil carbon sequestration, increase ecosystem carbon sequestration, and restore soil health. The use of resource conservation technologies has been proven to increase soil carbon and cropping intensity, which ultimately leads to higher crop productivity. Another way to adapt to climate variability and improve soil health is through crop diversification, which involves a proper cropping sequence. Preventing soil erosion, improving ecosystem C, and supporting sustainable agriculture can be achieved through improved land use from traditional jhum or improved jhum practices. Finally, integrating different components, such as crop and animal husbandry, fisheries, and silviculture, may be an adaptive measure to cope with climate variability and a sustainable farming system. The aim of this chapter is to offer various proven agricultural technologies for climate change adaptation in the hills of Mizoram.

Lungmuana (✉) · S. Choudhury · Lalhruaipuii · P. L. Lalrinsanga · B. Sailo
ICAR RC NEH Region, Mizoram Centre, Kolasib, Mizoram, India

R. Hauchhum · Vanlalruati · E. Lalparmawii · S. Doley
Department of Forestry, Mizoram University, Aizawl, Mizoram, India

J. K. Soni
ICAR-Directorate of Weed Research, Jabalpur, Madhya Pradesh, India

S. Mandal et al. (eds.), *Climate Change Vulnerability*, Climate Change Management, https://doi.org/10.1007/978-3-032-16783-5_16

Introduction

The north-east India comprises hills and intermontane valleys where different agricultural activities, from traditional to intensive, interplay. The region is a biodiversity hotspot and is a high rainfall zone. One of the regions in Meghalaya, Mawsynram, receives on average 11,963–12,550 mm (Kuttippurath et al. 2021). Due to the geographical settings in hill agriculture, sudden changes in rainfall patterns can lead to floods, droughts during the important vegetation, affecting agricultural production at large (Kumar et al 2026). A recent study by Chakraborty et al. (2025a, b) revealed that there has been an increasing variability and negative anomalies in monsoonal rainfall in NE India, which significantly affects the agricultural sector in the region. Temperature fluctuations and unpredictable rainfall patterns have been shown to have a detrimental impact on the yield of both food and non-food grains, endangering the livelihoods of small and marginal farmers (Kumar and Sharma 2022). Northeast India is experiencing shifting cultivation or jhum from the past, a traditional agriculture system where natural forests is being converted for cultivation. In NE India, shifting cultivation accounts for around 85% of these conversions (Singh et al. 2018). Because it entails the slashing and burning of natural forests, it would negatively affect the soil organic carbon (SOC) stock and overall soil health and make the ecosystem fragile (Sahoo et al. 2019; Verma et al. 2026).

Climate resilience is the capacity of ecosystems, communities, or systems to predict, adapt to, and recover from the consequences of climate change. Extreme weather disasters have shown how crucial resilience is to any comprehensive climate action program because climate change is a global and hyper-local issue. Agriculture and other land uses are major contributors to greenhouse gas (GHG) emissions, accounting for over 23% of anthropogenic emissions of CO_2, CH_4, and N_2O, according to the IPCC (2021). The growing season, crop duration, and soil moisture regime may all be altered by increasing temperature and changing precipitation patterns, which could induce drought and negatively affect crop yield (Calzadilla et al. 2013). One of the major causes of greenhouse gas (GHG) emissions, a dominant factor towards global warming, is the conversion of forest ecosystems into various land uses (Raihan et al. 2018). As climate change alters ecosystems, it affects the spread of diseases, with warmer climates expanding the range of pathogens, vectors, and parasites that affect animals. This not only threatens biodiversity but also impacts food security, as livestock face increased vulnerability to heat stress and disease outbreaks (Chaplin-Kramer et al. 2014; Tahir et al. 2025).

As the amount of carbon (C) reservoir in the soil exceeds the total C mass of the atmosphere and biomass, SOC is a major potential sink of atmospheric CO_2 (Yang et al. 2022). Even small changes in SOC stocks can have a significant impact on atmospheric CO_2 levels and climate change (Das et al. 2024). Thus, strategies to increase C sequestration in the soil may play a positive and adaptive role towards global warming. Through direct C sequestration, a well-designed, multifunctional plantation can reduce the strain on natural forests, restore some of the ecological services they offer, and slow down global warming (SiwSiwar et al. 2016). This calls for an adaptive strategy to increase system sustainability, C sequestration, and appropriate agronomic intervention while addressing the region's food security. In the hills of NE India including Mizoram, ecological degradation due to shifting cultivation may be conversion to permanent plantations, agroforestry and integrated farming system (IFS) to ensure ecological sustainability and farmers security (Hazarika et al. 2024). Plantation and conservation is becoming more and more important in order to restore degraded land, limit logging in wild forests, and C sequestration (Kelty 2006). Agro-advisory services could be crucial in protecting small and marginal farmers in the area from risky climate conditions. This chapter aims to provide an overview of the current situation and adaptation tactics for the hill farmers in Mizoram in response to different climate change-related concerns. The management and adaptation measures against climate aberrations would benefit greatly from these reviews.

Climate Variability in Mizoram

The dependence on rain-fed agriculture in Mizoram makes it extremely vulnerable to climatic variations, which in turn affects agricultural production and food security. The livelihoods of small and marginal farmers who control most of the region's agricultural landscape are at risk due to the negative effects of unfavourable climate conditions, such as rising temperatures and unpredictable rainfall patterns, on both food grain and non-food grain productivity (Kumar and Sharma 2022). Food insecurity is made worse by the overall effects of climate change, which raise the likelihood of crop failure and lower yields, even though some crops may exhibit positive growth trends (Kumar and Upadhyay 2019; Singh 2017). Additionally, the relationship between food security and agricultural production is clear since, especially in states where agriculture is the main economic activity, decreasing crop yields pose a direct threat to the population's nutritional status (Kumar and Nirmala 2023). In order to improve food security in Mizoram, it is imperative that climate variability be addressed through sustainable farming techniques and infrastructure upgrades.

Changes in rainfall patterns, temperature fluctuations, and major climatic oscillations such as the Indian Ocean Dipole (IOD) and the El Niño–Southern Oscillation (ENSO) are the main causes of the climate variability seen in Mizoram (WMO 2022). Case studies from North-East India provide a detailed examination of these influences: The 2009 El Niño caused a 73% monsoon shortfall, affecting paddy sowing in Mizoram and the region (IMD 2010). In contrast, the positive Indian Ocean Dipole (IOD) in 2019 increased rainfall in central and southern India, but the North-East was drier, registering 88% long period average (LPA), with the Nagaland–Manipur–Mizoram–Tripura subdivision classed as "deficient" (IMD 2019). Additionally, in 2023, El Niño circumstances caused a rainfall reduction of 82% in the North-East, highlighting the impact of ENSO on regional monsoon variability (IMD 2023). Since Mizoram relies on the southwest monsoon for agriculture, unpredictable rainfall patterns are possible. Increased unpredictability has led to seasonal droughts and postponed monsoons, which hurt rice agriculture, a vital crop (Tosato 2023). Elevated temperatures and humidity also increase forest fires. However, the overarching trend indicates a reduction in fires, while specific districts face heightened risk attributed to localised climatic conditions (Bhavsar et al. 2023). The impact of these drivers on the ecosystem is significant, leading to biodiversity loss, influencing water resources, and posing challenges to the sustainability of traditional agricultural practices. Resilience to climate change effects requires adaptive measures (Lalthanpuia et al. 2022; Tiwari 2006).

The LPA rainfall in Mizoram from 1990 to 2022 exhibits significant intra-annual variability when analysed alongside observations from 2013 to 2022 and 2018 to 2022 (Saha et al. 2016; Chakraborty et al. 2025a, b). The primary monsoon months (June–September), typically providing averagely 399.4–487.6 mm, indicate a declining trend. In the period from 2013 to 2022, June rainfall exceeded the LPA by 13.5%. However, in the subsequent period from 2018 to 2022, it fell short by 1.6%. Meanwhile, July's rainfall exhibited relative stability, fluctuating within a margin of ±1.9%. August exhibited the most significant deficits, measuring 406.3 mm from 2013 to 2022 (−16.7%) and 446.8 mm from 2018 to 2022 (−8.4%) compared to the 487.6 mm LPA (Saha et al. 2024). The post-monsoon period exhibited a notable increase: October rainfall surpassed the long-term average of 173.0 mm by 28.6% for the years 2013–2022 and by 20.5% for 2018–2022. Additionally, November totals approached double the normal of 27.1 mm, showing a 76.7% increase in 2018–2022 (Fig. 16.1). Significant wet anomalies were found in the monsoon season (June–September), with 4-month standardised Precipitation Index (SPI) values ranging from +2.0 to +3.1 ($p < 0.05$–0.01), and in the post-monsoon season (October–December), with 3-month SPI values ranging from +1.8 to +2.6 ($p < 0.01$), according to the SPI analysis. In contrast, the pre-monsoon months showed a significant dryness, indicated by a 3-month SPI of −2.6 ($p < 0.01$). The analysis of

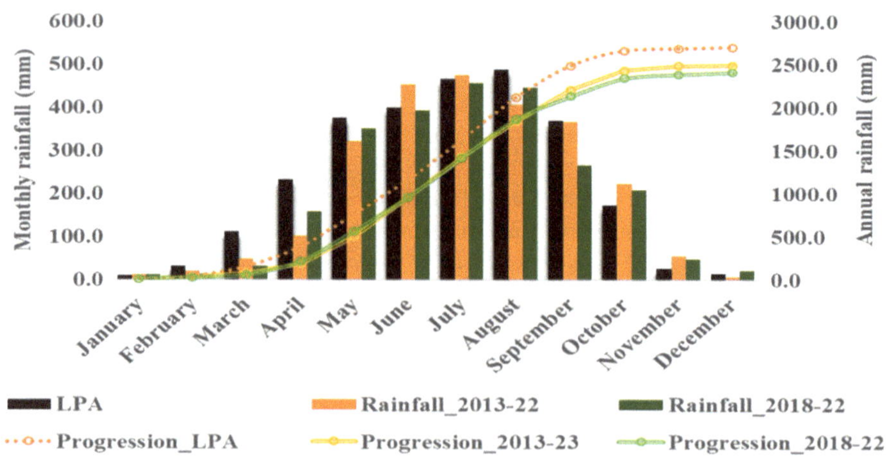

Fig. 16.1 Monthly and annual rainfall trends: LPA vs. recent decadal observations (2013–2022 and 2018–2022) (1990–2022)

seasonal contributions underscores this shift—pre-monsoon rainfall, previously accounting for approximately 22.47% of the annual cycle, has decreased by 23.9%. Similarly, the monsoon share, which stood at around 63.82%, has seen a reduction of 6.52%. In contrast, post-monsoon rainfall, which constitutes 7.96%, has risen by 18.25% above normal, with the months of October and November experiencing a notable increase of 25.3% (Fig. 16.1) (WMO 2022). The changes in seasonality have heightened the risks of waterlogging in lowland rice fields during the harvest period, highlighting a shift from the predominance of peak monsoon effects to more significant influences from the post-monsoon period.

The analysis of temperature trends in Mizoram indicates notable spatial and seasonal differences between Kolasib and Aizawl. In Kolasib, the annual mean temperatures exhibited a consistent upward trend from 1990 to 2024. Notably, there was a marked increase in February's minimum temperature ($Z = +3.14$, $p < 0.01$) and March's minimum temperature ($Z = -2.48$, $p < 0.05$), indicating a pattern of warming winters and elevated temperatures during the pre-monsoon months (Fig. 16.2), which aligns with the findings of Saha et al. (2016). In Aizawl, long-term IMD gridded datasets (1951–2010) reveal a consistent warming trend, with mean maximum temperatures rising at approximately 0.02–0.03 °C annually, and minimum temperatures increasing at around 0.01–0.02 °C per year, particularly noted during the winter and pre-monsoon seasons (Jhajharia and Singh 2011; Jain and Kumar 2012). Seasonal analysis indicates that Aizawl experiences a marginally stronger increase in temperature during the pre-monsoon months (March–May) in comparison to Kolasib, whereas winter warming appears to be similar between the two locations. Both locations illustrate the overarching trend in Northeast India of warming winters and heightened pre-monsoon heat. However, Aizawl's urban environment may exacerbate the increase in minimum temperatures due to urban heat island effects (Tahir et al. 2025).

Land Use Land Cover Change and Soil Organic C (SOC)

The global C cycle depends heavily on soil C, which also determines soil fertility and production (Van der Werf et al. 2009). Research conducted worldwide has revealed significant variation in the relationship between land use and soil C (Ali et al. 2017). In Mizoram, it is widely known that the transition from natural to agricultural practices caused the loss of SOC (Manpoong and Tripathi 2019; Lungmuana et al. 2023a). One of the main variables influencing the distribution patterns of nutrients in the soil (Islam and Weil 2000) is thought to be the alteration of the soil brought about by changes in land use types and patterns, which has resulted in a decline in soil productivity (Ayoubi et al. 2011) (Fig. 16.3). Soil C management is therefore essential to maintaining ecological balance and a climate change adaptation plan.

Shifting cultivation or jhum system has been reported to deteriorate the soil fertility and decrease the SOC in the first year cultivation. Lower SOC stock in young

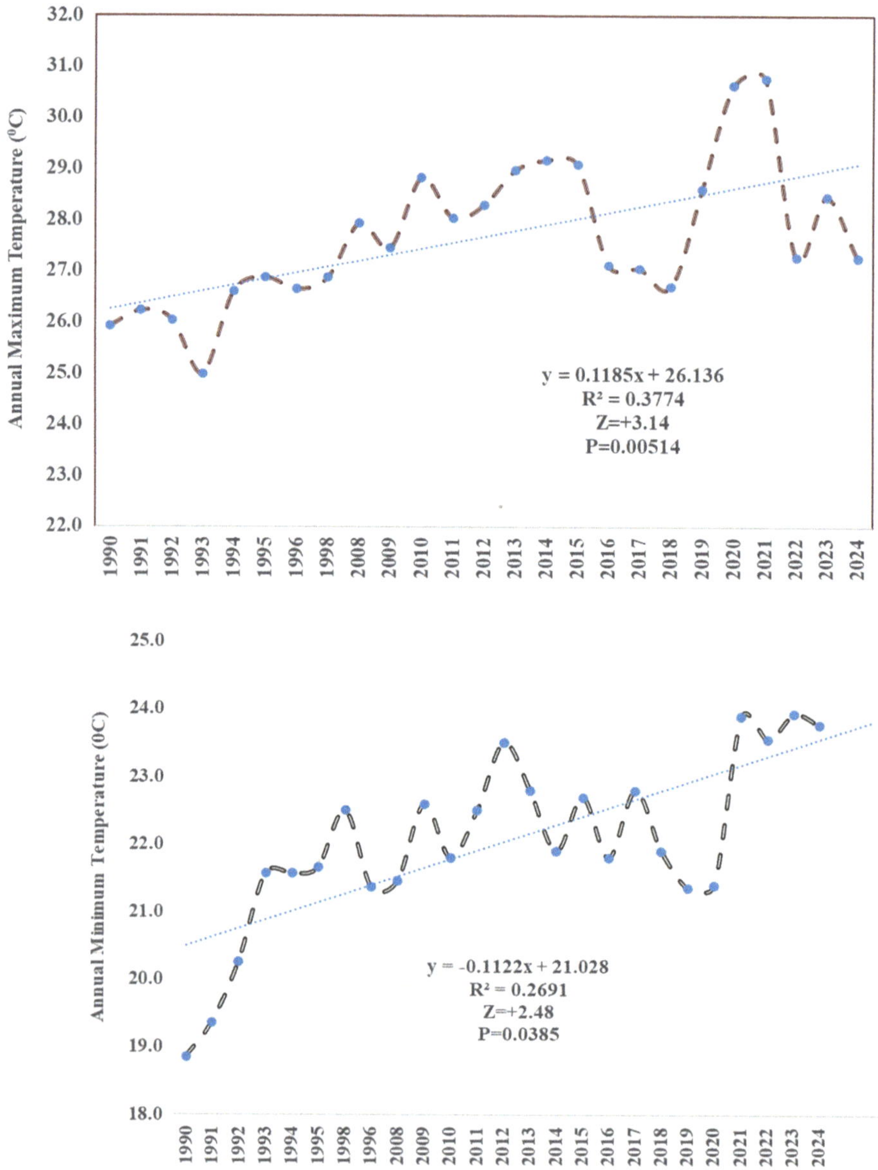

Fig. 16.2 Annual maximum and minimum temperature dynamics with trend analysis (1990–2024)

forest fallow (<10 years) compared to old forest fallow (>10 years) has also been observed (Lungmuana et al. 2018). Recently, shifting cultivation has slowly shifted towards plantations and has resulted to increase in SOC due to various factors (Lungmuana et al. 2019). The conversion of existing jhum to oil palm, teak, areca-nut, and home garden crops resulted in a positive rate of change in the C stock in all

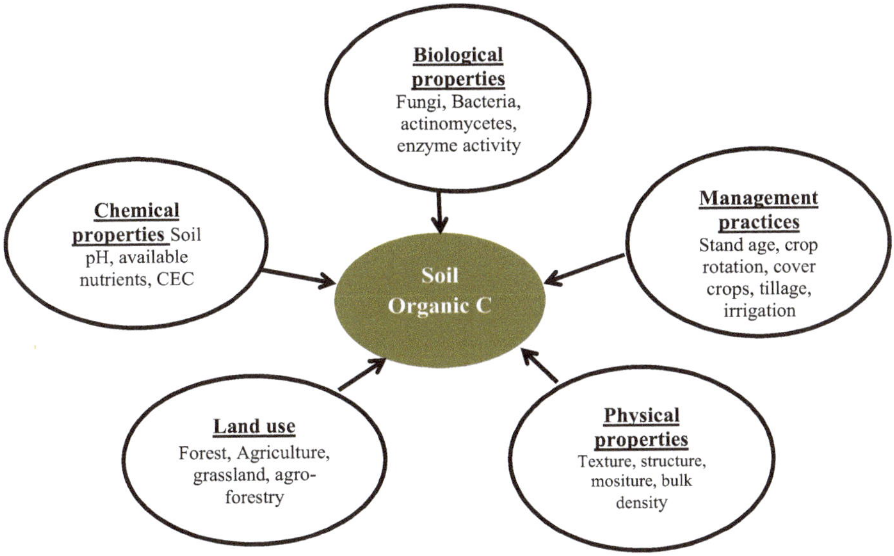

Fig. 16.3 Schematic presentation of factors affecting SOC

Table 16.1 Variations in soil organic carbon (OC) across various land uses in Mizoram

Land use system	Soil C stock (Mg ha^{-1})	References
Rubber	45.03–59.34	Manpoong and Tripathi (2021), Hrahsel and Sahoo (2024), Kenye et al. (2019), Sahoo et al. (2023), Manpoong and Tripathi (2019), Singh et al. (2018)
Teak	44.66–49.09	
Oil palm	26.05–43.35	
Arecanut	30.08–44.71	
Natural forest	52.74–67.90	
Bamboo	29.83–47.47	
Homegarden	46.82–50.85	
Jhum	27.87–29.67	
Fallow (5 and 10 years)	61.35–69.56	
Agroforestry	50.85–62.37	

Note: The soil C stock values represent up to 40 cm soil depth

the pools (Manpoong and Tripathi 2021; Hrahsel and Sahoo 2024). All land uses showed an increase in SOC stocks, however the magnitude of the changes varied depending on the vegetation age and land use (Table 16.1). Previous studies indicated that the analysis of SOC was calculated up to 30 cm during the period 2018–2023 expressed as C stock in Mg per ha. These indicate that the rate of SOC change may either increase (Lemma et al. 2006) or decrease (Farley et al. 2004) as a result of land use transitions. Additionally, this suggests the degrading effects of shifting cultivation in northeast India, where land becomes scarce and reduces the fallow periods (Uhlig et al. 1994).

Agroforestry or Tree-based systems have greatly enhanced the ecosystem's ability to store C and facilitate climate change adaptation and mitigation. This emphasises how important and vulnerable the vegetation biomass pool is, as the effects of land use management on changing soil and climate conditions have a substantial impact on its sequestration. Consequently, managing land use change requires focussing on finding workable solutions, such agroforestry systems, that would save species and eventually store soil C and tree biomass. The effectiveness varies widely depending on species composition, management practices and previous land use system, and poorly planned plantations can risk reducing rather than enhancing SOC (Lal 2004).

Strategies to Increase Soil C Sequestration in Mizoram

The schematic presentation of technological approaches to enhance SOC (Fig. 16.4) highlights the role of improved land management practices and biological inputs in promoting C sequestration and soil health. Inter-cropping and crop rotation increase the diversity of plant residues and root systems, thereby enhancing belowground C inputs and improving nutrient cycling (Cong et al. 2015).

Agroforestry systems integrate trees with crops or pastures, providing continuous organic matter inputs through litter fall, root turnover, and canopy protection, while also improving microclimatic conditions (Nair et al. 2009; Paustian et al. 2016). Appropriate agroforestry models on hill slopes could be crucial for raising SOC, enhancing ecosystem resilience, and enhancing farmer livelihoods. Tree-covered agroforestry systems can reduce soil erosion and promote the formation of soil aggregates by protecting the soil surface from erosive pressures, moisture, and direct sunlight. Additionally, agroforestry systems with deeply rooted trees can

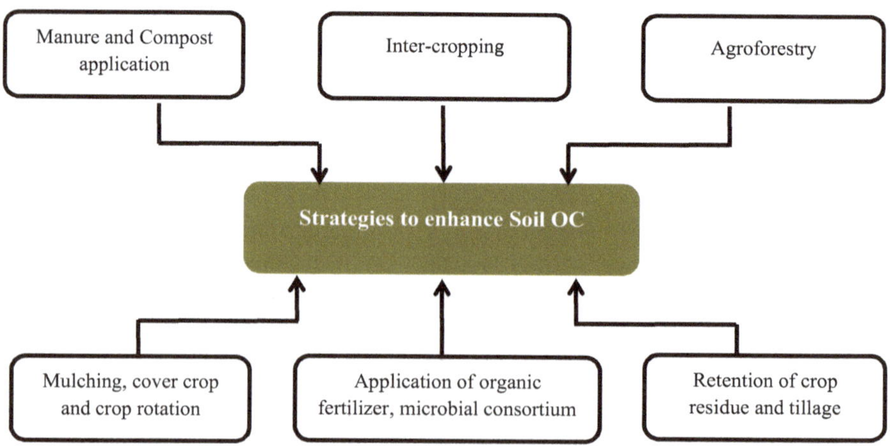

Fig. 16.4 Schematic presentation of technological approach to enhance soil OC

enhance soil structure by boosting soil porosity, aeration, and water penetration rates. This may facilitate the incorporation of organic matter into the soil matrix and raise the amount of SOC (Jose 2009). Agroforestry systems with N fixing leguminous trees such as tree bean could increase soil C storage than other land use in Manipur (NE India) (Ansari et al. 2022). Other than the sustainability effect on the soil C build up, suitable agroforestry systems are beneficial for farmer economic livelihood. The reduction and transformation of shifting cultivation to permanent agroforestry systems such as rubber, oil palm, teak, arecanut etc. in the NE states improves the livelihood of the farmers (Lungmuana et al. 2024).

The application of manure, compost, and organic fertilizers supplies stable organic matter and nutrients that stimulate microbial activity and long-term SOC accumulation (Lal 2004; Sommer and Bossio 2014). Cover cropping maintains living plant cover during fallow periods, preventing C losses through erosion and leaching, while simultaneously contributing biomass to the soil (Blanco-Canqui et al. 2015). North east India as a biodiversity hotspot is rich in indigenous legumes (N fixers) which can be used to improve degraded soils and increase soil C. A steady supply of nutrients for plant uptake is provided by the gradual release of nutrients from manures and compost into the soil through mineralization and microbial activity (Achu et al. 2013). The introduction of microbial consortia enhances decomposition, nutrient mineralization, and the stabilization of organic matter through microbial by-products and symbiotic interactions (Bhattacharyya et al. 2016).

Conservation tillage with residue retention can increase the quantity of organic matter that is added to the soil (Farmaha et al. 2022). Crop residues left over after harvest can serve as a protective layer, reducing soil erosion and increasing soil moisture, which both aid in the decomposition of organic residues (Giller et al. 2009). Conservation tillage or resource conservation farming methods increase the stability of soil organic matter leading to increased C sequestration in the soil. A study by Rangappa et al. (2024) in the north east hill region, observed that conservation tillage (minimum and no tillage) in rice followed by pea increased the physiological traits of pea and yield. Similarly, Das et al. (2020) has also found conservation tillage in rice followed by rapeseed increased the soil properties including SOC compared to conventional tillage. Inclusion of suitable legume (soybean) with maize as a sequence is another practice to increase soil health, diversification and cropping intensity (Lungmuana et al. 2023b) as mono cropping systems of maize and rice has to be diversified to increase cropping intensity. Technology such as direct seeded rice to save energy and time can also be promising in hills to adapt to climate change. As monsoon arrives earlier direct seeded rice can be used to utilize the early monsoon arrival and the field could also be used for successful *rabi* cultivation.

An Integrated Farming System (IFS) is a sustainable farming method that combines different elements including crops, animals, and other farming operations to maximise resource use, reduce environmental impact, and boost farm profitability overall. In addition to lowering greenhouse gas emissions, management techniques including employing legumes for nitrogen fixation, altering cultivation methods, and managing nutrients through crop residues and composting can make crops more resilient to climate change (Paramesh et al. 2022). Agroforestry, alley cropping, and intercropping improved nutrient availability and C sequestration, which mitigated the effects of climate change (Bell et al. 2014). Crop-livestock integration is a highly profitable, productive, and environmentally sustainable farming strategy (Gill et al. 2009; Yadav and Sharma 2013).

In order to reduce risks and increase resistance to climate change, IFS integrates several cropping systems and other agricultural operations to create a stable and sustainable production system (Singh et al. 2020). Climate uncertainty has made farming riskier, particularly for small and marginal farmers that lack resources and are unable to adopt new, improved technologies (Paramesh et al. 2022). Enhancement of fertiliser usage efficiency, nutrient recycling, and increased soil microbial activity can be attained when crops are integrated with livestock, fisheries, etc. (Sujatha and Bhat 2015). Adopting an environmentally friendly, comprehensive strategy is necessary. One such strategy is the IFS, which provides benefits for environmental conservation while also addressing the nutritional security of farmers' livelihoods (Fatima et al. 2023). In a study by Kumar et al. (2023) from Mizoram, four models of IFS were demonstrated in Kolasib and found to generate a gross income of 115.03–321.69% and B:C of 1.25 to 2.13 compared to the traditional mixed farming system adopted by the farmers.

Climate Change and Animal Health

The livestock sector in Mizoram is crucial for the rural economy, providing food, income, and protection against crop failures. However, it faces challenges such as low productivity among native breeds and increasing susceptibility to climate-sensitive and cross-border diseases. To establish a resilient agricultural framework, a strategic approach involving quantitative data analysis, proactive climate forecasts, and a combination of traditional and contemporary adaptation strategies is needed.

Regional productivity data (Figs. 16.5 and 16.6) shows a concerning trend of declining output in key livestock segments. The most significant occurrence is the extended outbreak of African Swine Fever (ASF), a transboundary viral illness with a 100% case fatality rate. The outbreak spread to all 11 districts of Mizoram, causing an estimated 70% depopulation of the susceptible pig population and subsequent culling efforts (CEIC 2025). The outbreak has led to the death of over 62,301 pigs, with an additional 49,411 culled to curb it (Table 16.2). The financial and social consequences have been significant, impacting over 11,000 households and causing an estimated financial loss of Rs 896.72 crore (Table 16.2). The mortality rate from March to July 2021 was estimated at 31.1%, with a case fatality rate of 92.1%. The livestock disease trend over the last 5 years (2019–2020 to 2024–2025) in Mizoram is highlighted in Table 16.3. This event has not only disrupted the state's economy, but it also underscores the significant vulnerability of a sector that is so dependent on pig farming.

Mizoram's livestock sector faces challenges from endemic diseases linked to climate, with 18% of cattle infected with haemoprotozoan diseases. These diseases

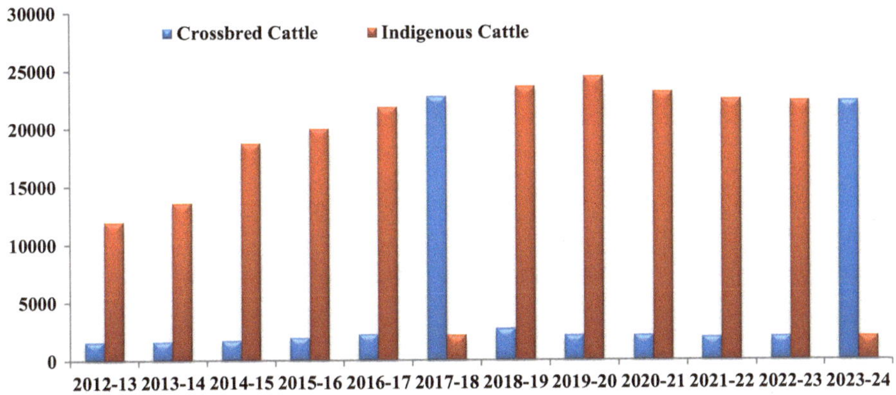

Fig. 16.5 Status of milk production (in tonnes) of Mizoram from 2012–2013 to 2023–2024. (Source: Department of Animal Husbandry and Veterinary, Govt. of Mizoram, India)

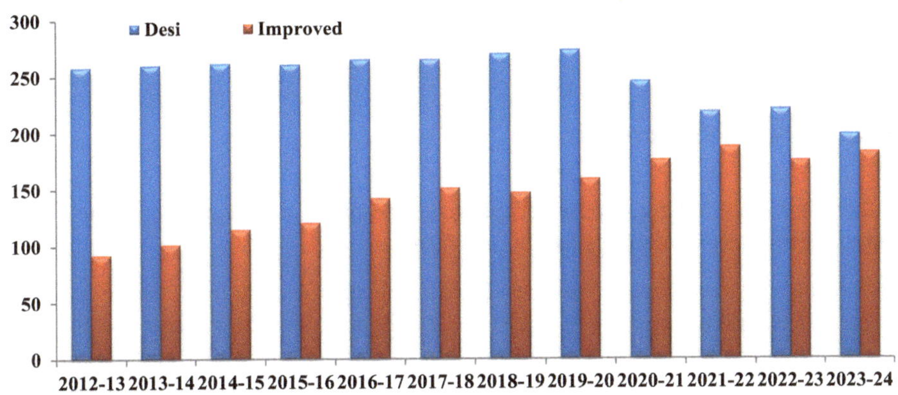

Fig. 16.6 Egg production (in tonnes) of Mizoram, India from 2012–2013 to 2023–2024. (Source: Department of Animal Husbandry and Veterinary, Govt. of Mizoram, India)

fluctuate with seasons, with monsoon (62.90%) and summer (29.62%) being key risk factors (Joane Mary et al. 2022). The life cycle of tick vectors, dependent on temperature, humidity, and rainfall, is linked to disease incidence. Climate projections suggest increased rainy days and longer rainfall conditions, leading to increased disease transmission and a higher year-round burden, affecting cattle health and productivity (Directorate of Science and Technology 2012).

Foundational and Innovative Adaptation Models: Mizoram is implementing a two-pronged adaptation strategy to address climate-related and economic challenges. The state is focusing on breed improvement and resilient housing and biosecurity. The Mithun breed, of great socio-economic and cultural importance in the Northeast, is being enhanced and preserved through initiatives at the Mithun Multiplication Centre and the Mithun Breed Development and Conservation Centre. Farmers are shifting to goat rearing as a viable alternative, and the state government seeks scientific guidance from the Indian Council of Agricultural Research (India

Table 16.2 African Swine Fever (ASF) scenario in Mizoram

Year	Nos. of pigs died	Nos. of pigs culled	Economic loss (Rs. In crore)	Nos. of villages affected
2021	33,417	12,568	334.14	272
2022	12,795	11,686	210.32	149
2023	1139	980	15.77	36
2024	14,950	24,177	336.49	258
Total	62,301	49,411	896.72	715

Source: Department of Animal Husbandry and Veterinary, Govt. of Mizoram, India

Today NE 2025). Resilient livestock housing is also being developed, using traditional materials like bamboo and straw. The Mizoram State Action Plan on Climate Change emphasizes the importance of developing climate-resilient animal sheds at community and institutional levels. These strategies aim to provide practical and sustainable solutions, setting Mizoram's livestock sector on a path towards a more resilient future (Directorate of Science and Technology 2012).

Strategies to Mitigate Risks

Although climate change poses serious risks to animal health, proactive measures can mitigate these challenges for the hill farmers.

Sustainable Agricultural Practices: Integrating agroforestry into farming systems provide shade and reduce heat stress for animals. Providing shade, clean water, and adjusting feeding schedules can help animals cope with rising temperatures to reduce the impact of heat stress on animal health. Rotational grazing helps prevent overgrazing and promotes biodiversity, while water-efficient irrigation systems conserve water in drought-prone areas (Thornton and Herrero 2015).

Implementing Proactive Measures: Proactive strategies like disease surveillance, biosecurity measures, and strengthening veterinary healthcare systems are vital for preventing disease outbreaks.

Habitat Management: Healthy habitats are vital for reducing disease susceptibility in animals. Protecting natural areas, restoring degraded habitats, and managing human-wildlife interactions help maintain ecosystems and reduce the risk of disease spread.

Weather Forecasting: By understanding the relationship between weather conditions and disease dynamics, pig farmers can adopt more effective biosecurity measures, adjustment of housing and ventilation systems, and vaccination scheduling and planning more strategically can mitigate the impact of adverse weather on pig health and productivity (Kumar and Singh 2022).

Table 16.3 Livestock disease trends over the last 5 years (2019–2020 to 2024–2025) in Mizoram

Name of disease	2020–2021			2021–2022			2022–2023			2023–2024			2024–2025		
	OB	AT	DT	OB	AT	DT	OB	AT	DT	OB	AT	DT	OB	AT	DT
Classical Swine Fever	1	8	3	–	–	–	–	–	–	–	–	–	–	–	–
Goat pox	–	–	–	1	82	63	–	–	–	–	–	–	–	–	–
African Swine Fever (in thousand)	–	–	–	0.27	45.99	33.42	0.15	23.49	12.8	0.043	2.1	1.14	0.26	39.13	14.95
Lumpy Skin Disease	–	–	–	–	–	–	–	–	–	1	56	4	–	–	–
PPR	–	–	–	–	–	–	–	–	–	2	229	199	–	–	–

Source: Department of Animal Husbandry and Veterinary, Govt. of Mizoram, India
OB outbreak, *AT* attack, *DT* death

Aquaculture and Climate Change

The provision of affordable animal protein, jobs, and revenue via aquaculture and fisheries has been essential to the livelihood and nutritional security of the people of Mizoram. Fish production in the state needs to be improved. At the moment, 5270.89 MT of fish were produced, with a productivity of 0.90 MT per hectare. The per capita availability of fish is still relatively low at 5.88 kg, leaving 6220 MT short of the recommended per capita requirement of 11 kg. Table 16.4 shows the trend of fish production in Mizoram, which shows low and falling productivity in the state. Of the 24,000 ha of possible fish farming land in the state of Mizoram, only 5888.08 ha (24.53%) have been used, suggesting that more agricultural land might be added to improve the state's overall fish production. According to Table 16.5, which shows the district-by-district status of fish farming practices in Mizoram, the state's entire fish production was contributed by Kolasib district (24.96%) and Mamit district (24.57%). The two districts in the state have the most potential for aquaculture growth.

Climate change may cause some fish species to grow more quickly, increasing production (Klinger et al. 2017). However, severe temperatures may also cause species to die from thermal stress (Reid et al. 2019), which is difficult to prevent. According to the FAO (2018), the biggest concern to aquaculture is the emergence of new diseases and invasive species brought on by the direct consequences of climate change. Fish growth, mortality, and reproduction will be indirectly impacted by the changes in their surroundings brought about by climate change, whereas feeding, migratory, and breeding behaviour will be directly impacted. Climate change will also lead to the degradation of habitat through catastrophic occurrences like cyclones, droughts, and floods. Farmers in the state of Mizoram have frequently reported instances of fish being destroyed by flooding. Additionally, eutrophication—which may be brought on by high temperatures and changed precipitation—is starting to pose a problem in several of the state's major waterways.

Table 16.4 Trend of fish production and productivity during 2015–2023

Year	Area (ha)	Production (MT)	Productivity (MT ha^{-1} $year^{-1}$)
2015	4850.00	6308.00	1.33
2016	5450.14	6980.00	1.28
2017	5468.34	7013.00	1.28
2018	5477.04	7024.13	1.28
2019	5492.04	7043.10	1.28
2020	5507.08	4304.69	0.78
2021	5507.08	4575.05	0.78
2022	5759.08	5029.85	0.87
2023	5888.08	5270.89	0.90

Source: Fisheries Dept., Govt of Mizoram (2024)

Table 16.5 District wise fish production and farming area in Mizoram during 2023–2024

District	Total area (ha)	Total production (MT)
Aizawl	457.71	320.397
Lunglei	620.61	372.366
Saiha	475.21	380.168
Kolasib	1196.21	1315.831
Mamit	1177.41	1295.151
Lawngtlai	816.89	653.512
Serchhip	442.69	354.152
Champhai	493.75	444.375
Saitual	72.2	46.93
Khawzawl	79.2	51.48
Hnahthial	56.2	36.53
Total	5888.08	5270.892

Source: Fisheries Dept., Govt of Mizoram (2024)

Agro Advisory Service (AAS): GKMS

Challenges and Barriers in GKMS Implementation

The Gramin Krishi Mausam Sewa (GKMS) project in Northeast India, particularly Mizoram, faces multiple geographic, linguistic, institutional, and socio-economic barriers that limit effective communication and adoption. The hilly terrain of the region and weak mobile networks create last-mile gaps, delaying SMS and app-based advisories in remote border villages (Chowdhury et al. 2022). Language diversity—over 22 ethnic groups—makes advisories in Hindi or English difficult to comprehend and trust (Saha et al. 2021). Even when accessible, recommendations are often perceived as generic and unsuitable for local systems such as jhum and terrace farming (Malik et al. 2020; Saha et al. 2021; Chowdhury et al. 2023).

ICT infrastructure remains limited, and poorly placed information centers reduce timely access to guidance (Nath et al. 2024). Low digital awareness further delays farmers' use of crop and pest management advisories (Chinlampianga 2011; Nath et al. 2024). Social and psychological barriers—resistance to new technologies, reliance on informal networks, and peer norms—also influence adoption. While older farmers continue to depend on traditional forecasting, many are beginning to blend indigenous knowledge with scientific forecasts. Confidence in GKMS advisories rises when modern predictions visibly align with lived experiences and acknowledge indigenous indicators (Saha et al. 2021; Irumva et al. 2023).

Advisory Dissemination and Effectiveness

The volatile weather of the region—flash floods, cloudbursts, and intense rainfall—often shifts outside these intervals, reducing real-time relevance of the Agro Advisory Bulletins (AABs) (Roy et al. 2021). In addition, advisory systems rarely integrate real-time crop–soil conditions, limiting their precision and location-specific usefulness (Nigam et al. 2023). Digital literacy gaps and weak feedback channels between farmers and officials further hinder smartphone-based advisories (Sharma 2024).

Despite these constraints, effectiveness improves with higher farmer exposure and human capital. Education, landholding size, and farming experience correlate positively with technology adoption (Choudhury et al. 2018; Mazhar 2024). Access to extension services and household literacy also enhance comprehension and application of advisories (Sati and Vangchhia 2017). Literate and semi-literate farmers are more adept at using SMS, Meghdoot/MAUSAM apps, and forecasts alongside field observations (Saha et al. 2021; Chowdhury et al. 2022). Connectivity issues persist in Mizoram, Nagaland, and Arunachal Pradesh, but multi-channel dissemination strategies—via social media, electronic media, KVKs, NGOs, and state directorates—have improved adoption. In Mizoram, 74% of farmers reported receiving timely advisories when such multi-agency approaches were employed (Chowdhury et al. 2023; Paul et al. 2025).

Role of GKMS in Risk-Prone Systems

GKMS (AMFU, Kolasib) is particularly critical in rain-fed systems such as jhum cultivation. Farmers use biweekly AABs and SMS alerts to plan burning, land preparation, sowing, and input use (Zothansanga and Beingachhi 2019; Amith et al. 2022). Currently, AMFU-Kolasib serves 85,890 registered farmers (~32% of operational holdings) (Agriculture Census, 2022–2023). Benefits are substantial: GKMS generates average returns of ₹28,500 per farmer annually, amounting to more than ₹35 crore per year in Mizoram alone (Chowdhury et al. 2023). Advisories provide essential guidance on drainage, mulching, irrigation scheduling, and varietal choice—key for resilience against floods, waterlogging, and dry spells. To address unforeseen extreme events, supplementary weather bulletins, weekly alerts, and Impact-Based Forecasts (IBFs) are occasionally disseminated via WhatsApp, Facebook, radio, TV, and mobile apps (Chowdhury et al. 2023).

Technological Innovations and Future Directions

Agro-climatic analogues highlight the potential of digital climate-informed advisory services (DCAS). By combining weather intelligence, participatory design, and digital platforms, GKMS advances climate-smart agriculture (Kumar et al.

2024). Emerging technologies—IoT, GIS, and AI/ML—enable real-time soil–crop monitoring, including parameters like moisture, pH, and nutrients. Wireless sensors provide timely alerts, while machine learning models such as Random Forest and KNN achieve up to 93% accuracy in crop suitability prediction (Kumar et al. 2018; Kushal et al. 2022; Manju et al. 2024; Jana et al. 2024).

Complementary digital initiatives, such as "Ran Vulhtute Thian," showcase the potential of mobile platforms in rural service delivery (Hmar et al. 2020). When integrated with IoT analytics for irrigation, pest control, and decision support, these innovations can significantly enhance precision, though adoption remains limited without full real-time integration (Sarma and Singh 2021).

Adoption Trends and Impacts

Comparisons over time highlight marked improvements. From 2014 to 2019, only 2051 bulletins were issued, reaching fewer than 60,000 farmers, with adoption below 10% and minimal benefits (3% seed savings; 8% cost savings) (Saha et al. 2021). By 2024–2025, outreach expanded to 343,695 farmers, with 4463 bulletins disseminated through diverse channels, including 728 crop bulletins, 1664 weather bulletins, 1664 district advisories, 388 block advisories, and 19 special bulletins. Mizoram recorded notable resource efficiency gains: 28.97% seed savings (41.52% cost reduction), 25.67% fertilizer/pesticide savings (31.50% cost savings), 20.34% manpower savings (19.62% cost savings), and 21.03% irrigation savings (31.96% cost savings) (Chowdhury et al. 2023; Paul et al. 2025).

Micro-level case studies underscore these benefits. In 2017, a farmer from Darlak saved ₹6147 by harvesting early after a rainfall warning, while another in 2016 saved ₹12,500 per acre by timely pesticide application. Similarly, a ginger farmer in Hmunpui earned ₹2200 more per bigha after applying fungicide against soft rot (Saha et al. 2021; Chowdhury et al. 2023). These examples demonstrate how context-specific, timely advisories translate into economic gains and strengthen resilience.

Conclusion

It was evident that the climate of the NEH region, including Mizoram, had a significant variability over the past years. The effect will slowly be observed in crop production and the animal sector. It is therefore an important task to deal with the adverse effects of a changing climate on agricultural production or adapt to resilient technologies for future food security. Awareness and acceptance from the farming community are a prerequisite strategy to adapting to resilient technologies. Diverse land use patterns in several sectors, including plantations, agroforestry, agriculture, and forests, have recently become common in the NE hills, including Mizoram.

These patterns are impacted by a number of factors, including resource scarcity, market opportunities, government policies and interventions, vulnerability to resource availability, and a shift in perspective. The overall ecosystem's capacity to store C is impacted by these land use changes in a variety of ways, and the management techniques used resulted in significant variations between C pools. For example, the shifting of land use from temporary shifting cultivation to settled agriculture, where farmers can improve their livelihood. Therefore, agroforestry systems and plantations with sustainable management practices could be adopted on a large scale for the restoration of degraded lands and to enhance soil OC in Mizoram, and most importantly livelihood of the farmers. Similarly, agronomic practices, IFS systems, and models etc., could also improve ecosystem sustainability. Agrometeorological advisory services like GKMS could serve as an important tool in forecasting, which could save farmers' crops from loss due to sudden extreme climate events, disease, and pest incidences promote sustainable agricultural practices in the hills of Mizoram.

References

Achu F, Kanmi N, Katzo C (2013) Effects of compost and organic green manure on soil fertility and nutrient uptake in wheat-rice cropping system. Int J Manures Fertil 2(10):407–412

Ali S, Begum F, Hayat R, Bohannan BJM (2017) Variation in soil organic carbon stock in different land uses and altitudes in Bagrot Valley, Northern Karakoram. Acta Agric Scand Sect B Soil Plant Sci 67:551–561. https://doi.org/10.1080/09064710.2017.1317829

Amith G, Ramesh, Avinash GB, Haroli M, Thimmareddy H, Dharani C (2022) Agromet advisory services for climate smart agriculture. J Exp Agric Int 44:1–7. https://doi.org/10.9734/jeai/2022/v44i430810

Ansari MA, Choudhury BU, Mandal S, Jat SL, Meitei CB (2022) Converting primary forests to cultivated lands: long-term effects on the vertical distribution of soil carbon and biological activity in the foothills of Eastern Himalaya. J Environ Manage 301:113886. https://doi.org/10.1016/j.jenvman.2021.113886

Ayoubi S, Khormali F, Sahrawat KL, Rodrigues de Lima AC (2011) Assessing impacts of land use change on soil quality indicators in a Loessial soil in Golestan Province, Iran. J Agric Sci Technol 13:727–742. 20.1001.1.16807073.2011.13.5.13.4

Bell LW, Moore AD, Kirkegaard JA (2014) Evolution in crop–livestock integration systems that improve farm productivity and environmental performance in Australia. Eur J Agron 57:10–20. https://doi.org/10.1016/j.eja.2013.04.007

Bhattacharyya P, Roy KS, Neogi S, Adhya TK, Rao KS, Manna MC (2016) Effects of microbial inoculants on soil organic carbon sequestration and greenhouse gas mitigation in agricultural soils. Appl Soil Ecol 100:124–137. https://doi.org/10.1016/j.apsoil.2015.12.011

Bhavsar D, Chakraborty K, Mohapatra J, Sarma K, Aggarwal SP (2023) Role of climatic variables on forest fire in the state of Mizoram. In: Disaster management and risk reduction: multidisciplinary perspectives and approaches in the Indian context. Springer, pp 238–309. https://doi.org/10.1007/978-981-99-6395-9_20

Blanco-Canqui H, Shaver TM, Lindquist JL, Shapiro CA, Elmore RW, Francis CA, Hergert GW (2015) Cover crops and ecosystem services: insights from studies in temperate soils. Agron J 107(6):2449–2474. https://doi.org/10.2134/agronj15.0086

Calzadilla A, Rehdanz K, Betts R, Falloon P, Wiltshire A, Tol RSJ (2013) Climate change impacts on global agriculture. In: Climate change. Springer, pp 357–334. https://hdl.handle.net/10419/32519

CEIC (2025) Livestock production: meat: cattle: Mizoram. https://www.ceicdata.com/en/india/livestock-production-meat/livestock-production-meat-cattle-mizoram. Accessed 5 Sept 2025

Chakraborty D, Roy A, Singh NU, Saha S, Das SK, Mridha N, Yumnam A, Paul P, Gowda C, Biam KP, Patra S, Amrutha T, Singh BP, Mishra VK (2025a) Assessing climate change impact on rainfall patterns in northeastern India and its consequences on water resources and rainfed agriculture. Earth 6(1):2. https://doi.org/10.3390/earth6010002

Chakraborty S, Sarkar A, Datye A, Praveen V, Burman DPK, Shivamurthy Y, Samal N, Saha SH, Yadava MG, Chattopadhyay R, Trivedi N, Trivedi RK, Sharma A (2025b) Precipitation isotopes and monsoon dynamics in the core monsoon zone of India. Sci Rep 15:6761. https://doi.org/10.1038/s41598-025-88640-x

Chaplin-Kramer R, Dombeck E, Gerber J, Knuth KA, Mueller ND, Mueller M, Ziv G, Klein AM (2014) Global malnutrition overlaps with pollinator-dependent micronutrient production. Proc R Soc Lond B Biol Sci 281:1471–2954. https://doi.org/10.1098/rspb.2014.1799

Chinlampianga M (2011) Traditional knowledge, weather prediction and bioindicators: a case study in Mizoram, northeastern India. Indian J Tradit Knowl 10(1):207–211. http://nopr.niscair.res.in/handle/123456789/11083

Choudhury BN, Das SC, Sarmah MC, Ahmed M (2018) A study on the influence of socio-economic factors on knowledge and technology adoption of sericulture farmers of Aizawl district of Mizoram. Int J Adv Res Ideas Innov Technol 4(3):001–004. https://www.ijariit.com/manuscripts/v4i3/V4I3-1137.pdf

Chowdhury S, Saha S, Shakuntala I, Chakraborty D, Lungmuana, Lalhruaipuii, Das A, Sethy BK, Mishra VK, Daschaudhuri D, Choudhury M, Lalmalsawmi, Vanthawmliana, Das S, Balasubramanian R, Ghosh K, Attri SD, Singh KK, Lalliansanga S (2022) In: Lalsiamliana J (ed) Forcible role of Gramin Krishi Mausam Sewa network for increasing hill farming resiliency in North East India. ICAR Research Complex for NEH Region, Mizoram Centre. ISBN 978-93-5620-372-3

Chowdhury S, Saha S, Doley S, Shakuntala I, Chakraborty D, Lungmuana, Lalhruaipuii, Hazarika S, Mishra VK, Lalmalsawmi, Vanthawmliana, Balasubramanian R, Ghosh K, Attri SD, Singh KK, Bhan SC, Lalliansanga S (2023) In: Lalsiamliana J (ed) Multi-stakeholder agro-advisory network: an efficient tool for promoting climate resilient agriculture in North East Indian states. ICAR Research Complex for NEH Region, Mizoram Centre. ISBN 978-93-5759-893-4

Cong WF, Hoffland E, Li L, Six J, Sun JH, Bao XG, Van Der Werf W (2015) Intercropping enhances soil carbon and nitrogen. Glob Chang Biol 21(4):1715–1726. https://doi.org/10.1111/gcb.12738

Das, A., Layek, J., Idapuganti, R.G., & Basavaraj, S. (2020). Conservation tillage and residue management improves soil properties under a upland rice–rapeseed system in the subtropical eastern Himalayas. Land Degrad Dev, 31(14):1775–1791. https://doi.org/10.1002/ldr.3568

Das S, Kim PJ, Nie M, Chabbi (2024) Soil organic matter in the Anthropocene: role in climate change mitigation, carbon sequestration, and food security. Agric Ecosyst Environ 375:109180. https://doi.org/10.1016/j.agee.2024.109180

Department of Fisheries, Government of Mizoram, India (2024). https://fisheries.mizoram.gov.in

Directorate of Science & Technology, Government of Mizoram (2012) Mizoram state action plan on climate change (2012–2017). Ministry of Environment, Forest and Climate Change. https://moef.gov.in/uploads/2017/09/MIZORAM.pdf

FAO (2018) Impacts of climate change on fisheries and aquaculture: synthesis of current knowledge, adaptation and mitigation options, FAO fisheries and aquaculture technical paper No. 627. FAO, Rome, Italy

Farley KA, Kelley EF, Hofstede RGM (2004) Soil organic carbon and water retention following conversion of grasslands to pine plantations in the Ecuadoran Andes. In: Ecosystems. Springer, pp 729–739. https://doi.org/10.1007/s10021-004-0047-5

Farmaha BS, Sekaran U, Franzluebbers AJ (2022) Cover cropping and conservation tillage improve soil health in the southeastern United States. Agron J 114(1):296–316. https://doi.org/10.1002/agj2.20865

Fatima A, Singh VK, Babu S, Singh RK, Upadhyay PK, Rathore SS, Kumar B, Hasanain M, Parween H (2023) Food production potential and environmental sustainability of different integrated farming system models in Northwest India. Front Sustain Food Syst 7:959464. https://doi.org/10.3389/fsufs.2023959464

Gill MS, Singh JP, Gangwar KS (2009) Integrated farming system and agriculture sustainability. Indian J Agron 54(2):128–139. https://doi.org/10.59797/ija.v54i2.4790

Giller KE, Witter E, Corbeels M, Tittonell P (2009) Conservation agriculture and smallholder farming in Africa: the heretics' view. Field Crop Res 114(1):23–34. https://doi.org/10.1016/j.fcr.2009.06.017

Hazarika A, Nath AJ, Reang D, Pandey R, Sileshi GW, Das AK (2024) Climate change vulnerability and adaption among farmers practicing shifting agriculture in the Indian Himalayas. Environ Sustain Indic 23:100430. https://doi.org/10.1016/j.indic.2024.100430

Hmar L, Rahman S, Tochhawng L (2020) Ran Vulhtute Thian: the smart extension services through smart phone in picturesque Hills of Mizoram. J Ext Syst 36(2):12–15. https://doi.org/10.48165/jes.2020.36203

Hrahsel L, Sahoo UK (2024) Variations of soil organic carbon in different plantations in Mizoram. Int J Ecol Environ Sci 50(2). https://doi.org/10.55863/ijees.2024.0036

IMD (2010) Annual report 2010. India Meteorological Department, pp 4–6

IMD (2019) Annual report 2019 (MoES/IMD/Annual Report 2019/2020/01), vol 9–16. India Meteorological Department

IMD, CRS, NCC (2023) Annual climate summary 2023. https://www.imdpune.gov.in/cmpg/Product/Annual_Climate_Summary/annual_summary_2023.pdf

India Today NE (2025) Mizoram hosts 3rd Mithun Day, CM Lalduhoma stresses integrated…. https://www.indiatodayne.in/mizoram/story/mizoram-cm-inaugurates-mithun-day-plans-thenzawl-peace-city-1271037-2025-09-02

IPCC (2021) Climate change 2021. In: Shukla PR, Skea J, Calvo Buendia E, Masson-Delmotte V, Pörtner H-O, Roberts DC, Zhai P, Slade R, Connors S, van Diemen R, Ferrat M, Haughey E, Luz S, Neogi S, Pathak M, Petzold J, Portugal Pereira J, Vyas P, Huntley E, Kissick K, Belkacemi M, Malley J (eds) IPCC special report on climate change, desertification, land degradation, sustainable land management, food security, and greenhouse gas fluxes in terrestrial ecosystems. https://doi.org/10.1017/9781009157988.004

Irumva O, Twagirayezu G, Uwimpaye F, Ntakiyimana C, Manzi HP, Nyirandayisabye R, Hakuzweyezu T, Nizeyimana JC, Itangishaka CA (2023) Amalgamating weather-related indigenous knowledge into modern forecasting knowledge adopted by farmers on climate prediction. In: Novel perspectives of geography, environment and earth sciences, vol 6. https://doi.org/10.9734/bpi/npgees/v6/6026A

Islam KR, Weil RR (2000) Land use effects on soil quality in a tropical forest ecosystem of Bangladesh. Agric Ecosyst Environ 79:9–16. https://doi.org/10.1016/S0167-8809(99)00145-0

Jain SK, Kumar V (2012) Trend analysis of rainfall and temperature data for India. Curr Sci 102(1):37–49

Jana S, Chatterjee D, Pal N, Pal K, Roy K, Bashak S (2024) AI in soil health monitoring: a data-driven approach. Int J Sci Technol Eng 12(10):1327–1335. https://doi.org/10.22214/ijraset.2024.64871

Jhajharia D, Singh VP (2011) Trends in temperature, diurnal temperature range and sunshine duration in Northeast India. Int J Climatol 31(9):1353–1367. https://doi.org/10.1002/joc.2164

Joane Mary HC, Rajesh JB, Prasad H, Sarma K, Roychoudhury P, Ali A, Deka D, Bayan H, Ravindran R (2022) Prevalence and characteristics of haemoprotozoan infections of cattle in Mizoram. Indian J Anim Sci 92(4):417–421. https://doi.org/10.56093/ijans.v92i4.124017

Jose S (2009) Agroforestry for ecosystem services and environmental benefits: an overview. Agrofor Syst 76(1):1–10. https://doi.org/10.1007/s10457-009-9229-7

Kelty MJ (2006) The role of species mixtures in plantation forestry. J Econ Manag 233(2–3):195–204. https://doi.org/10.1016/j.foreco.2006.05.011

Kenye A, Sahoo UK, Singh SL, Gogoi A (2019) Soil organic carbon of different land uses of Mizoram, Northeast India. AIMS Geosci 5(1):25–40. https://doi.org/10.3934/geosci.2019.1.25

Klinger DH, Levin SA, Watson JR (2017) The growth of finfish in global open-ocean aquaculture under climate change. Proc R Soc Lond B Biol Sci 284:20170834. https://doi.org/10.1098/rspb.2017.0834

Kumar N, Nirmala MM (2023) Climatic variations on Indian agricultural productivity. In: Artificial intelligence, Internet of Things, and Society 5.0. Springer Nature, pp 71–80. https://doi.org/10.1007/978-3-031-43300-9_7

Kumar A, Sharma P (2022) Impact of climate variation on agricultural productivity and food security in rural India. Soc Sci Res Netw 39. https://hdl.handle.net/10419/81545

Kumar, A., Singh, K. K., & Lalnuntluanga. (2022). Emerging diseases in pigs: diagnosis and management in Northeast India. New Delhi: Springer

Kumar S, Upadhyay SK (2019) Impact of climate change on agricultural productivity and food security in India: a state level analysis. Indian J Agric Res 53(2):133–142. https://doi.org/10.18805/A-5134

Kumar P, Motia S, Reddy SRN (2018) Integrating wireless sensing and decision support technologies for real-time farmland monitoring and support for effective decision making. Int J Inf Technol:1–19. https://doi.org/10.1007/S41870-018-0218-9

Kumar P, Chakraborty A, Sagar AK et al. (2026). Investigating Heatwave Features: Creating an Intensity-Duration-Frequency Model for India's Principal Climate Zones. Earth Syst Environ. https://doi.org/10.1007/s41748-025-00979-3

Kumar YB, Lalramhlimi B, Lalhruaipuii, Lalrinsanga PL, Lungmuana, Soni JK, Doley S (2023) Success of integrated farming system for enhancing farmer's income in Mizoram. Indian Farm 73(08):39–43

Kushal BJ, Jain NS, Raaju NS, Gowda KGV, Rani AKP, Nath GS (2022) Real time crop prediction based on soil analysis using Internet of Things and machine learning. In: 2022 international conference on edge computing and applications (ICECAA). IEEE, pp 1–6. https://doi.org/10.1109/ICECAA55415.2022.9936417

Kuttippurath J, Murasingh S, Stott PA, Sarojini BB, Jha MK, Kumar P, Nair PJ, Varikoden H, Raj S, Francis PA (2021) Observed rainfall changes in the past century (1901–2019) over the wettest place on earth. Environ Res Lett 16:024018

Lal R (2004) Soil carbon sequestration impacts on global climate change and food security. Science 304(5677):1623–1627. https://doi.org/10.1126/science.1097396

Lalthanpuia H, Lalmalsawma S, Lalnunzira J (2022) Assessment of district level climate vulnerability of Mizoram, India: water resources approach. J Clim Change 8(3):21–29. https://doi.org/10.3233/JCC220018

Lemma B, Kleja DB, Nilsson I, Olsson M (2006) Soil carbon sequestration under different exotic tree species in the Southwestern Highlands of Ethopia. Geoderma 136:886–898. https://doi.org/10.1016/j.geoderma.2006.06.008

Lungmuana, Choudhury BU, Saha S, Das A, Buragohain J, Dayal V, Singh AR, Boopathi T, Dutta S (2018) Impact of postburn jhum agriculture on soil carbon pools in the north-eastern Himalayan region of India. Soil Res 56(6):615–622. https://doi.org/10.1071/SR18031

Lungmuana, Singh SB, Choudhury BU, Vanthawmliana, Saha S, Hnamte V (2019) Transforming jhum to plantations: effect on soil microbiological and biochemical properties in the foot hills of north Eastern Himalayas, India. Catena 117:84–91. https://doi.org/10.1016/j.catena.2019.02.008

Lungmuana, Hauchhum R, Lalremsang P (2023a) Soil microbial pools as an indicator of soil health in different land use Systems of North India. In: Mishra G, Giri K, Nath AJ, Francaviglia R (eds) Soil carbon dynamics in Indian Himalayan Region. Springer Nature, Singapore, pp 189–204. https://doi.org/10.1007/978-981-99-3303-7_10

Lungmuana, Soni JK, Doley S (2023b) Improving growth, yield and soil health of maize soybean sequence by balance application of lime and fertilizers in acidic soil of Mizoram. J Indian Soc Soil Sci 71(4):448–453. https://doi.org/10.5958/0974-0228.2024.00013.6

Lungmuana, Sangma CBK, Angami T, Shakuntala I (2024) Shifting cultivation to settled agriculture for sustaining soil health and agriculture productivity in North-East India. In: Tripathi SK, Brearley FQ (eds) Shifting cultivation systems: from indigenous practices to soil fertility in North-East India. Springer Nature, Switzerland, pp 143–154. https://doi.org/10.1007/978-3-031-70388-1_11

Malik M, Sehgal M, Kanojia AK, Singh R (2020) Use of information and communication tools for north-Eastern region of India: a paradigm shift. J Pharmacognosy Phytochem 9:595–598. https://www.phytojournal.com/special-issue/2020.v9.i4S.12228/use-of-information-and-communication-tools-for-north-eastern-region-of-india-a-paradigm-shift

Manju G, Kishor S, Binson VA (2024) An IoT-enabled real-time crop prediction system using soil fertility analysis. Eng 5(4):2496–2510. https://doi.org/10.3390/eng5040130

Manpoong C, Tripathi SK (2019) Soil properties un-der different land use system of Mizoram, North East India. J Multidiscip Appl Nat Sci 11(1):121–125. https://doi.org/10.31018/jans.v11i1.1999

Manpoong C, Wapongnungsang, Tripathi SK (2021) Soil carbon stock in different land-use systems in the hill terrain of Mizoram, Northeast India. J Multidiscip Appl Nat Sci 13(2):723–728. https://doi.org/10.31018/jans.v13i2.2615

Mazhar SH (2024) Adoption level of advanced cocoon production techniques in Aizawl, district of Mizoram. Int J Agric Ext Soc Dev 7(8):244–248. https://doi.org/10.33545/26180723.2024.v7.i8d.919

Nair PKR, Kumar BM, Nair VD (2009) Agroforestry as a strategy for carbon sequestration. J Plant Nutr Soil Sci 172(1):10–23. https://doi.org/10.1002/jpln.200800030

Nath D, Tamuk Y, Gohain I (2024) Constraints encountered by the KVK beneficiaries of Manipur and their suggestions. J Exp Agric Int 46(9):1061–1067. https://doi.org/10.9734/jeai/2024/v46i92902

Nigam R, Bhattacharya B, Pandya MR (2023) Satellite agromet products and their adaptation for advisory services to Indian farming community. J Agrometeorol 25(1):42–50. https://doi.org/10.54386/jam.v25i1.2084

Paramesh V, Ravisankar N, Behera U, Arunachalam V, Kumar P, Rajkumar RS, Misra SD, Kumar RM, Prusty AK, Jacob D, Panwar AS, Mayenkar T, Reddy VK, Rajkumar S (2022) Integrated farming system approaches to achieve food and nutritional security for enhancing profitability, employment, and climate resilience in India. Food Energy Secur 11(2). https://doi.org/10.1002/fes3.321

Paul S, Chakraborty D, Tripathi AK (2025) Frontline extension services as a buffer against social vulnerability to climate change: A case study of shifting cultivators in Northeast India. J Environ Manag 377:124607. https://doi.org/10.1016/j.jenvman.2025.124607

Paustian K, Lehmann J, Ogle S, Reay D, Robertson GP, Smith P (2016) Climate-smart soils. Nature 532(7597):49–57. https://doi.org/10.1038/nature17174

Raihan A, Begum RA, Said MNM, Abdullah SMS (2018) Climate change mitigation options in the forestry sector of Malaysia. J Kejuruteraan 1:89–98. https://doi.org/10.17576/jkukm-2018-si1(6)-11

Rangappa K, Das A, Layek J, Basavaraj S, Debnath S, Bhupenchandra I, Devi AG, Mohapatra KP, Choudhury BU, Mishra VK (2024) Conservation tillage and residue management practices in rice improves stress tolerance of succeeding vegetable pea by regulating physiological traits in Eastern Himalayas. Sci Hortic 327:112842. https://doi.org/10.1016/j.scienta.2024.112842

Reid GK, Gurney-Smith HJ, Flaherty M, Garber AF, Forster I, Brewer-Dalton K, Moccia RD (2019) Climate change and aquaculture: considering adaptation potential. Aquac Environ Interact 11:603–624. https://www.jstor.org/stable/26920866

Roy A, Kolady D, Paudel B, Yumnam A, Mridha N, Chakraborty D, Singh NU (2021) Recent trends and impacts of climate change in North-Eastern region of India: a review. J Environ Biol 42(6):1415–1424. https://doi.org/10.22438/jeb/42/6/MRN-1701

Saha S, Chakraborty D, Singh SB, Chowdhury S, Syiem EK, Dutta SK, Lungmuana, Choudhury BU, Boopathi T, Singh AR, Ramakrishna Y, Roy A (2016) Analyzing the trend in thermal discomfort and other bioclimatic indices at Kolasib, Mizoram. J Agrometeorol 18(1):57–61

Saha S, Chowdhury S, Shakuntala I, Chakraborty D, Mishra VK, Lalliansanga S, Lalhmachhuana P, Balasubramanian R, Ghosh K, Baxla AK, Singh KK, Lungmuana, Sunani S, Das A, Saithantluanga H (2021) Agro advisory services for managing weather related risks in hill agriculture of Mizoram. ICAR Research Complex for NEH Region, Umiam, Meghalaya. ISBN: 978-93-5408-945-9

Saha S, Pandey P, Kumar A, Babu S, Layek J, Chaudhury S, Das A (2024) Agro-ecosystem services in hill agriculture: a North East Indian perspectives. J Agric Phys 24(2nd JAP special issue):S135–S155

Sahoo UK, Singh SL, Gogoi A, Kenye A, Sahoo SS (2019) Active and passive soil organic carbon pools as affected by different land use types in Mizoram, Northeast India. PLoS One 14(7):1–16. https://doi.org/10.1371/journal.pone.0219969

Sahoo UK, Ahirwal J, Giri K, Mishra G, Francaviglia R (2023) Modelling land use and climate change effects on soil organic carbon storage under different plantation systems in Mizoram, Northeast India. Agriculture 13:1332. https://doi.org/10.3390/agriculture13071332

Sarma D, Singh R (2021) Adoption of Agromet Advisory Services in shifting cultivation systems of Northeast India. Agric Syst 190:103106. https://doi.org/10.1016/j.agsy.2021.103106

Sati VP, Vangchhia L (eds) (2017) Population and socio-economic development. Springer, Cham, pp 35–49. https://doi.org/10.1007/978-3-319-45623-2_4

Sharma R (2024) Identification of various constraints faced by farmers in adopting government schemes in the state of Rajasthan, India. Asian J Agric Ext Econ Sociol 42(4):123–130. https://doi.org/10.9734/ajaees/2024/v42i42402

Singh AK (2017) Climate change and productivity of food-grain and cash crops in India. LAP LAMBERT Academic Publishing

Singh SL, Sahoo UK, Gogoi A, Kenye A (2018) Effect of land use changes on carbon stock dynamics in major land use sectors of Mizoram, Northeast India. Indian J Environ Prot 9:1262–1285. https://doi.org/10.4236/jep.2018.99057

Singh VK, Rathore SS, Singh RK, Upadhyay PK, Shekhawat K (2020) Integrated farming system approach for enhanced farm productivity, climate resilience and doubling farmers' income. Indian J Agric Sci 90:1378–1388. https://doi.org/10.56093/ijas.v90i8.105884

SiwSiwar C, Chinade AA, Mohamad S, Isahak A (2016) Economic valuation of soil carbon sequestration services in Malaysia's forest sector: a review of possible approaches. J Sustain Sci Manag 11:14–28. https://agris.upm.edu.my:0/17676

Sommer R, Bossio D (2014) Dynamics and climate change mitigation potential of soil organic carbon sequestration. J Environ Manag 144:83–87. https://doi.org/10.1016/j.jenvman.2014.05.017

Sujatha S, Bhat R (2015) Resource use and benefits of mixed farming approach in arecanut ecosystem in India. Agric Syst 141:126–137. https://doi.org/10.1016/j.agsy.2015.10.005

Tahir Z, Haseeb M, Mahmood SA et al. (2025). From Green to Grey: Assessing the Impact of Urbanization on Land Surface Temperature and Thermal Comfort in Multan City. Earth Syst Environ 9, 777–800. https://doi.org/10.1007/s41748-025-00598-y

Thornton PK, Herrero M (2015) Adapting to climate change in the mixed crop and livestock farming systems in sub-Saharan Africa. Nat Clim Chang 5(9):830–836. https://doi.org/10.1038/nclimate2754

Tiwari RC (2006) Analytical study on variation of climatic parameters at Aizawl, Mizoram (India). Bull Arunachal For Res 22(1&2):33–39

Tosato T (2023) Climate change and agroecosystems in the hill and mountain regions of Northeast India. In: Sustainable food value chain development. Springer, pp 37–60. https://doi.org/10.1007/978-981-19-6454-1_3

Uhlig J, Hall CAS, Nyo T (1994) Changing patterns of shifting cultivation in selected countries in Southeast Asia and their effect on the global carbon cycle. In: Effects of land-

use change on atmospheric CO_2 concentrations. Springer, pp 145–200. https://doi.org/10.1007/978-1-4613-8363-5_5
Van der Werf GR, Morton DC, De Fries RS, Olivier JGJ, Kasibhatla PS, Jackson RB, Collatz GJ, Randerson JT (2009) CO_2 emissions from forest loss. Nat Geosci 2:737–738. https://doi.org/10.1038/ngeo671
Verma P, Kaur J, Arora MS et al. (2026). Modeling the Impact of Direct Air Capture on Forest Biomass and Population Dynamics. Earth Syst Environ 10, 407–427. https://doi.org/10.1007/s41748-025-00599-x
WMO (2022) State of the climate in Africa 2021 (WMO–No. 1300). World Meteorological Organization
Yadav CM, Sharma RK (2013) Crop-livestock integrated farming system for the marginal farmers in rainfed regions of Bhilwara District in Rajasthan. Indian J Dryland Agric Res Dev 28:74–76. https://doi.org/10.3389/fsufs.2024.1338299
Yang Y, Shi Y, Sun W, Chang J, Zhu J, Chen L, Fang J (2022) Terrestrial carbon sinks in China and around the world and their contribution to carbon neutrality. Sci China Life Sci 65(5). https://doi.org/10.1007/s11427-021-2045-5
Zothansanga D, Beingachhi B (2019) The new land use policy: a panacea for shifting cultivation in Mizoram. Senhri J Multidiscip Stud 4(1):21–29. https://doi.org/10.36110/SJMS.2019.04.01.003

Chapter 17
Climate Action Through WASH: Addressing SDG Implementation Gaps in Selected Rural Areas of South West Bengal Through the Lens of SDG 13

Lopamudra Ganguly, Subhadip Mondal, and Saptadipa Chakraborty

Abstract Climate change is worsening the water, sanitation and hygiene (WASH) situation in South 24 Parganas, a highly vulnerable delta region of West Bengal. This study explored how cyclones, salinity intrusion and rising temperatures are impacting WASH systems while considering variations in gendered vulnerabilities to these climate extremes. The study also explored adaptation strategies related to SDG 6 and SDG 13. A mixed-methods design was used, involving household surveys ($n = 95$), infrastructure audits ($n = 50$), six focus group discussions, 12 key informant interviews and remote sensing analysis using the Normalized Difference Salinity Index (NDSI). Quantitative data were analyzed using Pearson correlation, regression modelling and spatial analysis in QGIS. Cyclones Amphan (2020), Yaas (2021) and Remal (2024) damaged or destroyed 70–80% of sanitation infrastructure and contaminated more than 40,000 tube wells. The area affected by salinity increased by 24.07 km^2 between 2019 and 2024. Compared to 2020, the temperature rise (+1.2 °C) over this five-year period reduced access to safe water by 12% ($p < 0.01$) and 53% of women reported their sanitation insecurity increased after disasters occurred. Climate extremes are removing the viability of WASH systems in deltaic West Bengal. Climate-resilient interventions such as raised toilets, decentralized desalination and gender-responsive planning are needed to ensure that SDG 6 and SDG 13 may be achieved. Integrative policies that link disaster management, water governance and gender equity were desperately needed.

Abbreviations

C2 L2/C2 L1	Collection 2 level 1/level 2 (Remote sensing data processing standards)
FGDs	Focus Group Discussions

L. Ganguly (✉) · S. Mondal · S. Chakraborty
Department of Geography, Vedanta College, Kolkata, West Bengal, India

S. Mandal et al. (eds.), *Climate Change Vulnerability*, Climate Change Management, https://doi.org/10.1007/978-3-032-16783-5_17

Ha	Hectares
IMD	India Meteorological Department
INCOIS	Indian National Centre for Ocean Information Services
IPCC	Intergovernmental Panel on Climate Change
KIIs	Key informant interviews
km/h	Kilometres per hour
km^2	Square kilometers
NAPCC	National Action Plan on Climate Change
NDMA	National Disaster Management Authority
NDSI	Normalised Difference Salinity Index
NGO	Non-Governmental Organization
OLI/TIRS	Operational Land Imager/Thermal Infrared Sensor
OSM	OpenStreetMap
PRA	Participatory rural appraisal
QGIS	Quantum Geographic Information System
SDG	Sustainable Development Goal
UNICEF	United Nations International Children's Emergency Fund
USGS	United States Geological Survey
WASH	Water, sanitation and hygiene
WHO	World Health Organisation

Introduction

Access to safe water, sanitation, and hygiene (WASH) is a foundation of public health, human dignity, and sustainable development, serving as key enablers of SDG 3 (Good Health and Well-being), SDG 5 (Gender Equality), SDG 6 (Clean Water and Sanitation), and SDG 13 (Climate Action). However, climate change threatens the stability of WASH systems through rising sea levels, salinization, inconsistent rainfall, and increased frequency of cyclones. As warned by the Intergovernmental Panel on Climate Change (2022), these threats are increasingly accentuated in low-lying deltaic regions where fragile infrastructure is regularly disrupted, aggravating already serious health and equity challenges. South 24 Parganas, located in the Sundarbans delta in West Bengal, is representative of these vulnerabilities. The multiple storm surges, embankment breaches, and salinization experienced by the district over time have made access to safe water and sanitation exceedingly precarious. Cyclones, namely Aila (2009), Bulbul (2019), Amphan (2020), Yaas (2021), and Remal (2024), demonstrate the magnitude of these risks. Amphan alone contaminated over 40,000 tube wells and left over 70% of households without functioning toilets. Yaas destroyed nearly 60% of latrines in some villages, while Remal destroyed community sanitation facilities in 75% of surveyed villages, forcing households to access contaminated pond water. These multifaceted and repeated shocks demonstrate the underlying structural weaknesses in WASH infrastructure and recovery processes.

The pressures of WASH insecurity are not universally felt. Women and adolescent girls, as the predominant water collectors and managers of domestic hygiene, encounter multiplied challenges. In periods of drought, in areas like South 24 Parganas, women account for 3–5 h a day in searches for water and in flood periods, the potential risk of injury, disease, and gender-based violence (GBV) increases. Although potentially not considered, menstrual hygiene management is near impossible in disaster shelter settings, a trend that has seen some examination in the literature across the world. Caruso et al. (2017) identified that sanitation insecurity was a particularly troubling stressor for women in disaster-prone areas, while Sommer et al. (2016) noted that limited menstrual hygiene facilities were correlated with absenteeism from schools, resulting in long-term negative psychosocial and sexual health outcomes. It follows then that disruptions to WASH due to climate change increase the severity of existing gender inequities. Similar cases from around the globe lend support to these findings. In Bangladesh, Jeandron et al. (2019) showed that storm surges disrupted fresh water supplies for months, with immediate effects felt on women and children. In the Mekong Delta, Mishra et al. (2022) showed that a decentralized adaptation of solar-powered water filtration reduced salinity effects by 40% in fact. UNESCO (2021) reported that sea-level rises have already rendered traditional WASH systems based on groundwater sources impossible in several Pacific Island nations. These examples show that WASH insecurity in South 24 Parganas is indicative of a much larger global spectrum of climate-induced WASH stress, but must be understood in the context of a district-level crisis as well. Despite India's policy frameworks, such as the National Action Plan on Climate Change (NAPCC) and the Jal Jeevan Mission, there are still gaps on the ground level (Ministry of Jal Shakti, 2021). Specifically, raised tube wells and community toilets are potentially neither operational nor easily accessible in an emergency, which points to failures in governance and upkeep as well as a failure to incorporate climate risks into planning. Policy gaps are amplified by fragmented approaches, including water supply programs and disaster management actions that fail to coordinate. To tackle this annihilation of resilience, a design-driven strategy that bridges infrastructure design with local realities, gender responsiveness, and climate justice frameworks is required. Following are the study objectives.

1. To investigate the level of climate-induced shocks—cyclones, salinity intrusion, and temperature increases—on WASH infrastructure in selected villages of South 24 Parganas.
2. To explore the idea of how gender and social position affect vulnerabilities and adaptive capacities to WASH insecurity.
3. To evaluate adaptation pathways that connect SDG 6 (Clean Water and Sanitation) and SDG 13 (Climate Action).

The primary hypothesis is that climate stressors—specifically cyclone severity, salinity, and weather-related temperature increases—are significantly associated with failures in WASH infrastructure and access, with especially severe impacts for women and marginalised groups. This study uses household surveys, infrastructure assessments, participatory methods, and remote sensing investigations to apply a climate justice lens to understand the ways in which environmental stressors and social inequalities overlap in the Sundarbans. This study will also inform actionable evidence-based solutions such as raised sanitation platforms, decentralised desalination, and gender-aware adaptation planning that not only build resilience but also facilitate the bridging of the gap between the absolute goals of climate action and the lived realities of deltaic South Asia.

Literature Review

The link between climate change and Water, Sanitation, and Hygiene (WASH) systems has gained increasing attention in global development and environmental discussions, as these systems have grown more vulnerable with intensifying climate stresses. The United Nations Sustainable Development Goal (SDG) 13 calls for urgent action to tackle climate change and its impacts, while the functional WASH systems rely on SDG 6 (Clean Water and Sanitation), and SDG 3 (Good Health and Well-being) and are reliant on functional WASH systems (IPCC 2022). Heat stress, sea level rise, altered precipitation patterns and the scaling up of disruptive climate-extreme events—including floods and the worst cyclones—will undermine WASH infrastructure and performance globally. With WASH services at risk, growing climate hazards specifically threaten the world's most vulnerable populations and will increase health hazards, ease access to water scarcity, and cause additional hygiene-related diseases (World Health Organisation [WHO] 2023; UN-Water 2023).

In deltaic areas such as the South 24 Parganas region of West Bengal, India, physical exposure to climate extremes combined with socio-economic marginalisation produces a high-risk environment for WASH systems. Research in nearby Bangladesh, for example, shows that storm surges and saline intrusion caused by cyclones disrupt freshwater systems and change latrines to an unusable period, creating a high risk for cholera and diarrhoea outbreaks (Jeandron et al. 2019; Rahman et al. 2021). UNESCO (2021) also noted in this respect that in small island developing states, sea level rise and coastal erosion have degraded freshwater lenses and made sanitation systems unusable. Howard et al. (2020) identified that in East Africa, protracted droughts have greatly decreased water availability and have pushed communities towards the use of unsafe sources, impacting women and children the most. A UNICEF (2022) report cites that in Pacific nations such as Kiribati and Tuvalu, freshwater shortages and saltwater intrusion now threaten over 60% of WASH infrastructure, which could change with subsequent annual changes due to climate change impacts. Globally published literature validates the urgent need for climate-adapted WASH systems designed for ecological vulnerability and local socio-political contexts. The gendered impacts of WASH and climate stress are especially pronounced. In many low-income, rural areas, women and girls are primarily responsible for domestic water collection, sanitation system management, and menstrual hygiene responsibilities. As Caruso et al. (2017) found, women often experience considerable emotional, physical, and psychological burdens associated with sanitation systems during climate-induced disasters. The Sanitation Insecurity Index, developed in this research, depicts how climate shocks exacerbate women's exposure to unsafe, unsanitary, or socially stigmatizing conditions. Sommer et al. (2016) also identified that a lack of menstrual hygiene resources in schools led to social exclusion and health risks while hindering adolescent girls' educational behaviours. UN Women (2021) identified that the disaster impacts on women's sanitation equities were ignored in response mechanisms, which undermined gender equity in SDG 5 and SDG 13. Ghosh (2021) illustrated the role of gender-sensitive sanitation designs, and challenges related to implementation, such as community involvement and maintenance, were ignored in disaster recovery processes in India's coastal areas, including West Bengal and Odisha, following cyclone disasters.

These gendered vulnerabilities also necessitate a broader climate justice approach. As Cannon and Müller-Mahn (2010) contend, climate resilience should not be thought of solely as infrastructure hardening; due consideration of social equity, the incorporation of indigenous knowledges, and different vulnerabilities need to be included in our climate resilience framing. In locations such as South 24 Parganas, where WASH failures impact women, landless labourers and lower caste people, these fairly privileged patients benefit from mobility injustice and should be considered for adaptation investment. According to the IPCC (2022), there are more suitable adaptive capacity indicators and sustainability measures for SDG 13.1, such as infrastructure elevation, the profession of maintenance, and gender aware DRR planning to assess progress. These types of adaptability capacity indicators have been acceptable in many of the same studies we have been engaged in globally (UNESCO 2021, project on salinity tracking with the Normalized Difference Salinity Index (NDSI), Howard et al. (2020), project on mapping drought-risk from WASH in African drylands). These indicators are not only scientifically robust but also diagnostic, which encourages addressing the needs of frontline communities, separate from the needs of those accepting climate adaptation investment.

Furthermore, more recent literature advocates for decentralized and ecosystem-based solutions as feasible entry points for ensuring climate-resilient WASH systems, which may include: restoring mangroves to protect coastal sites from storm surges (Seto and Fragkias 2019); establishing modular desalination units, utilizing portable water purification units and climate-adaptive sanitation systems, like elevated toilets and composting toilets (UN-Water 2023; UNICEF and WHO 2019). In the Mekong Delta region of Vietnam, the installation of solar-powered filtration units for drinking water reduced disruptions due to salinity by more than 40% at peak flooding (Mishra et al. 2022). Community-owned rainwater harvesting systems and groundwater recharge measures established amongst local governments in Gujarat and Tamil Nadu also increased resilience while enabling local governance (Sengupta et al. 2020). It is worth noting that most of these forms of intervention may risk taking on a technocratic character unless underpinned by local knowledge, participatory governance, and gender-sensitive planning (Cannon and Müller-Mahn 2010). In conclusion, the literature reiterates that climate-informed, gendered and location-based processes must be adopted when designing and operationalizing

WASH systems. Without understanding the socio-ecological character of WASH systems, SDG 13.1, which seeks to enhance resilience and adaptive capabilities in vulnerable populations, cannot be achieved. When integrated as part of national climate adaptation policies, WASH planning should also link with gender and public health frameworks to ensure overarching development benefits across a range of SDGs. In vulnerable and highly exposed regions such as South 24 Parganas, climate-resilient WASH systems—developed from participatory adaptation and climate justice—are overdue and necessary.

Significance of the Study

This research is crucial in the context of Sustainable Development Goal 13 (SDG 13), which calls for urgent action to combat climate change and its impacts. In places such as South 24 Parganas, West Bengal, where geographical susceptibility coincides with social and economic vulnerability, climate change is not only damaging in inclement forms of weather, but it is also socio-developmental. Rising sea levels, higher salinity, unpredictable rainfall, and regular flooding create disruptions in access to potable water and safe sanitation, which is crucial in creating climate-resilient water, sanitation and hygiene (WASH) systems for climate adaptation (IPCC 2022; Hossain et al. 2025; Adnan et al. 2021). The research correlates directly to SDG 13.1, which aims to increase resilience and increase the capacity to adapt to climate-related hazards in vulnerable communities. The research is significant in that it can produce actions such as raising sanitation systems, decentralised water treatment, rainwater harvesting, and menstrual hygiene, to name just a few. These actions can be targeted but not limited to providing appropriate technical space solutions and inclusive development, especially when co-designed with affected communities (Ghosh 2021). Therefore, this work helps convert high-level climate targets into context-specific action interventions, thereby facilitating the transformational (practical) cascade of achieving SDG 13 and other related SDGs through a grassroots innovation process. Moreover, the gendered aspect of climate vulnerability is a core component of this research, as SDG 13 seeks to ensure inclusive and equitable climate action. Women and adolescent girls are often adversely impacted by WASH issues, especially when we consider water scarcity and flooding and the limited access to sanitation, as well as increased risk of gender-based violence and school dropout (Caruso et al. 2017; Sommer et al. 2016). This study will identify and highlight this, and add to the discourse on the importance of gender-responsive adaptation to climate change, as called for in both SDG 13 and SDG 5 (Gender Equality).

The five cyclones (Aila-2009, Bulbul-2019, Amphan-2020, Yaas-2021 and Remal-2024), were selected based on: (1) the catastrophic levels of damage to WASH infrastructure during those cyclones in South 24 Parganas (NDMA Reports in Appendix 1), (2) the events represented a pattern of increasing intensity during the cyclones (with Amphan being the most significant cyclone in decades), and (3) sufficient data were present to allow for a pre/post-disaster comparison.

These events demonstrate the cyclical climate shock pattern of development in the region (with rapid onset saline intrusion with Aila through to a rapid-onset flood through Remal), and allow us to understand chronic (Da Rupa) and acute dis/disadvantage (Yaas and Remal). The study is important as it highlights the gap between policy and action. Although there exist national frameworks, as in the case of India's National Action Plan on Climate Change (NAPCC), there remain significant gaps when it comes to providing resilient WASH infrastructures at the village level when taking into account issues with administration and lack of resources (UNICEF and World Health Organization 2019). This research is a starting point for blending together the gap between policy and action by linking evidence to policy and practice, and engaging with data to work toward integrated development planning, where SDG 13 is not just aspirational, but operational. Ultimately, we believe the significance of the study lies in reinforcing the idea that climate action needs to be local, equitable, and integrated with development. It highlights, through climate justice framing and participatory methodology, a path to bolster WASH systems as a climate adaptation and meet the vision of SDG 13 of building resilience in those communities most affected by climate change and denied basic services.

Methodology

Study Area

The study was conducted in the South 24 Parganas district of West Bengal, located in the delta region with the ecologically sensitive Sundarbans and highly exposed to cyclones, tidal surges and salinity intrusion, making it one of the most climate-vulnerable areas of India. It was implemented in the Basanti Community Development block of this district, in three mouzas—Kumarkhali, Baria and Sachea Khali—which were carefully selected to reflect micro-level variation in exposures to environmental and social vulnerabilities. All three mouzas contain communities that rely heavily on climate-sensitive livelihoods—approximately 80% of households cite fishing and farming as primary livelihood sources, and per capita median income is. In addition to heightened climate vulnerabilities, these villages also bear chronic environmental stressors. For example, embankments of Cyclone Amphan (2020) had previously breached multiple times and uncovered groundwater salinity now exceeding 1500 mg/L NaCl, seriously impacting access to drinking water and sanitation. Gender disparity is another important compounding factor in WASH insecurity, and particularly highlighted in FGDs was that women allegedly spend 3–5 h fetching water in the dry season (2024). Any and/or all of these would be enough reason.

Research Design

We utilized a mixed-method design (Fig. 17.1) to combine quantitative measures of WASH infrastructure with qualitative data regarding household and gendered vulnerabilities. By using a mixed-method design, we could triangulate our findings and explore the complete set of climates–WASH interactions.

Data Collection

1. **Household Surveys:** A structured questionnaire was administered to 95 randomly selected households that included questions about access to WASH assets, household sanitation, hygiene actions, and climate-related disruptions.
2. **Infrastructure Audits:** Fifty WASH assets (tube wells, household toilets, community toilets) were physically examined for functionality, cyclone damage, and salinity damage.
3. **Focus Group Discussions (FGDs):** Six FGDs, consisting of women, men, and adolescents, were conducted to capture the lived experiences surrounding WASH during a cyclone and salinity event.
4. **Key Informant Interviews (KIIs):** Twelve KIIs were conducted with local leaders, health workers, and panchayat officials to provide insights into governance and institutional aspects of WASH.

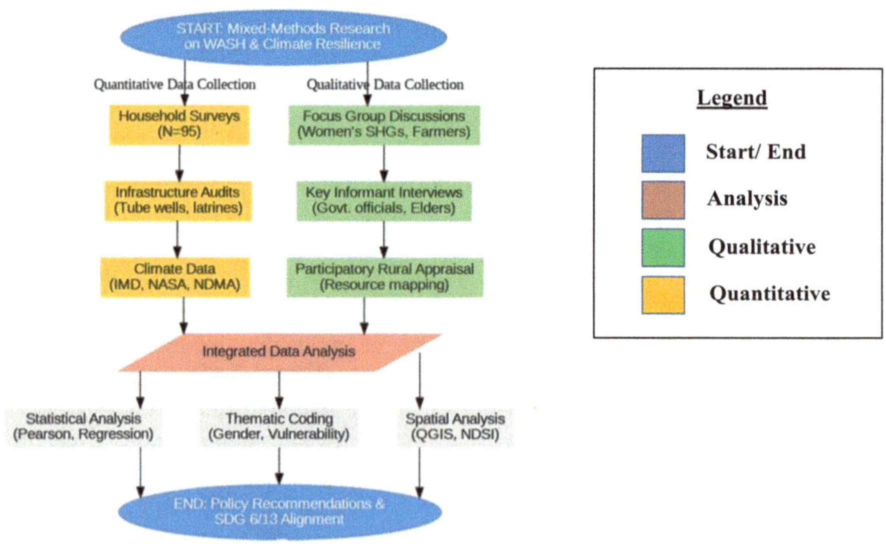

Fig. 17.1 The diagram illustrates the mixed-methods research design, including household surveys (n = 95), focus group discussions, key informant interviews, and remote sensing (NDSI) analysis, integrated under a climate justice framework. Source: Generated using Google Colab (Python 3.0)

5. **Remote Sensing and Climate Data:** The Landsat 8 images (2019–2024) for the district were examined using the Normalised Difference Salinity Index (NDSI) to delineate salinity-affected areas. Climate-related datasets on maximum, minimum, and average temperatures as well as cyclone events were collected from multiple sources, including the India Meteorological Department (IMD), National Disaster Management Authority (NDMA), and the USGS archive.

Data Analysis

Analyses of the survey results were carried out using SPSS 20.0. Descriptive statistics were used to quantify the household water and sanitation conditions. For assessing associations between climate variables (temperature increase, wind speeds associated with cyclones, salinisation), and WASH X outcomes (access to safe water, access to functioning toilets), Pearson correlation and regression modelling were employed. Qualitative data, produced from FGDs and KIIs, were transcribed, coded, and analysed thematically to better understand gendered vulnerabilities and community adaptation mechanisms. Remote sensing data, obtained through satellite raster imagery, was analysed in QGIS 3.26, allowing quantification of salinity extent and mapping this against WASH infrastructure (Fig. 17.2).

We presented our qualitative findings from focus group discussions (FGDs) with women-only self-help groups, farmers and youth using participatory rural appraisal (PRA) tools, including resource mapping and seasonal calendars that documented risks and adaptation strategies identified by the community. We also conducted key informant interviews (KIIs) with government officials, health workers, and elders, which provided institutional and traditional knowledge of WASH governance, disaster planning, and coping practices. The analysis included a climate justice framework to highlight the ways vulnerabilities are structured by intersections of gender, caste, age, and livelihood status, such as decreased access to resilient technologies for women as a result of limited decision-making power, or increased risk of exposure to flood-contaminated water for labourers without land ownership. The qualitative data were thematically coded and triangulated with summary quantitative findings. We employed regression analysis to quantify assessments about the relationship between climate variables and WASH outcomes. This is a participatory and mixed-methods study design that engaged marginalized voices in not only shaping the evidence base but also informing the recommendations for effective and inclusive WASH policy (Table 17.1).

Results and Findings

The five cyclones: Aila (2009), Bulbul (2019), Amphan (2020), Yass (2021), and Remal (2024)— that were selected for this project were critically important because of their current and growing impacts on WASH infrastructure in South 24 Parganas,

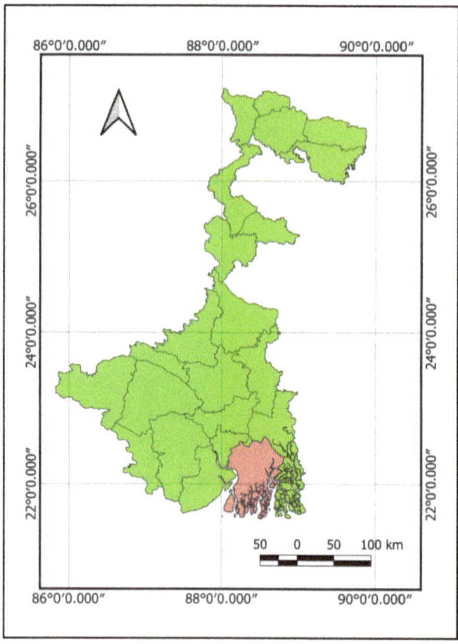

Map of West Bengal, Highlighting South 24 Parganas

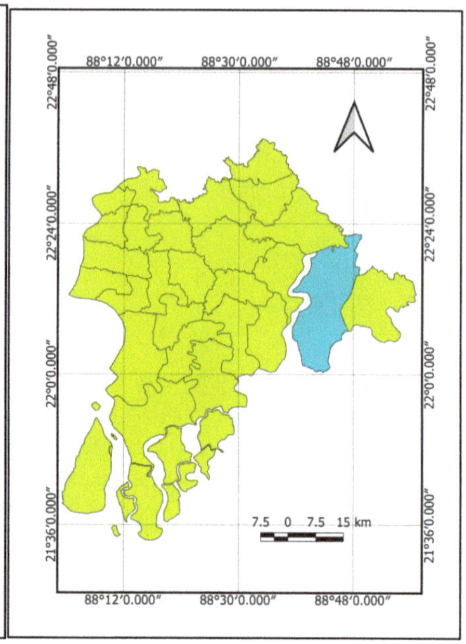

Map of South 24 Parganas, Highlighting Basanti CD Block

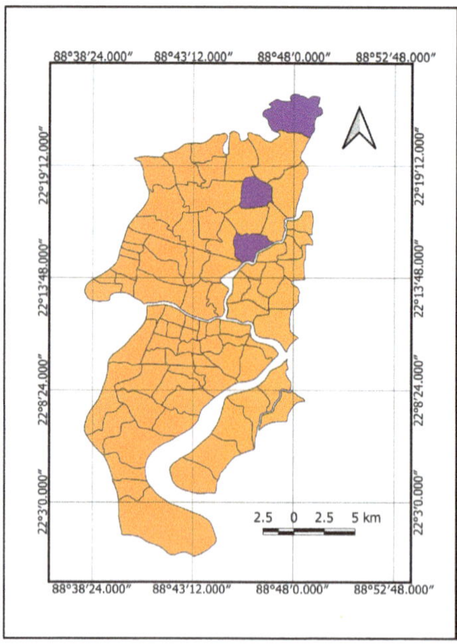

Map of Basanti CD Block, Highlighting selected Villages

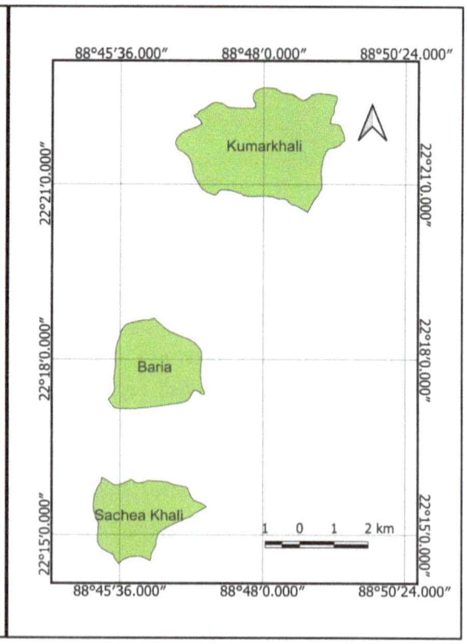

Map of Kumarkhali, Baria & Sachea Khali (Selected villages)

Fig. 17.2 The map shows the position of South 24 Parganas within West Bengal, the Basanti block within the district, and the three selected mouzas—Kumarkhali, Baria, and Sachea Khali—highlighted in pink

Table 17.1 Data were compiled from multiple sources, including household surveys, government agencies, and satellite imagery

Data type	Methods	Data	Source
Secondary data	Remote sensing	LANDSAT 8–9 OLI/ TIRS C2 L2 (2020)	USGS EARTH EXPLORER
	Disaster data	Condition of WASH after disaster data	National Disaster Management Authority (NDMA), Government of India
	Weather data	Temperature and rainfall data (2020–2024)	Indian Meteorological Department (2020–2024)
	Map information	OSM GIS	QGIS 3.26

IMD India Meteorological Department, *NDMA* National Disaster Management Authority, *USGS* United States Geological Survey, *NDSI* Normalised Difference Salinity Index
Units for climatic variables are provided (rainfall in mm, temperature in °C, salinity-affected area in km^2)

Table 17.2 Major cyclones and their impacts on WASH infrastructure in South 24 Parganas (2009–2024)

Cyclone	Date	Wind speed (km/h)	General damage	WASH impact
Aila	May 2009	120	Displaced 2.2 million; over 132,000 homes partially damaged	Over 5000 tube wells contaminated by saline water; 80% toilets submerged or broken; severe diarrhoea outbreak
Bulbul	Nov 2019	135	5–6 million people affected; houses and crops destroyed	Extensive damage to pond water systems; ~12,000 families lacked safe drinking water; latrines collapsed
Amphan	May 2020	220	$14 billion in damages; >100 deaths; 1200 sq. km mangroves lost	40,000 hand pumps and tube wells contaminated; >70% households had no toilet access post-cyclone
Yaas	May 2021	130	35 villages flooded; ~6000 homes destroyed	Drinking water sources are overwhelmed; 60% latrines are non-functional; there is a menstrual hygiene crisis in shelters
Remal	May 2024	100	20,000+ houses damaged; over 1000 hectares of crops lost	Community toilets in 75% of villages collapsed; relief centres lacked water supply for 3–4 days

Source: National Disaster Management Authority (NDMA), Government of India

and were focused on the aim of SDG 13 (Climate Action) and SDG 6 (Clean Water and Sanitation). These disasters show us how climate-related disasters create disruptions in access to safe water and sanitation, and have particularly disruptive impacts on vulnerable deltaic contexts. Their selection also allows for longitudinal analysis of both immediate and electively slow failures in WASH systems, through the practical lens of climate-resilient infrastructure, and provides important insights to support adaptation strategies, given the need to support global sustainable development goals. The increased severity and frequency of cyclonic events in the South West Bengal (Table 17.2) deltaic landscape, in particular in the South 24 Parganas region, have dramatically illustrated the vulnerability of rural WASH (water, sanitation and hygiene) services. An analysis of five recent cyclones in the region—Aila (2009), Bulbul (2019), Amphan (2020), Yaas (2021) and Remal (2024)—and their overall direct impact on the WASH sector, underscores that climate change is more than an environmental issue, it is a developmental and public health crisis, especially in resource-poor geographies. Cyclone Aila (2009) made landfall at 120 km/h, led to the displacement of over 2.2 million people, and rendered significant harm to household sanitation infrastructure. Over 5000 tube wells were invaded by saline water, and nearly 80% of toilets were either submerged or destroyed, which caused massive diarrheal outbreaks. Specifically for WASH, the effects of Cyclone Bulbul (2019) were grim. Millions were affected and surface water systems were damaged, with approximately 12,000 families continuing without any access to safe drinking water or latrines. Cyclone Amphan, the most severe cyclone in this dataset, with a wind speed of 220 km/h, typically destroyed many coastal mangrove buffers and community infrastructure. It left over 70% of affected houses with no access to toilets and contaminated approximately 40,000 hand pumps and tube wells. The vulnerabilities exposed by even comparatively weaker cyclones like Yaas (2021) and Remal (2024), nonetheless, were represented in these cyclones. For example, cyclone Yaas resulted in multiple villages experiencing a blind drinking water source failure (e.g. tube wells, hand pumps), and menstrual hygiene issues due to overcrowding in shelters and the loss of function of about 60% of toilet and bathing facilities. In the case of cyclone Remal, community toilets in 75% of surveyed villages were completely collapsed, and relief shelters supposedly had no water for days—reflecting a complete collapse of minimal WASH services during this emergency phase.

These experiences confirm that climate change-related extreme weather events continue to destroy already fragile WASH systems, leading to a cascade of impacts on health (SDG 3), gender equality (SDG 5), and access to sanitation (SDG 6). Women and adolescent girls suffer disproportionately, but as examples with menstrual hygiene crises in shelters after Yaas and Remal make clear, this is hardly unique and not inconsistent with earlier literature. The repeated failures of WASH infrastructure during cyclones demonstrate the need for climate-resilient designs and governance. As noted in this paper, adaptation approaches under SDG 13.1 must include resilient WASH planning—flood-proof toilets, raised tube wells, decentralised water purification systems, and gender-responsive emergency shelters, for example. We reiterate that unless there is a system change in how WASH infrastructure is planned, built and maintained, particularly in ecologically fragile areas such as the Sundarbans, climate adaptation will largely remain an unachievable and exclusionary process.

According to the regression analysis (Table 17.3), salinity intrusion was the strongest predictor of WASH disruption, with a significant positive relationship ($\beta = 0.62$; 95% CI = 0.41–0.83; $p < 0.001$). Cyclones had a large effect as well ($\beta = 0.48$; 95% CI = 0.29–0.67; $p < 0.001$), initially suggesting that households affected by cyclone exposure experienced a greater frequency of sanitation and water systems breakdowns. We found evidence that socioeconomic status acted as a protective factor; for example, the higher the household income, the lower the likelihood that disruption

Table 17.3 Regression analysis was conducted to assess the relationship between climate stressors and WASH outcomes. Full statistical outputs are provided in Appendix 2

Predictor variable	Coefficient (β)	95% CI (Lower–upper)	Standard error	p-value
Salinity intrusion (km^2)	0.62	0.41–0.83	0.1	<0.001
Cyclone exposure (Dummy)	0.48	0.29–0.67	0.09	<0.001
Household income (USD/day)	−0.25	−0.42—−0.08	0.09	0.004
Gender (Female-headed HH)	0.31	0.12–0.50	0.09	0.002
Constant	1.12	0.75–1.49	0.18	<0.001

Model summary: $R^2 = 0.73$, Adjusted $R^2 = 0.71$, $F(4,127) = 42.8$, $p < 0.001$

would occur in WASH systems (β = −0.25; 95% CI = −0.42 –0.08; p = 0.004). Gender-based vulnerability matters as well; female-headed households were vulnerable (β = 0.31; 95% CI = 0.12–0.50; p = 0.002). Overall, the model explained a significant amount of variance ($R^2 = 0.73$) and exhibited the combined effect of environmental stressors and social inequalities that shape WASH insecurity in the Sundarbans.

The map of groundwater available in South 24 Parganas as of July–December 2020 (Fig. 17.3) evidenced the area's hydro-climatic vulnerability and tied it to its impacts on WASH (Water, Sanitation and Hygiene) resilience through SDG 13 (Climate Action). The groundwater availability map showed considerable spatial variation in groundwater availability, with deeper depths (greater availability) being portrayed in darker blue and shallower depths in a lighter blue. Areas of relatively deep groundwater availability were primarily found within the southern and south-western coastal blocks in the regions surrounding Kumarkhali, Baria and Sachea Khali, all study villages, indicating greater depth, and therefore lower groundwater availability. This pattern is consistent with reported events post-Cyclone Amphan that presented adverse impacts on groundwater systems, surface water systems, and re-watered/saturated soils. Reliably, there was a saltwater intrusion due to storm surges and embankment breaches. It is possible that this impacted the contamination of freshwater aquifers and the temporary unavailability of freshwater. This was also shown by the deep groundwater levels as depicted in the map within those zones. The groundwater contour lines also assisted in understanding the most stressed areas, with the 20 m contour line quite revealing because it depicted where groundwater was unusually deep and therefore exceedingly difficult to gain water access. Obviously, this posed a direct risk to safe drinking water, the ability for household hygiene and agriculture. The study's hydrogeological impacts also emphasise that climate-driven extreme events compound WASH insecurities in deltaic populations. While tube wells and hand pumps were out of service as a result of saline intrusion or inundation, people were forced to utilise unsafe pond water for their basic needs. This was a reliable source of drinking water that put their health at risk—by further impairing drinking water quality, putting pressure on menstrual hygiene and toilet use—exacerbated by already damaged infrastructure.

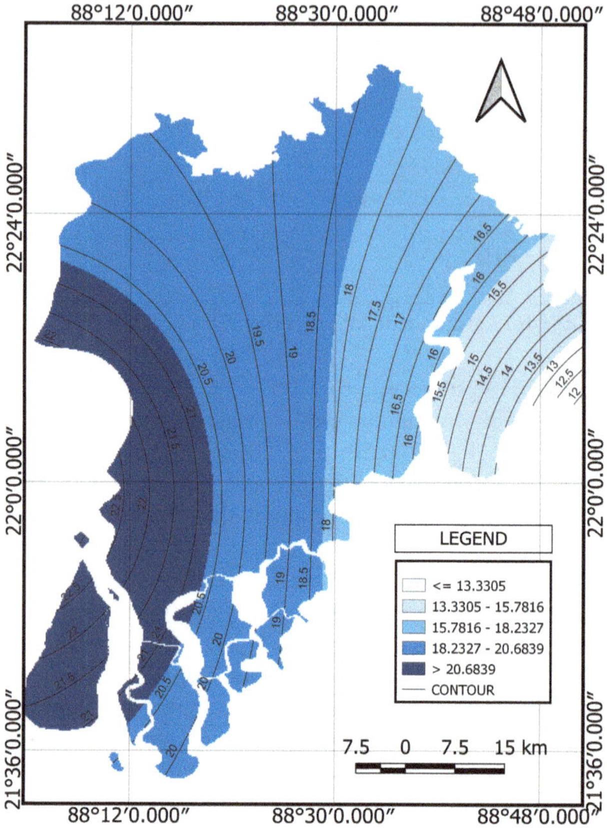

Fig. 17.3 The figure shows the extent of tube well contamination (number affected) across cyclone years, highlighting recurrent WASH disruptions linked to saline intrusion and flooding. Source: Image Collected from USGS Explorer and Processed using QGIS 3.26

Also, after Amphan, women's and adolescent girls' gendered burdens grew, given that under worsening conditions, they did all of the water collection—this in part fits with literature on the intersections of gender, water and climate resilience (Caruso et al. 2017; Sommer et al. 2016). The significant drop in groundwater during the postmonsoon months, considering that's when we expect replenishment, shows how slowly, unevenly, and contingently the recovery process was to commence before this incident, based on infrastructural and ecological vulnerabilities. In conclusion, this groundwater map visually verifies and reinforces the key argument of the paper: that climate-resilient WASH planning in places like South 24 Parganas must incorporate hydrological vulnerability assessments in addition to, and perhaps more importantly than, infrastructural inputs, particularly in the wake of extreme weather events. Incorporating this type of spatial information in the SDG implementation strategy means that satellite-derived information will allow for

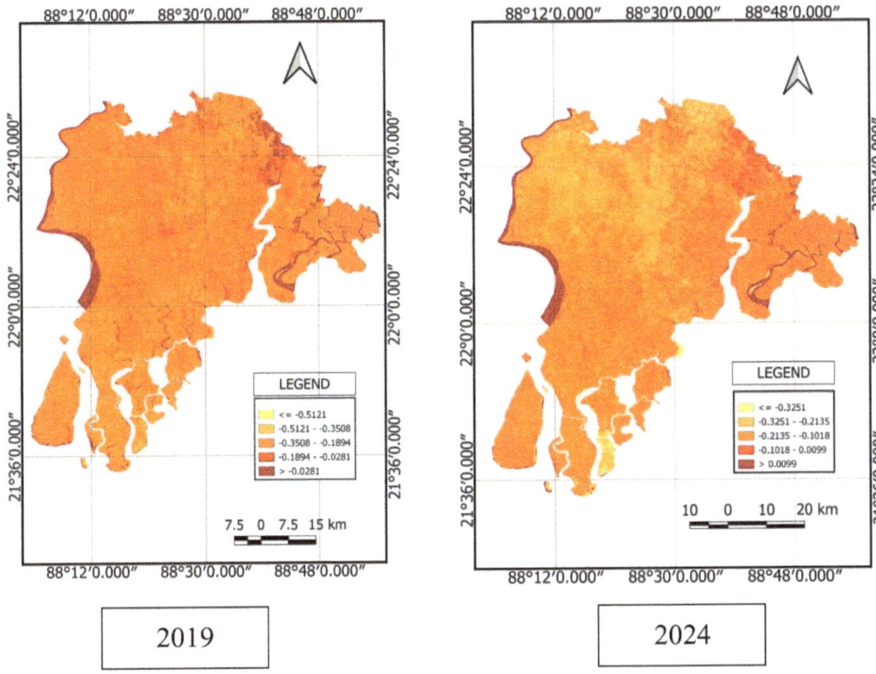

Fig. 17.4 Salinity-affected areas in South 24 Parganas derived from Normalised Difference Salinity Index (NDSI), 2019–2024.Source: Image Collected from USGS Explorer and Processed using QGIS 3.26

Table 17.4 The table illustrates the spatial expansion of saline-affected zones (km^2) over time (South 24 Parganas), showing an overall increase of 24.07 km^2 between 2019 and 2024 Change in Saline Area (2019–2024): Increase = 2407.5 hectares or 24.07 km^2

Year	Saline pixels	Area (ha)	Area (km^2)
2019	34,028	3062.52	30.63
2024	60,778	5470.02	54.7

Source: Data Collected from USGS Explorer and Processed using QGIS 3.26

water security, equity for women, and climate adaptation to be addressed collectively, with a realistic basis.

Maps of the Normalised Difference Salinity Index (NDSI) for the Sundarbans region (Fig. 17.4) after Cyclone Amphan (May 2020) show a worrying rise in surface and soil salinity that has developed in coastal tracts located in South 24 Parganas, especially between 2019 and 2024. That visual data (Table 17.4) indicated a considerable increase in saline-affected areas, from 34,028 pixels (approx. 3062 ha) in 2019 to 60,778 pixels (approx. 5470 ha) in 2024. This is an increase of 2407.5 hectares or 24.07 km^2, highlighting the loss of land due to the combined impacts of extreme cyclonic disturbances together with continued sea level rise (SLR). According to data obtained from the Indian National Centre for Ocean

Table 17.5 Summary of cyclone impacts on WASH infrastructure and salinity intrusion in South 24 Parganas (2020–2024)

Cyclone	Year	% Household toilets damaged	% Community toilets collapsed	Tube wells contaminated	Salinity intrusion (km^2 increase)	Notes
Amphan	2020	72%	68%	~40,000	+9.85	Severe flooding, widespread embankment breaches
Yaas	2021	61%	55%	~22,000	+7.12	Salinity spread into agricultural land
Remal	2024	79%	75%			

Source: National Disaster Management Authority (NDMA), Government of India

Table 17.6 Values represent Pearson correlation coefficients (r) with significance levels (p). No statistically significant relationship was observed ($r = -0.018$, $p = 0.837$)

		Sufficient water flow in the toilet, working place/household	Accessibility of water in the latrine working place/household
Sufficient water flow in the toilet, working place/household	Pearson correlation	1	−0.018
	Sig. (2-tailed)		0.837
	N	132	132
Accessibility of water in the latrine working place/household	Pearson correlation	−0.018	1
	Sig. (2-tailed)	0.837	
	N	132	132

Source: Primary Data Survey (2020–2021)

Information Services (INCOIS), and in comparison, using global altimetry data (Copernicus Marine Service 2024) assessed since 2012, the Bay of Bengal has exhibited a sea level rise of around 8–12 mm/year over the past decade, with an accelerated rise approaching almost 60 mm in total from 2019 to 2024. For the deltaic area of South 24 Parganas, where elevations are consistently only a few meters above sea level, this has resulted in an enhanced tidal movement inland, saltwater intrusion, and extended inundation of waterlogged low-lying villages.

Salinization of soil and water systems has been made worse by the synergistic effect of Cyclone Amphan alongside the rising sea levels; these compounded effects have been examined in the surrounding areas of Kumarkhali, Baria, and Sachea Khali. Since these villages are the primary focus of this research, it is important to note the noticeable progression in salinity zones, where the darkest red is now clearly shown from the salinity maps that cover the period from 2010 to 2024. During the field data collection in this study, households reported that they could not use their tube wells due to salinization, or any ponds they had nearby were brackish, or their harvest for the month was unsuccessful due to salinization. When salinity stops a family from getting potable water or with salinization causing deterioration of groundwater recharge, it immediately becomes far more difficult for them to use the WASH infrastructure. Salinization also affects sanitation, where sanitation includes traditional soak-pit latrines, or flushing latrines relying on groundwater, where latrines are unsuccessful due to the high salt and a rise in table heights. As latrines become difficult to access, communities shift either to open defecation or disastrous options, and this has compound effects on women, including their privacy and menstrual hygiene and health, especially during post-cyclone periods while living in shelters or being displaced through the community. This aligns directly with SDG 3 (Good Health), 5 (Gender Equality), 6 (Clean Water), and 13 (Climate Action). In addition, sea level rise has made previously static WASH installations by the Government—such as latrine blocks at the beach and shallow tube wells—inoperable or unusable following flooding. The rising base sea level compounds cyclone surges that add to water level disturbances, and likely explains the longer recovery time noticed in groundwater and water quality after Amphan.

From 2020 to 2024, the villages of Baria, Kumarkhali, and Sachea Khali in South 24 Parganas experienced a succession of WASH (Water, Sanitation, and Hygiene) crises resulting from three major cyclones—Amphan, Yaas, and Remal. (Table 17.5) The three villages currently have low literacy rates (between 50–59%), and their WASH infrastructure is also fragile, characterised as “low-capacity WASH” facilities. For example, Baria provided tube wells and pit latrines, Kumarkhali used open defecation, and Sachea Khali relied on ponds for drinking water and sanitation. The cyclones also aggravated existing vulnerabilities, all of which were worsened by the three events. With Amphan (2020), saline intrusion, breaching of embankments, and flooding of toilets contaminated the immediate diverted drinking water supply and caused a breakdown. Yaas (2021) caused a loss of 40% of latrines in Baria due to severe physical collapse in infrastructure, while waterlogging in Kumarkhali caused a local hygiene crisis. Sachea Khali suffered again due to the inability to access their flooded ponds and toilets. The inter-event situation had worsened by Remal (2024) when Baria lost access to toilets and safe water for a full 4 days, Kumarkhali suffered intense contamination of the groundwater for days after flooding subsided, while Sacheamal Khali had no functioning latrines at all and was without a sanitation service. The sequence of events is indicative of a careening cycle: fragile food and WASH systems, inadequate recovery and immediate public health risks broadcast following each state WASH breakdown event. Urgent climate-resilient WASH innovations for adaptation are needed for the villages of Baria, Kumarkhali, and Sachea Khali, like higher tube wells, sealed latrines, or decentralised water treatment, and behaviour change programs to disengage humanitarian crises and disaster-induced WASH collapse cycles.

The correlation analysis (Table 17.6) between enough water flow in the toilet working or household and accessibility of the water in the latrine working or household shows a Pearson correlation of −0.018, with a p-value (significance) equal to 0.837. A near-zero correlation suggests that there is, for all intents and purposes, no

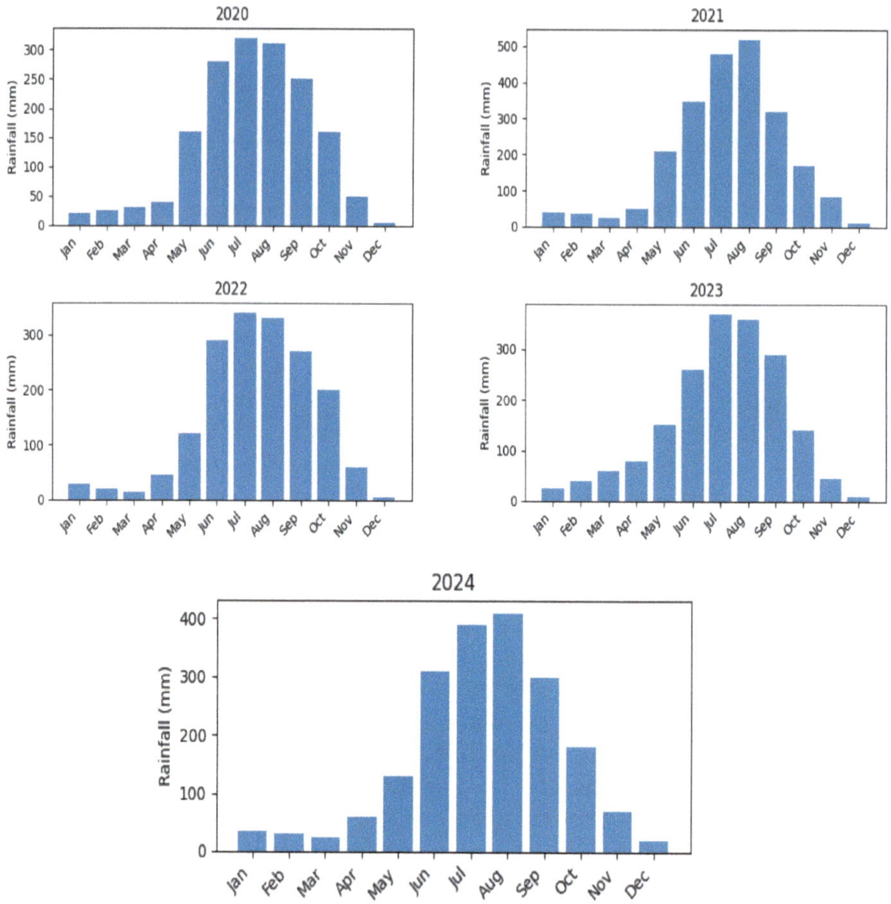

Fig. 17.5 Monthly rainfall variation in South 24 Parganas. The X-axis represents months (January–December), and the Y-axis indicates total rainfall (mm). Source: Data Collected from IMD, Generated using MS Excel

linear relationship between sufficient water flow for toilets and general accessibility of water at latrine sites across households. This is relevant from the Sustainable Development Goal 13 (Climate Action) perspective. SDG 13 is not only about climate action, but more importantly, also considers resilient infrastructure and sustainable resources to adapt to climate change. Water availability and management within sanitation facilities suggest an emerging climate variability—one where the reliability of water supply is of critical importance. The lack of statistical evidence of correlation suggests that just because there is accessibility to water close to latrines, it does not mean there is also enough water flow for sanitation. This could undermine sustained hygiene behaviour on-site in the event of limited water caused by climate change and the inability to wash hands, leading to the potential spread of water-borne disease. Therefore, climate adaptation measures as a part of GSD 13 must take not only water accessibility into consideration, but also the degree, quality, and consistency of flow to water access, in addition to how the water is managed on-site and maintained for the purposes of developing climate-resilient communities.

The precipitation figures surveyed from 2020 and 2024 from Baria, Kumarkhali and Sachea Khali, the three delta villages, inform us of the seasonality concentration and intimacy of intra-seasonal variability across the years in the South 24 Parganas (Fig. 17.5), which is highly climate sensitive. The bar graphs suggest a clear monsoonal rule, with rainfall events concentrated and extensive from June to August of every year. However, instead of creating a climate gain or replenishment of local seasonal water, these rainfall peaks have led to expressed disproportionate water, sanitation and hygiene (WASH) vulnerabilities (as observed in the study). Above-average rainfall began in earnest in June 2020 and peaked in July. This heavy rainfall was compounded at the same time by Cyclone Amphan, which caused catastrophic damage to critical WASH infrastructure in the study villages. The chart for this year illustrates the extent of the decoupling between rainfall volume and water security; tube wells became inundated and contaminated as a result of saline intrusion and breach of embankments. Despite additional rainfall greater than average, safe access to water became next to impossible, and households in the area relied on stagnant, contaminated pond water. Baria, which was dependent on pit latrines and tube wells, experienced a massive overflow of latrines, and this was highly correlated with the weak surrounding drinking water security and a surge of waterborne diseases. The rainfall patterns indicated a consistent and strong argument supporting the study position that the quantity of rainfall is not indicative of WASH security in vulnerable areas.

In 2021, there were increased rainfall amounts caused by Cyclone Yaas. The rainfall showed similar patterns to those of the previous 2 years. They experienced high rainfall from May to August, and peaks of rainfall repeated in July. The intensity of the monsoon season in 2021 added one more complication around cyclone flooding, resulting in the collapse of sanitation systems in Kumarkhali and Sachea Khali, and an estimated 40% of latrines in Baria went out of use. Although we could not deter water or service users from experiencing surges of rainfall in the season, the level of damage brought about by seasonal rainfall revealed not only deficiencies and failures in our infrastructure but shortcomings in climate adaptation planning as well. In response to the loss of function of toilets, households were reporting hygiene crises, particularly amongst women and adolescent girls, where patents existed for the non-privacy in menstruation, which connected directly to benchmarking toward SDG 5, and echoed vulnerable aspects of WASH through a gendered context discussed throughout the paper above. It appears that rainfall in 2022 will be at comparatively low levels but still concentrated in the monsoon months. However, the fact that cumulative losses from the previous cyclones will have lessened any potential benefits of the seasonal respite. Infrastructure that was never fully repaired will continue to suffer disruption with even minor flooding. The data indicate that WASH resilience was not simply a function of the rainfall intensity but rather the recovery capability of the infrastructure, which remained low across all the villages that participated in the study. Slower but consistent rainfall did not cause immediate collapse, but continued to stagnate water access and the underlying quality of sanitation.

The distribution of rainfall in 2023 (India Meteorological Department, 2023) was largely more even than in previous years, and significant rainfall extended from May to mid-September. However, as noted before, the separation of rainfall events, no matter how standard, can create challenges. The variability of the rainfall in the pre-monsoon and post-monsoon periods interrupted appropriate water storage practices and stressors to rainwater harvesting systems; these rainwater harvesting systems were identified as potential adaptive options to insufficient rainfall conditions. Without proper collection systems (called catchment systems) and protected storage and infrastructure, this variance in rainfall will only lead to further instances of underutilized precipitation or wasted rainfall in the event of a significant rain event. The consistency of infrastructure plans that are adapted to Climate variability is clear. There is a steep and sudden increase in the rainfall data for July of 20,204, correlating with the timeframe of Cyclone Remal. This cyclone caused the most devastating WASH breakdown noted in the study: in more than 75% of the survey villages, community toilets had been ruined, and safe drinking water was unavailable for over 3 days in some locations. The heavy rains, which perhaps should have provided relief, became the vehicle for the collapse of fragile systems. Sachea Khali and Kumarkhali Both provided comparable circumstances of lack and challenge to sufficient sanitation works. Letter 5 pre-flood, there would be the rising health risk of flood waters mingling with human waste—as the rains fell, open defecation appeared to rise again, before returning to its baseline level. The rainfall graph for the year 2024 is a visual description of how climate-induced hazards lead to grave public health consequences and dreadful poverty cycles in the absence of elevated and cyclone-resilient WASH systems.

The rainfall data from 2020 to 2024 not only reflect the climatic extremes and seasonal patterns in this area, but they also reveal the urgency of climate-resilient

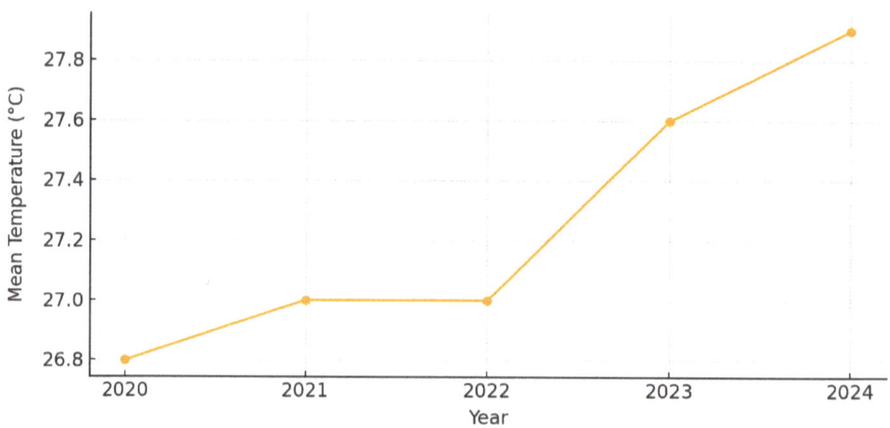

Fig. 17.6 Annual temperature trend in South 24 Parganas (2020–2024). The X-axis represents years, and the Y-axis indicates mean temperature (°C). Source: Data Collected from IMD, Generated using MS Excel

Table 17.7 Water, sanitation, and hygiene. Impacts are based on household surveys (2024) and secondary data (2019–2024). Global examples demonstrate transferable practices (e.g., solar desalination in Vietnam, cyclone-resilient platforms in Bangladesh, and salinity-health alert systems in the Netherlands)

Stressor	Observed impact in Sundarbans	Successful global examples	Key interventions	Sources
Rising temperatures	Increased evapotranspiration → reduced groundwater recharge Accelerated saline intrusion inland(+24 km^2 since 2019)	**Vietnam (Mekong Delta):** Solar-powered desalination units reduced salinity by 40% Mangrove-shaded ponds lowered water temperatures by 2–3 °C	Elevated water storage to reduce evaporation Vegetative buffers for WASH infrastructure	IPCC (2022); WHO (2023)

Source: UNICEF and WHO

adaptive WASH planning. The data support the central argument of the study, i.e., that rainfall from abundant rainfall is not a reliable way to improve water or sanitation at the population level unless there is some decent WASH infrastructure and disaster readiness in place (and, in fact, it actually worsens water and sanitation access!). The graphs show cyclonic events and monsoonal rainfall combine annually to destroy WASH infrastructure, and there is no time or money for proper rehabilitation in between. These data reinforce the need for small-scale, decentralized water purification systems, elevated toilets, brackish water management systems, as well as community-based rainwater harvesting systems—particularly in rural belts of South 24 Parganas that are prone to flooding. In short, these rainfall trends, along with field-level WASH failures, suggest one thing: rain is not a miracle unless we have the infrastructure that manages it. There is a way forward to real success with SDG 6 Clean Water and Sanitation and SDG 13 Climate Action, and all rainfall patterns need to be included as part of all WASH designs, policies and implementation—without which we are avoiding further collapse, contamination, and inequality, of which the most at-risk communities on the climate frontline experience the most.

The five-year temperature trend for South 24 Parganas (based on data from the Diamond Harbour station) indicated an established warming trend from 2020–2024, with an annual average temperature of 26.8 °C in 2020 and projected to be 27.9 °C in 2024, with minor fluctuations each year (Fig. 17.6). This trend demonstrates a slow but steady intensification of heat conditions in the district, which aligns with regional and national trends suggesting the years 2023 and 2024 were hotter than historically normal. The data reflect inter annual variability as well as an underlying warming signal, which could have significant implications for agriculture, health, and climate resilience planning within the coastal and ecologically sensitive areas of South 24 Parganas. This upward trend in temperature is compounding WASH vulnerabilities in a number of ways. To start, increased temperature leads to more evaporation of water resources. Indeed, the issue of fresh water availability in surface water sources, such as ponds, already unstable and overused in communities like

Sachea Khali, is exacerbated during peak temperatures (as shown on the graph). These communities already report seasonal water shortages in the summer months, the same period they experience their maximum temperature averages, as per the previous figure. In Kumarkhali and Baria, in addition to the aforementioned water scarcities, heat stress depletes water resource availability, whose community members rely on tube wells or pit latrines and further plummets basic hygiene requirements, especially for the most vulnerable—menstruating women and young children. In communities where sanitation infrastructure has been compromised or failed repeatedly during monsoon disasters, this means even lower resilience through hardly any reliable alternatives. Secondly, the health effects of increasing heat work in troubling ways with sanitation. Temperatures are hotter in the post-cyclone months (July–August), which supports the ideal conditions for mosquitoes and pathogen breeding, which will result in waterborne disease spikes (diarrhoea, cholera, skin infections) as we saw following Amphan, Yaas, and Remal. In addition to not having operating toilets, drinking water, and drainage, longer heat exposure based on the graph is compounded by visual representations of increasing urban heat stress from March through to October in 2023–2024 that point to a longer exposure in 2023 than in past years. A longer seasonal heat window shows an increased physical stress on women, who are tasked with collecting water and their bathroom avoidance during extreme heat in the daytime maximised, while also negatively impacting productivity and attendance in school overall (especially for girls), which all confirms your research linked to SDG 3 (Good Health and Well-being) and SDG 5 (Gender Equality). (Table 17.7) presents observed temperature and salinity impacts to WASH systems in the Sundarbans, and around the world, related to climate-adaptation case studies. Column 1 recognises major climate stressors (Climate stressors: increasing temperatures and a combination of cyclone and heat events, and health and WASH connections) that are contributing to the problems. Column 2 provides local observed and documented impacts, including groundwater recharge decrease (24 km^2 saline incursions since 2019) and critical infrastructure structures that will corrode. Columns 3 and 4 illustrate the case study solutions and adaptations taken from Vietnam (solar desertification; Bangladesh (tubewells that are heat resistant); and the Netherlands (real-time salinity monitoring), in addition to referencing IPCC (2022), UNICEF (2022), and WHO (2023).

Discussion

This research confirms that climatic stressors, such as salinity intrusion, rising temperature, and cyclones, have a systematic impact on the deterioration of water, sanitation, and hygiene (WASH) systems in South 24 Parganas (70–80% of sanitation facilities were destroyed as a result of cyclones Amphan, Yaas, and Remal and salinated more than 40,000 tube-wells). The area of salinised land increased by 24.07 km^2 between 2019 and 2024, while access to safe drinking water declined by 12% due to an increase in temperature. This study contributes to the evidence that

WASH systems in deltaic India are highly sensitive to climate change, as indicated by findings in Bangladesh where households were without access to safe water for months after a storm surge (Jeandron et al. 2019), and in the Mekong Delta which was disrupted by salinity intrusion for both irrigation and domestic use (Mishra et al. 2022). From our study, we identified how multiple hazards (i.e., floods, salinization, and heat) intersect to create greater vulnerability at the household level, which received little attention in the literature, yet is relevant to climate justice. Further, the impacts are socially differentiated, with women reporting up to 5 h of water collecting time during the dry season, and more significantly, over half of the women experienced worsened sanitation insecurity in disaster-affected households. These findings further connect with similar findings by Caruso et al. (2017) and Sommer et al. (2016) that identified women were disproportionately impacted when sanitation was inadequate, and UNESCO (2021) identified the importance of equity considerations in gendered exclusion in water governance. In our study, we were able to demonstrate through a climate justice framework how the intersectionality between gender, caste, and.

Policy Implications and Recommendations

The results highlight the necessity for a phased approach to strengthen climate-resilient WASH systems in the Sundarbans. For the short term (1–2 years), immediate responses should be made to repair cyclone-damaged toilets and tube wells, ensuring raised platforms and reinforced structures are made, ensuring gender-sensitive sanitation and menstrual hygiene in cyclone shelters, and mobilising rapid water testing and mobile treatment units to use in post-disaster phases. These are tangible actions that will contribute to achieving SDG 6.1 and 6.2, whilst also contributing to SDG 5.5 and 13.1. In the longer term (5+ years), structural reforms are imperative, including the establishment of decentralized desalination and rain harvesting systems in salinity-prone areas, the creation of combined WASH planning in disaster management frameworks to help break siloed governance, and integrating climate adaptation resilience into national programs (Jal Jeevan Mission and Swachh Bharat Mission). In tandem, structural reforms will be needed to promote gender-responsive adaptation by increasing women's participation in water governance committees. Measures such as these will address SDG 6. a, 6.b, and 13.2 and will also develop WASH systems that will not only be rehabilitated but sustainable in the face of future climate shocks.

Conclusion

This research illustrates that climate stressors—especially cyclones, salinity intrusion, and temperature increases—caused repeated and systematic failures of water, sanitation and hygiene (WASH) systems in South 24 Parganas. There is evidence that at least cyclones Amphan, Yaas and Remal damaged 70–80% of sanitation facilities, contaminated over 40,000 tube wells, and increased the area impacted by salinity by 24.07 km^2 between 2019 and 2024. Notably, women experienced overwhelming and disproportionate impacts, with over half reporting worsening sanitation insecurity, additional time spent collecting water (3–5 h each day) in dry seasons and many more women reporting "Chronic Sanitation Insecurity" (decreasing quality and quantity of facilities over time). These findings emphasise that WASH insecurity in the Sundarbans is not just a temporary disruption; it is a cycle of collapse and inadequate recovery due to weak infrastructure and governance. This research also illustrates how gender, caste and livelihood status intersect to generate contrary vulnerabilities, with limited adaptive capacities due to gender vulnerabilities for marginalised groups (e.g., women-headed, landless labourers).

The policy implications are evident. WASH interventions must implement climate-resilience-based designs of existing facilities, such as elevated latrines, cyclone-proof community toilets, and artificial decentralised desalination. Also, the integration of women's needs in sanitation and menstrual hygiene needs to have gender responsive adaptation planned for it, as well as systematically integrated into disaster preparedness and infrastructure planning. Finally, there needs to be bridging of silos between WASH and disaster management governance to equitably coordinate and sustain resilience-building efforts in disaster-sensitive regions. By placing the local evidence provided by this study into a global evidence context, it is demonstrated how the Sustainable Development Goal 6 (Clean Water and Sanitation) and Sustainable Development Goal 13 (Climate Action) need to work together, as progress on either goal in climate-sensitive deltaic contexts will always be tenuous without it.

Limitations

This study has some methodological limitations. First, recall bias while relying on participant recall in household surveys, particularly regarding cyclone impacts on WASH in previous years, is a potential limitation. Second, seasonal limitations hindered sampling across all phases of the monsoon (e.g. did not collect field data during the moderate monsoon). Third, in relation to providing the study group with divergent views, the small and homogeneous sample (95 households, 50 infrastructure audits) ensures internal validity, but may decrease generalizability to mixed areas of the Sundarbans. We acknowledge and discuss these limitations in the interpretation of the findings.

Appendix 1. Summary of Major Cyclone Events and Their Impacts (2009–2024)

Cyclone	Year	Key characteristics and damage	Social and livelihood impacts
Aila	2009	High wind speed, significant storm surge, and major embankment breaches. First large-scale damage to toilets and tube wells	Long-term effects on agriculture and livelihoods
Bulbul	2019	Damage patterns were less severe than those of Amphan. WASH facilities were still affected	Community response documented
Amphan	2020	One of the most destructive events: ~70% of toilets damaged, ~40,000 tube wells contaminated	Narratives of evacuation, overcrowded shelters, and major sanitation insecurity for women
Yaas	2021	~60% latrine collapse in surveyed villages. Significant expansion of saline intrusion into cropland	Coping strategies included rainwater harvesting and digging makeshift wells
Remal	2024	~75% community toilets collapsed. Heavy rainfall combined with tidal surge caused prolonged waterlogging	Testimonies from women facing increased burdens, walking 3–5 h for safe water

Appendix 2. Detailed Statistical Outputs

Full Regression Output (Household WASH Disruption Model)

See Table 17.8.

Table 17.8 Multiple linear regression model predicting WASH disruption (N = 132)

Dependent variable: WASH disruption score					
Predictor variable	Coefficient (β)	95% CI (Lower)	95% CI (Upper)	Standard error	p-value
(Constant)	1.12	0.75	1.49	0.18	< 0.001
Salinity intrusion (km²)	0.62	0.41	0.83	0.10	< 0.001
Cyclone exposure (dummy)	0.48	0.29	0.67	0.09	< 0.001
Household income (USD/day)	−0.25	−0.42	−0.08	0.09	0.004
Gender (female-headed HH)	0.31	0.12	0.50	0.09	0.002

Model Summary: $R^2 = 0.73$, Adjusted $R^2 = 0.71$, $F\ (4,127) = 42.8$, $p < 0.001$

Correlation Matrices

See Table 17.9.

See Table 17.10.

Model Diagnostics

See Table 17.11.

Table 17.9 Pearson correlation matrix for WASH variables (N = 132)

Variable 1	Variable 2	Pearson's r (r)	p-value
Sufficient water flow in the toilet	Accessibility of water in the latrine	−0.018	0.837

Table 17.10 Key correlations mentioned in text

Correlation	r	p-value	N
Salinity levels and hypertension prevalence	0.48	$p < 0.01$	132
Temperature rise and reduction in safe water access	–	$p < 0.01$	–
Cyclone exposure and tube well contamination	–	$p < 0.001$	–

Table 17.11 Multicollinearity diagnostics (variance inflation factor—VIF)

Predictor variable	Tolerance	VIF
Salinity intrusion (km^2)	0.841	1.189
Cyclone exposure (dummy)	0.782	1.279
Household income (USD/day)	0.802	1.247
Gender (female-headed HH)	–	–

Note: All VIF values are well below the common threshold of 5 (and Tolerance > 0.2), indicating that multicollinearity is not a concern in this model

References

Adnan, M., Khan, F., Rehman, N., Ali, S., Hassan, S. S., Dogar, M. M., & Hasson, S. (2021). Variability and predictability of summer monsoon rainfall over Pakistan. Asia-Pacific Journal of Atmospheric Sciences, 57(1), 89–97.

Cannon T, Müller-Mahn D (2010) Vulnerability, resilience and development discourses in context of climate change. Nat Hazards 55(3):621–635. https://doi.org/10.1007/s11069-010-9499-4

Caruso BA, Clasen TF, Yount KM, Cooper HL, Hadley C, Haardörfer R (2017) Assessing women's negative sanitation experiences and concerns: the development of a novel sanitation insecurity measure. Int J Environ Res Public Health 14(7):755. https://doi.org/10.3390/ijerph14070755

Copernicus Marine Service. (2024) Global sea level rise monitoring indicator. European Union. https://marine.copernicus.eu

Ghosh S (2021) Community-based adaptation strategies for climate resilience in Indian coastal villages. J Environ Manage 287:112325. https://doi.org/10.1016/j.jenvman.2021.112325

Howard G, Calow R, Macdonald A, & Bartram J (2020) Climate change and water and sanitation: Likely impacts and emerging trends for action. Annual Review of Environment and Resources, 45, 253–276. https://doi.org/10.1146/annurev-environ-012320-083852

Hossain MA, Begum M, Akhtar MN et al. (2025). Predicting water scarcity in northern Bangladesh using deep learning and climate data. npj Clim Atmos Sci 8, 348. https://doi.org/10.1038/s41612-025-01030-y

India Meteorological Department (2023) Climate Data Services Portal. https://cdsp.imd.gov.in

Intergovernmental Panel on Climate Change. (2022). Climate change 2022: Impacts, adaptation, and vulnerability. Contribution of Working Group II to the Sixth Assessment Report of the Intergovernmental Panel on Climate Change (H.-O. Pörtner et al., Eds.). Cambridge University Press. https://doi.org/10.1017/9781009325844

Jeandron, A., Saidi, JM, Kapama, A, Burhole, M., Birembano, F., Vandevelde, T., & Ensink, JHJ (2019). Water supply interruptions and waterborne disease outbreaks in low-income settings. PLoS Medicine, 16(5), e1002756. https://doi.org/10.1371/journal.pmed.1002756

Ministry of Jal Shakti (2021) Annual report 2020–21. Government of India

Mishra BK, Kumar P, Saraswat C, Chakraborty S, & Gautam A (2022) Solar-powered decentralized water treatment systems for salinity-affected delta regions: A case study from the Mekong Delta. Water, 14(3), 412. https://doi.org/10.3390/w14030412

Rahman, MS, Islam ARMT, Bodrud-Doza M, & Muhib MI (2021) Salinity intrusion and its impact on public health in coastal Bangladesh. Environmental Health Insights, 15, 1–12. https://doi.org/10.1177/117863022110xxxx

Seto KC, & Fragkias M. (2019) Mangrove protection and coastal resilience in the face of climate change. Current Opinion in Environmental Sustainability, 36, 83–89. https://doi.org/10.1016/j.cosust.2018.11.006

Sengupta, A., Mukherjee, S., & Pal, S. (2020). Community-based groundwater recharge and climate resilience in semi-arid India. Sustainability, 12(18), 7456. https://doi.org/10.3390/su12187456

Sommer M, Caruso BA, Sahin M, Calderon T, Cavill S, Mahon T, Phillips-Howard PA (2016) A time for global action: addressing girls' menstrual hygiene management needs in schools. PLoS Med 13(2):e1001962. https://doi.org/10.1371/journal.pmed.1001962

UNICEF & World Health Organization (2019) Progress on household drinking water, sanitation and hygiene 2000–2017: special focus on inequalities. UNICEF and WHO

United Nations Educational, Scientific and Cultural Organization (2021) Water security and climate change: Regional perspectives from small island developing states. UNESCO Publishing

United Nations Children's Fund. (2022). Water under fire: UNICEF global water security report 2022. UNICEF

United Nations Children's Fund (UNICEF), & World Health Organization (2019) Progress on household drinking water, sanitation and hygiene 2000–2017: Special focus on inequalities. WHO & UNICEF Joint Monitoring Programme (JMP).

UN Women (2021) Explainer: How climate change disproportionately impacts women and girls. UN Women. https://www.unwomen.org

UN-Water (2023) United Nations world water development report 2023: Partnerships and cooperation for water. UNESCO

World Health Organization (2023) Climate change and health: Key facts. WHO. https://www.who.int/news-room/fact-sheets/detail/climate-change-and-health

Chapter 18
Traditional Ecological Knowledge and Climate Resilience: A Study of PVTGs in the Western Ghats

K. Jayarajan, Sreelu Sreepadi, A S Sana, T. Sreelesh, and V Lakshana

Abstract Traditional Ecological Knowledge (TEK) plays a crucial role in enhancing climate resilience among indigenous communities, the Particularly Vulnerable Tribal Groups (PVTGs) in the Western Ghats region of India. This study explores the TEK and adaptive practices of the Cholanaickan communities using focus group discussions centred on five thematic variables: environmental perception, traditional practices, biodiversity knowledge, risk awareness, and threats to knowledge systems. The findings reveal the persistence of traditional practices such as water storage, season-responsive water use, and mixed cropping. However, concerns arise regarding the erosion of TEK among youth and the lack of institutional support. The study employs descriptive statistics and multivariate analyses to identify patterns and relationships among the variables. Results underscore the effectiveness of TEK in mitigating climate impacts and the need for integrating indigenous knowledge with scientific approaches. The study also highlights the challenges posed by the marginalization of TEK in formal decision-making processes and the importance of inclusive, community-based frameworks. Policy recommendations emphasize the need for empowering indigenous communities as key stakeholders in climate adaptation strategies and the development of culturally appropriate capacity-building initiatives. Future research should prioritize longitudinal studies, comparative analyses across PVTGs, and participatory action research to ensure tangible, community-driven outcomes. The findings contribute to the development of sustainable and locally grounded approaches to climate resilience that respect and preserve the intellectual and cultural rights of indigenous communities.

K. Jayarajan (✉) · A. S. Sana · V. Lakshana
Department of Geography, Government College, Chittur, Palakkad, Kerala, India

Sreelu Sreepadi
HSST Geography, Kozhikkode, Kerala, India

HSST Geography, GHSS Kayanna, Kozhikkode, India

T. Sreelesh
Department of Geography, Government College, Nilambur, Kerala, India

S. Mandal et al. (eds.), *Climate Change Vulnerability*, Climate Change Management, https://doi.org/10.1007/978-3-032-16783-5_18

Introduction

Climate change disproportionately affects Indigenous communities worldwide, threatening their traditional ways of life and cultural practices. Many Indigenous peoples rely on natural resources and ecosystems that are increasingly vulnerable to environmental shifts (Dogar and Almazroui 2022). As a result, these communities face challenges in maintaining their livelihoods, food security, and spiritual connections to the land. Climate change is expected to have significant impacts on the biodiversity and ecosystems of the Western Ghats (Dogar et al. 2019). Rising temperatures and changing precipitation patterns may lead to shifts in species distributions and alterations in vegetation communities. These changes could potentially disrupt delicate ecological balances and threaten endemic species found nowhere else on Earth. Climate change poses significant challenges to Particularly Vulnerable Tribal Groups (PVTGs) due to their heavy reliance on natural resources and traditional livelihoods. These communities often lack the adaptive capacity and resources to cope with changing environmental conditions. As a result, PVTGs face increased risks of food insecurity, water scarcity, and displacement from their ancestral lands.

Indigenous communities have developed unique strategies for adapting to climate change over generations. These strategies are often based on traditional ecological knowledge and sustainable practices that have proven effective in maintaining ecological balance. Studying and incorporating these indigenous approaches into modern climate resilience efforts can provide valuable insights and innovative solutions for addressing global environmental challenges.

The Western Ghats, a biodiversity hotspot in India, is home to several Particularly Vulnerable Tribal Groups (PVTGs) who have lived in harmony with nature for centuries. These indigenous communities possess a wealth of Traditional Ecological Knowledge (TEK) that has been passed down through generations. Their intimate understanding of local ecosystems, weather patterns, and natural resource management practices offers valuable insights for climate change adaptation and resilience strategies. Traditional Ecological Knowledge (TEK) plays a crucial role in enhancing climate resilience and adaptive capacity of indigenous communities. TEK encompasses a wide range of practices, including management of customary land and resources, social networks, and cultural traditions that help communities observe and respond to local climate change (Hosen et al. 2020). This knowledge system is particularly valuable for fostering adaptation and resilience to environmental and climate change, as it provides practical tools for communities to adapt to climate risks and promotes socio-ecological resilience (Dogar et al. 2017a, 2017b; Dogar and Sato 2018, 2019; Hosen et al. 2020; Mcmillen et al. 2016). While there is a generalized worldwide trend of TEK erosion, substantial pockets of TEK persist in both developing and developed countries (Gómez-Baggethun et al. 2013). A common trend observed is hybridization, where traditional knowledge merges with novel forms of knowledge and technologies to create new knowledge systems (Gomez-Baggethun et al. 2013). However, the dynamics of TEK loss in response to ecosystem change and disruption to social-ecological systems remain poorly understood (Turvey et al. 2018).

TEK serves as a reservoir of experiential knowledge that can provide important insights for designing adaptation and mitigation strategies to cope with global environmental change (Gomez-Baggethun et al. 2013). It enables pastoralists and other indigenous communities to navigate social-ecological changes, thereby increasing the sustainability of their practices and fostering social-ecological resilience (Sharifian et al. 2022). By facilitating the combination of experiential with experimental knowledge and fostering complementarity of different knowledge systems, TEK can increasingly contribute to more resilient social-ecological outcomes in natural resource management (Prober et al. 2011).

Traditional Ecological Knowledge plays a crucial role in helping indigenous communities adapt to and build resilience against climate change impacts. TEK encompasses a wide range of practices, including sustainable land and resource management, food security strategies, and social networks that enable collective action during calamities (Hosen et al. 2020). Indigenous communities often observe and respond to local climate changes through their TEK, which helps them manage customary lands and resources to ensure food and resource security (Hosen et al. 2020). The perception of time and climate change can differ between Western science and indigenous knowledge systems. Indigenous communities tend to focus on relationality, noticing interactional changes among humans and other species, and linking changes to a broader suite of socio-political phenomena, including the long arc of colonialism (Chisholm Hatfield et al. 2018). This unique perspective can provide valuable insights for climate change adaptation strategies. While specific information about PVTGs in the Western Ghats is not available, the importance of TEK in building climate resilience is well-documented across various indigenous communities worldwide. Preserving and integrating TEK into climate change adaptation policies could be crucial for supporting vulnerable tribal groups in facing environmental challenges. Further research specific to the Western Ghats region would be necessary to understand the unique TEK and climate resilience strategies of PVTGs in that area. Traditional ecological knowledge (TEK) arose from thousands of years of direct interaction with nature. Traditional ecological knowledge is also known as indigenous knowledge or native science. This comprises peasant, indigenous, and rural communities' knowledge, practice, and beliefs that have evolved via cultural transmission and are more helpful in improving the effectiveness of species and habitat decision making (Withanage and Lakmali 2023).

The term "tribe" refers to a socially cohesive unit associated with a territory (Haseena 2014). 'Adivasi,' is a term for tribes commonly used in India. It originated from the Sanskrit language: 'adi' meaning 'original' and 'vasi' meaning 'resident'. The term "indigenous people" also refers to tribes and is extensively used worldwide (Peters and Mika 2017). There are 476 million indigenous people globally, scattered across 90 nations, who belong to over 5000 indigenous peoples and speak over 4000 languages (Amnesty International 2022). Indigenous people represent about 5% of the world's population, and around 80% of people live in Asia, Africa, Latin America, and the Caribbean (LAC) (Congressional Research Service 2021).

Particularly Vulnerable Tribal Groups (PVTGs) in India

The tribal communities in India are particularly susceptible to challenges due to low literacy rates, limited access to technology, and economic difficulties. Certain subgroups have remained isolated and have not benefited from governmental social programs. In 1973, the Dhebar Commission identified Primitive Tribal Groups (PTGs) as a distinct category for the less developed tribal groups. In 2006, the Government of India reclassified these groups as Particularly Vulnerable Tribal Groups (PVTGs). According to MoTA (2014), seventy-five tribal groups have been designated as PVTGs based on their social, economic, and educational conditions. These PVTGs employ pre-agricultural technologies, exhibit stable or declining populations, have extremely low literacy rates, and rely on a subsistence economy (SCSTRTI 2015). PVTGs are located across 18 states and one union territory (Revised Scheme of Development of Particularly Vulnerable Groups 2015). Among the 75 listed PVTGs, Odisha has the highest number with 13 groups, followed by Andhra Pradesh with 12, Bihar including Jharkhand with 9, Madhya Pradesh including Chhattisgarh with 7, Tamil Nadu with 6, and both Kerala and Gujarat with five groups each (Census 2011).

Scheduled Tribes and Particularly Vulnerable Tribal Groups in Kerala

Kerala is home to 4,84,839 tribal individuals across 1,19,788 family units, representing 1.43 percent of the state's total population (Rajasenan et al. 2019). The tribal communities in Kerala predominantly reside in the forests and mountainous regions of the Western Ghats, adjacent to Karnataka and Tamil Nadu (KIRTADS 2017). Among the 36 officially recognized communities, the Paniyan tribe is the most populous, comprising 88,450 individuals or 18.24% of the state's total tribal population. Six additional tribes, namely Malai Arayan (33,216), Mavilan (30,867), Kurumans (24,505), Muthuvan (23,746), Irular (23,721), and Kanikkar (21,251), collectively account for 57.94% of the entire tribal population (Census 2011). The Cholanaickan tribe is the smallest, with a population with the overall tribal population. The sex ratio of the total Scheduled Tribe population in Kerala is 1035, indicating a majority of females and significantly exceeding the national average. Wayanad district has the highest concentration of Scheduled Tribe population (18.5%), followed by Idukki (4.8%), Palakkad (1.7%), and Kasaragod (3.7%) (STDD 2013). A total of 78% of Scheduled Tribe families in Kerala are settled in five districts: Wayanad, Idukki, Palakkad, Kasaragod, and Kannur. The district of Alappuzha has the lowest Scheduled Tribe population in the state, contributing 0.3% to the total district population.

Five Particularly Vulnerable Tribal Groups (PVTGs) are prevalent in Kerala., namely Cholanaickan, Kadar, Kattunayakan, Kurumbas, and Koraga. There are

6771 PVTG families within a population of 26,273, comprising 13,073 men and 13,200 women (STDD 2013). PVTGs constitute 6.17% of the total Scheduled Tribes in Kerala. The Cholanaickan tribe is located in the Ghats region of the Karulai and Vazhikkadavu forest ranges in the Nilambur valley of Malappuram district. The Kadar, also referred to as Wayanad Kadar, are found in Thrissur, Palakkad, and Kozhikode, with a predominant presence in the Wayanad district, accounting for 0.608% of the overall Scheduled Tribe population. The Koragas are situated in the Kasaragod district and the surrounding areas of Karnataka state, comprising approximately 0.326% of 390 households. The Kurumbas primarily reside in the Attappady and Palakkad districts (KIRTADS 2016).

The Cholanayikans the Cave men of Kerala" inhabit in the Reserve forests of Karulai Range, and the Chungathara Range of the vested forests of Nilambur valley. Particularly during the Monsoon, they are food gatherer. Cholanaikan, the cave man of Kerala still subsists exclusive on food gathering and honey collection. Cholanaikan, of Kerala is one of the most primitive communities with a strength of 409 persons, who normally resides in caves and do not know any agricultural practices. The name Shola Naikan"s first appeared in Edgard Thurston's work (1975). He mentioned it as a synonym for the Katunaykans. The natural ecology of the Cholanaickan has been largely encroached upon by the migrant plainsmen and brought within the threshold of the cultural heritage of the peasant society. However, despite their regular cultural contact with the non-tribal migrant people, the Cholanaickan did not get an acceptable status in the cultural milieu of the local peasant society. As a matter of fact, they still practice primitive economy and adopt a nomadic life (Mathur 1996). The Cholanaickan refer to themselves as "Malanaikan" or "Cholanaickan," a name that means "kings of the forest." The term "Chola" or "shoals" refers to the deep evergreen forest, and "naikan" means king. They are called Cholanaickan because they inhabit the inner parts of dense forests. According to tradition, they were once warriors who migrated from Mysore in Karnataka along with a king. More than a century ago, due to war in the region, they are believed to have taken refuge in the forests. The Cholanaickan are generally of short stature with sturdy, well-built bodies. Their complexion ranges from dark to light brown. They typically have round or oval faces with a depressed nasal root, a medium nasal bridge, and a straight profile. Their lips are thin to medium, and their hair is usually long and curly. Both men and women are short in height and share similar physical features. They live in rock shelters known as "Kallu alai" or in temporary campsites made of leaves. They reside in small groups consisting of two to seven primary families. Each group is called a "Chemmem." The Cholanaickan strictly follow rules established by their ancestors to maintain the territories of each Chemmem. Their dialect is a mixture of Kannada, Tamil, and Malayalam, and it has no written script. The Chemmem are widely scattered across forest ranges. The Cholanaickan depend on food gathering, hunting, and the collection of minor forest produce for their livelihood. They use various tools, weapons, household articles, clothing, and ornaments in their daily lives. Rice is their staple food, supplemented by tubers, seeds, fruits, and meat. Their livelihood is entirely dependent on the forest, and the collection and sale of minor forest produce serve as their primary source of income.

Many traditional customs, practices, and taboos continue to be observed within the community. The habitat of the Cholanaickan reflects their close relationship with nature. Often referred to as the "cavemen of Kerala," they live in different parts of the forest within well-defined territories called Chemmem. The boundaries of each Chemmem are usually marked by natural features such as rivers or hills. Patrilineally related groups of two to ten families live in rock shelters (alai) and huts (manai) located along the banks of rivers and streams. The alai used for habitation are typically rocks that open toward the river and are naturally protected, offering safety from wild animals such as elephants.

The Kattunayakan, recognized as the largest Particularly Vulnerable Tribal Group (PVTG) in Kerala, predominantly reside in the dense forest and hilly regions. They are also referred to as Then Kurumar, Kaattu Kurumar, and Jenu Kurumbar (Tharuvana 2020). Their subsistence is primarily derived from forest-based activities, including food gathering, the collection of Non-Timber Forest Products (NTFP), agriculture, forest labour, and small-scale cultivation within their settlements (Kakkoth 2005). These indigenous communities are regarded as the custodians of the forest. The term 'Kattunayakan' translates to 'Lord of the Forest,' originating from the words 'katu' meaning forest, and 'nayakan' or 'naicken' meaning lord (Kakkoth 2005). The Kattunayakan tribes in Kerala are distributed across the districts of Wayanad, Kozhikode, Malappuram, Palakkad, and Idukki (SCSTRTI 2015). The tribe is descended from a primitive hunter-gatherer group that traditionally inhabited hilltops and caves, and they are known for their skills in honey collection and hunting (Shylajan and Mythili 2007). Of the total 5137 Kattunayakan families, the majority are located in Wayanad (4369), followed by Malappuram (517), Palakkad (218), and Kozhikode (32). These families are settled across 49 Gram Panchayats, with 22 in Wayanad, followed by Malappuram (16), Kozhikode (5), Palakkad (5), and Idukki (1) districts.

Methods

This study investigates Traditional Ecological Knowledge (TEK) and climate resilience among the Cholanaickan communities, Particularly Vulnerable Tribal Groups (PVTGs) residing in the Western Ghats region of Kerala, India. According to the Socio-Economic Status Report (2013), there are 409 Cholanaickan-inhabited areas distributed across multiple panchayaths. In Karulai Panchayath, the settlements include Mancheri, Mannala, Thalippuzha, Pulimunda, Kuppamala, Nagamala, Panappuzha, Vilakkumala, Meenmutty, Poochappara, and Tannipoyil. In Vazhikkadavu Panchayath, the settlement is located in Alakkal, while in Amarambalam Panchayath, it is situated in Achanala. For the present study, a stratified sampling method was adopted to ensure adequate representation across these settlements. A total of 75 samples, constituting 18% of the population, were collected.These habitations are geographically remote and difficult to access, often requiring trekking distances of 10–15 kilometres. During the monsoon season, swollen rivers frequently isolate these areas, further restricting accessibility. Seasonal mobility characterizes the lifestyle of the Cholanaickan community: they reside in temporary shelters along riverbanks and caves in the interior forest areas.

Data collection was carried out primarily through focus group discussions (FGDs), which provided an in-depth platform to explore community-based ecological knowledge, perceptions of climate change, and adaptive practices. The FGDs were structured around five thematic variables, each reflecting essential dimensions of TEK (Fig. 18.1).

1. Environmental Perception and Ecological Indicators
2. Traditional Practices and Resource Management
3. Biodiversity Knowledge and Conservation Ethics
4. Risk Awareness and Adaptive Capacity
5. Threats to Traditional Knowledge Systems

Each variable was operationalized through a series of five culturally pertinent and clearly articulated statements, enabling participants to indicate their levels of agreement or disagreement. These variables were derived from initial ethnographic insights and refined in collaboration with local resource persons. The collected data were coded and entered into SPSS version 30. Descriptive statistics were employed to summarize community responses, while multivariate analyses, including factor analysis and cluster analysis, were utilized to identify patterns and relationships among the variables. This analytical approach facilitated the examination of how

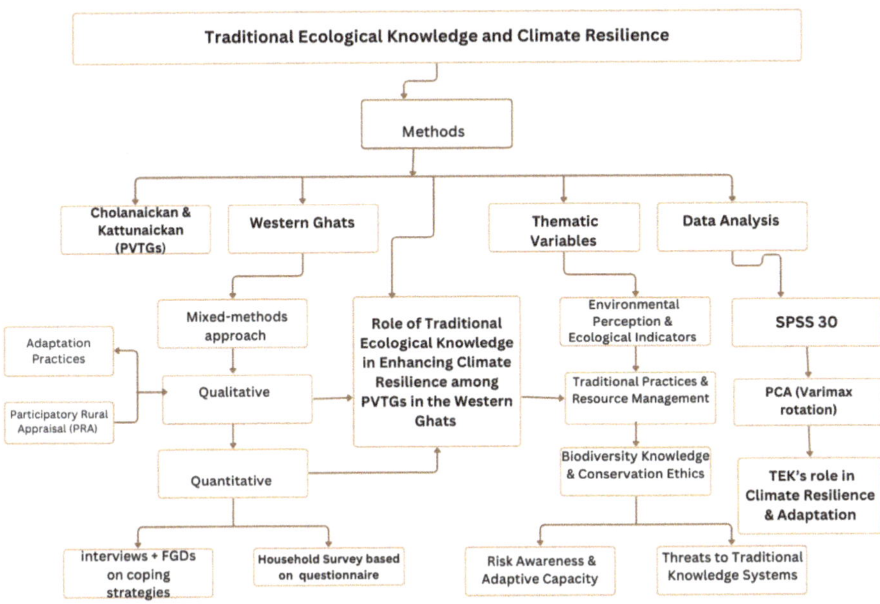

Fig. 18.1 Graphical representation of the overall methodology of traditional ecological knowledge and climate resilience

Traditional Ecological Knowledge (TEK) contributes to adaptive capacity in the context of climate change, as well as how socio-cultural and policy dynamics influence the persistence of these knowledge systems.

Results

Traditional Ecological Knowledge (TEK) of the Cholanaickan in Climate Risk Context

The results of the study provide a comprehensive understanding of how the Cholanaickan communities utilize Traditional Ecological Knowledge (TEK) to perceive, interpret, and respond to climate-related risks (Tables 18.1, 18.2, 18.3, 18.4, 18.5, 18.6, and 18.7). Drawing upon observations of natural indicators—such as plant phenology, animal and insect behaviour, wind patterns, and lunar cycles—these communities have developed intricate systems for forecasting weather, managing resources, and sustaining livelihoods in fragile forest ecosystems. The analysis highlights both the strengths of these practices, including sustainable water management, cultural-ritual taboos, and biodiversity preservation, as well as the emerging challenges posed by modernization, policy neglect, and intergenerational gaps in knowledge transfer. By combining descriptive insights with factor analysis, the results illustrate that TEK operates as a multifaceted and dynamic framework, balancing ecological observation with cultural values, while simultaneously adapting to the pressures of climate change and socio-economic transitions.

Traditional Ecological Knowledge (TEK) of the Cholanaickan—Climate Risk Adaptation Framework

Adaptive Strategies and Cultural-Spiritual Resilience of the Cholanaickan

The Traditional Ecological Knowledge (TEK) of the Kattunaickan and Cholanaickan communities exemplifies how indigenous knowledge systems function as essential frameworks for climate risk adaptation. Their perception of time is structured around natural indicators such as the position of the sun, the length of shadows, and seasonal changes, thereby replacing mechanical timekeeping with ecological rhythms. Seasonal awareness is informed by bird calls, insect behaviour, and plant phenology, including the flowering of neem, jackfruit, and mango, which signal shifts in weather and agricultural cycles. Sensory acuity—through smell, taste, and sound—facilitates survival by enabling the identification of edible plants,

Table 18.1 Traditional ecological knowledge (TEK) of the Cholanaickan in climate risk context

Category	Description	Indicators
Time perception	Knowledge of time based on natural cues rather than mechanical clocks	Measuring time by shadow size (noon = smallest shadow)—planning activities (fruit, bamboo collection) by sun position, heat intensity, and shadow length - "Neru" means sun; divided into five parts: (1) *Ottheru*—Morning, (2) *Ucheru*—Noon, (3) *Chanjeru*—Evening, (4) *Eedu*—Night, (5) *Oomdhchokku*—Midnight
Seasonal awareness	Recognition of seasons through ecological and sensory cues	Cricket and bird sounds differ by season—yellow butterflies appear at monsoon's end, signalling dry weather—Three main seasons: *Kodaikkalai* (rainy), *Cheetheikalai* (cold), *Bechekalai* (summer)—Transitional periods: *Paatukalai, Maruthuvkalai, Jhenukalai, Benguuvkalai*—Maruthu (*Terminalia tomentosa*) flowering = start of summer—Venga (*Pterocarpus marsupium*) flowering = seasonal shift—mushroom blooming = onset of monsoon—"Mootiikai" (*Baccaurea courtallensis*) blooming = onset of summer
Sensory acuity	Use of smell, taste, and sound for survival, safety, and foraging	Smelling/hearing animals from a distance for hunting/ avoiding danger—Identifying edible mushrooms by smell (elders train children)—Tasting honey to identify plants
Phenological knowledge	Tracking plant life cycles for climate prediction and livelihood planning	Neem flowering = pre-monsoon—Jackfruit fruiting = summer—Mango blooms= rising temperature and rains
Animal behaviour	Reading animal actions as weather/ environmental signals	*Pattukappei* frog calls = rain—Ground-nesting birds moving higher = flood warning—Animal silence = storm warning—*Cheedalu* (Malabar whistling thrush) = summer indicator—*Chuchuulal* (Hoopoe) = accident warning—Cicadas (*Oore*): 1. *Kodaioore* = post-monsoon onset 2. *Maluluoore* = winter–summer transition 3. *Bellioore* = buzzing after sunset
Insect behaviour	Insects as predictors of rain and floods	Winged termites enmasse = monsoon onset—Black ants moving eggs = flood warning—*Jyanukwinni* (*Apis dorsata*—honey bee) high honey production before monsoon
Wind patterns	Local winds named and tracked to anticipate weather	*Kadakathu kaatu* (dry wind) = seasonal shift/poor rains
Cloud formations	Cloud observation as short-term rain forecast	Tall cumulus from specific directions = rain in hours/days
Cultural embedding	TEK is transmitted orally through stories, songs, and rituals	Rituals aligned with fruiting/flowering cycles—*Beede* (musical instrument) made with honeybee wax
Climate adaptation	Immediate responses to ecological signals for survival and livelihood security	Covering food, shifting shelters, or moving livestock based on signals—Elephant movement before landslides (*Naadu Barddh*)—Yellow butterflies (*Aresanu Bandu*) = end of monsoon

Table 18.2 Wind patterns, cloud formations, and insect indicators

Sub-category	Description	Examples/observations	Community actions
Wind patterns	Local winds are named and linked to weather outcomes	*Kadakathu kaatu* (dry wind) = dry spell moist winds = incoming rain	Modify travel routes prepare farming/ collection tasks secure shelters
Cloud formations	Cloud shape, height, and direction interpreted for short-term weather forecasting	Tall cumulus clouds from certain directions = imminent rainfall	Delay outdoor work collect/store water adjust forest-based activities
Insect behaviour	Insects act as indicators for rain or flood risks	Winged termites swarming = onset of monsoon ants moving eggs = flood warning	Cover/stores food secure huts/shelters move livestock shift camps to safer ground

Table 18.3 Seasonal calendars

Sub-category	Description	Indicators/markers	Usage/purpose
Lunar and ecological cues	Calendar guided by moon phases and environmental changes	Moon phases and humidity—Bird migrations—Flora/fauna changes —Position change of the sun (*Neru*)—Lunar month (*Dingeikalai*) used for tracking menstrual cycle and delivery time—*Eedudingai* = new moon	Identifying weeks, months, and seasonal transitions—Planning ceremonial timings—Guiding personal and community life cycles
Forest livelihood timing	Seasonal cycles align forest-based livelihood activities	Flowering of *Terminalia*—Ground moisture = yam collection—Mushroom flush post-rain—Specific tuber crops: *Kavale* (used in marriages), *Chuulai* (used for purification after delivery or menstrual period)	Honey collection—Wild yam harvesting—Herb gathering—Ensuring sustainable use of resources
Cultural ceremonies	Rituals and festivals are timed with ecological markers	Pre-monsoon = rain-invoking rituals—Full moon = ancestor worship—Solstices = forest deity offerings—Seasonal festival *Beidh* at onset of summer (day of no hunting)—Hills revered as gods (*Maladaivai*)	Aligning spiritual and social life with ecological rhythms—Strengthening cultural identity and community cohesion
Climate sensitivity	Oral knowledge adapts to ecological changes	Unpredictable rainfall—Altered flowering/blooming times	Elders reinterpret seasonal timing—younger generations adapt through oral transmission and flexible practices

Table 18.4 Water management practices

Sub-category	Description	Natural indicators	Practices/actions
Water conservation	No formal irrigation; water sourced and preserved via TEK	Location of sacred groves, stream behaviour, vegetative density	Build check dams, maintain forest springs, preserve sacred groves as hydrological buffers
Aquifer identification	Identification of underground water through flora and fauna behaviour	Moisture-loving trees (Ficus, Myristica), moss growth, animal trails to hidden water points	Use for seasonal water access; protect and respect natural water sources

Table 18.5 Adaptive strategies in changing climate

Strategy	Description	Examples/actions
Water practice revival	Reintroduction of traditional water conservation methods	Reactivating old check dams, expanding sacred groves
Rotational foraging and resting	Allowing land to regenerate naturally	Avoiding overuse of patches, letting soil and water sources recover

Table 18.6 Spiritual and cultural practices

Sub-category	Theme	Description/practice	Purpose/effect
Weather-related rituals	Rain invocation	Rituals conducted during droughts or delayed monsoons; seasonal celebration *Beidh* at summer onset, where no hunting of animals or birds is allowed	Appeasement of nature spirits; reinforcement of community cohesion
Forest and animal health	Ecosystem balance	Ceremonies during disease outbreaks or ecological stress; silent periods observed for forest regeneration. Belief that killing snakes is a sin; snake poison (*aavu maddu*) recognized as harmful. Bird watching used for signals (e.g., *chuchulal* hoopoe signaling storms)	Restores ecological balance; halts overexploitation; enhances ecological awareness
Ecological taboos	Cultural restrictions	Restrictions on cutting sacred trees, plucking flowers during bloom, or entering forests during nesting. Elephant called *dodai janthuve* (equal to god). Honey collection banned in monsoon to protect bees. Hunting/killing animals or birds only for food	Protects biodiversity and regeneration cycles; promotes respect for species and ecosystems
Ecological ethics	Subsistence and moderation	Harvest limited to need; waste discouraged. Mushrooms preserved for junglefowl (*Gallus sonneratii*)	Encourages sustainability, moderation, and ecological balance
Reciprocity with nature	Spiritual exchange	Offerings (flowers, smoke, rice) made to forest spirits as thanks for resources	Reinforces humility, gratitude, and responsible use of natural resources
Collective consent	Communal decision-making	Elders consulted before hunting or land clearance	Ensures actions are ecologically, socially, and spiritually appropriate

Table 18.7 Oral histories and intergenerational knowledge

Sub-category	Theme	Content/practice	Function/impact
Memory of past climate events	Community climatic archives	Stories and songs about floods, droughts, or storms	Provide historical benchmarks; guide preparedness for future events
Songlines and seasonal markers	Ecological learning through music	Songs linking flowering, animal calls, and rain patterns. Example: *Jyanupattu* (honeysong) sung to prevent bee stings or pain	Ingrain ecological patterns in memory; support timing of resource use
Elders as climate archives	Long-term landscape observers	Elders recall forest thinning, river drying, species loss, or cold fogs	Provide baseline data for monitoring ecological change
Intergenerational dialogue	Experiential learning	Storytelling, forest walks, and rituals used to teach youth	Embed restraint, reciprocity, and continuity of TEK
Recognizing climatic shifts	Local ecological monitoring	Observing changes in monsoon winds, flowering cycles, and fog disappearance. Example: Fewer animal attacks noted at noon (*ucheru*)	Support adaptive farming, foraging, and ritual scheduling
Climate resilience role	Oral knowledge as climate data	Oral histories reflect micro-changes often overlooked by instruments	Fill scientific data gaps; provide culturally appropriate insights for policymaking and adaptation

mushrooms, and animals. Animal and insect behaviours, such as frog calls, cicada buzzing, or ant movements, serve as predictive markers for rain, floods, or storms, prompting timely community responses like relocating livestock or securing shelters. Similarly, local winds and cloud formations provide short-term forecasts that guide farming, water storage, and travel. TEK is embedded in cultural practices, rituals, and taboos, ensuring that ecological knowledge is transmitted through oral traditions, festivals, and spiritual beliefs, such as the summer onset celebration "Beidh", which prohibits hunting. Beyond observation, TEK fosters climate resilience through adaptive strategies, including rotational foraging, water conservation, and aquifer identification. Oral histories and songs function as community climate archives, offering intergenerational learning and ecological ethics that promote restraint and reciprocity. Collectively, these practices demonstrate that TEK is not only a cultural heritage but also a dynamic, evolving system that integrates ecological understanding with practical adaptation, making it highly relevant in addressing contemporary climate challenges. Building on this factor analysis, it becomes evident that the Cholanaickan's extensive traditional ecological knowledge is not fragmented but operates as an integrated system of observation, interpretation, and response. Their perceptions of climate-related changes—such as altered rainfall patterns, shifting flowering cycles, and irregular animal behaviours—are consistently linked to adaptive practices that ensure both immediate survival and long-term sustainability. For instance, changes in insect behaviour or cloud formations are swiftly translated into decisions about travel, food storage, and foraging, while oral calendars and ritual cycles are recalibrated by elders to account for emerging climatic irregularities. Such adaptive responses reveal not only the resilience of their ecological worldview but also the dynamic capacity of TEK to evolve under climate stress. The factor analysis thus underscores how ecological indicators, cultural practices, and adaptive strategies converge to create a holistic resilience framework, positioning the Cholanaickan community's knowledge as a valuable complement to scientific climate monitoring and adaptation planning.

Table 18.8 shows the analysis of the variables of TEK (Traditional Ecological Knowledge) among 75 respondents highlights both the strengths and challenges of indigenous adaptive practices. The results show that awareness of climate variability is quite strong, as a majority reported noticing local weather changes (M = 0.68) and acknowledged the predictive role of elders (M = 0.64). Ecological indicators such as flowering and fruiting patterns (M = 0.61), animal behaviour (M = 0.63), and traditional water practices (M = 0.67) continue to be valued for forecasting and seasonal planning. Water security emerges as a significant area, with high agreement on traditional storage (M = 0.73) and conservation practices, while shifting cultivation, mixed cropping, and indigenous seeds are also recognized for their role in building resilience (M = 0.65, M = 0.51).

Community-level practices, including sacred grove preservation (M = 0.65), biodiversity knowledge (M = 0.68), sustainable harvesting (M = 0.69), and traditional taboos (M = 0.59), demonstrate an embedded ethic of sustainability. At the same time, collective responses such as decision-making (M = 0.44) and reliance on animal or insect behaviour (M = 0.44) showed lower agreement, suggesting possible erosion of these practices. Social dimensions remain important, as women and elders' role in knowledge-sharing during crises (M = 0.60) and the recognition of forest signs for environmental health (M = 0.57) continue to support resilience.

However, modernization poses clear challenges. Many participants expressed that young people are losing interest in TEK (M = 0.68), government policies neglect traditional systems (M = 0.60), and modern education undermines ecological practices (M = 0.68). These findings indicate that while TEK remains central in weather prediction, resource management, and disaster awareness, intergenerational continuity is at risk, making revitalization efforts and policy support essential for sustaining indigenous resilience frameworks.

The Total Variance Explained in the factor analysis of Traditional Ecological Knowledge (TEK) among 75 respondents elucidates the structuring of indigenous adaptive practices around multiple, interrelated dimensions. The extracted factors collectively encapsulate both the enduring strengths of traditional knowledge—such as weather forecasting through ecological indicators, sustainable resource management, and cultural-ritual practices—and the challenges posed by modernization, including declining youth interest and policy neglect. This distribution of variance indicates that TEK is not a singular construct but a multifaceted system, wherein ecological observation, social institutions, and cultural values interact to

Table 18.8 Thematic dimensions of traditional ecological knowledge (TEK) identified through FGDs

Sl no	Variables of TEK	Mean	Std. deviation	Analysis N
Var1	I have noticed changes in local weather patterns over the years	0.68	0.470	75
Var2	Elders in our community can predict weather using traditional methods	0.64	0.483	75
Var3	Flowering and fruiting patterns of local plants help us predict rainfall	0.61	0.490	75
Var4	Bird or insect behaviour is used by our community to foresee climate changes	0.44	0.500	75
Var5	We observe animal migration or behaviour to understand environmental changes	0.63	0.487	75
Var6	We use traditional water practices that respond to seasonal variations	0.67	0.475	75
Var7	Shifting cultivation or mixed cropping helps with drought/flood risks	0.65	0.479	75
Var8	Traditional water collection methods are still in use	0.65	0.479	75
Var9	Indigenous seed varieties are more resilient to climate change	0.51	0.503	75
Var10	Our community stores water traditionally to manage scarcity	0.73	0.445	75
Var11	Community members use forest signs to assess environmental health	0.57	0.498	75
Var12	Sacred groves and forest patches are preserved for ecological reasons	0.65	0.479	75
Var13	Knowledge of local biodiversity helps us adapt to environmental stress	0.68	0.470	75
Var14	We harvest forest resources in ways that ensure future availability	0.69	0.464	75
Var15	Traditional taboos prevent overexploitation of natural resources	0.59	0.496	75
Var16	Awareness of risks related to floods, droughts, and landslides	0.61	0.490	75
Var17	Local practices/rituals protect from natural disasters	0.65	0.479	75
Var18	Collective decision-making helps respond to climate threats	0.44	0.500	75
Var19	Women and elders share knowledge during crises	0.60	0.493	75
Var20	Young people are losing interest in traditional knowledge	0.68	0.470	75
Var21	Government policies ignore our traditional knowledge	0.60	0.493	75
Var22	Modern education reduces traditional ecological practices	0.68	0.470	75

support resilience. Concurrently, the relatively lower loadings on collective decision-making and modern influences underscore vulnerabilities in the continuity of these practices. Thus, the variance explained reflects both the robustness of TEK as an adaptive framework and the pressures it faces in contemporary contexts. The results are further corroborated by the scree plot, which confirmed the factor structure, while the extraction method of Principal Component Analysis (PCA) with Kaiser Normalization and rotation ensured a clearer interpretation of the underlying dimensions of TEK.

Table 18.9 shows the component analysis reveals that nine components have initial eigenvalues greater than 1, following Kaiser's criterion, and together they account for approximately 62.1% of the total variance in the dataset. This suggests a moderately complex underlying structure, with multiple dimensions contributing to the observed data patterns. After extraction and varimax rotation, the variance explained becomes more evenly distributed across components. The rotated solution shows that the first component explains 8.51%, the second 8.35%, and so on, with the ninth component contributing 5.73%. The cumulative variance explained after rotation reaches 62.1%, indicating a reasonably good representation of the original variables across the nine components. However, the relatively low percentage of variance explained by each individual factor (none exceeding 9%) and the slow decline in eigenvalues suggest that the data lacks strong, dominant factors (Fig. 18.2).

The component analysis reveals that nine components have initial eigenvalues greater than 1, following Kaiser's criterion, and together they account for approximately 62.1% of the total variance in the dataset. This suggests a moderately complex underlying structure, with multiple dimensions contributing to the observed data patterns. After extraction and varimax rotation, the variance explained becomes more evenly distributed across components. The rotated solution shows that the first component explains 8.51%, the second 8.35%, and so on, with the ninth component contributing 5.73%. The cumulative variance explained after rotation reaches 62.1%, indicating a reasonably good representation of the original variables across the nine components. However, the relatively low percentage of variance explained by each individual factor (none exceeding 9%) and the slow decline in eigenvalues suggest that the data lacks strong, dominant factors.

The specified dimension underscores the influence of traditional ecological knowledge (TEK) in enhancing community awareness and resilience to climate-related risks (Table 18.10). Variables such as the capacity of elders to forecast weather through traditional methods, alongside the dissemination of knowledge by women and elders during crises, underscore the importance of intergenerational knowledge systems. These practices enhance local awareness of environmental hazards, including floods, droughts, and landslides. Nevertheless, the pronounced negative loading of the variable indicating a decline in young people's interest in traditional knowledge points to a concerning generational gap in the transmission of this wisdom. This dimension can thus be interpreted as representing "Intergenerational Knowledge and Community-Based Climate Awareness," underscoring both the

Table 18.9 Total variance explained

Component	Initial eigenvalues			Extraction sums of squared loadings			Rotation sums of squared loadings		
	Total	% of variance	Cumulative %	Total	% of Variance	Cumulative %	Total	% of Variance	Cumulative %
1	2.262	10.282	10.282	2.262	10.282	10.282	1.872	8.511	8.511
2	2.071	9.413	19.695	2.071	9.413	19.695	1.836	8.346	16.857
3	1.595	7.252	26.947	1.595	7.252	26.947	1.648	7.490	24.348
4	1.548	7.037	33.984	1.548	7.037	33.984	1.523	6.922	31.270
5	1.508	6.856	40.840	1.508	6.856	40.840	1.434	6.519	37.788
6	1.342	6.100	46.940	1.342	6.100	46.940	1.433	6.515	44.303
7	1.213	5.512	52.452	1.213	5.512	52.452	1.368	6.217	50.521
8	1.091	4.957	57.409	1.091	4.957	57.409	1.288	5.854	56.374
9	1.033	4.696	62.105	1.033	4.696	62.105	1.261	5.730	62.105
10	0.979	4.452	66.557						
11	0.954	4.336	70.893						
12	0.846	3.843	74.737						
13	0.834	3.792	78.529						
14	0.803	3.650	82.179						
15	0.733	3.334	85.512						
16	0.614	2.790	88.302						
17	0.556	2.528	90.830						
18	0.491	2.230	93.060						
19	0.436	1.981	95.041						
20	0.424	1.926	96.967						
21	0.382	1.737	98.704						
22	0.285	1.296	100.000						

Extraction Method: Principal Component Analysis

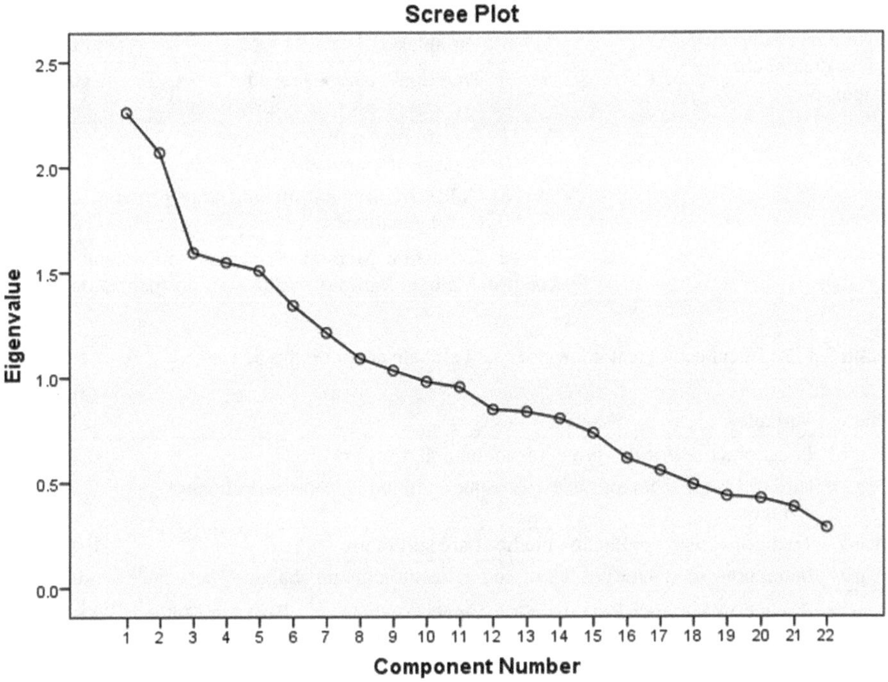

Fig. 18.2 Scree plot shows principal components to retain

Table 18.10 Intergenerational knowledge and community-based climate awareness

Sl no	Variable	Eigen value
Var2	Elders in our community can predict weather using traditional methods	0.589
Var19	Women and elders share knowledge during crises	0.585
Var16	Awareness of risks related to floods, droughts, and landslides	0.574
Var20	Young people are losing interest in traditional knowledge	−0.690

Source: Extraction Method: Principal Component Analysis and Rotation Method: Varimax with Kaiser normalization

value of traditional knowledge in climate adaptation and the urgent need to engage younger generations in preserving and applying this heritage.

This dimension underscores the tensions between traditional knowledge systems and contemporary governance within the framework of environmental stewardship and climate response (Table 18.11). The strong positive loading for the statement that traditional taboos prevent overexploitation of natural resources reflects the role of cultural norms and indigenous practices in promoting sustainable resource management. Additionally, the moderate positive association with the belief that

Table 18.11 Traditional governance vs. policy integration in climate response

Sl no	Variables	Eigen value
Var15	Traditional taboos prevent overexploitation of natural resources	0.655
Var21	Government policies ignore our traditional knowledge	0.497
Var18	Collective decision-making helps respond to climate threats	−0.793

Source: Extraction Method: Principal Component Analysis. Rotation Method: Varimax with Kaiser normalization

Table 18.12 Local ecological knowledge and climate adaptation practices

Var	Variable	Eigen value
Var17	Local practices/rituals protect from natural disasters	0.757
Var4	Bird or insect behaviour is used by our community to foresee climate changes	0.449
Var8	Traditional water collection methods are still in use	0.411
Var9	Indigenous seed varieties are more resilient to climate change	−0.387

Source: Extraction Method: Principal Component Analysis and Rotation Method: Varimax with Kaiser normalization

government policies ignore traditional knowledge points to a perceived marginalization of indigenous wisdom in formal decision-making processes. Conversely, the high negative loading for collective decision-making helps respond to climate threats suggests a disconnect or lack of alignment between traditional practices and inclusive, community-based decision-making frameworks. Collectively, these insights suggest that although traditional systems offer ecological benefits, they are frequently marginalized by policy and may not be effectively integrated into broader participatory governance frameworks.

This dimension underscores the ongoing significance and implementation of traditional ecological practices in community-based climate adaptation strategies (Table 18.12). The strong positive loading for the statement that local practices and rituals protect from natural disasters emphasizes the community's reliance on culturally embedded coping mechanisms to mitigate environmental risks. The use of animal behaviour, such as birds or insects, to anticipate climate changes, and the persistence of traditional water collection methods, further illustrate the embeddedness of local ecological knowledge in everyday life. However, the negative loading on the belief that indigenous seed varieties are more resilient to climate change may suggest a decline in confidence or usage of these traditional agricultural resources, potentially due to external pressures such as the introduction of hybrid seeds or modern agricultural practices. Overall, this dimension captures the value and partial erosion of place-based ecological practices in the context of modern climate challenges.

Table 18.13 shows that dimension centres on how communities perceive and respond to environmental changes, particularly through traditional practices. The

Table 18.13 Ecological perception and traditional water adaptation

Sl no	Variables	Eigen value
Var6	We use traditional water practices that respond to seasonal variations	0.797
Var1	I have noticed changes in local weather patterns over the years	0.450
Var5	We observe animal migration or behaviour to understand environmental changes	−0.535

Source: Extraction Method: Principal Component Analysis and Rotation Method: Varimax with Kaiser normalization

high positive loading on the use of traditional water practices that respond to seasonal variations reflects the adaptive nature of indigenous knowledge in managing natural resources. Similarly, the moderate positive loading for awareness of changes in local weather patterns over the years indicates a strong observational understanding of environmental shifts based on lived experience. However, the negative loading on using animal migration or behaviour to understand environmental changes may suggest a decline in this specific form of ecological observation, possibly due to reduced reliance on or visibility of fauna-based indicators. Overall, this dimension highlights a blend of continuity and decline in traditional climate responsiveness, with water management practices remaining strong, while other ecological cues like animal behaviour may be diminishing in importance.

The three variables presented suggest a complex relationship between traditional ecological knowledge and perceptions of environmental change among the Cholanaickan community. The strong positive loading for Var6 (*We use traditional water practices that respond to seasonal variations*, 0.797) indicates that traditional water management remains a significant and active component of the community's adaptive practices. Var1 (*I have noticed changes in local weather patterns over the years*, 0.450) also loads positively, though moderately, suggesting that individuals are increasingly aware of shifting climatic conditions, likely through direct observation and long-term familiarity with their environment. However, the negative loading of Var5 (*We observe animal migration or behaviour to understand environmental changes*, −0.535) points to a possible decline or shift in reliance on faunal indicators, which may reflect ecological disruptions or a weakening of certain traditional observation systems. Collectively, these variables reveal a dimension of "Ecological Perception and Traditional Water Adaptation", where community members actively draw on culturally embedded practices (like water management) to respond to seasonal changes, while also grappling with evolving environmental cues that may challenge or alter older interpretive frameworks.

This set of variables reveals a nuanced tension between traditional ecological knowledge and the influence of modernity, particularly formal education. Table 18.14 Var22 (*Modern education reduces traditional ecological practices*, 0.773) shows a strong positive loading, indicating a widespread perception that the spread of modern education is leading to a decline in the use of Indigenous ecological knowledge. Var11 (*Community members use forest signs to assess environmental health*, 0.441) and Var12 (*Sacred groves and forest patches are preserved for ecological reasons*, 0.386) suggest that traditional knowledge systems and cultural

Table 18.14 Tensions between tradition and modernity in ecological knowledge

Sl no	Variables	Eigen value
Var22	Modern education reduces traditional ecological practices	0.773
Var11	Community members use forest signs to assess environmental health	0.441
Var12	Sacred groves and forest patches are preserved for ecological reasons	0.386
Var8	Traditional water collection methods are still in use	−0.405

Source: Extraction Method: Principal Component Analysis and Rotation Method: Varimax with Kaiser normalization

Fig. 18.3 Traditional ecological knowledge, intergenerational transmission, and community climate resilience: navigating continuity, adaptation, and modernity

values still play a meaningful role in environmental monitoring and conservation. However, the negative loading of Var8 (*Traditional water collection methods are still in use*, −0.405) may imply that some practical aspects of traditional practices—such as water collection—are being abandoned or are less prevalent. Together, these variables define a dimension that can be titled "Tensions Between Tradition and Modernity in Ecological Knowledge", capturing the community's struggle to maintain cultural and environmental practices in the face of changing educational and social influences (Fig. 18.3).

Discussions

Traditional Ecological Knowledge (TEK) has been shown to be highly effective in enhancing climate resilience across various communities worldwide. TEK systems provide valuable insights for adaptation strategies, fostering social-ecological resilience, and promoting sustainable resource management (Hosen et al. 2020; Mcmillen et al. 2016). Indigenous communities utilize TEK to predict weather events, manage resources, and respond to local climate change impacts (Filho et al. 2023; Kupika et al. 2019). For instance, in Sarawak, Malaysian Borneo, indigenous communities use TEK to manage customary land and resources, ensuring food and resource security in the face of climate-induced challenges (Hosen et al. 2020).

While some traditional coping strategies have decreased in importance over time (e.g., forecasting, storage, and mobility), others have adapted and maintained their relevance (e.g., livelihood diversification, knowledge transmission, and cultural identity) (Tahir et al. 2025; Mcmillen et al. 2016). This highlights the dynamic nature of TEK and its ability to evolve in response to changing environmental conditions. However, despite its acknowledged value, there are still significant gaps in effectively integrating TEK systems within legal frameworks and national adaptation policies, particularly in Africa (Filho et al. 2023). TEK plays a crucial role in climate change adaptation and resilience-building, especially for indigenous, rural, and place-based communities (Parrey et al. 2025; Dorji et al. 2024; Mcmillen et al. 2016). By combining TEK with modern climate change strategies, communities can develop more effective and culturally appropriate adaptation plans (Filho et al. 2023). The integration of TEK in natural resource management and climate change monitoring can lead to more resilient social-ecological outcomes (Prober et al. 2011). However, to maximize the benefits of TEK, there is a need for further research and efforts to bridge it with scientific knowledge while ensuring the holistic integrity of traditional knowledge systems (Dorji et al. 2024; Shawoo and Thornton 2019).

Intergenerational knowledge sharing and community-based approaches play a crucial role in fostering climate awareness and action across different age groups. Research indicates that both older and younger generations can contribute valuable perspectives and experiences to address climate change challenges (Ayalon et al. 2022). Place-based civic science (PBCS) projects involving intergenerational partnerships have shown promise in engaging youth from marginalized communities in environmental understanding and action, helping to build agency and efficacy while addressing emotionally heavy issues like climate change (Gallay et al. 2021).While generational diversity can initially hinder knowledge sharing in workplace settings, creating an intergenerational climate free from ageist attitudes and employing boundary-spanning leadership can facilitate knowledge exchange and improve group performance (Hans et al. 2023). This highlights the importance of fostering inclusive environments for effective intergenerational collaboration on climate issues.

Community-based participatory research (CBPR) approaches have demonstrated success in promoting intergenerational dialogue and cultural knowledge transmission, particularly in indigenous communities (Wexler 2011). Such methods can be adapted to address climate change awareness and action at the local level. Additionally, after-school programs using participatory methods have shown effectiveness in increasing children's climate change knowledge and motivation for action, emphasizing the potential of informal learning environments for children's intergenerational and political influence (Trott 2019). By fostering intergenerational partnerships and community-based initiatives, we can harness collective wisdom and inspire informed local action to address complex challenges like climate change (George et al. 2011).

Traditional Ecological Knowledge (TEK) of indigenous people plays a crucial role in cultural sustainability and addressing governance gaps. TEK encompasses a wealth of practices, beliefs, and knowledge systems that have been developed over generations through direct interaction with the environment (Hosen et al. 2020). This knowledge is particularly valuable for fostering adaptation and resilience to environmental and climate change (Mcmillen et al. 2016). Indigenous communities have observed significant changes in their local environments, such as increased temperatures and uncertain weather patterns, which have impacted their livelihoods. However, they have responded by managing their customary lands and resources to ensure food and resource security, providing examples of sustainable management of terrestrial and inland ecosystems (Hosen et al. 2020). The social networks and institutions of indigenous communities enable collective action, strengthening reciprocal relationships that they rely on during calamities (Hosen et al. 2020). Whereas TEK is recognized as a powerful tool for sustainable resource management, there are challenges in integrating it with Western scientific approaches. The governance of socio-ecological change requires a shift in the understanding of 'innovation' to emphasize contributions to societal goals rather than economic growth or technological modernization (Ludwig and Macnaghten 2019). TEK is a practical tool that helps indigenous communities adapt to climate risks and promotes socio-ecological resilience, upholding social empowerment and sustainable resource management (Hosen et al. 2020). However, to effectively bridge the governance gaps, there is a need for a more inclusive approach that respects indigenous worldviews and self-determination. This may involve developing Indigenous-driven co-governance models that provide better prospects for integrating TEK and Western science for the sustainability of social-ecological systems (Hill et al. 2012).

Local Ecological Knowledge (LEK) and Traditional Ecological Knowledge (TEK) play crucial roles in climate change adaptation practices, particularly for indigenous and rural communities (Parrey et al. 2025). These knowledge systems provide valuable insights into environmental changes, coping strategies, and resilience-building measures (Kupika et al. 2019; Mcmillen et al. 2016). Communities utilize LEK and TEK to observe climate change indicators such as decreasing rainfall, increasing temperatures, and changes in vegetation phenology (Kupika et al. 2019). This knowledge informs adaptation strategies like crop diversification, shifting to drought-resistant crops, and adjusting planting schedules

(Mugambiwa 2018; Yang et al. 2019). For instance, in Taxkorgan, indigenous elders use ecological cues like grass height and migratory bird arrivals to determine the timing of seasonal activities (Yang et al. 2019). While some traditional coping strategies have decreased in importance (e.g., forecasting, storage, and mobility), others have adapted and maintained their relevance, such as livelihood diversification, knowledge transmission, and cultural identity preservation (Mcmillen et al. 2016). This highlights the dynamic nature of LEK and TEK in response to changing environmental conditions. While integrating LEK and TEK into climate adaptation strategies can enhance community resilience and provide culturally appropriate solutions (Filho et al. 2023; Nalau et al. 2018). However, there are challenges in effectively mainstreaming these knowledge systems into formal adaptation policies and practices (Filho et al. 2023). To maximize the potential of LEK and TEK, it is crucial to blend adaptation strategies from individual households to community and landscape levels (Adhikari et al. 2018) and to complement traditional knowledge with modern climate change strategies (Filho et al. 2023).

Traditional ecological knowledge (TEK) plays a crucial role in understanding and managing ecosystems, particularly in the context of climate change and water management. In Colca Valley, Southern Peru, climate change has led to uncertainties in water provision, prompting a focus on material practices and knowledge practices to understand climate change effects and water management responses (Stensrud 2014). This approach emphasizes the importance of examining water practices and the realities they create, as well as the relationships between humans, environment, infrastructure, and other-than-human beings.

Apparently, there are tensions between traditional and modern approaches to ecological knowledge and water management. In Ethiopia, for example, the Borana and Konso communities have managed their water sources (wells and ponds, respectively) for over five centuries using traditional methods. However, modern water management systems often ignore these traditional practices, despite the fact that they incorporate concepts such as cost recovery, ownership, equity, and enforcement (Behailu et al. 2016). This highlights the need to integrate indigenous practices and thoughts when planning water management projects.

The importance of TEK in biodiversity conservation and ecosystem management is evident across various studies. In northern Canada, the Anishnaabe (Ojibwa) people's traditional practices and cultural landscapes contribute to temporal and spatial biodiversity (Berkes and Davidson-Hunt 2006). Similarly, in the Hani Rice Terraces System in China, TEK has enabled sustainable management of terraces and natural resources for over a thousand years (Yuan et al. 2014). However, there is a concerning trend of declining TEK among younger generations, which may affect the sustainability of these systems. To address this, it is crucial to combine historical data with contemporary biomonitoring and invest in training new taxonomists to ensure intergenerational connectivity and the transmission of cultural and environmental heritage (Gravina et al. 2020).

Incorporating traditional ecological knowledge (TEK) into climate policies presents significant opportunities for collaborative and effective climate adaptation strategies. Studies suggest that embracing indigenous knowledge and local institutions in climate change adaptation planning can enhance the adoption and scaling of climate-smart agriculture innovations in smallholder farming (Makate 2019).

To effectively incorporate TEK into climate policies, several recommendations emerge:

1. Adopt a rights-based framing and give greater decision-making power to Indigenous peoples within climate policy structures (Shawoo and Thornton 2019). This addresses the need to recognize and counteract the marginalization of Indigenous peoples due to unequal power relations and colonialism.
2. Develop mechanisms to ensure the holistic integrity of TEK and build overall resilience of climate mitigation and adaptation systems (Shawoo and Thornton 2019). This involves moving beyond simply "integrating" knowledge to truly valuing and preserving Indigenous worldviews.
3. Engage local communities and their indigenous institutions as active stakeholders in designing, planning, and implementing climate adaptation programs (Makate 2019). This participatory approach enhances input from rural people and improves the scaling of climate-smart innovations.
4. Create a policy framework that uses a holistic approach, complementing scientific rigor with the wealth of TEK (Ingty 2017). This can lead to improved policy responses to climate change by bridging Western and Indigenous knowledge systems.
5. Incorporate TEK into climate research, education, and resource planning at national, regional, and local levels (Vinyeta and Lynn 2013). This integration can inform the detection of environmental changes, development of adaptation strategies, and implementation of sustainable land-management principles.

To protect and promote indigenous knowledge, policies should focus on empowering Indigenous communities as key stakeholders in climate adaptation decision-making (Datta and Kairy 2024). This involves shifting from top-down, colonial approaches towards more inclusive, culturally sensitive strategies that recognize the value of Indigenous voices in climate discourse.

Capacity building needs for TEK-based climate resilience include:

1. Facilitating assessments from local to national scales to support collaborative efforts in protecting health from current and future climate hazards (Berry et al. 2018).
2. Investing in promoting indigenous knowledge and providing access to credit for climate-resilient infrastructure (Okoronkwo et al. 2024).
3. Fostering collaborative research as a cornerstone for sustainable rural development (Okoronkwo et al. 2024).
4. Continuous learning and improvement through capacity-building workshops and progress monitoring (Okoronkwo et al. 2024).
5. By addressing these policy implications and recommendations, climate adaptation strategies can become more inclusive, effective, and sustainable, leveraging the valuable insights and practices of Indigenous communities in the face of climate change.

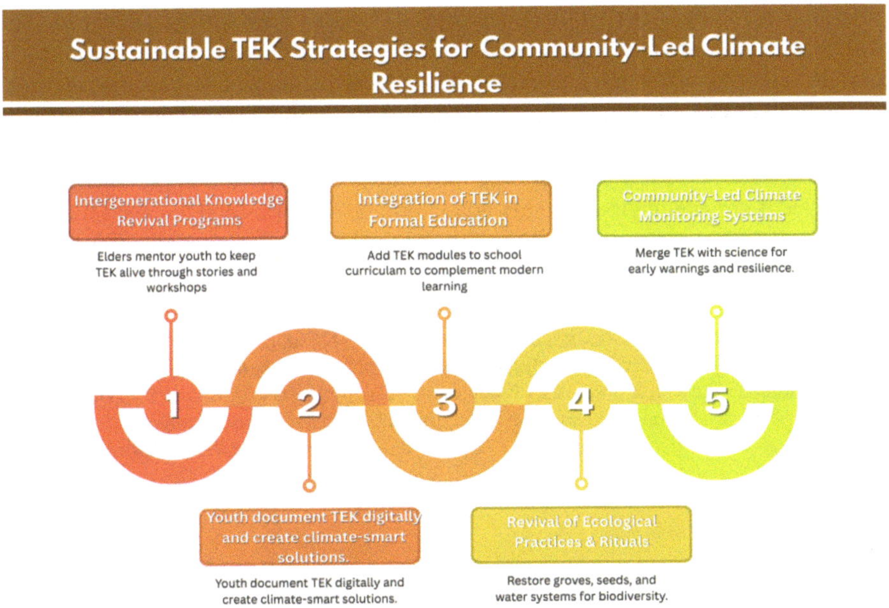

Fig. 18.4 Sustainable TEK strategies for community-led climate resilience

Integrating Traditional Ecological Knowledge (TEK) into local initiatives is essential for building climate resilience within communities (Fig. 18.4). Initially, programs aimed at reviving intergenerational knowledge should facilitate connections between elders, women, and youth through storytelling and seasonal observation walks, thereby ensuring the preservation and transmission of TEK. Furthermore, it is imperative to incorporate TEK into school and college curricula, allowing modern education to complement rather than supplant indigenous wisdom concerning weather patterns, sacred groves, and water practices. Additionally, participatory climate monitoring systems that integrate TEK indicators—such as avian behaviour, flowering cycles, and water rituals—with contemporary climate data can enhance the efficacy of early warning systems. Moreover, the restoration of sacred groves, the revival of resilient indigenous seeds, and the implementation of traditional water harvesting methods can significantly improve biodiversity and ecosystem health. Finally, empowering youth to digitally document TEK and develop climate-smart agricultural and water management practices will facilitate the integration of tradition and technology, thereby ensure sustainability and fostering local ownership.

Conclusion

This study examines the role of Traditional Ecological Knowledge (TEK) in enhancing climate resilience among Particularly Vulnerable Tribal Groups (PVTGs) in the Western Ghats region of India. The research focuses on the Cholanaickan communities, using focus group discussions to explore their ecological knowledge, climate perceptions, and adaptive practices. The findings reveal that traditional practices such as water storage, biodiversity knowledge, season-responsive water use, and mixed cropping remain relevant and actively practiced. However, concerns arise regarding the erosion of traditional knowledge among youth and the lack of institutional support for these practices. The study highlights the importance of integrating local knowledge into climate adaptation frameworks and the need for inclusive policies that empower indigenous communities as key stakeholders in climate resilience efforts.

Acknowledgement The authors gratefully acknowledge the support of the Indian Council of Social Science Research (ICSSR) for facilitating this research, undertaken as part of the ICSSR Multi-Disciplinary Research Studies on Particularly Vulnerable Tribal Groups (PVTGs) of India, 2024–2025. This work contributes to the major project titled *"Climate Change and Indigenous Food Security: Adaptation Strategies of the Cholanaikan Tribe in the Western Ghats.*

References

Adhikari S, Baral H, Nitschke C (2018) Adaptation to climate change in Panchase Mountain ecological regions of Nepal. Environments 5(3):42. https://doi.org/10.3390/environments5030042

Amnesty International (2022) Indigenous peoples. Amnesty International. https://doi.org/10.4324/9781315265797-30

Ayalon L, Roy S, Aloni O, Keating N (2022) A scoping review of research on older people and intergenerational relations in the context of climate change. Gerontologist 63(5):945–958. https://doi.org/10.1093/geront/gnac028

Behailu BM, Pietilä PE, Katko TS (2016) Indigenous practices of water management for sustainable services. SAGE Open 6(4):215824401668229. https://doi.org/10.1177/2158244016682292

Berkes F, Davidson-Hunt IJ (2006) Biodiversity, traditional management systems, and cultural landscapes: examples from the boreal forest of Canada. Int Soc Sci J 58(187):35–47. https://doi.org/10.1111/j.1468-2451.2006.00605.x

Berry P, Campbell-Lendrum D, Shumake-Guillemot J, Enright PM, Villalobos Prats E (2018) Assessing health vulnerabilities and adaptation to climate change: a review of international Progress. Int J Environ Res Public Health 15(12):2626. https://doi.org/10.3390/ijerph15122626

Chisholm Hatfield S, Marino E, Whyte KP, Dello KD, Mote PW (2018) Indian time: time, seasonality, and culture in traditional ecological knowledge of climate change. Ecol Process 7(1). https://doi.org/10.1186/s13717-018-0136-6

Census of India (2011) Population statistics 2011 census. Government of India

Congressional Research Service (2021) Indigenous peoples in Latin America. CRS. https://doi.org/10.4324/9780429499869

Datta R, Kairy B (2024) Decolonizing climate change adaptations from indigenous perspectives: learning reflections from Munda indigenous communities, Coastal Areas in Bangladesh. Sustainability 16(2):769. https://doi.org/10.3390/su16020769

Dogar MM, Sato T (2018) Analysis of climate trends and leading modes of climate variability for MENA region. J Geophys Res Atmos 123(23):13–074

Dogar MM, Sato T (2019) Regional climate response of Middle Eastern, African, and South Asian monsoon regions to explosive volcanism and ENSO forcing. J Geophys Res Atmos 124(14):7580–7598

Dogar MMA, & Almazroui M (2022). Revisiting the strong and weak ENSO teleconnection impacts using a high-resolution atmospheric model. Atmospheric Environment, 270, 118866. https://doi.org/10.1016/j.atmosenv.2021.118866
Dogar MM, Kucharski F, Sato T, Mehmood S, Ali S, Gong Z, ... & Arraut J (2019). Towards understanding the global and regional climatic impacts of Modoki magnitude. Global and Planetary Change, 172, 223–241. https://doi.org/10.1016/j.gloplacha.2018.10.004
Dogar MM, Kucharski F, Azharuddin S (2017a) Study of the global and regional climatic impacts of ENSO magnitude using SPEEDY AGCM. J Earth Syst Sci 126(2):30
Dogar MM, Stenchikov G, Osipov S, Wyman B, Zhao M (2017b) Sensitivity of the regional climate in the Middle East and North Africa to volcanic perturbations. J Geophys Res Atmos 122(15):7922–7948
Dorji T, Rinchen K, Morrison-Saunders A, Blake D, Banham V, Pelden S (2024) Understanding how Indigenous knowledge contributes to climate change adaptation and resilience: a systematic literature review. Environ Manag 74:1101–1123. https://doi.org/10.1007/s00267-024-02032-x
Filho WL, Kalangu JW, Mucova SAR, Efitre J, Adelekan I, Donkor FK, Sanni M, Balogun A, Ayal DY, Simpson NP, Totin E, Zvobgo L, Musiyiwa K, Wolf F (2023) Is indigenous knowledge serving climate adaptation? Evidence from various African regions. Dev Policy Rev 41(2). https://doi.org/10.1111/dpr.12664
Gallay E, Furlan Brighente M, Flanagan C, Lowenstein E (2021) Place-based civic science-collective environmental action and solidarity for eco-resilience. Child Adolesc Mental Health 27(1):39–46. https://doi.org/10.1111/camh.12537
George D, Whitehouse C, Whitehouse P (2011) A model of Intergenerativity: how the intergenerational school is bringing the generations together to Foster collective wisdom and community health. J Intergenerational Relatsh 9(4):389–404. https://doi.org/10.1080/15350770.2011.619922
Gomez-Baggethun E, Reyes-García V, Corbera E (2013) Traditional ecological knowledge and global environmental change: research findings and policy implications. Ecol Soc 18(4). https://doi.org/10.5751/es-06288-180472
Gravina MF, Giampaoletti J, Lezzi M, Giangrande A, Ventura D, Del Pasqua M, Bonifazi A (2020) Perception of changes in marine benthic habitats: the relevance of taxonomic and ecological memory. Diversity 12(12):480. https://doi.org/10.3390/d12120480
Hans S, Nayeem AM, Mikkilineni S, Gupta R (2023) Exploring the relationship between generational diversity and knowledge sharing: the moderating role of workplace intergenerational climate, boundary-spanning leadership and respect. Employee Relat 45(6):1437–1454. https://doi.org/10.1108/er-11-2022-0507
Haseena Ajims M (2014) Scope of education and dropout among tribal students in Kerala: a study of Scheduled Tribes in Attappady. Int J Sci Res Publ 4(1):1–13. http://www.ijsrp.org
Hill R, George M, Jackson S, Abel N, Robinson CJ, Grant C (2012) A typology of indigenous engagement in Australian environmental management: implications for knowledge integration and social-ecological system sustainability. Ecol Soc 17(1). https://doi.org/10.5751/es-04587-170123
Hosen N, Hamzah A, Nakamura H (2020) Adaptation to climate change: does traditional ecological knowledge hold the key? Sustainability 12(2):676. https://doi.org/10.3390/su12020676
Ingty T (2017) High mountain communities and climate change: adaptation, traditional ecological knowledge, and institutions. Clim Change 145(1–2):41–55. https://doi.org/10.1007/s10584-017-2080-3
Kakkoth S (2005) The primitive tribal groups of Kerala: a situational appraisal. Stud Tribes Tribals 3(1):47–55. https://doi.org/10.1080/0972639X.2005.11886519
Kerala Institute for Research, Training and Development Studies of Scheduled Castes and Scheduled Tribes (KIRTADS) (2016) Tribals in Kerala. Government of Kerala
Kerala Institute for Research, Training and Development Studies of Scheduled Castes and Scheduled Tribes (KIRTADS) (2017) Scheduled tribes Kerala. Government of Kerala

Kupika OL, Nhamo G, Kativu S, Gandiwa E (2019) Local ecological knowledge on climate change and ecosystem-based adaptation strategies promote resilience in the middle Zambezi biosphere reserve, Zimbabwe. Scientifica 2019(1):1–15. https://doi.org/10.1155/2019/3069254

Ludwig D, Macnaghten P (2019) Traditional ecological knowledge in innovation governance: a framework for responsible and just innovation. J Responsible Innov 7(1):26–44. https://doi.org/10.1080/23299460.2019.1676686

Mathur PRG (1996) Tribal situation in Kerala. Kerala Historical Society

Mugambiwa SS (2018) Adaptation measures to sustain indigenous practices and the use of indigenous knowledge systems to adapt to climate change in Mutoko rural district of Zimbabwe. Jamba 10(1):388. https://doi.org/10.4102/jamba.v10i1.388. PMID: 29955251; PMCID: PMC6014056

Makate C (2019) Local institutions and indigenous knowledge in adoption and scaling of climate-smart agricultural innovations among sub-Saharan smallholder farmers. Int J Clim Change Strateg Manag 12(2):270–287. https://doi.org/10.1108/ijccsm-07-2018-0055

Mcmillen H, Ticktin T, Springer HK (2016) The future is behind us: traditional ecological knowledge and resilience over time on Hawai'I Island. Reg Environ Chang 17(2):579–592. https://doi.org/10.1007/s10113-016-1032-1

Nalau J, Becken S, Mackey B (2018) Ecosystem-based adaptation: a review of the constraints. Environ Sci Policy 89:357–364. https://doi.org/10.1016/j.envsci.2018.08.014

Okoronkwo DJ, Ugwoke RU, Okoro GG, Ugwu CH, Mbah EC, Ozioko RI, Nwobodo C, Nwagbo UV (2024) Climate smart agriculture? Adaptation strategies of traditional agriculture to climate change in sub-Saharan Africa. Front Clim 6. https://doi.org/10.3389/fclim.2024.1272320

Parrey HA, Bhat MS, Shah SA, Sumira M, Thoker IA, Dogar MM et al (2025) Evaluation of the perception of climate change and its implications across various geo-physical regions of Kashmir Himalayas. Nat Hazards 121(16):18685–18712

Peters MA, Mika CT (2017) Aborigine, Indian, indigenous or first nations? Educ Philos Theory 49(13):1229–1234. https://doi.org/10.1080/00131857.2017.1279879

Prober SM, O'Connor MH, Walsh FJ (2011) Australian aboriginal peoples' seasonal knowledge: a potential basis for shared understanding in environmental management. Ecol Soc 16(2). https://doi.org/10.5751/es-04023-160212

Rajasenan D (2019) Tribal populations in Kerala's development process: an impact evaluation of policies and schemes. Revista Venezolana de Análisis de Coyuntura 15(2):85–110

Scheduled Castes & Scheduled Tribes Research and Training Institute (SCSTRTI) (2015) Particularly vulnerable tribal groups of Odisha, vol I. (Ota AB, Mohanty SC, eds). SCSTRTI.

Scheduled Tribes Development Department (STDD) (2013) Scheduled tribes of Kerala: report on the socio-economic status (Issue November). Government of Kerala

Shylajan CS, Mythili G (2007) Community dependence on nontimber forest products: a household analysis and its implication for forest conservation (WP-2007-00). Indira Gandhi Institute of Development Research

Sharifian A, Cabeza M, Molnár Z, Fernández-Llamazares Á, Wario HT (2022) Dynamics of pastoral traditional ecological knowledge: a global state-of-the-art review. Ecol Soc 27(1). https://doi.org/10.5751/es-12918-270114

Shawoo Z, Thornton TF (2019) The UN local communities and indigenous peoples' platform: a traditional ecological knowledge-based evaluation. WIREs Clim Change 10(3). https://doi.org/10.1002/wcc.575

Stensrud AB (2014) Climate change, water practices and relational worlds in the Andes. Ethnos 81(1):75–98. https://doi.org/10.1080/00141844.2014.929597

Tahir Z, Haseeb M, Mahmood SA, Hassan NE, Dogar MM, Hassan SZ (2025) From green to grey: assessing the impact of urbanization on land surface temperature and thermal comfort in Multan City. Earth Syst Environ 9(2):777–800

Tharuvana A (2020) Wayanattile Aadivasikal: Charitravum varthamanavum. Kerala Bhasha Institute

Trott CD (2019) Children's constructive climate change engagement: empowering awareness, agency, and action. Environ Educ Res 26(4):532–554. https://doi.org/10.1080/13504622.2019.1675594

Thurston E (1975) Castes and tribes of Southern India, reprint, Delhi, vol 6, pp 71–72

Turvey ST, Bryant JV, Mcclune KA (2018) Differential loss of components of traditional ecological knowledge following a primate extinction event. R Soc Open Sci 5(6):172352. https://doi.org/10.1098/rsos.172352

Vinyeta K, Lynn K (2013) Exploring the role of traditional ecological knowledge in climate change initiatives, vol 879. U. S. of Agriculture Forest Service Pacific Northwest Research Station. https://doi.org/10.2737/pnw-gtr-879

Wexler L (2011) Intergenerational dialogue exchange and action: introducing a community-based participatory approach to connect youth, adults and elders in an Alaskan native community. Int J Qual Methods 10(3):248–264. https://doi.org/10.1177/160940691101000305

Withanage WKNC, Gunathilaka MDKL (2023) Theoretical framework and approaches of traditional ecological knowledge. In: Traditional ecological knowledge of resource management in Asia. Springer, pp 27–43. https://doi.org/10.1007/978-3-031-16840-6_3

Yang H, Xu J, Wang Z, Ranjitkar S, Salim MA, Goldberg SD, Zhong M, Jiang Y, Zhai D (2019) Role of traditional ecological knowledge and seasonal calendars in the context of climate change: a case study from China. Sustainability 11(12):3243. https://doi.org/10.3390/su11123243

Yuan Z, Li W, Fuller A, He L, Cheng S, Bai Y, Cao Z, Lun F, Liu M, Min Q (2014) Exploring the state of retention of traditional ecological knowledge (TEK) in a Hani Rice Terrace Village, Southwest China. Sustainability 6(7):4497–4513. https://doi.org/10.3390/su6074497

Chapter 19
Climate-Resilient Aquaculture and Farming Systems for Northeast India: Strategies for Adaptation and Livelihood Enhancement

Lopamudra Sahoo, Huirem Bharati, Vinay Singh, Janmejay Parhi, Abhijit Singha, and B. U. Choudhury

Abstract Northeast India, comprising eight states and home to over 40 million people, more than 82% of whom live in rural areas, is highly vulnerable to climate-induced stresses such as rising temperatures (projected increase of ~1.7 °C by 2030) and intensified rainfall (expected to rise in 73% of districts), resulting in more frequent flooding. These changes threaten key livelihood sectors like agriculture and aquaculture, especially among indigenous and marginal communities. This paper presents strategies for climate-resilient adaptation that focus on low-input aquaculture systems, which are ideal for environments that are ecologically sensitive and have limited resources. Notable interventions include floodplain-based integrated farming systems utilizing raised beds and water-retaining lowlands. Species diversification with hardy, native fish such as Clarias magur, Amblypharyngodon mola, and Labeo calbasu enhances year-round productivity and food security. Composite Fish Culture with Small Indigenous Species (SIS), which enables rural households to earn up to Rs. 10,000/month, contributing nearly 50% of total household income, while improving nutritional intake through vitamin A- and iron-rich species like *Amblypharyngodon mola* and *Puntius sophore*; Low-Cost Mushroom Cultivation, using a bamboo-based microclimate-regulated house, which yields 441 kg/year of oyster mushrooms and generates a sustainable livelihood, offering protein-rich food and recycling farm waste; and Low-Cost Backyard Farming Systems, integrating vegetables, SIS-based aquaculture, and Khaki Campbell ducks on 100–400 m^2 plots, generating Rs. 20,000–30,000/year in surplus income while enhancing household nutrition and resource recycling. These integrated strategies not only increase household-level fish production by up to 2–3 times but also offer scalable, low-cost solutions for building adaptive capacity among over 30 million rural inhabitants

L. Sahoo (✉) · H. Bharati · V. Singh · A. Singha · B. U. Choudhury
ICAR Research Complex for NEH Region, Tripura Centre, Lembucherra, West Tripura, India

J. Parhi
ICAR-CIFA, Bhubaneswar, India

S. Mandal et al. (eds.), *Climate Change Vulnerability*, Climate Change Management, https://doi.org/10.1007/978-3-032-16783-5_19

across the region. In particular, these technologies empower rural and tribal women, reduce external input dependency, and conserve biodiversity, thereby contributing to climate resilience, food security, and gender-inclusive rural development.

Introduction

The North East region of India, comprising eight states, is a unique geographical area characterized by diverse landscapes, rich biodiversity, and a predominantly rural population. This region, covering 26.2 million hectares and home to 40 million people, is distinguished by its low population density, significant tribal presence, and extensive forest cover. The area is dominated by two major river basins, the Brahmaputra and Barak, and its population heavily relies on natural resources for sustenance. The North East region of India, comprising eight states, is a unique geographical area characterized by diverse landscapes, rich biodiversity, and a predominantly rural population. This region, covering 26.2 million hectares and home to 40 million people, is distinguished by its low population density, significant tribal presence, and extensive forest cover. The area is dominated by two major river basins, the Brahmaputra and Barak, and its population heavily relies on natural resources for sustenance. Climate change poses a significant threat to this ecologically sensitive region (Azharuddin and Dogar 2016; Dogar and Shahid 2019). Recent studies project an average temperature increase of 1.7 °C across almost all districts, with some areas potentially experiencing rises exceeding 2 °C. Rainfall patterns are also expected to shift dramatically, with 57 out of 78 districts likely to see increased precipitation, some by as much as 25%. The frequency of extreme rainfall events is predicted to rise by 26%, exacerbating the risk of floods. The region's unique topography and location contribute to distinct precipitation and drainage patterns, with thunderstorms and monsoon rains playing crucial roles in the annual water cycle (Dogar et al. 2017, 2019). However, climate change is rendering these patterns increasingly unpredictable and erratic, leading to devastating impacts on local communities, agriculture, and livestock.

The vulnerability of the North East region is further compounded by widespread poverty, particularly in remote areas, and the prevalence of traditional farming practices. Many farmers lack formal education and resources, making it challenging for them to adapt to the changing climate independently. This situation underscores the urgent need for developing and implementing effective coping strategies to help farmers adapt to the impending challenges posed by climate change. The North East region's unique ecological and cultural characteristics make it particularly susceptible to the impacts of climate change. Its diverse ecosystems, including tropical and subtropical forests, grasslands, and wetlands, support a wide array of flora and fauna, many of which are endemic to the region. These ecosystems not only contribute to global biodiversity but also provide essential ecosystem services to local communities, such as water regulation, soil conservation, and carbon sequestration. The region's agricultural sector, which forms the backbone of the rural economy, is highly vulnerable to climate change impacts. Traditional farming practices, often based on indigenous knowledge systems, are becoming less reliable due to changing weather patterns. Crops that have been cultivated for generations may no longer be suitable under new climatic conditions, necessitating a shift in agricultural practices and crop choices. Furthermore, the North East region's geopolitical significance adds another layer of complexity to its climate change challenges. Sharing international borders with several countries, including Bangladesh, Bhutan, China, and Myanmar, the region's climate-related issues have transboundary implications (Rasul 2015). Changes in water availability, for instance, could potentially affect international relations and regional stability. The socio-economic fabric of the North East is also at risk because of climate change. The region's high dependence on natural resources for livelihoods makes communities particularly vulnerable to environmental change. Traditional occupations, such as fishing, forestry, and small-scale agriculture, are likely to be severely impacted, potentially leading to increased poverty, food insecurity, and migration. In light of these multifaceted challenges, there is an urgent need for comprehensive and context-specific climate change adaptation strategies in the northeastern region.

These strategies must not only address the immediate impacts of climate change but also consider the long-term sustainability of the region's unique ecosystems and its cultural heritage. Integrating traditional knowledge with modern scientific approaches, enhancing community resilience, and promoting sustainable development practices are crucial for building the adaptive capacity of the region's population. Climate change poses a significant threat to this ecologically sensitive region. Recent studies project an average temperature increase of 1.7 °C across almost all districts, with some areas potentially experiencing rises exceeding 2 °C (Barik and Baidya Roy 2023). Rainfall patterns are also expected to shift dramatically, with 57 out of 78 districts likely to see increased precipitation, some by as much as 25%. The frequency of extreme rainfall events is predicted to rise by 26%, exacerbating the risk of floods. The region's unique topography and location contribute to distinct precipitation and drainage patterns, with thunderstorms and monsoon rains playing crucial roles in the annual water cycle. However, climate change is rendering these patterns increasingly unpredictable and erratic, leading to devastating impacts on local communities, agriculture, and livestock. The vulnerability of the North East region is further compounded by widespread poverty, particularly in remote areas, and the prevalence of traditional farming practices. Many farmers lack formal education and resources, making it challenging for them to adapt to the changing climate independently. This situation underscores the urgent need for developing and implementing effective coping strategies to help farmers adapt to the impending challenges posed by climate change.

Several key technologies are recommended to enhance farmers' adaptability to climate change scenarios. These include the implementation of climate-smart agricultural practices, such as conservation tillage, crop rotation, and integrated pest management. Additionally, the adoption of drought-resistant and heat-tolerant crop varieties can significantly improve resilience to changing weather patterns in the region. Precision farming techniques that utilize remote sensing and GPS technology allow for more efficient resource management and targeted interventions.

Water management technologies play a crucial role in adaptation strategies. These encompass efficient irrigation systems, such as drip irrigation and rainwater harvesting techniques. The use of weather forecasting tools and early warning systems can help farmers make informed decisions regarding planting and harvesting times. Finally, the integration of renewable energy sources, such as solar-powered irrigation pumps, can reduce dependence on fossil fuels and increase overall farm sustainability. When implemented in combination, these technologies can significantly bolster farmers' ability to adapt to and mitigate the impacts of climate change on agricultural production. Climate change impacts on aquatic ecosystems and associated livelihoods are growing, and we have tried to outline a few adaptation activities that can aid in enhancing the adaptation level of resource-poor farmers of the region.

Species Diversification for Climate-Resilient Aquaculture

The North East Indian region is blessed with many potential indigenous and endemic fish species that can be utilized for aquaculture. In the present context of climate change, species diversification in aquaculture through the introduction of various indigenous fish species specific to Northeast India can prove to be a sustainable option for enhancing fish production and paving the way for climate-smart and resilient aquaculture. The adoption of a greater number of fish species for culture will also reduce the adverse effects of climate change. The development of culture technologies and the introduction of potential indigenous species as new candidate species for aquaculture would not only help improve the region's economy through enhanced fish production but also facilitate the protection and conservation of these valuable fishery resources.

Aquaculture in the northeastern region of India is concentrated mostly on carp culture, which is largely focused on six species, *viz*., the three Indian 'major carps,' catla (*Catla catla*), rohu (*Labeo rohita),* and mrigal (*Cirrhinus mrigala)*; and the three exotic or 'Chinese' carps, grass carp (*Ctenopharyngodon idella*), silver carp (*Hypopthalmicthyes molitrix*), and common carp (*Cyprinus carpio*) (Jayasankar 2018). Expansion and diversification of aquaculture through species diversification is the need of the hour for a sustained and climate-resilient mode of fish production despite the changing climate scenario. Medium and minor carps, small-sized indigenous fishes (SIFs), air-breathing fishes, and catfishes are potential candidates for diversification as they are in great demand among the local population of the region.

Medium and Minor Carps

Some of the medium and minor carps, such as *Osteobrama belangeri, Bangana dero, Labeo gonus, Labeo calbasu, Labeo bata, Cirrhinus reba, Puntius gonionotus,* etc., are potential indigenous candidates for diversification. These carps have an initial higher growth rate and market acceptability at 300–400 g, which makes them ideal species for polyculture with major carps. These species are suitable candidates for the utilization of seasonal ponds with water retention of 5–6 months.

Small Indigenous Fishes (SIFs)

Maximum diversity of small indigenous fishes (SIFs) has been recorded from the North East region of India (Roy et al. 2023). Many SIFs have been associated with the social and cultural rituals of the region. Besides this, the SIFs play a vital role in the livelihood, health and the general well-being of the poor populations of the region. Many SIF species have potential for culture with high demand and high nutritional value, and thus, can be introduced as candidate species in freshwater aquaculture systems in the region. Some of the potential SIF species are *Amblypharyngdon mola, Notopterus notopterus, Puntius sarana, Puntius ticto, Puntius chola, Puntius sophore, Salmostoma bacaila, Esomus danricus, Glossogobius giuris, Danio devario, Parambassis ranga,* and *Chanda nama,* etc.

Catfishes

Owing to the high market demand and consumer preference for catfishes and the availability of huge resources in the form of swamps and derelict water bodies, catfishes have a lot of prospects in the region as candidate species for aquaculture. Some of the important high-value catfish species in Northeast India are *Clarius magur* (magur), *Heteropneustes fossilis* (singhi), *Ompok bimaculatus, Ompok pabda, Mystus* spp., etc. With the standardisation of seed production and culture technology of *Clarius magur*, *Heteropneustes fossilis,* and *Ompok bimaculatus*, their farming can be expanded throughout the entire region.

Air-Breathing Fishes

Air-breathing fishes such as murrels (*Channa striatus* and *Channa marulius*) and *Anabas testudineus* have great demand as live fish, particularly in the northeast Indian region, and fetch a higher price compared to the Indian major carps and

Chinese carps. These species are potential candidates for culture as they have high economic value, consumer preference, and high nutritional value. These fish species can be cultured in the waterlogged low low-lying swampy areas and shallow water bodies available in the region.

Alternative Fish Species Other than IMCs and Chinese Carps for Culture in Tripura

Significance of Species Diversification

Fish production can be made all year round by the addition of more compatible fish species with the IMCs and the Chinese carps. Most of the medium and minor carps and SIFs are compatible with Indian major carps for polyculture systems. They can also be cultured in pen enclosures, as most of them are omnivores/herbivores and can easily digest the plant protein sources. These species can be cultured with minimum maintenance cost. The initial growth rate of many minor and medium carps is faster than the IMCs. Marketable size of these carps is around 100–300 g compared to that of IMCs (700–800 g). These attributes make them good candidates for short-duration culture practice in seasonal water bodies. The medium and minor carps are also high fecund fish, like the IMCs, and are suitable for seed production and can help in meeting the demand for fish seeds in the region. A culture of self-recruiting fish can be done in less manageable ponds with shallow water depth, with minimum maintenance cost. Fishes such as Notopterus sp., Ompok sp., Magur, and Singhi can be cultured together with small-sized self-recruiting SIFs such as mola, which will eventually enhance income (Table 19.1, Fig. 19.1). Most of the medium and minor carp species, air-breathing and catfish species like Ompok bimaculatus, O. pabda, Clarius magur, Heteropneustes fossilis, etc., fetch a higher price than the IMCs. The small-sized fish also fetch a good price in the market, making their culture

Table 19.1 Species and their feeding habit

Species	Feeding habit
Ompok bimaculatus	Carnivorous/omnivorous
Notopterus notopterus	Omnivorous
Clarius magur	Carnivorous/omnivorous
Heteropneustes fossilis	Carnivorous/omnivorous
Labeo gonius	Omnivorous
Labeo calbasu	Omnivorous
Labeo bata	Omnivorous
Puntius gonionotus	Plankton and soft aquatic weeds
Esomus dandricus	Insectivorous
Amblypharyngodon mola	Plankton

Labeo rohita (Rohu) *Catla catla* (catla)

Cirrhinus mrigala (Mrigal) *Ctenopharyngodon idella* (grass carp)

Hypopthalmicthyes molitrix (silver carp) *Cyprinus carpio* (common carp)

Notopterus notopterus (kandla) *Ompok bimaculatus* (pabda)

Labeo calbasu (kalbasu) *Labeo gonius* (gonius)

Labeo bata (bata) *Puntius gonionotus*

Amblypharyngodon mola (moka) *Esomus dandricus*

Fig. 19.1 Different types of fishes

profitable. Cultivation of indigenous species will contribute to the development of local communities, as well as the protection and conservation of indigenous resources.

Composite Fish Culture with SIS (Small Indigenous Species)

(Technology developed at ICAR, Tripura)

Three species of Indian major carps, viz., catla (*Catla catla*), rohu (*Labeo rohita*), and mrigal (*Cirrhinus mrigala*), and exotic species such as silver carp (*Hypophthalmichthys molitrix*), grass carp (*Ctenopharyngodon idella*), and common carp (*Cyprinus carpio*) are presently cultured in Tripura. The majority of fish eaten by the rural poor people is the small indigenous fish species (SIS), which have been defined as species that grow to a maximum length of about 25 cm. Sixteen small indigenous species of fish have been identified for culture in India. Among these, mola (*Amblypharyngodon mola*) and punti (*Puntius sophore*) are important because of their high nutritional value. Integrating SIS into composite fish culture systems with major carps offers multiple benefits. Firstly, it can enhance the nutritional intake of rural communities by providing a diverse range of fish species in their diet. Secondly, as SIS commands higher market prices, their inclusion can potentially increase the income of farm families. This approach to aquaculture not only addresses nutritional needs but also contributes to economic sustainability in rural areas, making it a promising strategy for both food security and livelihood improvement (Filipski and Belton 2018).

Livelihood Potential

The tribal farm families are still dependent on traditional practices, depending more on fish capture rather than culture. With the introduction of the scientific method of fish culture, the dependence on natural resources will reduce considerably. Implantation of the stocks of big and small fish will strengthen their food & nutritional security, which will in turn support sustainable livelihood activities. Pisciculture will not only add to their net family income but also provide them with an alternative source of income generation (Ponnusamy and Devi 2017). Based on our calculations, after we intervene with the initial supply of inputs, this technology of alternative livelihood can contribute to 50% of the family income (taking into consideration that the average family income is Rs. 5000/month/household and after our intervention, there is an additional income of 10,000/month) (Table 19.2). So, it can be recommended that if they save 50% of their profit, they can easily make the system sustainable by stocking fish for composite fish culture from next year onwards. This sustainable approach to pisciculture can lead to long-term economic benefits for tribal farm families. By adopting scientific fish culture methods, they

Table 19.2 Economics of fish farming in composite culture (with SIS) and traditional practice

Parameter	Composite culture with SIS (Rs.)	Traditional farming (Rs.)
Total cost of production (Rs./ha)	1,40,000	42,666
Gross profit (Rs./ha)	2,60,000	58,386
Net returns (Rs./ha)	1,20,000	15,720
Benefit-cost ratio	1.85	1.36

can reduce their reliance on traditional capture practices and increase their overall productivity. Furthermore, this shift towards sustainable aquaculture can potentially improve the ecological balance of local water bodies while providing a steady source of income and nutrition for the community.

Sale prices of fish during the period were Rs. 100/-respectively. Mola and Puti were sold at Rs. 400/kg.

Conservation Potential

It is important to prioritize SIFS for conservation and restoration programs. About 23 per cent of small indigenous fish species provide food and nutrition, and are of economic value as ornamental fish, supporting livelihoods and nutrition, especially for people of the eastern and north-eastern parts of India (Roy et al. 2023). The north-eastern parts of India are the biodiversity hotspots for SIFFS. In terms of resilience, 62% of SIFFS are highly resilient (in which the population doubling time is 2 years or less), while 38 per cent have medium resilience (with a population doubling time of over 5 years), making it important to consider conservation measures for these species. Though there is a lot of traditional knowledge on the benefits of SIFFS, this is poorly documented. Considering the nutritional value of SIS is essential to maintain ecological, nutritional, and socioeconomic equilibrium with respect to the conservation and management of species. Through this technology, we aim to introduce two species for incorporation initially (mola and Puti), and the addition of other important cultivable species will be increased over the following years (Fig. 19.2 and Table 19.3). This will lead to conservation of these minor species along with additional income and food security.

Economics

Fish farming is one of the most profitable and dependable sources of income in an agricultural production system. The adopted farmers under this program will benefit both in terms of fish production and additional revenue. The income from the total additional fish production (2500 kg harvest multiplied by Rs. 100/kg) was calculated to be Rs. 2,50,000/ha.

Small indigenous fish

Indian Major Carps

Fig. 19.2 Small indigenous fish, and Indian Major Carps

Table 19.3 Major carps and their SIS

Sl. no.	Parameters	Major carps	SIS
1	Survival (%)	85%	As they are self-recruiting, they will multiply year-round
2	Average size at harvest (kg)	1	1–5 g
3	Total production (kg)	2500	Along with it, there will be partial harvesting of mola and puti of about 6 kg every 3 months
4	Farm gate sale price (Rs.)	100	300–400 for SIS
5	Gross income (Rs.)	2,50,000	An additional 10,000–20,000 after household consumption

Feasibility to Be Produced by Local Women

Traditionally, women were not involved in fishing practices as it was considered to be a male domain. But our technology concentrates not only on large farmers but also caters to the needs of small and marginal farmers who have only a backyard pond for their subsistence. For a fish harvesting farm, workers need to be incorporated. But the SIS can be harvested easily by the farm women with the use of traditional traps. The perception of rural women is that when the smaller species were cooked, they were assured of a share. This was not necessarily the case when the bigger species were cooked. The intra-household distribution of fish, depending on the kind of fish cooked, is another important dimension that upholds the involvement of farm women to take up fish culture as they are assured of adequate nutrition for their family (Beveridge et al. 2013). In the case of small species, unlike the larger ones, they will be marketed locally within the villages, especially by women and children. SIS catches when sold to local markets would also provide a source of income, whenever needed—a supplementary annual income of about 10,000–15,000 Rs. Hence, this technology will motivate the village women to take up fish culture for food, nutrition, and livelihood security.

Utilization of Shallow Water Bodies for Fingerling Production

(Technology developed at ICAR, Tripura)

Most of the ponds in NE India are shallow and seasonal, with water holding capacity for 6–8 months. Post-winter, the water level in these ponds falls drastically, thereby forcing the farmers to sell off their undeveloped produce. Fishes as small as 200–500 g are sold in local markets at a very marginal price (Fig. 19.3). This whole scenario is uneconomic and non-viable. These ponds can be used as a grow-out pond, but will not be beneficial as the growth time for the fish will be very less. Hence, these ponds can be used for stunted fingerling production. The fingerlings can be reared in shallow ponds with no feeding. Fingerling will be stunted and will thrive on the natural productivity of the system. The stunted fingerling can be sold at a remunerative price the following year once the rain starts.

Multiple Stocking and Multiple Harvesting with Stunted Fingerling

(Technology refined at ICAR Tripura Centre)

Stunted fingerlings are suitable stocking material for carp culture because of their higher survival rate. They can reach marketable size within a shorter time period, i.e., grow up to 700–800 g within 2.5–3 months, leading to higher yield &

Fig. 19.3 Shallow pond utilized for fingerling production and fingerlings

Fig. 19.4 Stocking of stunted fingerling and harvest in the farmers' field

income. These have good market demand as they utilize seasonal grow-out ponds efficiently, and the fish can be sold at a higher price, too. Multiple stocking and multiple harvesting (MSMH) is a method of composite fish farming, through which fish production can be increased by many folds (6–10 tons/hectare/year). This technology is a climate-resilient technology, as the majority of farm ponds are seasonal. Regular stocking with stunted fingerlings gives better growth and survival of fish in the ponds compared to stocking with spawn (Fig. 19.4). Multiple harvesting releases the obnoxious gases and mixing of bottom nutrients, which enhances the primary productivity of the system. This also gives remuneration every 4 months to the farmers, which they can utilise for their household and other agricultural operations. Through this technology, farmers can get higher yields & income compared to traditional systems.

Floodplain-Based Farming System

(Technology developed at ICAR, Tripura)

This system is targeted for the low-lying flood-prone zones. The zone can be divided into raised beds and lowland. The raised bed and the lowlands can be used for paddy cultivation during the rainy season. In addition, lowland can also be used for fish culture, where in high demand local fishes like magur, singhi, koi, etc. can be cultured (Fig. 19.5). In addition, the raised beds can be used for vegetable cultivation during the winter and summer seasons. A trench is dug at one side of the system so that the fish can accumulate once the water level goes down. In addition, the trench will also ease the harvesting of fish. This system will be very beneficial to the farmers because if the flood destroys the paddy crop in the lowland, there will be an additional upland paddy to overcome distress. In addition, the fish will generate household nutritional security as well as additional income.

Fig. 19.5 Raised bed and low-land model and deep Trench for fish on one side of the system

Trap Pond Management

This is a technology suggested by FAO for flood-prone areas of Bangladesh and can also be adapted to these areas. On seasonally flooded land, bushes and branches can be used to build traps for flood periods, and traditional bamboo traps can be modified to reduce catch per unit effort. The traps also provide habitat for young native species. As this cultivated land is unavailable for several months of the year, this increased catch provides additional food and income sources.

Small-Scale Homestead Pens

This technology is also suggested by FAO for flood-prone areas of Bangladesh and can also be adapted to these areas. Bamboo pens with trap doors can be built next to homes. These can be stocked with some fish, and then when seasonal floods come, the fish will not be washed away, and new fish will be introduced via floodwater. Households can eat the larger fish, and the smaller fish washed can be restocked in the pens.

Low-Cost Backyard Farming System

(Technology developed at ICAR, Tripura)

Out of a 36-lakh population in Tripura, 11 lakhs are landless labourers. They are the workforce in agriculture and other sectors. They come below the Poverty Line and are the most vulnerable group in terms of food and nutritional security. The average family size is 5–6 members. Their only possession is a small house with 100–400 m^2 land adjacent to it. Usually, the male member goes to work, and the female and the children stay back. To improve their food and nutritional security, we conceptualized a program on backyard farming for these landless laborers, taking into consideration the available area and the time a lady can spare after her normal household work (Sahoo et al. 2023).

Livelihood Potential

Based on ICMR recommendations, the following table cites the requirements of the different food components of a household. With this backdrop, we developed a farming system with a concept of "By the women, for the women and of the women" for ensuring sustainable food and nutritional security of the household. The following table cites the requirements and production/week of different food components from the farming system. Seasonal vegetables with local preference were produced throughout the year. Compost pits were dug so that the waste produced from the system can be recycled back to the system. Mola (*Amblypharyngodon mola) and* local puti (*Puntius sophore*) were the 2 species that were introduced into the ponds. Mola and puti breeds in the pond and therefore enables poor households to increase their consumption of fish. A 50 kg fish was harvested from the 200 m^2 pond, along with partial harvesting of mola and puti every month. Mola and puti have 16% protein and around 7% lipid. Along with that, they provide essential nutrients like Vit A, Ca, Fe, and Zn. Mola is very rich in vitamin A as retinol, and making mola part of the diet can help those suffering from malnutrition and micronutrient deficiency. Ducks like Khaki Campbell can produce 200 eggs annually. Apart from that, 10 ducks can produce 1Kg manure/day, which can be used for Fishery ponds and vegetable production. With our intervention, the lady, after devoting 3 h per day, could ensure sufficient food items for her household to ensure food and nutritional security. After that, she had some surplus items in terms of mainly vegetables and a few eggs, which she can sell in the local market to ensure some sort of livelihood security for her family. The male member was required to purchase rice, some pulses, oil, etc., through his daily wages.

Conservation Potential

Our technology of backyard farming has the potential to manage diversity in agriculture and aquaculture. Often, a woman's influence in society seems to be restricted to the boundaries of her home, but a closer look, especially within the rural communities, reveals that they have multi-tier responsibilities within and beyond the confines of their house and home garden. They make decisions about which crops and varieties to grow, eat, and sell. These women farmers can be considered visionaries who play a significant role in managing agricultural biodiversity. Women are very much aware of the opportunities that can be gained through diverse vegetable cultivation, which they consider to be a more practical income generation option given the limited resources within their reach, such as land and finances. Women are also instrumental in ensuring the dietary components that enter the household food basket. Much of what is grown on the home farm contributes to the household's dietary intake, which is complemented with purchases from the market, which are usually bought by the men. These women are very eager to increase their knowledge on the nutritional content of different food items and are aware of the importance of having a balanced diet for their household. Our stakeholders for this technology are farm women who will, in fact, play an active role in the management of agricultural biodiversity, which can contribute to the resilience of their farming systems, improve nutrition and increase incomes, and, in turn, empower the women within the communities.

Economics

The average income of landless labourers is Rs. 300–500/day without the security of year year-round income. Through our technology, we can empower the farm woman to have assured food and nutritional security throughout the year. There will still be some surplus that can be sold in local markets for income generation. Through this technology, they can earn an additional amount of Rs. 20,000–30,000/year.

Low-Cost Mushroom Cultivation

Empowerment in the context of women's development is a way of defining, challenging and overcoming barriers in a woman's life through which she increases her ability to shape her life and environment. Mushroom growing is one agricultural activity in which women can play a vital role without sacrificing their household responsibilities. Mushroom cultivation is simple, low-cost, labour-intensive and suitable for rural areas, which can provide employment in both the semi-urban and rural areas. This cultivation will improve the socio-economic condition of farmers, families and solve employment problems of both literate and illiterate, especially women. It can also be a good tool for rural development. Mushrooms contain 20–30% protein, which has been recognized as an alternate source of good quality protein, higher than vegetables and fruits, and is of high quality. ICAR Tripura Centre has come up with a low-cost technology for mushroom farming with a view to uplift the living standards of the poor women and also to add high-quality protein in their daily diets to eradicate malnutrition problems. In Tripura, the major crop being grown in 90% of the cultivable area is paddy, which assures a year-round availability of paddy straw, which is a suitable substrate for mushroom cultivation. There is abundant growth of bamboo, which is available at affordable rates to the farmers, and can be used for the construction of the low-cost mushroom house. Mushroom cultivation can be practised easily by all farmers, especially women of the household, simultaneously along with their regular occupation, as it is less labor-intensive.

Livelihood Potential

Mushroom farming has the capacity to enhance the livelihood of a farm family in a short period of time. This will motivate them to continue their production. Oyster mushroom has the potential to be cultured around the year under Tripura conditions. Our technology has the potential to generate an additional income of Rs. 61,200/ household/year under the backyard system. Per capita consumption of mushrooms will be increased from almost nil to around 50 g/day.

Income of Oyster Mushrooms

August–March total 9 crops can be harvested @ 140 beds/crop = 140 × 9 = 1260 beds
0.35 kg fresh mushrooms can be harvested from each bed = 1260 × 0.35 = 441 kg
1 kg of fresh mushrooms is sold in the market @ Rs. 200 = 441 × 200 = Rs. 88,200
Profit = Rs. 88,200 – Rs. 27,000 = Rs. 61,200

Conservation Potential

Any agricultural and farm waste materials, viz., paddy straw, wheat straw, sawdust, farm yard trashes, etc., can easily be used for the cultivation of oyster mushroom. Use of such waste for mushroom production is a better, and profitable, and eco-friendly way of waste disposal. Hence, mushroom farming is an ecologically friendly practice that generates food of high nutritional value from farm waste materials.

Economics

To promote the mushroom cultivation and to encourage the farmers to take mushroom cultivation as one of the components of the Integrated Farming System, a low-cost mushroom house (Rs. 12,060/–), which is affordable to the small and marginal farmers, was constructed at ICAR Research Complex for NEH Region, Tripura centre as a demonstration unit (Technology developed and demonstrated at ICAR Tripura Centre). The detailed cost of construction and the economics (Table 19.4) were worked out for the benefit of the farmers. In addition, as the

Table 19.4 Expenditure on low-cost mushroom house L × B × H (12 × 10 × 7 ft.)

Sl. no	Materials used	Quantity	Rate/unit	Amount in rupees
1.	Bamboo	200	12	2400
2.	Barrack bamboo	12	100	1200
3.	Polythene sheets	5 m	140	700
4.	Nylon wires	10 m	120	120
5.	Bricks	100	10	1000
6.	Sand		15	90
7.	Labor	25-man days	262	6550
			Total	12,060

bamboo unit creates a cooler environment, favourable for the good yield of the mushroom. The house will be demonstrated to the farmers with the emphasis on integrating mushroom cultivation as one of the components in their Integrated Farming System to improve their livelihood, along with the economic and nutritional security of their family.

Expenditure

Mushroom house (12 × 10 × 7) L × B × H can accommodate 140 beds in a hanging system

1. 100 kgs of Paddy Straw@ Rs. 10 = Rs. 1000
2. 70 Spawn packets @ Rs. 10 = Rs. 700
3. Chemicals = Rs. 500
4. Wires, Polthene bags = Rs. 300
5. Miscellaneous = Rs 500
6. Total = Rs. 3000
7. Total expenditure = Rs. 3000 × 9 (nine crops can be taken within a season) = Rs. 27,000

Feasibility to Be Practiced by Farm Women

Mushrooms can provide employment and income to rural women who are confined within the family and have leisure time for productive pursuits (Bijla and Sharma 2023). It needs less land area, less investment, and gives early returns. Mushroom finds a ready market in hotels and homes. Rural women can be involved in this vocation for self-employment by using this simple technology. This particular micro-enterprise helps the women not only in supplementing their income but also in lessening the family burden on them on account of food and nutrition. Another important aspect is that the farm women have easy access to farm by-products, the paddy straw, which constitutes the major bulk of the raw material. Being a part-time employment avenue, women can easily combine mushroom farming with household and other farming activities.

Conclusion

The climate-resilient strategies and technologies presented in this paper offer promising solutions for enhancing the adaptive capacity and livelihoods of rural communities in Northeast India in the face of climate change. Key interventions such as species diversification in aquaculture, composite fish culture with small indigenous species, floodplain-based integrated farming systems, low-cost mushroom cultivation, and backyard farming systems provide multiple benefits:

1. Increased resilience: These approaches help farmers and fishers adapt to changing climate conditions through diversified production systems and climate-smart practices.
2. Enhanced food and nutrition security: The integration of diverse fish species, vegetables, and mushrooms improves household nutrition and dietary diversity.
3. Improved livelihoods: The technologies generate additional income streams, with potential earnings of Rs. 20,000–30,000 per year from backyard systems and up to Rs. 61,200 per year from mushroom cultivation.
4. Women's empowerment: Many of the interventions can be managed by women, providing opportunities for income generation and decision-making.
5. Resource conservation: The approaches promote efficient use of land and water resources while conserving indigenous fish biodiversity.
6. Low-cost and scalable: The technologies are designed to be affordable and implementable by small-scale farmers with limited resources.

By combining traditional knowledge with modern scientific approaches, these integrated strategies offer scalable solutions for building climate resilience, food security, and sustainable rural development across Northeast India's diverse landscapes. Further research and extension efforts are needed to refine and widely disseminate these promising technologies to benefit the over 30 million rural inhabitants of the region. Policymakers and development agencies should prioritize support for implementing and scaling up these climate-resilient farming and aquaculture systems as part of broader climate adaptation and rural development initiatives in Northeast India.

References

Azharuddin S, & Dogar M M (2016). Decadal climate variability and predictability. CURRENT SCIENCE, 110(8), 1397. http://www.currentscience.ac.in/Volumes/110/08/1397.pdf

Beveridge MCM, Metian M, Phillips MJ, Troell M, Hall SJ, Thilsted SH (2013) Meeting the food and nutrition needs of the poor: the role of fish and the opportunities and challenges emerging from the rise of aquaculture[a]. J Fish Biol 83(4):1067–1084. https://doi.org/10.1111/jfb.12187

Bijla S, Sharma VP (2023) Status of mushroom production: global and national scenario. Mushroom Res 32(2):91–98. https://doi.org/10.36036/mr.32.2.2023.146647

Dogar MMA, & Shahid A (2019). Revisiting the climatic impacts of ENSO Modoki varying magnitude. In AGU Fall Meeting Abstracts (Vol. 2019, pp. GC31D-1221).

Dogar MM, Kucharski F & Azharuddin S (2017). Study of the global and regional climatic impacts of ENSO magnitude using SPEEDY AGCM. J Earth Syst Sci 126, 30. https://doi.org/10.1007/s12040-017-0804-4

Dogar MM, Kucharski F, Sato T, Mehmood S, Ali S, Gong, Z, ... & Arraut J (2019). Towards understanding the global and regional climatic impacts of Modoki magnitude. Global and Planetary Change, 172, 223–241. https://doi.org/10.1016/j.gloplacha.2018.10.004

Filipski M, Belton B (2018) Give a man a fishpond: modeling the impacts of aquaculture in the rural economy. World Dev 110:205–223. https://doi.org/10.1016/j.worlddev.2018.05.023

Jayasankar P (2018) Present status of freshwater aquaculture in India - a review. Indian J Fisher 65(4):157–165. https://doi.org/10.21077/ijf.2018.65.4.81300-20

Sahoo L, Kandpal BK, Das A, Debnath C, Singh V, Devi HL, Bharati H, Parhi J, Singha A, Datta J, Das B, Mishra VK (2023) Improving the nutritional and livelihood security of landless laborer through the backyard farming system. Front Sustain Food Syst 7:1206367. https://doi.org/10.3389/fsufs.2023.1206367

Ponnusamy K, Devi MK (2017) Impact of integrated farming system approach on doubling farmers' income. Agric Econ Res Rev 30(conf):233. https://doi.org/10.5958/0974-0279.2017.00037.4

Rasul G (2015) Water for growth and development in the Ganges, Brahmaputra, and Meghna basins: an economic perspective. Int J River Basin Manag 13(3):387–400. https://doi.org/10.1080/15715124.2015.1012518

Roy A, Manna RK, Das BK, Naskar M, Ekka A, Sahu SK, Sinha A (2023) Determinants influencing fishermen's willingness-to-participate and willingness-to-pay for conservation of small indigenous fishes: a model-based insight from Indian Sundarbans. Front Sustain Food Syst 7:1215091. https://doi.org/10.3389/fsufs.2023.1215091

Chapter 20
Integrated Strategies for Flood Risk Mitigation in Tripura: A Comprehensive Approach to Resilience and Sustainability

Ganesh Das and B. U. Choudhury

Abstract The state of Tripura is becoming more susceptible to recurrent flooding because of the combination of intense monsoonal rainfall, complex terrain, and inadequate drainage and infrastructure systems. Unprecedented rainfall in 2024 caused Tripura to experience one of the most severe flood disasters in recent history, resulting in extensive inundation across the state. The 2024 floods are examined in this paper through the use of data from government reports, humanitarian agencies, and disaster monitoring platforms to examine their causes, impacts, and response strategies. More than 17 lakh people were affected by the floods and over 65,000 individuals were displaced to relief camps. The disaster caused a significant number of deaths, damage to homes and public infrastructure, widespread power outages, and major disruptions to water and sanitation services. Over 1.60 lakh hectares of farmland were submerged, impacting both paddy and vegetable crops, leading to long-term livelihood and food security concerns. Coordinated efforts of state authorities, national disaster response teams, and community volunteers resulted in the mobilization of rescue and relief operations. Although efforts were made, the disaster highlighted critical shortcomings in early warning dissemination, infrastructure resilience, and community preparedness. This study stresses the importance of having an integrated, multi-sector flood risk mitigation plan that incorporates structural and non-structural approaches, improved forecasting systems, and sustainable development planning to foster long-term resilience in the state.

G. Das (✉)
ICAR-Krishi Vigyan Kendra West Tripura, ICAR RC for NEH Region, Belbari, West Tripura, India

B. U. Choudhury
ICAR Research Complex for NEH Region, Tripura Centre, Lembucherra, West Tripura, India

S. Mandal et al. (eds.), *Climate Change Vulnerability*, Climate Change Management, https://doi.org/10.1007/978-3-032-16783-5_20

Introduction

The Northeastern region (NER), encompassing states such as Arunachal Pradesh, Assam, Manipur, Meghalaya, Mizoram, Nagaland, Sikkim, and Tripura, lies between 21°50′ and 29°34′ N latitude and 85°34′ and 97°50′ E longitude. It spans 26.2 million hectares, approximately 8% of India's geographical area, and has a population of approximately 50 million. Extreme weather events are common in Tripura due to its topography and climatic patterns. The rainfall is primarily caused by the southwest monsoon, which starts in mid-May and lasts until October. Flooding and erosion are common in Tripura's plains area (Choudhury et al. 2022). Furthermore, Himalayan glaciers such as the Gangotri glacier are retreating at approximately 28 meters per year. This contributes to erratic hydrological flows in rivers, increasing flood risk during the monsoon and potentially reducing water availability in the long term, including Tripura (Das et al. 2009). Tripura, part of India's Northeastern region, is particularly vulnerable to flood hazards due to its unique geological and climatological conditions. Frequent flooding, soil erosion, and waterlogging are caused by the region's erosive landforms, dense forest cover, and heavy monsoonal rainfall (Choudhury et al. 2022). Climate change has intensified these challenges, necessitating a comprehensive and integrated flood risk mitigation strategy that aligns sustainability with resilience (Dogar and Sato 2019; Parrey et al. 2025; Dogar et al. 2025; Hossain et al. 2025; Choudhury et al. 2024). Unprecedented flooding in Tripura (2024) was caused by continuous and intense rainfall starting from August 19, 2024, as reported by the Government of Tripura (2024a, b). The situation quickly escalated, with districts like Gomati, South Tripura, Unakoti, and West Tripura suffering the most. It is worth noting that Bokafa in South Tripura experienced an extreme rainfall of 493.6 mm in a single day. Over 17 lakh people were affected by August 22, and over 65,400 individuals were accommodated in 450 relief camps throughout the state. There was a severe impact on agriculture.

According to early estimates, approximately 5000 hectares of vegetable crops and 1.20 lakh hectares of paddy fields were still submerged, resulting in significant crop losses. Additionally, all educational institutions have been closed until further notice. The impact of the 2024 flood was particularly severe. According to official assessments, the agriculture sector alone suffered losses worth INR 1284.0 crore, while allied sectors incurred damages worth INR 740.0 crore. Over 2.66 lakh farmers were affected by the floods, causing damage to 1.2 lakh hectares of cropped land. The human impact was also significant, resulting in the loss of 36 lives and the displacement of over 1 lakh people across the state. To respond, the government distributed INR 20.0 crore as interim relief and submitted a request for INR 128.8 crore in assistance, which includes INR 113.0 crore for agriculture and INR 15.80 crore for horticulture and plantation crops (Business Standard 2024). Unprecedented rainfall took place in the state, with West Tripura receiving 182 mm, Khowai 157.6 mm, and Gomati 153.1 mm in 72 h. Eight districts experienced widespread flooding as a result of the intense downpour.

The floods caused significant disruption to infrastructure and essential services. Over 1396 houses were partially damaged, 368 were severely damaged, and 184 were completely destroyed, particularly in South Tripura, Gomati, and Unakoti districts. Relief and mobility efforts were severely hindered by landslides at over 2032 locations and road erosion at 1952 sites, with only partial restoration completed at 579 locations. The power infrastructure also suffered a heavy blow, with 844 poles broken, 151 transformers malfunctioning, and 310 kilometers of power lines damaged, while two substations remained non-functional (Relief Web 2024a, b). According to Sphere India (2024), the floods in Tripura in August 2024 triggered a humanitarian emergency marked by significant loss of life, mass displacement, widespread infrastructure damage, and severe disruptions to food security and livelihoods. The disaster highlighted the region's growing exposure and vulnerability to climate-related hazards.

Human Impact and Displacement

The floods resulted in 32 deaths, three injuries, and the disappearance of one person. About 1,49,087 individuals were displaced and relocated to 821 relief camps across the state, resulting in the displacement of over 38,000 families. In Udaipur (Gomati district) and parts of South Tripura, floodwaters made it extremely difficult to access essential services.

Infrastructure Damage

The physical infrastructure was extensively destroyed by the floods. More than 9000 homes were completely destroyed, and nearly 38,000 homes suffered partial damage. The mobility and relief efforts were severely hindered by over 2500 landslides that blocked vital roadways. Thousands of electric poles and hundreds of transformers were damaged, resulting in significant damage to the power sector. Furthermore, water pipelines and sanitation systems suffered significant damage, causing a reduction in access to clean water and basic hygiene.

Agriculture and Livelihoods

Tripura's agricultural economy was severely affected as approximately 1.60 lakh hectares of farmland were submerged, affecting both paddy fields and vegetable crops. The disaster impacted livestock rearing and local businesses, resulting in 64% of surveyed households experiencing livelihood crises. The Dhalai district's highest livelihood impact affected 93% of households. The economic vulnerability was highlighted by 56% of households reporting monthly incomes below INR 10,000.

Food Security and Nutrition

The flood caused alarming damage to food systems. 52% of households experienced disruptions in food supplies, with South Tripura reporting the highest incidence at 70%. Damage to household food stocks was widespread, with 81% of households reporting losses. The impact of damage on household food stocks was widespread, with 81% of households reporting losses. Alarmingly, 86% of families stated that their food supplies would last only 1–2 weeks, and merely 2% had reserves for over 4 weeks. For 69% of respondents, fresh fruits and vegetables were not accessible, and there was a decline in meal frequency post-disaster, especially in the most affected districts. Flooding in Tripura is not just an environmental issue, but also a socio-economic crisis that requires immediate attention. Floods that disrupt lives, damage infrastructure, and derail local economies are common in the state, which is marked by its diverse topography and intersecting river systems. The region's reliance on rain-fed agriculture increases the vulnerability of its economy to climate-induced disruptions. Floods regularly affect about 60% of the population and 70% of the state land area, particularly low-lying riverine regions (Bhowmik and Das 2017). The floods are caused by both natural and human factors, such as intense monsoonal rainfall and the region's unique topography, as well as deforestation, rapid urbanization, and unplanned infrastructure development. Multifaceted mitigation strategies are required, as evidenced by flood-prone regions worldwide. The problem in Tripura is made worse by the absence of drainage infrastructure, insufficient early warning systems, and a lack of community preparedness. Though Gramin Kisan Mousam Seva (GKMS) from ICAR Tripura center had a significant impact on flood warning by reaching out to state departments, NGO, FPC, social media, television, and news channels. The flood risks in the district reflect wider national issues, as the northeast of India consistently experiences the highest annual rainfall in the country (Debnath et al. 2023).

The purpose of this paper is to identify effective strategies for reducing floods in Tripura. The insights presented here aim to provide information to both local authorities and national policymakers and global stakeholders who are interested in addressing the shared challenge of flood resilience. The significance of dealing with flooding in Tripura extends beyond emergency disaster management. Long-term regional development, safeguarding livelihoods, and protecting critical ecosystems can be achieved through effective mitigation. This study emphasizes the need to implement robust flood management strategies tailored to local contexts as climate change continues to increase the frequency and severity of extreme weather events (Dogar et al. 2017a, 2017b, Dogar and Almazroui 2022; Dogar 2020).

Table 20.1 Rainfall data for Tripura, 2024 (January–December)

Month	Jan	Feb	Mar	Apr	May	Jun	Jul	**Aug**	Sep	Oct	Nov	Dec
Rain (mm)	0	28	89.7	21	453.8	480.8	285.4	**714**	131.9	150.6	53.4	0

Sources: ICAR RC for NEH Region, Tripura Centre, 2025, Rain: Monthly Total Rainfall (mm)

Flood Characteristics in Tripura

Tripura is highly susceptible to various types of flooding due to its undulating topography of hills and valleys, which intersects several river systems. These floods manifest through various mechanisms.

1. **Intense Rainfall:** The region experiences some of the heaviest monsoonal downpours in the country (Table 20.1), which generate rapid accumulation of water and surface runoff. This overloads the district's limited drainage system, leading to widespread flooding in urban and rural areas (ReliefWeb 2024a, b).
2. **Riverine Floods:** Tripura is traversed by key rivers, including the Haora, Lohar Nala, and Gumti. Low-lying downstream areas are significantly impacted by overflow from these rivers due to prolonged rainfall or upstream activities. Poor embankments and silt in riverbeds are often the cause of this type of flooding (Bhowmik and Das 2017).
3. **Flash Floods:** The sudden, short-duration events are caused by localized, high-intensity rainfall. The piedmont zones of the district experience flash floods frequently due to steep slopes that cause water to cascade rapidly into settlements and agricultural fields, leaving little time for evacuation (Bhowmik and Das 2017).
4. **Urban Flooding:** Poorly planned urbanization has caused waterlogging during heavy rainfall. Inadequate stormwater drainage systems, the encroachment of natural waterways, and the loss of greenery are the causes of this problem.

Contributing Factors to Flood Vulnerability

(i) **Topographical Constraints:** Natural bottlenecks for water flow are created by the district's alternating hills and valleys. Depressed areas, especially in river basins, tend to retain water for extended periods, which can increase the severity of floods.

(ii) **Soil and Geology:** Many parts of Tripura have loose, sandy soils, which decrease infiltration rates and result in higher surface runoff during rains.

(iii) **Deforestation and Land Use Change:** The natural capacity of forests to absorb rainfall has been reduced by extensive deforestation for agriculture and urbanization, which has led to an increase in the likelihood of floods.

(iv) **Climate Change Impacts:** The increase in temperature has resulted in increased rainfall patterns, resulting in monsoons that are heavier and less predictable. In recent years, flooding has become more frequent and severe due to this (Choudhury et al. 2024).

Socio-Economic Implications

Both rural and urban populations are affected by the significant socio-economic implications of floods in Tripura. Communities' challenges are exacerbated by the immediate disruption of essential services, such as transportation, healthcare, and education. Roads, bridges, and public transport systems frequently become impassable, preventing critical services from being accessible. Healthcare facilities become overstretched, and education is disrupted. The submersion of farmlands is causing the destruction of crops, livestock, and fisheries. Food production and the livelihoods of many rural families depend on these sectors. The region's food insecurity is further exacerbated by the long-term effects on food production and the loss of income for farmers due to washed away crops. The economic impact is also significant in urban areas. Flooding paralyzes trade and commercial activities, forcing businesses to close due to inundation or infrastructure damage. The local economy relies heavily on small and medium enterprises, but they are experiencing significant financial losses and may not be able to reopen. Job losses are caused by this, which further exacerbates the poverty levels in urban areas.

Economic hardship is exacerbated by the paralysis of the economy and the destruction of urban infrastructure like roads, markets, and factories. The combination of lost income, destruction of property and rising costs of goods forces many families into deeper poverty. Due to crop failure and damage, many people move to urban centers to find work and shelter as rural areas become uninhabitable. The migration is putting more pressure on cities, which are already struggling with overcrowding and a lack of basic infrastructure. The emergence of informal settlements, which often lack sanitation, healthcare, and clean drinking water, leads to a rise in waterborne diseases and further health risks. The socio-psychological effects of flooding are also significant. Trauma, grief, and uncertainty are common experiences for communities affected by floods. The loss of life and property, along with the displacement of families, can lead to increased mental health issues, such as anxiety and depression. Without adequate support systems, these problems could persist, further hindering recovery. Furthermore, the financial burden on both local and national governments is significant. The expense of disaster response, recovery, and rebuilding drains resources, leading to a decrease in funding for development projects. The region's vulnerability to future floods is exacerbated by the lack of comprehensive flood management strategies and climate adaptation measures, which perpetuate the cycle of disaster, recovery, and loss.

Historical Flood Events

The frequency and severity of flood events in Tripura has increased, demonstrating the urgent need for better flood management strategies.

1. **August 2024 Floods:** Over 77,000 residents were displaced across Tripura due to record-breaking rainfall during the latest flood event. The key infrastructure was severely damaged, with 1396 homes destroyed and 151 transformers damaged, resulting in widespread power outages. Many communities were left in distress due to the destruction of homes and critical infrastructure, which disrupted daily life and essential services (Government of Tripura 2024a, b).
2. **2018 Floods:** The state experienced significant agricultural losses as a result of this event, which submerged approximately 12,000 hectares of farmland. The economic losses were made worse by the damage to rural roads and bridges, which prevented relief operations for weeks.
3. **2004 Flash Floods:** The piedmont zones were affected by flash floods caused by intense localized rainfall, with particular impact on the villages along the Lohar Nala Basin. The sudden onset of flooding led to the loss of over 50 lives and highlighted the need for more robust early warning systems.
4. **1958 Cyclonic Floods:** A devastating cyclonic storm that caused widespread flooding across northern Tripura was described in historical accounts from 1958. The state's nascent infrastructure was disrupted by this event (Chaudhury 1964).

Patterns and Trends

The analysis of flood patterns over the decades reveals several crucial trends:

(a) **Increased Frequency:** Annual flood reports have shown a significant increase in flood occurrences since the 1990s, which coincides with rapid urbanization and deforestation.
(b) **Greater Intensity:** The intensity of rain during monsoons has increased (Table 20.1), with events like the 2024 flood causing record-breaking downpours in short timeframes.

GIS-Based Flood Hazard Mapping

Flood risk assessment has been transformed by the use of Geographic Information Systems (GIS) and Remote Sensing (RS). The integration of multiple datasets is possible with these tools, leading to high-resolution hazard maps that pinpoint flood-prone areas with precision. Key findings and methodologies are outlined below:

(i) **Lohar Nala Basin:** Approximately 20–25% of the basin is flood-prone. A composite flood hazard index identified population density, proximity to rivers, and agricultural losses as critical risk factors (Bhowmik and Das 2014). Remote sensing data, combined with field surveys, revealed that seasonal siltation exacerbates flooding.
(ii) **Manu-Deo River Basin:** Analytical Hierarchical Process (AHP) modeling was employed to prioritize flood-prone areas based on elevation, drainage density, and rainfall intensity. The results pinpointed Kailashahar and Kanchanpur as high-risk hotspots (Debnath et al. 2023).
(iii) **Urban Flood Mapping in Agartala:** GIS applications have been utilized to track urban expansion and its impact on drainage patterns. By overlaying historical flood data with urban land use maps, researchers identified areas of recurrent waterlogging and proposed targeted drainage upgrades.

Methodological Advancements

I. **Data Integration:** The integration of satellite imagery (e.g., Landsat, Sentinel) with hydrological models has enhanced the accuracy of flood predictions.
II. **Thematic Mapping:** Layers depicting land use, soil types, and elevation have been instrumental in identifying flood triggers and vulnerabilities.
III. **Community Validation:** Ground-truthing exercises involving local stakeholders have ensured the relevance and applicability of GIS findings.

Policy and Practical Applications

(a) **Zoning and Land Use Regulation:** Flood hazard maps inform zoning laws to restrict development in high-risk areas, ensuring safer urban planning.
(b) **Resource Allocation:** Hazard maps guide the prioritization of investments in embankments, flood shelters, and drainage systems.
(c) **Disaster Preparedness:** Maps have been incorporated into early warning systems, helping communities and local authorities to prepare more effectively.

Mitigation Strategies

Structural Measures

I. **Drainage Systems***:* Enhanced urban drainage systems are critical for mitigating waterlogging in Agartala and other urban areas. Underground pipelines, flood gates, and pumping stations have proven effective in expediting water removal after heavy rainfall. Regular maintenance of these systems is equally vital to prevent blockages caused by silt and debris.

II. **Riverbank Embankments and Reinforcement:** Embankments along the Haora, Lohar Nala, and Gumti rivers have been strengthened to reduce overflow risks during monsoons. Innovative solutions, such as bioengineering techniques using vegetation to stabilize embankments, can further enhance these structures.

III. **Flood Shelters:** Elevated community shelters equipped with food, water, and sanitation facilities provide safe havens during floods. These shelters must be strategically located to ensure accessibility within the shortest possible time.

IV. **Retention Ponds and Reservoirs:** Creating artificial ponds and reservoirs can help in storing excess rainwater, thereby reducing the volume of runoff reaching rivers and drainage systems. This approach not only mitigates flooding but also contributes to groundwater recharge.

Non-Structural Measures

(i) **Early Warning Systems:** Automatic weather stations and hydrological monitoring networks can predict flood risks with higher accuracy. Alerts disseminated via mobile networks and community radio stations can ensure timely evacuation and preparedness.

(ii) **Land Use Regulation:** Implementing zoning laws to prevent construction in flood-prone areas is essential. Designating floodplains as protected zones and promoting afforestation efforts can reduce runoff and improve soil absorption.

(iii) **Capacity Building:** Training programs like the Aapda Mitra initiative empower local communities with skills in flood response and rescue operations. Mock drills, first-aid training, and disaster education campaigns enhance overall community resilience.

(iv) **Flood Insurance:** Introducing subsidized insurance schemes can mitigate the financial impact of floods on vulnerable populations. Policies tailored to cover property, crops, and livestock losses can provide much-needed economic relief.

Policy Recommendations

(a) **Integrated Water Resources Management (IWRM):** A district-level IWRM framework that aligns with state and national disaster management plans can streamline efforts. This approach ensures that water resources are managed sustainably while addressing flood risks.

(b) **Public-Private Partnerships (PPPs):** Engaging private entities in infrastructure development, such as constructing drainage systems and flood shelters, can mobilize resources and accelerate project implementation.

(c) **Climate Adaptation Strategies:** Given the increasing unpredictability of rainfall patterns, incorporating climate adaptation measures into planning processes is imperative. Projects like elevated road networks, climate-resilient crops, and renewable energy integration can reduce flood impacts over time.

Challenges and Opportunities

Challenges

Flood mitigation in Tripura faces a multitude of challenges that complicate planning and implementation. These include:

I. **Climate Change and Variability:** Climate change has exacerbated weather patterns, making rainfall more intense and less predictable. Rising temperatures increase the likelihood of extreme weather events, stretching the capacity of existing flood management systems.

II. **Inadequate Infrastructure:** Ageing and poorly maintained drainage systems, embankments, and flood shelters fail to cope with the increasing volume and intensity of water flow. Urban centers like Agartala are particularly vulnerable to waterlogging due to inadequate stormwater management.

III. **Resource Constraints:** Limited financial and human resources hinder the development and maintenance of critical infrastructure. The high cost of advanced technologies and extensive flood defenses poses a significant barrier to their widespread adoption.

IV. **Institutional Gaps:** Coordination among local, state, and national agencies remains weak. Overlapping mandates and bureaucratic inefficiencies delay the implementation of flood mitigation projects and the dissemination of early warnings.

V. **Unplanned Urbanization:** Rapid urban expansion, often encroaching on natural waterways and floodplains, exacerbates flood risks. Poor zoning laws and enforcement contribute to the development of high-risk areas.

VI. **Community Vulnerability:** Many residents, particularly in rural and economically disadvantaged areas, lack the resources and knowledge to prepare for and recover from floods. Socio-economic factors further amplify vulnerabilities, leaving marginalized groups at greater risk.

Opportunities

Despite the challenges, there are plenty of opportunities to improve flood management and build resilience in Tripura.

(i) **Advances in Technology:**
GIS and remote sensing tools enable detailed flood risk assessments, predictive modeling, and real-time monitoring of flood-prone areas. Early Warning Systems is investing in automated weather stations and mobile alert systems that can provide communities with timely warnings and reduce casualties and property damage.

(ii) **Policy Innovation:**
Implementing Integrated Water Resource Management (IWRM) ensures that water resources are sustainably managed across sectors. Enforcing zoning laws to restrict development in flood-prone areas and designating floodplains as protected zones can mitigate risks.

(iii) **Public-Private Partnerships (PPPs):**
Collaboration with private entities can mobilize financial and technical resources for flood mitigation projects, such as building resilient infrastructure or introducing insurance schemes for vulnerable populations.

(iv) **Community-Based Approaches:**
Initiatives like the *Aapda Mitra* program, which trains local volunteers in disaster response, empower communities to act as first responders during emergencies. Awareness campaigns and educational programs can enhance community preparedness and resilience.

(v) **Capacity Building:**
Training programs for local authorities and planners in flood risk management can bridge knowledge gaps. Collaborative efforts with academic institutions can support research and innovation in flood mitigation strategies.

(vi) **Climate Adaptation Strategies:**
Promoting the use of climate-resilient crops and agricultural practices reduces economic losses in flood-affected areas. Integrating green infrastructure, such as rain gardens and permeable pavements, in urban planning enhances water absorption and reduces runoff.

(vii) **Funding Mechanisms:**
Accessing international funds, such as the Green Climate Fund, can support large-scale infrastructure projects and community-based adaptation initiatives. Introducing microfinance schemes for rural communities can support small-scale, locally driven mitigation projects.

Integrated Flood Risk Mitigation Strategies

(i) **Land and Soil Management**

- Minimum and Zero Tillage: These practices help in retaining soil moisture and reducing runoff, thereby lowering flood vulnerability (Das et al. 2009).
- Contour Bunding and Mulching: In-situ soil and water conservation techniques that prevent surface erosion and improve water absorption capacity.

- Land Restoration: Restoring degraded lands through cover cropping and vegetative barriers helps in stabilizing slopes and controlling sediment flow.

(ii) **Water Resource Management**

- Rainwater Harvesting: Both in-situ (mulching, land shaping) and ex-situ (ponds, micro-harvesting structures like *Jalkund*) methods can help in reducing surface runoff and managing excess rainfall (Choudhury 2023).
- Watershed Development: Integrated approaches involving afforestation, check dams, and drainage planning reduce the risk of flash floods.
- Agroforestry: Introducing multipurpose trees and hedgerows along waterways enhances hydrological regulation and reduces erosion.

(iii) **Climate-Adapted Agriculture**
Drought-Resistant, Short-Duration Varieties: Selection of climate-resilient crops aids in minimizing losses from unpredictable rainfall and flood episodes (Das et al. 2009). SRI and Aerobic Rice Cultivation: These water-efficient techniques reduce greenhouse gas emissions and increase flood resilience. Revised Crop Calendars: Adjusting sowing dates in line with changing rainfall patterns can significantly reduce crop damage from floods.

(iv) **Ecosystem-Based Approaches**
Forest and Wetland Conservation: Forests and natural wetlands act as buffers during heavy rainfall and absorb excess water, helping regulate river discharge. Biomass Management in Shifting Cultivation: Replacing slash-and-burn with in-situ biomass recycling enhances soil organic matter and reduces CO_2 emissions (Das et al. 2009; Devi and Choudhury 2013).

(v) **Community Infrastructure and Planning**
Flood-Resilient Infrastructure: Building raised storage facilities and flood-safe community shelters enhances local coping capacity. Early Warning Systems: Timely weather updates and flood alerts help communities prepare for extreme events. Integrated Farming Systems: Combining livestock, fisheries, and crop production improves resource use efficiency and household resilience (Das et al. 2021). To build long-term resilience against flooding in Tripura, a multi-sectoral and ecosystem-based approach is essential. By integrating modern techniques with traditional knowledge and focusing on land, water, and community-based solutions, the region can move towards sustainable flood risk reduction. This comprehensive strategy aligns well with national and global goals for climate adaptation and sustainable development.

Conclusion

Addressing floods in Tripura requires a holistic approach that integrates structural interventions with community-based initiatives. Advanced technologies like GIS and early warning systems are crucial for preparedness, but community participation and education are necessary to ensure that these measures reach their full potential. In addition, promoting collaboration among multiple stakeholders, including public-private partnerships, can bridge resource gaps and speed up the implementation of projects. The integration of climate adaptation strategies into flood management plans ensures that solutions are sustainable and resilient to future climate variability. Policymakers must ensure equitable access to resources and opportunities for vulnerable populations while prioritizing long-term investments in infrastructure, technology, and human resources to achieve these goals. Transforming policies into actionable outcomes requires strong institutional frameworks and capacity building at the local level. The collective dedication of government agencies, communities, and private stakeholders will ultimately determine the success of flood mitigation efforts in Tripura. By adopting an inclusive and adaptive approach, the region can transform the challenge of recurring flooding into an opportunity for sustainable development and resilience building.

References

Bhowmik M, Das N (2014) Flood hazard mapping of the Lohar Nala Basin, West Tripura. Arch Appl Sci Res 6(3):142–153

Bhowmik M, Das N (2017) Rainfall frequency analysis to predict flood in West Tripura District, Tripura, North-East India. Int J Geomatics Geosci 7(3):310–312

Business Standard (2024, August 26) Floods in Tripura cause INR 2024 crore loss to agriculture and allied sectors; 2.66 lakh farmers affected. Business Standard India. https://www.business-standard.com/

Chaudhury AK (1964) Flood in Tripura in October 1958. Mausam 15(2):239–246

Choudhury BU (2023) Soil moisture conservation in water-deficit hill agriculture: affordable to small landholdings. Indian Farming 73(04):13–17

Choudhury BU, Nengzouzam G, Ansari MA, Islam A (2022) Causes and consequences of soil erosion in northeastern Himalaya, India (a review paper). Curr Sci 122(7):772–789

Choudhury BU, Daschaudhuri D, Ghosh K (2024) Tracking the long-term progression of weather variables in West Tripura through the lens of climate change. J Agric Phys 24(2):151–167

Das A, Ghosh PK, Choudhury BU, Patel DP, Munda GC, Ngachan SV, Chowdhury P (2009) Impact of climate change on agriculture. In: ISPRS Archives XXXVIII-8/W3 Workshop Proceedings: Ahmedabad, India

Das A, Datta D, Samajdar T, Idapuganti RG, Islam M, Choudhury BU, Mohapatra KP, Layek J, Babu S, Yadav GS (2021) Livelihood security of small holder farmers in eastern Himalayas, India: pond based integrated farming system a sustainable approach. Curr Res Environ Sustain 3:100076. https://doi.org/10.1016/j.crsust.2021.100076

Debnath S, Sethi LN, Gautam VK, Nath NK (2023) Flood-prone area mapping of Manu-Deo River Basin, Tripura using GIS and RS. Indian J Ecol 50(2):539–545

Devi NL, Choudhury BU (2013) Soil fertility status in relation to fallow cycles and land-use practices in shifting cultivated areas of Chandel District Manipur, India. IOSR J Agric Vet Sci 4(4):1–9

Dogar MM, & Sato T (2019). Regional climate response of Middle Eastern, African, and South Asian monsoon regions to explosive volcanism and ENSO forcing. Journal of Geophysical Research: Atmospheres, 124(14), 7580–7598.

Dogar MM, Kucharski F, and Azharuddin S (2017a). Study of the global and regional climatic impacts of ENSO magnitude using SPEEDY AGCM. Journal of Earth System Science, 126(2), 30.

Dogar MM, Stenchikov G, Osipov S, Wyman B, & Zhao M (2017b). Sensitivity of the regional climate in the Middle East and North Africa to volcanic perturbations. Journal of Geophysical Research: Atmospheres, 122(15), 7922-7948.

Dogar MMA (2020). Study of the regional climatic impacts of tropical explosive volcanism in the Middle East and North Africa region (Doctoral dissertation, Hokkaido University).

Dogar MMA, & Almazroui M (2022). Revisiting the strong and weak ENSO teleconnection impacts using a high-resolution atmospheric model. Atmospheric Environment, 270, 118866.

Dogar, M.M., Watanabe, S., Fujiwara, M. et al. Tropical volcanic impacts on MENA climate via ENSO and NAO dynamics in a high-top model. npj Clim Atmos Sci 8, 330 (2025). https://doi.org/10.1038/s41612-025-01201-x

Government of Tripura (2024a) Updates on flood situation in the state on 22-08-24. Directorate of Information and Cultural Affairs

Government of Tripura (2024b, August 22) Updates on flood situation in state on 22-08-2024: Review of Chief Minister on relief and response measures [Press release]. Directorate of Information and Cultural Affairs, Government of Tripura

Hossain, M.A., Begum, M., Akhtar, M.N. et al. Predicting water scarcity in northern Bangladesh using deep learning and climate data. npj Clim Atmos Sci 8, 348 (2025). https://doi.org/10.1038/s41612-025-01030-y

Parrey, H.A., Bhat, M.S., Shah, S.A. et al. Evaluation of the perception of climate change and its implications across various geo-physical regions of Kashmir Himalayas. Nat Hazards 121, 18685–18712 (2025). https://doi.org/10.1007/s11069-025-07534-3

ReliefWeb (2024a) Tripura floods and landslides: joint rapid needs assessment. https://reliefweb.int/report/india/tripura-floods-and-landslides-joint-rapid-needs-assessment-september-2024

ReliefWeb (2024b, August 23) India: floods and landslides in Tripura—situation report as of 23 Aug 2024. ReliefWeb. https://reliefweb.int/

Sphere India (2024, August) Joint rapid needs assessment: Tripura floods 2024. Inter-Agency Coordination Committee, Sphere India. https://www.sphereindia.org.in/

Chapter 21
Sustainable Horticultural Practices for Climate Change Adaptation and Mitigation

Bapi Das, Burhan Uddin Choudhury, Biswajit Das, Samik Choudhury, Parag Majumder, Bipasha Saha, Pradip Kumar Sarkar, Rahul Kumar, Abhijit Debnath, P. Barman, and Ashima Suklabaidya

Abstract The agricultural industry is confronted with a substantial risk as a result of climate change, resulting in a rise in temperatures, shifting precipitation patterns, and an increase in the number of extreme weather events. All of these factors pose a threat to the quality of crops, their production, and long-term viability of agricultural systems as a whole. Climate-resilient horticulture uses innovative techniques and state-of-the-art technologies to tackle these escalating problems. The creation of crop types that are more tolerant to heat, drought, and salinity is a crucial component of this strategy since it allows them to thrive in challenging environmental circumstances. Crops like pineapple, strawberry, banana, dragon fruit, chilli, cherry tomato, and ridge gourd require proper soil moisture, which can only be maintained by water conservation measures like drip irrigation, rainwater harvesting, and different mulching techniques like weed mat, black polythene, and bio-mulching. Soil fertility and crop resilience are further enhanced by enhancing soil health using sustainable methods like conservation tillage and organic additions. Precision horticulture and other technological innovations maximise input utilisation and minimise waste, while controlled environment agriculture, which includes soilless culture, hydroponics, and aeroponics, provides stability in growth conditions by providing protection from unpredictable weather. Moreover, advances in biotechnology and genetic engineering offer promising avenues for creating plant species that are more resilient. Despite the advancements, there are still obstacles such as costly implementation costs, informational barriers, and restrictive regulations. For climate-smart horticulture to reach its full potential and to ensure long-term

B. Das (✉) · B. U. Choudhury · B. Das · P. Majumder · B. Saha · P. K. Sarkar · R. Kumar · P. Barman · A. Suklabaidya
ICAR Research Complex for NEH Region, Tripura Centre, Lembucherra, West Tripura, India

S. Choudhury
ICAR Research Complex for NEH Region, Mizoram Centre, Kolashib, Mizoram, India

A. Debnath
Krishi Vigyan Kendra, Dhalai, Tripura, India

S. Mandal et al. (eds.), *Climate Change Vulnerability*, Climate Change Management, https://doi.org/10.1007/978-3-032-16783-5_21

agricultural sustainability and food security in a warming world, these obstacles must be removed.

Introduction

Climate change significantly affects horticulture by affecting crop quality, productivity, and growth. In addition to causing heat stress, rising temperatures can interfere with pollination, blooming, and fruit setting. Changes in rainfall patterns, on the other hand, can cause droughts or floods that harm crops and weaken soil. Production in horticulture is further endangered by the increasing occurrence of severe weather phenomena like hail and storms. Furthermore, different temperature zones are changing the areas where specific crops, pests, and illnesses can flourish, which presents growers with new difficulties. Changes in climate have a significant impact on pest populations in particular, which can alter the dynamics of ecosystems. Horticultural systems' long-term sustainability and resilience are under jeopardy due to these effects. The ability of a system to foresee, endure, and bounce back from unfavourable occurrences while preserving or improving its fundamental operations and structure is referred to as resilience (IPCC 2012). Building resilient horticultural systems that can withstand climate disturbances and recover well when conditions improve is crucial, as this idea emphasises.

Importance of Climate-Resilient Practices

Climate-Smart Horticulture (CSH) is a proactive, integrated method to making horticultural systems more adaptable, sustainable, and resilient around climate change (Tahir et al. 2025). Innovative technologies, eco-friendly growing methods, and flexible management strategies are all brought together to make better use of resources, slow down environmental damage, and guarantee the long-term viability of horticultural production. Adding carbon to soil makes plants better able to handle problems caused by climate change (Verma et al. 2026). This is a very important part of making sure there is enough food for everyone, helping rural businesses grow, and making sure people get enough nutrition. Integrating digital tools like precision farming and real-time weather tracking gives farmers even more power to handle changing weather conditions (Mandal et al. 2018, 2020). Adopting climate-resilient practices, farmers can protect their crops and incomes and also help reach larger sustainability goals. This will keep horticulture viable and useful in a climate that is becoming less predictable.

Climate Change

Long-term changes in climate conditions backed by statistical evidence of significant variations in mean states or climate system variability are referred to as climate change by the Intergovernmental Panel on Climate Change (2007). Only when these changes continue over long periods of time—usually several decades or longer—are they deemed to be part of climate change.

Current Perspectives on the Impact and Adaptation of Horticultural Crops

Fruit Crops

Perennial fruit crops demand a large initial investment to establish orchards, but their yearly yields are seriously threatened by the growing unpredictability of climate conditions. Long-term decisions made by farmers on crop selection and economic viability are greatly impacted by these uncertainties (Rymbai et al. 2024). Seasonal losses that occur one after the other can deter growers from continuing to cultivate these crops, frequently leading to their complete abandonment. Significant climate change in Himachal Pradesh, including warmer temperatures, longer summer droughts, and less winter snowfall, is encouraging farmers to look for more financially and climate-resilient options. Many in the lower parts of Kullu and Mandi districts are turning to vegetable cultivation, especially the production of tomatoes and peas, as a result of the decline in output in areas that were once fairly conducive to the production of apples. Peach cultivation has supplanted apple orchards in the Sirmaur district's Rajgarh area. Furthermore, guava and mango are becoming more and more susceptible to drastic temperature changes brought on by heat and cold waves (Table 21.1). In addition to high temperatures during blooming and fruit development, extended dry spell, and irregular rainfall at water stress times affect citrus crops. In banana plants, 31–32 °C speeds maturation and bunch growth (Singh et al. 1966; Turner et al. 2007) (Table 21.2).

Vegetable Crop

Studies have revealed that early growth of tomato crops under soil moisture stress can lead to a 26% decline in yield (Verma et al. 2022, 2024). Likewise, Srinivasa Rao (1995) found that when temperatures rose above 28 ° C and plants experienced water stress, tomato yields dropped by 30–45% across various cultivars. To combat moisture stress, drought- and rainfall-resistant cultivars have been introduced such as tomato variety Arka Vikas, onion variety Arka Kalyan, Bhima Shakti, Bhima

Table 21.1 Effect of climate change on the nutritiousness of fruit

Conditions	Effect	Reason
Elevated CO_2 levels	Increases carbohydrate (sugar/starch) content but reduces protein and essential minerals like iron, zinc, and magnesium	Plants photosynthesize more under higher CO_2, leading to "carbon dilution "more sugars but fewer nutrients per unit of mass
Higher temperatures	Alters fruit development, often leading to lower concentrations of vitamins (like vitamin C) and phytochemicals	Heat stress can interfere with enzyme activity and biochemical pathways that synthesize these nutrients
Water stress/ drought	Leads to smaller fruit size and can reduce water-soluble nutrients like vitamin C and some B vitamins	Drought affects nutrient uptake from soil and plant metabolism
Extreme weather events	Damage to crops reduces quality; may lead to early or delayed ripening, which can change nutrient profiles	Storms, floods, and hail can damage fruit tissue, disrupt ripening, and reduce shelf life and nutrient content
Pests and diseases	Increased pest pressure under warmer climates can lead to more chemical use or damaged fruits with altered quality	Stress from pests reduces the plant's ability to allocate energy toward producing healthy, nutrient-rich fruit

Table 21.2 Varieties tolerant to moisture stress

Fruit crop	Rootstock
Almond	*Prunas xerophila* and *P. amygdaliflora*
Apple	MM111
Pear	*Pyrus betulaefolia* and *Pyrus calleryan*
Sapota	Khirni (Manilkara)
Citrus	Alemow, Swingle
Guava	*Psidium cujavilis*
Ber	*Ziziphus rotundifolia* and *Z. nummularia*
Fig	Gular (*Ficus glumerata*)

Source: Adopted from Nimbolkar et al. (2016)

Super (Das et al. 2025), and chilli variety Arka Lohit. In tomatoes, high temperatures negatively affect fruit quality and size, reduce fruit set, and cause substantial yield losses. Geisenberg and Stewart (1986) identified the optimal daily average temperature range for successful tomato fruit set as 21–24 ° C. The pre-anthesis stage in tomatoes is particularly temperature-sensitive. Fruit set in peppers is hampered by high post-pollination temperatures, suggesting that fertilisation is extremely temperature-dependent. Similarly, temperature affects the sex expression of cucumber flowers. While female flower production is favoured by lower temperatures, male flower development is encouraged by higher temperatures. High temperatures in onions reduce the growing season, which eventually lowers output.

Spices and Plantation Crops

In some areas, including the Maidan region of Karnataka and Coimbatore in Tamil Nadu, as a result of recurrent droughts, coconut production has recently declined. The production of coconuts has, however, typically increased over the previous 50 years, with the exception of a four-year period during which production fell by about 300 million nuts each year. Since cashews are mostly grown when it rains, they are especially vulnerable to droughts that are caused by changes in moisture levels and other weather changes (Table 21.3). Unseasonal rainfall and heavy dew during flowering and fruiting stages, alongside reduced pollinator activity, negatively impact nut set by increasing the risk of pest infestations and disease outbreaks. Rainfall-induced soaking during the harvesting process affects the nuts' quality and shelf life.

Effects of climate change impact both conventional and organic horticulture, but the nature and severity of these effects and how each system responds can differ:

Table 21.3 The impact of climate change on organic and conventional horticulture

Conditions	Organic horticulture	Conventional agriculture
Temperature increases	Organic systems may be more resilient due to better soil health and biodiversity, but can still face yield drops from extreme heat without synthetic inputs to compensate	Higher temperatures may lead to greater pest pressure and heat stress in plants, potentially increasing the need for synthetic pesticides and irrigation
Water scarcity	Heavily reliant on irrigation, with water-intensive crops and chemical fertilizers that can degrade soil's water-holding capacity	Builds soil organic matter, which enhances water retention, making it better suited to cope with drought, though limited irrigation infrastructure can still be a challenge
Pest and disease pressure	Emphasizes natural pest control (e.g., beneficial insects, crop rotation), but climate change may outpace these biological methods unless adaptively managed	Increased pests due to warming often lead to higher chemical pesticide use, which can result in resistance and environmental harm
Soil health	Prioritizes compost, cover crops, and reduced tillage practices that enhance soil stability and carbon sequestration	Often depletes soil structure through monoculture and chemical inputs, making soils more vulnerable to erosion from heavy rains
Crop yields	May experience lower yields initially, but demonstrates more stability over time due to greater ecosystem resilience	Short-term yields may be maintained through technology, but with higher environmental costs and diminishing returns under extreme conditions

Source: Adopted from Manyaku et al. (2024)

Adaptive and Innovative Methods for Sustainable Horticulture in a Changing Climate

Despite the ongoing progression of climate change, atmospheric concentrations of CO_2 and methane are projected to continue rising, potentially contributing to temperature increases, heightened water demand, and more severe biotic and abiotic stresses. While high temperatures can cause lowers yields and raises the chance of crop failure, inadequate chilling in mid-hill areas may prevent optimal flowering in apple agriculture. The possible dangers that climate variability may pose are highlighted by these worries. However, technical advancements have frequently been made in response to historical climate fluctuations, helping to lessen these difficulties. Conventionally, crops that are sensitive to temperature, such cauliflower, potatoes, tomatoes, and cabbage, can only flourish in areas with long daylight hours (Table 21.4). However, developments like the creation of suitable cultivars and better management of production systems have now made it feasible to attain large yields even in subtropical and slightly subtropical environments.

Heat Resistant and Drought Tolerant Varieties

Numerous robust cultivars have been created to withstand these unfavourable environmental circumstances through focused breeding and selection initiatives (Rymbai et al. 2024). The ability of these cultivars to retain yield and quality in the face of heat and water stress has led to their specific selection or breeding.

Table 21.4 Varieties and rootstock tolerant to drought

Fruit crop	Rootstock/variety
Pomegranate	*Passiflora incarnata*
Annona	Arka Sahan
Fig	Deanna, excel
Guava	*Psidium cujavilis*
Cherry	Mahaleb
Citrus	Rangpur lime
Ber	*Ziziphus nummularia*
Sapota	Khirni
Passion fruit	*Passiflora incarnata*
Ber	*Ziziphus rotundifolia*
Grape	110R, Dogridge

Source: Adopted from Bose et al. (2001) and Singh et al. (2009a)

Salt Tolerant Varieties

The capacity to endure salinity is a multifaceted physiological and genetic characteristic. Traditional breeding methods have yielded limited advancements in enhancing this feature (Flowers 2004). Successful salt tolerance breeding requires effective screening techniques, sufficient genetic diversity, and the ability to transfer desired traits into target species. Field-based screening is generally discouraged due to inconsistent salinity levels in field soils. In Tunisia, the Bendi pepper cultivar outperformed the other evaluated types in terms of yield when subjected to salt stress. *S. esculentum* accession PI17426 has been associated with quick germination in tomatoes under saline conditions (Foolad and Jones 1991) (Table 21.5).

Water Management Strategies

The necessity for high-quality water presents a considerable challenge to vegetable production. Water management significantly impacts vegetable yield and quality (Verma et al. 2024; Hossain et al. 2025). Poor irrigation, whether excessive or insufficient, can result in uneven plant growth, heightened vulnerability to diseases, and nutritional imbalances. Timely water management strategies like soil moisture conservation are critical agronomic practices for sustaining yields during drought stress in regions with limited and unpredictable water resources (Khan et al. 2025; Verma et al. 2022). The choice of irrigation method is influenced by various factors, including crop variety, water availability, soil properties, and geography. Surface irrigation is employed on over 80% of the irrigated land globally; however, its field-level application efficiency often falls below 40% to 50% (Von and Chieng 2004).

Drip Irrigation/Micro-Irrigation

Unlike traditional surface irrigation methods, this system provides precise control over water and fertiliser application, significantly reducing runoff and deep percolation losses. Drip irrigation systems have been shown to save between 50 and 80% of

Table 21.5 Varieties/rootstock tolerant to salinity

Fruit crop	Rootstock/variety
Mango	Nileswar dwarf, Kurukram
Mango	Bappakai
Grape	110R, Dogridge
Fig	F. glomerata
Lime	Rangpur lime
Citrus	Cleomerata

Source: Adopted from Singh et al. (2009a, b)

Fig. 21.1 Drip irrigation in Dragon Fruit cultivation

water. In the cultivation of chilli peppers, for example, water consumption was significantly decreased when drip irrigation took the place of furrow irrigation, particularly when high-efficiency delivery regimes were used. According to Sanchez-Guerrero et al. (2009), drip irrigation reduced nitrous oxide (N_2O) emissions in melon crops by 70%. Subsurface drip systems can even achieve higher water consumption efficiency than surface systems (Sharma et al. 2011). Additionally, a drip system that is kept up to date improves the quality of tomato fruit while also saving water (Rudich et al. 1977). In order to satisfy the unique needs of tomato plants, irrigation plans must be customised to account for their varying water requirements at different stages of growth (Fig. 21.1).

Rainwater Harvesting

Water harvesting is a traditional approach to addressing water scarcity across different climatic regions (Balusamy et al. 2024). Considering climate change, the conservation of floodwater and rainfall can enhance agricultural yields and reduce failure rates, thereby improving the productivity of lands typically regarded as marginal. In India, cashew is commonly grown as a rainfed crop on underutilised land unsuitable for alternative agricultural practices. Collecting rainwater, implementing in situ soil and water conservation, and ensuring a consistent supply of cashew seedlings are critical strategies to tackle this issue. Cashew plantations have been effectively established on steep and arid sites through techniques including modified concrete bunds and the burial of coconut husks in staggered trenches along slopes (Rejani and Yadukumar 2010).

Rainwater harvesting can be done in two main ways: surface storage in structures such as check dams, ponds, jalkund, and underground tanks, or through groundwater recharge. The latter involves various techniques aimed at restoring aquifer levels by capturing rainfall where it falls known as on-site water collection. Farmers also use soil moisture management methods to enhance rainwater retention and contribute to groundwater recharge. Public participation is vital in water conservation efforts, and stakeholders must be properly trained to improve their knowledge and implementation of sustainable water management practices.

Approaches to Soil Management

Mulching Practices in Horticulture

Crop management techniques such as mulching, raised beds, protective shelters are critical for maintaining soil moisture and protecting crops from environmental stresses, including heavy rainfall, and extreme temperature fluctuations (Verma et al. 2022). Vegetable and fruits cultivation can utilise both synthetic and organic mulches to enhance soil health and crop performance. These mulches serve multiple functions, such as minimizing runoff water and soil erosion, regulating soil temperature, minimising evaporation (Choudhury et al. 2022), managing weed growth, and preventing fruits from contacting the soil surface. Mulching can reduce irrigation water requirements by as much as 25%. Organic mulches enhance soil fertility, soil structure, and overall soil health. Rice straw and dry plant waste is readily available and recommended as mulch for summer tomato cultivation in tropical rice-producing areas. Organic mulches effectively suppress weed growth, significantly influenced by soil moisture and temperature (Choudhury et al. 2022; Verma et al. 2024). Jodaugienė et al. (2006) assessed the effects of various organic mulches, such as wood chips, peat, and chopped wheat straw, applied in layers of 5 and 10 cm over an eight-year period. Wood chips, straw, and peat were the most effective forms; significantly reducing weed emergence, particularly during peak germination periods. The application of mulching significantly enhances the growth of crops such as brinjal, okra, tomato, capsicum, king-chilli, and cucurbits in India, compared to plots without mulch (Pandita and Singh 1992; Verma et al. 2024). The weed mat technique for high-quality pineapple production effectively eliminates weeds, resulting in a 25–30% reduction in weed presence, while also preserving adequate soil moisture by allowing water to permeate through the micro-pores of the sheet (Das et al. 2024). Furthermore, the combination of drip irrigation and black polyethylene mulch resulted in a 57.87 t/ha increase in tomato yields (Singh et al. 2009a, b). In tropical lowland regions, it is recommended to integrate rice straw with dark-coloured plastic mulch. The dark plastic inhibits sunlight penetration to the soil, thereby further suppressing weed growth. In hot, rainy seasons, significant rainfall often leads to waterlogging and damage to fruit, consequently diminishing agricultural yields, such as those of tomatoes. Clear plastic rain shelters have been shown to enhance crop yields by protecting plants from rainfall and excess water (Midmore et al. 1992) (Fig. 21.2).

Fig. 21.2 Weed-mat technology in pineapple cultivation

Conservation Tillage

Conservation tillage is essential for promoting minimal tillage and maintaining stable soil cover, as these practices enhance soil organic matter and reduce the effects of wind, flooding, erosion, drought, and heavy rainfall. Further research could be conducted in the domains of conservation agriculture, risk management production techniques, and organic farming practices. Aerobic mineralisation and excessive disturbance from extensive soil tillage reduce soil organic matter. Research indicates that these methods enhance the resilience of agricultural systems and bolster farmers' ability to adapt to climate change (Lal 1987). These techniques can concurrently decrease carbon emissions associated with ploughing while enhancing carbon absorption through diminished erosion and increased residue retention. Farmers' community of Niger Delta have employed various strategies to adapt to climate change. 75.8% focused on seed preservation, 77.2% participated in mixed farming, and 80.2% initiated planting early in the wet season. Additionally, 67.5% of farmers reported an increase in weeding frequency, 71.8% implemented cover crops, 72.0% utilised organic fertilisers, and 74.2% employed inorganic fertilisers. Notable adaptation activities that mitigate the effects of climate change include low or zero tillage (57.2%), education on climate change (59.2%), watershed protection/mulching (64.5%), (Nzeadibe et al. 2011).

Improving Plant Survival under Stress Through Grafting

Plant performance under stress can be improved by grafting sensitive plant varieties (scions) onto more resilient rootstocks, especially in settings that are impacted by drought and salinity. Grafting brinjal was the first use of this technique in year 1950, 1960 and 1970; it was expanded to cucumbers and tomatoes (Edelstein 2004). According to research by Romero et al. (1997), melons grafted onto hybrid squash rootstocks showed noticeably higher levels of salt tolerance than their non-grafted counterparts.

Most vegetable crops are susceptible to excessive soil moisture, with tomatoes being particularly vulnerable (Verma et al. 2022). Despite ongoing efforts, limited genetic variability for waterlogging tolerance has made it difficult to prevent yield losses. However, certain eggplant accessions have shown strong resistance to flooding (Midmore et al. 1997). Therefore, flood-tolerant eggplant rootstocks that are also compatible with tomato scions can be used to improve tomato resilience under saturated soil conditions. Moreover, some eggplant genotypes possess drought tolerance, offering dual protection against both waterlogging and moisture deficit stress when used as rootstocks.

Technological Advancements in Climate-Resilient Horticulture

Precision Horticulture

Precision farming is a viable approach to boost agricultural output while reducing environmental damage and dependency on inputs like pesticides and fertilisers (Narzari et al. 2025). This approach, which emphasises the control of variability across both spatial and temporal scales, is closely related to the socioeconomic and environmental aspects of farming systems. According to Narzari et al. (2025), its strength is in encouraging resource efficiency, which raises agricultural income while reducing ecological harm. Precision agriculture is becoming a more and more important instrument for promoting sustainable farming methods as technology advances. Smallholder farmers in underdeveloped nations benefit most from it since it enables them to produce higher yields at cheaper costs and with fewer inputs.

Remote Sensing

Remote sensing is a set of techniques that collect field data without direct physical contact by detecting light reflected or emitted by vegetation and soil (Singhal et al. 2024). In Pennsylvania, researchers have used both solar and artificial light reflectance to calculate vegetation indices. The Normalised Difference Vegetation indicator (NDVI), which is the most widely used indicator, is especially helpful for examining plant canopies and identifying low amounts of chlorophyll in fruit. Several other indicators, which can also be derived from light reflection data, show strong correlation with chemically measured leaf chlorophyll (Singhal et al. 2024). Data captured electronically via satellites or aircraft which measures crop reflectance and electromagnetic emissions can provide critical insights into soil quality, plant growth stages, weed presence, and more. Such information is cost-effective and highly valuable for site-specific crop management strategies. As a result, this technology plays a key role in precision agriculture by enabling easy access to essential field data (Singhal et al. 2024).

Genetic Biotechnology Engineering and Biotechnology

Biotechnology has far-reaching applications, such as enhancing crop yields and stability by developing plant varieties that are more resistant to pests, insects, and environmental stressors like drought, cold temperatures, and acidic soils. It also contributes to enhancing the nutritional quality of food.

Molecular markers are very useful for speeding up the breeding of crops that can withstand stress and produce large yields. This is particularly crucial for farmers in poor nations since these markers enable the stacking of several genes linked to complex features and aid in introducing advantageous alleles from wild species into farmed crops. Several QTLs (quantitative trait loci) linked to stress tolerance have been found in tomatoes, for instance. Marker-assisted selection (MAS) is a promising method for improving salt tolerance because a small number of key QTLs explain the majority of the observed variance in characteristics.

Micropropagation is a key biotechnological tool used to produce disease-free plants, and it has already proven effective in the propagation of healthy plantlets of crops like bananas, strawberries, and stevia.

Weather Forecasting and Agricultural Insurance for Farmers

Weather forecasting and agricultural insurance are essential resources that help farmers better manage the uncertainties brought on by climate variability. In parallel, Geographic Information System (GIS) technology is becoming increasingly important in agriculture, especially in the cultivation of fruit crops. By generating climate suitability maps, GIS assists farmers in selecting the most appropriate areas for planting based on specific environmental factors (Mandal et al. 2020). There is also a rising demand for the development of mathematical models for tropical fruit trees to anticipate the effects of future climate scenarios such as drought, flooding, and elevated atmospheric CO_2 levels. Identifying and cultivating resilient cultivars that can thrive under these conditions will be vital. Additionally, tracking fruit tree responses using phenological scales which mark key stages of plant development provides critical data for understanding and adapting to climate-related changes, making them a valuable component of climate-smart horticulture.

Use of Drones and Sensors

Drones, formally known as Dynamic Remotely Operated Navigation Equipment (DRONE) or Unmanned Aerial Vehicles (UAVs) can be controlled manually using remote devices or smartphone applications, or they can operate autonomously via GPS-enabled autopilot systems (Singhal et al. 2024). Drone integration into agriculture has been a top priority for the Indian government in recent years. The Union Budget for 2022–2023 included plans to use "Kisan Drones" to modernise and fortify the agriculture industry (Narzari et al. 2025). The ability of high-capacity drones to deliver perishable goods like mature fruits, green vegetables, and fish direct from fields to markets is making them seem like game-changers. According to Choudhury et al. (2019), farmers can still use drone-captured aerial data to efficiently monitor and manage their crops even in the absence of widespread precision agriculture adoption. These agricultural drones are designed to monitor crop health and growth in order to increase productivity. Real-time insights into field conditions are provided by a combination of sophisticated sensors and high-resolution imaging. Drone-based data collection has a lot of promise to improve crop yields and farm performance overall (Narzari et al. 2025).

Controlled Environment Agriculture (CEA)

CEA systems can tremendously increase crop yield and nutritional value by carefully controlling important growing parameters like light, temperature, carbon dioxide levels, humidity, and nutrient supply (Verma et al. 2022). These technologies make it possible to produce food continuously throughout the year, regardless of changes in the outside weather, because growing is done within under controlled conditions. Kalantari et al. (2018) point out that putting these farms closer to consumers can significantly cut down on the distance food must travel, reducing transportation-related environmental effects.

Greenhouse Effect

Greenhouse gases (CO_2, CH_4, CO, CFCs, and N_2O) absorb long-wave radiation emitted from the earth surface, thereby trapping heat and contributing to global warming. This phenomenon is referred to as the greenhouse effect. These gases contribute to temperature regulation on Earth by retaining heat reflected from the surface, rather than substantially absorbing incoming solar radiation (Dogar 2019). The greenhouse effect can be categorised into two types: the natural greenhouse effect, which maintains a stable and hospitable climate essential for life, and the enhanced greenhouse effect, resulting from human activities like fossil fuel combustion and deforestation, which increases greenhouse gas concentrations and intensifies global warming. CO_2 is the primary gas regulating the Earth's temperature. This natural process has been expedited by the rise in atmospheric greenhouse gases due to human activities, resulting in increased heat retention and climate change.

Influence of Temperature

High temperatures can negatively impact pollination processes, often resulting in increased flower drop and fruit abortion. These conditions also tend to delay fruit ripening and reduce overall fruit quality, including poor coloration. Under prolonged heat stress, fruit crops are more prone to physiological disorders such as spongy tissue in mangoes, fruit cracking, and black spot in custard apples. In temperature-sensitive crops like pome and stone fruits, elevated temperatures can alter chilling requirements, leading to premature dormancy break. Mango trees have a low tolerance for cold temperature, however they may tolerate temperatures as high as 48 ° C for short periods of time (Mukherjee 1953). Mango trees prefer vegetative growth by nature, and when temperatures rise, more leaves are produced, which affects how they flower. Environmental variables like temperature, humidity, and the kind and intensity of light have a significant impact on the differences in dragon fruit's nutritional makeup, physical attributes, and biochemical properties (Parmar and Karetha 2020).

Citrus crops are grown from coastal regions to elevations as high as 2100 m, with optimal growth occurring between 2 °C and 30 °C. Temperatures below 0 °C can cause tree damage, while growth efficiency declines when temperatures dip under 13 °C. Dry root rot has become an increasingly serious global issue, especially under stress conditions linked to climate change, has a significant threat to citrus production worldwide (Said et al. 2022).

Influence of Irrigation Water

By 2025, most irrigated areas in India are anticipated to face higher water demand. On a global scale, net irrigation requirements are expected to increase by around 3.5–5% by 2025 and 6–8% by 2075, relative to conditions without climate change. In Himachal Pradesh's apple-growing regions, an estimated 80% reduction in apple yields has been linked to insufficient irrigation water, with the remaining 20% attributed to elevated evaporation rates. Furthermore, these areas saw a reduction in chill unit hours an essential factor for successful apple cultivation (Singh et al. 2016a).

Impact of Relative Humidity

Changes in relative humidity can negatively impact fruit set, leading to increased fruit drop in citrus varieties like oranges and mandarins, as well as other temperate and subtropical fruit crops. Both low and high humidity levels may hinder pollen germination by causing the stigmatic fluid to dry out, thereby reducing fruit set. Moreover, high humidity levels have been linked to a rise in mango hopper infestations and a greater occurrence of powdery mildew. Dates and grapes can be seriously harmed by pre-monsoon showers, which can occasionally result in a complete harvest loss. Pollen may be washed off flower stigmas by rainfall during the flowering period, which could lead to a poor or unsuccessful fruit set. In Gujarat, an 80–90% decline in mango output was caused by heavy dew after unseasonable rain during the flowering season.

Patil et al. (2015) reported that rainfall generally showed a negative correlation with the production of most fruits, with the exception of grapes and apples, which exhibited a positive relationship. Regarding minimum temperature, sapota and guava production, were positively correlated by 38.1% and 35.2% respectively, whereas grapes and pineapples showed negative correlations of 65% and 44%. As for maximum temperature, banana and mango yields had the strongest positive correlations at 39.5% and 39.4%, while grapes showed a strong negative correlation of 66.9%.

Influence of Frost

In temperate regions, spring frosts present a major risk to plants by damaging blossoms, which directly affects fruit set and development. This damage typically occurs in plant parts closest to the ground, where temperatures are coldest. Freezing can cause internal sap-carrying tissues in the young bark to rupture, often leading to the death of bark in young trees (Choudhury et al. 2019). Yadollahi (2011) examined both passive and active frost protection methods for reducing frost-related damage in grapevines grown in Shahrekord, Iran. The study found that heater treatments were particularly effective during the first year, primarily due to the absence of severe frost events. However, under freezing conditions, the application of Bordeaux mixture proved to be the most effective in minimizing damage to grape growth.

Influence of CO_2 on Fruit Crops

Investigation indicates that many peststhrive under increased carbon dioxide (CO_2) levels and changing climate conditions. Such environmental shifts also increase the risk of invasive pest species establishing themselves. In strawberries, exposure to elevated CO_2 and higher temperatures caused fruit yields to drop by 12% under low nitrogen and by 35% under high nitrogen conditions. This decrease is linked to a reduced number of inflorescences and smaller umbel sizes during the critical flower induction phase when plants are under stress.

Soil Less Culture

Soilless culture refers to cultivating plants without traditional soil, using instead nutrient solutions or inert substrates like sand, perlite, vermiculite, rockwool, or cocopeat. This approach is commonly employed in controlled settings such as greenhouses and is especially beneficial in regions where soil quality is poor or contaminated (Table 21.6).

Table 21.6 Different soil less culture techniques and their benefits

Different soil less culture technique	Crop	Benefits	Reference
Aeroponics	Potato	• Traditional methods only produce about 8 daughter tubers every year, however a single potato plant in an aeroponic system can produce over 100 mini tubers per row	Otazu (2008), Hussey and Stacey (1981) and CIP (2008)
		• Improved root aeration and more effective nutrient absorption from the fine mist of the atomised nutrient solution created by this procedure can lead to improved growth in potato seed	Barak et al. (1996) and Nichols (2005)
		• Aeroponics speeds up overall growth, improves root development, boosts plant survival rates, and reduces maturity time	Stoner (1983)
		• Average tuber yield achieved through aeroponics is higher compared to the yield obtained using traditional cultivation methods with the same plant material	Otazu (2008) and Tsoka et al. (2008)
Hydroponics	Tomato	• Tomatoes may grow up to 50% faster than in traditional soil-based methods, as they absorb all necessary nutrients directly through their root systems • Under optimal growing conditions, tomato yields can be up to 30% higher than those achieved with soil-based cultivation • Water usage can be cut by as much as 90%, as the nutrient solution is recycled instead of seeping away into the soil	Das (2023)

Table 21.6 (continued)

Different soil less culture technique	Crop	Benefits	Reference
Hydroponic	Lettuce	• Hydroponic lettuce reduces the risk of soil-borne infections by growing in a nutrient-rich water solution, doing away with the requirement for soil • The labour-intensive chores of weeding and soil preparation are removed with this technique, providing a more effective gardening option • Gardeners can precisely regulate elements such as light, temperature, and nutrients, which guarantees the best possible growth and well-being for their lettuce crops • Regardless of the season or location, hydroponic farming offers a consistent supply of fresh, premium lettuce throughout the year • Controlled environmental conditions allow growers to get consistent harvests and faster growth rates	Stewart (2024)
Hydroponics	Pakchoi	• Key benefits of hydroponics is its ability to be set up easily in clean, healthy environments with limited space, while still delivering high yields	Muslimah et al. (2024)

Hydroponics

Hydroponics is a modern cultivation method that employs a nutrient enriches liquid solution to cultivate plants without soil. Plant roots in this system are either submerged directly in the fluid or held up by inert media like sawdust, peat moss, rockwool, perlite, sand, gravel, vermiculite, or coconut coir. In the absence of soil, these substrates offer physical support. They are chosen because of their high porosity, which guarantees the ideal retention of water and air for strong root development. There are various kinds of hydroponic systems, such as nutrient film technology (NFT), drip, wick, ebb and flow, and water culture. Water culture is one of the most straightforward and often used techniques among these. Using this method, plants are put in pots with holes in the bottom so that their roots can grow into a reservoir that holds the fertiliser solution. An air pump that creates bubbles in the reservoir continuously aerates the nutrient solution to ensure there is enough oxygen present and to avoid root asphyxia.

Aeroponics

This technique is a soilless cultivation system where crop grow suspended in the air without any solid growing medium. The idea of growing plants without soil dates back to the 1920s, when botanists first used simple aeroponic setups to study root systems, primarily for research rather than commercial production (Barker 1922). In 1942, Carter developed air-culture methods involving mist to observe roots, and in 1957, Went officially coined the term "aeroponics" to describe this mist-based root growth technique. Aeroponics, which entails enclosing the roots in a dark chamber and misting them with nutrients, works well for a variety of horticultural crops, such as lettuce, tomato, cucumber (Cho et al. 1996; Gysi and von Allmen 1997; Biddinger et al. 1998; Park et al. 1997), and ornamentals like poinsettias and chrysanthemums (Scoggins and Mills 1998; Molitor et al. 1999).

Micropropagation

In vitro asexual reproduction techniques, such as micropropagation, facilitate the continuous and rapid multiplication system for disease free planting material production throughout the year. This approach works well for satisfying the steady, high demand for high-quality seed material. Potato plants have multiplied on a vast scale in an exceptionally short amount of time as a result of its use in pre-basic seed generation (Naik et al. 2000). According to Pruski (2001), the method is renowned for its adaptability and ability to spread quickly, leading to a high rate of multiplication.

Challenges to Implementing Resilient Horticulture

Climate change on horticulture has created a number of difficulties, highlighting the pressing need for technologically advanced, creative, and sustainable solutions. The scarcity of suitable plant cultivars those are climate resilient and the inadequacy of water management systems are major technological problems. Adoption is further impeded by financial obstacles such high initial investment costs, uncertain returns, and inadequate finance. Social and cultural challenges including gender inequality, entrenched traditional practices, and resistance to change create significant obstacles to effective community participation. Agricultural productivity is further threatened by diminishing water supplies, loss of arable land, and insufficient access to critical inputs like fertilizers. Additionally, climate change has altered pest behaviours and distribution, heightening risks to both crop yields and biodiversity.

Inconsistent government support, inadequately supported extension services, and precarious market access also impede the wider adoption of improved agricultural methods. Environmental stresses like erratic weather and soil erosion make these problems worse and further impeding progress are a lack of research and a sluggish transfer of technology. Addressing these challenges requires robust policy support, enhanced market connectivity, and the development of innovative financial solutions.

Future Research Directions

Given the ongoing and complex effects of climate change on horticulture, continuous research is essential to enhance resilience, develop innovative solutions, and ensure sustainable food systems. Key areas for future study include:

(a) **Breeding Climate-Resilient Crop Varieties:** Prioritizing the genetic improvement of horticultural crops to withstand heat, drought, pests, and diseases, while adapting to changing environments and improving nutritional quality (Dwivedi et al. 2016).
(b) **Advancing Precision Agriculture and Technology:** Investigating the incorporation of advanced technology, including artificial intelligence, remote sensing, and robotics, in precision agriculture to enhance resource utilisation and sustain productivity among climatic challenges (Wang and Zhang 2010).
(c) **Investigating Agroecological Approaches:** Deepening knowledge of diversified cropping systems, organic farming, and IPM to foster sustainable horticultural practices (Altieri and Nicholls 2017).
(d) **Soil Health and Microbiome Studies:** Understanding how climate change effects soil micro-organism and nutrient cycles, and how these contribute to plant resilience (Berg et al. 2014).
(e) **Promoting Climate-Smart Horticulture:** Identifying cost-effective, locally suitable practices that reduce vulnerability and boost productivity in the face of climate challenges (Lipper et al. 2014).
(f) **Enhancing Local Agroecosystem Resilience:** Conducting interdisciplinary research on the ecological and social capacities of local agroecosystems to support community-based resilience and sustainable horticulture (Lin 2011).
(g) **Improving Climate Information Services:** Developing reliable, accessible climate data and early warning systems to empower farmers with timely information for better decision-making (Morton and Wecker 2010).
(h) **Strengthening Governance, Policy, and Institutional Frameworks:** Assessing existing policies and exploring innovative governance models that enable adaptive, climate-resilient horticultural practices (Eriksen et al. 2011).

Conclusion

Climate resilient horticulture is essential for food security and sustaining horticultural productivity amid increasing climate unpredictability. This plan involves modifying horticultural techniques and implementing technologies to address extreme weather events, shifting climatic patterns, and increasing environmental stressors. Developing resilient crop varieties capable of enduring drought, flooding, and pest pressures is a primary strategy. Implementing effective water management strategies like micro irrigation and rainwater harvesting, is essential. Greenhouses and shade nets exemplify protective farming techniques that provide additional safeguarding against adverse conditions. Precision farming and integrated pest management (IPM) exemplify advanced technologies that improve crop resilience and optimise resource utilisation. Equally vital are efforts to enhance farmer education, promote community engagement, and establish supportive policy frameworks to facilitate widespread adoption. Despite challenges including high initial costs, limited awareness, and cultural barriers, shifting towards climate-resilient horticulture is indispensable for securing the agricultural sector's sustainability and adaptability in the long run.

References

Altieri MA, Nicholls CI (2017) Agroecology and the design of climate change-resilient farming systems. Agron Sustain Dev 37(1):36. https://doi.org/10.1007/s13593-016-0407-0

Balusamy A, Ramesh T, Moirangthem P, Talang HD, Chanu LJ, Gopalakrishnan B, Islam M, Hazarika S, Kandpal BK, Choudhury BU (2024) Indigenous wisdom for sustainable mountain agriculture in the eastern Himalayas (Meghalaya). Indian J Soil Conserv 52(1):46–57

Barak P, Smith JD, Krueger AR, Peterson LA (1996) Measurement of short-term nutrient uptake rates in cranberry by aeroponics. Plant Cell Environ 19(3):237–242. https://doi.org/10.1111/j.1365-3040.1996.tb00244.x

Barker BTP (1922) Studies on root development. Long Ashton Res Stn Annu Rep 1921:9–57

Berg G, Grube M, Schloter M, Smalla K (2014) Unraveling the plant microbiome: looking back and future perspectives. Front Microbiol 5:148. https://doi.org/10.3389/fmicb.2014.00148

Biddinger EJ, Liu CM, Joly RJ, Raghothama KG (1998) Physiological and molecular responses of aeroponically grown tomato plants to phosphorus deficiency. J Am Soc Hortic Sci 123(2):330–333. https://doi.org/10.21273/JASHS.123.2.330

Bose TK, Mitra SK, Sanyal D (2001) Fruits: tropical and subtropical, vol 1, 3rd edn. Naya Udyog

Cho YD, Kang SG, Kim YD, Shin GH, Kim KT (1996) Effects of culture systems on growth and yield of cherry tomatoes in hydroponics. RDA J Agr Sci 38:563–567

Choudhury BU, Webster R, Sharma V, Goswami J, Meetei TT, Krishnappa R, Raju PL (2019) Frost damage to maize in Northeast India: assessment and estimated loss of yield by hyper spectral proximal remote sensing. J Appl Remote Sens 13(4):044527–044527. https://doi.org/10.1117/1.JRS.13.044527

Choudhury BU, Nengzouzam G, Islam A (2022) Runoff and soil erosion in the integrated farming systems based on micro-watersheds under projected climate change scenarios and adaptation strategies in the Eastern Himalayan mountain ecosystem (India). J Environ Manage 309:114667. https://doi.org/10.1016/j.jenvman.2022.114667

Das A (2023) Hydroponic tomato farming: wonderful tomato cultivation. Times of Agriculture: A Resonance of Agriculture

Das B, Devi HL, Mahajan V, Sahoo L, Kumar R, Choudhury BU, Mishra VK (2025) Onion variety and season interaction for yield, storage and disease-insect dynamics under North East Indian agro-climatic condition. J Environ Biol 46(4):497–507

Dogar MMA (2019) The Sensitivity of DTR to the increased CO2 over Mid-latitude Semi-dry Regions. JOJ Hortic Arboric, 2(2):555584.

Dwivedi SL, Sahrawat KL, Upadhyaya HD, Ortiz R, Varshney RK (2016) Climate-resilient crop breeding. Curr Opin Plant Biol 31:172–180. https://doi.org/10.1016/j.pbi.2016.04.002

Edelstein M (2004) Grafting vegetable-crop plants: pros and cons. Acta Hortic 6(635):235–246. https://doi.org/10.17660/ActaHortic.2004.635.30

Eriksen S, Aldunce P, Bahinipati CS, Martins RD, Molefe JI, Nhemachena C, Ulsrud K (2011) When not every response to climate change is a good one: identifying principles for sustainable adaptation. Clim Dev 3(1):7–20. https://doi.org/10.3763/cdev.2010.0060

Flowers TJ (2004) Improving crop salt tolerance. J Exp Bot 55(396):307–319. https://doi.org/10.1093/jxb/erh003

Foolad MR, Jones RA (1991) Genetic analysis of salt tolerance during germination in *Lycopersicon*. Theor Appl Genet 81(3):321–326. https://doi.org/10.1007/BF00228677

Geisenberg C, Stewart K (1986) Field crop management. In: Atherton JG, Rudich J (eds) The tomato crop: a scientific basis for improvement. Chapman and Hall, pp 511–557

Gysi C, von Allmen F (1997) Balance of water and nutrients in tomatoes grown on soilless systems. Agrarforschung 4(1):1–4

Hossain MA, Begum M, Akhtar MN, Hossain MA, Islam MM, Almazroui M & Rahman M (2025) Predicting water scarcity in northern Bangladesh using deep learning and climate data. npj Climate and Atmospheric Science, 8(1):348

Hussey G, Stacey NJ (1981) In vitro propagation of potato (*Solanum tuberosum* L.). Ann Bot 48(6):787–796. https://doi.org/10.1093/oxfordjournals.aob.a086194

Intergovernmental Panel on Climate Change (IPCC) (2007) Climate change 2007: the fourth assessment report of the intergovernmental panel on climate change. Cambridge University Press

Intergovernmental Panel on Climate Change (IPCC) (2012) Managing the risks of extreme events and disasters to advance climate change adaptation: summary for policymakers. Cambridge University Press. https://www.ipcc.ch/report/srex

Jodaugienė D, Pupalienė R, Urbonienė M, Pranckietis V, Pranckietienė I (2006) The impact of different types of organic mulches on weed emergence. Agron Res 4(Special Issue):197–201

Kalantari F, Tahir O, Mohd J, Raheleh A, Fatemi E (2018) Opportunities and challenges in sustainability of vertical farming: a review. J Landsc Ecol 11(1):35–60. https://doi.org/10.1515/jlecol-2017-001

Khan A, Mandal S, Satpati L, Dogar MM, Hasanean H & Salah AM (2025) Analyzing drought: the impact of climate change on precipitation patterns in the Shilabati river Basin, West Bengal, India. Earth Systems and Environment, 9(3):1585–1614.

Lin BB (2011) Resilience in agriculture through crop diversification: adaptive management for environmental change. Bioscience 61(3):183–193. https://doi.org/10.1525/bio.2011.61.3.4

Lipper L, Thornton P, Campbell BM, Baedeker T, Braimoh A, Bwalya M, Herrero M (2014) Climate-smart agriculture for food security. Nat Clim Chang 4(12):1068–1072. https://doi.org/10.1038/nclimate2437

Mandal S, Choudhury BU, Satpati L (2020) Soil site suitability analysis using geo-statistical and visualization techniques for selected winter crops in Sagar Island, India. Appl Geogr 122:102249. https://doi.org/10.1016/j.apgeog.2020.102249

Manyaku A, Witbooi H, Laubscher CP (2024) The significance of organic horticulture in mitigating climate change and promoting the production of healthier fruits and vegetables. Appl Sci 14(2):2–18. https://doi.org/10.3390/app14020218

Midmore DJ, Roan YC, Wu MH (1992) Management of moisture and heat stress for tomato and hot pepper production in the tropics. In: Kuo CG (ed) Adaptation of food crops to temperature and water stress. AVRDC, pp 453–460

Midmore DJ, Roan YC, Wu DL (1997) Management practices to improve lowland subtropical summer tomato production: yields, economic returns, and risk. Exp Agric 33(2):125–137. https://doi.org/10.1017/S0014479797004057

Molitor HD, Fischer M, Popadopoulos AP (1999) Effect of several parameters on the growth of chrysanthemum stock plants in aeroponics. Acta Hortic 481(1):179–186. https://doi.org/10.17660/ActaHortic.1999.481.21

Morton JF, Wecker D (2010) Policies for horticulture and poverty alleviation in sub-Saharan Africa. Hortic Int J 3(2):215–226

Mukherjee SK (1953) The mango: its botany, cultivation, uses, and future improvements, especially as observed in India. Econ Bot 7(2):130–162. https://doi.org/10.1007/BF02930733

Muslimah Y, Harahap EJ, Lizmah SF, Siregar MPA, Martunis, Yuszikri (2024) Nutrients and growing media effect on growth and physiological traits of *Pak choi* plants (*Brassica rapa* L.). SABRAO J Breed Genet 56(2):652–659. https://doi.org/10.54910/sabrao2024.56.2.652

Naik PS, Chakrabarti SK, Sarkar D, Birman RK (2000) Potato biotechnology: Indian perspective. In: Khurana SMP, Shekhawat GS, Singh BP, Pandey SK (eds) Potato, global research and development, vol 1. Indian Potato Association, pp 194–211

Narzari R, Choudhury BU, Singhal G, Choudhary KK (2025) A critical review of how UAVs can transform precision agriculture in the realm of agroecology. Discover Soil 2(1):28. https://doi.org/10.1007/s44378-025-00055-2

Nichols MA (2005) Aeroponics and potatoes. In: Proceedings of the First International Symposium on Root and Tuber Crops: Food Down Under. International Society for Horticultural Science, pp 201–206

Nimbolkar PK, Awachare C, Reddy YTN, Chander S, Hussain F (2016) Role of rootstocks in fruit production: a review. J Agric Eng Food Technol 3(3):183–188

Nzeadibe TC, Egbule CL, Chukwuone N, Agu V (2011) Farmers' perceptions of climate change governance and adaptation constraints in Niger Delta Region of Nigeria. African Network Reports 7

Otazu, V (2008) Quality seed potato production using aeroponics: a potato production manual. International Potato Centre (CIP)

Pandita ML, Singh N (1992) Vegetable production under water stress conditions in rainfed areas. In: Kuo CG (ed) Adaptation of food crops to temperature and water stress. AVRDC, pp 467–472

Park HS, Chiang MH, Park HS (1997) Effects of form and concentration of nitrogen in aeroponic solution on growth, chlorophyll, nitrogen contents, and enzyme activities in *Cucumis sativus* L. plant. J Korean Soc Hortic Sci 38:642–646

Parmar VM, Karetha KM (2020) Physical and biochemical analysis of dragon fruit species from different regions of Gujarat. J Pharmacogn Phytochem 9(5):2863–2866

Patil NA, Yeldhalli RA, Patil BO, Laxmi TN (2015) Impact of climate change on major fruits in India. Asian J Environ Sci:34–38

Pruski K (2001) Micropropagation technology in early phases of commercial seed potato production (Doctoral dissertation). Wageningen University, Netherlands

Rejani R, Yadukumar N (2010) Soil and water conservation techniques in cashew grown along steep hill slopes. Sci Hortic 126(3):371–378. https://doi.org/10.1016/j.scienta.2010.07.032

Romero L, Belakbir A, Ragala L, Ruiz MJ (1997) Response of plant yield and leaf pigments to saline conditions: effectiveness of different rootstocks in melon (*Cucumis melo* L.). Soil Sci Plant Nutr 3(5):855–862

Rudich J, Zamski E, Regev Y (1977) Genotype variation for sensitivity to high temperature in the tomato: pollination and fruit set. Euphytica 26(1):193–200

Rymbai H, Deshmukh NA, Ramesh T, Verma VK, Baiswar P, Mawlein J, Talang HD, Dutta SK, Choudhury BU, Hazarika S, Mishra VK (2024) DRIS standards for identifying yield-limiting nutrients in Khasi Mandarin and their relationship with fruit yield on acidic soil of the eastern Himalayas, India. J Plant Nutr 47(8):1331–1346. https://doi.org/10.1080/01904167.2024.2308193

Said E, Nabil R, Abdessalem T, Zakaria E-H, Fouad M, Goksel O et al (2022) Dry root rot disease, an emerging threat to the citrus industry worldwide under climate change: a review. Physiol Mol Plant Pathol 117:1–19. https://doi.org/10.1016/j.pmpp.2021.101754

Sanchez-Guerrero MC, Lorenzo P, Medrano E, Baille A, Castilla N (2009) Effects of EC-based irrigation scheduling and CO_2 enrichment on water use efficiency of a greenhouse cucumber crop. Agric Water Manage 96(3):429–436. https://doi.org/10.1016/j.agwat.2008.09.017

Scoggins HL, Mills HA (1998) Poinsettia growth, tissue nutrient concentration, and nutrient uptake as influenced by nitrogen form and stage of growth. J Plant Nutr 21(1):191–198. https://doi.org/10.1080/01904169809365403

Sharma J, Upadhyay AK, Bande D, Patil SD (2011) Susceptibility of Thompson seedless grapevines raised on different rootstocks to leaf blackening and necrosis under saline irrigation. J Plant Nutr 34(11):1711–1722

Singh RN, Mujumder PK, Sharma DK (1966) Sex expression in mango (*Mangifera indica* L.) concerning prevailing temperature. Proc Am Soc Hortic Sci *89*:228

Singh HP, Shukla S, Malhotra SK (2009a) Ensuring quality planting material in horticultural crops. In: Lead papers of the 9th Agricultural Science Congress, pp 22–24

Singh R, Kumar S, Nangare DD, Meena MS (2009b) Drip irrigation and black polyethylene mulch influence on growth, yield, and water-use efficiency of tomato. Afr J Agric Res 4(12):1427–1430

Singh N, Sharma DP, Chand H (2016a) Impact of climate change on apple production in India: a review. Curr World Environ 11(1):251–259

Singh RK, Dubey SK, Singh RK, Buckseth T, Singh AK (2016) Farmer participatory seed potato production through Krishi Vigyan Kendra networking: an action research for enhanced availability of seed potatoes in India. Potato J 43(2):193–199

Singhal G, Choudhury BU, Singh N, Goswami J (2024) An enhanced chlorophyll estimation model with a canopy structural trait in maize crops: use of multi-spectral UAV images and machine learning algorithm. Eco Inform 83:102811. https://doi.org/10.1016/j.ecoinf.2024.102811

Srinivasa Rao NK (1995) Management of heat, moisture, and other physical stress factors in tomato and chilli in India. In: Collaborative vegetable research in South Asia: Proceedings of the SAVERNET Midterm Review Workshop. AVRDC, pp 232–240

Stewart, M. (2024). Hydroponic lettuce: how to grow crisp and healthy lettuce in water. FEI Hydroponics

Stoner RJ (1983) Aeroponics versus bed and hydroponic propagation. Floriculture Review 173(1):4477

Tahir Z, Haseeb M, Mahmood SA. et al. (2025). From Green to Grey: Assessing the Impact of Urbanization on Land Surface Temperature and Thermal Comfort in Multan City. Earth Syst Environ 9, 777–800. https://doi.org/10.1007/s41748-025-00598-y

Tsoka O, Demo P, Nyende AB, Kamau N (2008) Seed production of selected potato *(Solanum tuberosum L.)* clones under aeroponic conditions (Master's thesis). Department of Horticulture, Faculty of Agriculture, Jomo Kenyatta University of Agriculture and Technology

Turner DW, Fortescue JA, Thomas DS (2007) Environmental physiology of bananas (*Musa* spp.). Braz J Plant Physiol 19(4):463–484. https://doi.org/10.1590/S1677-04202007000400013

Verma P, Kaur J, Arora MS, et al. (2026) Modeling the Impact of Direct Air Capture on Forest Biomass and Population Dynamics. Earth Syst Environ 10, 407–427. https://doi.org/10.1007/s41748-025-00599-x

Verma VK, Choudhury BU, Jha AK (2022) Effect of soil amelioration on high value vegetables grown under protected conditions. Indian J Agric Sci 92(12):1464–1468

Verma VK, Choudhury BU, Rymbai H, Jha AK, Hazarika S, Mishra VK (2024) Irrigation scheduling for high value vegetable crops grown under protected cultivation in the hilly ecosystem of north-East India. Indian J Agric Sci 94(7):767–773. https://doi.org/10.56093/ijas.v94i7.147126

Von WS, Chieng SS (2004) A comparison between low-cost drip irrigation, conventional drip irrigation, and hand watering in Nepal. Agric Water Manag 64(2):143–160. https://doi.org/10.1016/S0378-3774(03)00205-6

Wang N, Zhang N (2010) Precision agriculture: a worldwide overview. Comput Electron Agric 74(1):1–4. https://doi.org/10.1016/j.compag.2010.08.009

Yadollahi A (2011) Evaluation of reduction approaches on frost damages of grapes grown in moderate cold climate. Afr J Agric Res 6(29):6289–6295. https://doi.org/10.5897/AJAR11.1029

Part VII
The Future of Rural Livelihoods in a Changing Climate

Chapter 22
The Future of Rural Livelihoods in a Changing Climate

Subhadwip Ghorai, Soumik Dey Roy, Sk Md Asif, Rajdeep Mohanta, Sahely Kanthal, Ishanu Mandal, and Pabitra Kumar Ghosh

Abstract Climate change poses a critical livelihood threat in rural South Asia, where reliance on climate-sensitive agriculture, fisheries, and natural resources subjects populations to the vagaries of increasing temperatures, unpredictable rainfall, flooding, and cyclones. In India's Sundarbans delta, repeated cyclones and sea-level rise have inundated agricultural land and relocated households, with farmers losing close to 100% of their arable land in some cases, underlining the compelling connections between environmental change and livelihood security. This chapter integrates evidence across 25–30 peer-reviewed articles (2015–2025) to evaluate the climate change impacts on rural livelihoods, determine adaptation options, and draw policy-relevant conclusions, with a focus on South West Bengal and comparable agro-ecological regions. A systematic literature review process was adopted, utilizing academic databases and grey literature, and studies were grouped thematically into impacts, adaptation options, and facilitating factors. Quantitative trends were extracted where possible. The results indicate pervasive threats: model projections indicate 20–30% yield losses of rice and wheat in mid-century warming projections, 70–80% of coastal households in West Bengal experience frequent storm-surge inundation, and salinity intrusion is making extensive areas of arable land unproductive. Adaptation practices documented are crop diversification, water harvesting at the community level, adoption of renewable energy, diversification of livelihoods, and the use of electronic advisories; for instance, surveys indicate more than 50% of smallholders in similar areas have adopted one or more climate-smart practices, although adoption in India is limited by poor extension services and seed availability. Generally, the evidence suggests that enhancing adaptive capacity through saline-resistant crops, renewable energy, social protection, and mainstreaming of indigenous knowledge and hybridized modern technologies is imperative, while harmony between climate and agricultural policy remains imperative to creating sustainable and resilient rural livelihoods.

S. Ghorai · S. D. Roy · S. M. Asif · R. Mohanta · S. Kanthal · P. K. Ghosh (✉)
Department of Agriculture, Brainware University, Barasat, Kolkata, West Bengal, India

I. Mandal
Faculty of Agriculture, JIS University, Kolkata, West Bengal, India

S. Mandal et al. (eds.), *Climate Change Vulnerability*, Climate Change Management, https://doi.org/10.1007/978-3-032-16783-5_22

Introduction

Rural livelihoods form the bedrock of most developing economies, sustaining billions of individuals on agriculture, livestock production, fisheries, aquaculture, and forest-based livelihoods (Dey et al. 2025a; Halder et al. 2022; Mandal et al. 2013; Mandal and Choudhury 2014). These are resource-based activities, dependent on land, water, forests, and biodiversity, and their productivity is directly connected to climatic stability. In South Asia, more than 60% of the population still depends on agriculture, rendering the region highly exposed to environmental shocks (Aryal et al. 2020; Shahzad et al. 2024; Dey et al. 2025b). Although traditional practices and farming have long allowed communities to prosper in a wide range of productive agro-ecological zones (Kokou et al. 2024; Mandal et al. 2015, 2018), the increasingly rapid rate of climate change now jeopardizes the sustainability of these systems.

The evidence of the past decade indicates that climate change has already started compromising agricultural productivity and household welfare. For example, crop modelling research estimates 20–30% staple yield reduction in rice and wheat in mid-century climate warming scenarios, with rainfed systems being most at risk (Köck-Schulmeyer et al. 2021). Temperature increase and changing precipitation patterns are also increasing drought frequency and infestations, while glacier loss and sea-level rise further threaten freshwater security (Muzammal et al. 2024; Halder et al. 2023; Khan et al. 2025). In South Asia, the number of individuals affected each year due to climate-related hazards has risen consistently, with floods, droughts, and cyclones being responsible for significant livelihood losses (de Heer and Choudhury 2025; Dogar et al. 2017, 2019; Dogar and Almazroui 2022). These effects spill over into agriculture, degrading fisheries, forestry, and other livelihoods that are resource-based.

Smallholder farmers, who predominate in the rural areas, continue to be disproportionately exposed due to their restricted access to credit, irrigation, markets, and technical information (Ngubane et al. 2025). Consequently, adaptive measures tend to be incremental in nature, e.g., adjustments to sowing times, low-level crop diversification, or short-distance migration instead of revolutionary. Constraints like high input prices, small landholdings, and poor extension systems limit the adoption of climate-smart technology, even when they are accessible (Atta-Aidoo et al. 2022). Socioeconomic limitations amplify the threats of food insecurity, poverty, and inequality, disproportionately falling on women and marginalized groups.

The Sundarbans delta in South West Bengal is a case in point. This vulnerable ecosystem, with millions of inhabitants, is being further exposed to cyclones, tidal surges, and saline intrusion that destroy cropland and fisheries. Empirical evidence indicates extensive income loss, out-migration, and livelihood diversification in the face of repeated shocks (Banerjee et al. 2023; Ghosh and Mistri 2023). Such facts point to the necessity for context-specific adaptation trajectories. Against this context, the chapter synthesizes results of around 30 peer-reviewed articles (2015–2025) to (i) outline the dominant means through which climate change is transforming rural livelihoods, (ii) evaluate coping and adaptation mechanisms, (iii) examine enabling and constraint factors that determine adaptive capacity, and (iv) offer policy-relevant suggestions to construct resilient and sustainable livelihoods in South Asia, with a focus on the Sundarbans region of West Bengal.

Methodology

This chapter is supported by a synthesis of 25–30 peer-reviewed reports and articles (2015–2025) on climate impacts, vulnerability, and adaptation in rural livelihoods. We accepted studies from throughout South Asia and similar developing contexts. Keywords were combinations of "climate change", "rural livelihoods", "adaptation", "vulnerability", and names of regions (e.g., "India", "South Asia", "Sundarbans"). Sources were predominant databases (Google Scholar, Web of Science) and institutional reports. Articles were chosen if they dealt with climate-related threats to agriculture/fisheries or discussed adaptation measures (institutional, technological, community-based).

We grouped the literature into thematic categories: (a) Impacts/Vulnerabilities (e.g., yield variation, exposure to disasters, health hazard); (b) Adaptation Strategies (e.g., agronomic changes, management of resources, diversification); (c) Socio-economic and Policy Context (e.g., access to credit/finance, extension, local institutions). Within each, we coded quantitative results where available (e.g., percentage change in yield, rates of adoption). We further documented regional examples (South Asia, India, Sundarbans) highlighted in case studies. The narrative review adhered to general PRISMA guidelines, with systematic search and thematic synthesis, albeit no meta-analysis. Overall, our analytical strategy was to synthesize disparate findings, highlighting repetitive trends, ranges of outcomes (e.g., 15–30% yield loss), and agreement regarding drivers and solutions.

Climate Change and its Impact on Rural Livelihoods

Agricultural Productivity and Food Security

Climate change has already altered weather patterns, reducing crop productivity across South Asia. Several studies project that rising temperatures and more erratic monsoons will lower yields of key staples by ~20–30% by mid-century if current trends continue (Hamer et al. 2025). For instance, modelling of South Asian cropping systems indicates that without adaptation, rice and wheat yields could decline sharply under +2 °C to +4 °C warming scenarios (Hossain et al. 2025). Extreme weather exacerbates this: heatwaves and droughts shorten growing seasons and boost pest/disease outbreaks, while unseasonal rains damage tender crops.

In India's coastal plains, warmer, wetter winters and hotter summers have already shifted sowing dates and reduced yields of maize, pulses, and oilseeds (Kumar et al. 2025). Conversely, some rainfed areas (e.g., parts of Odisha, West Bengal) see slight gains when increased monsoon rains fall on previously water-limited systems, but these gains are fragile and uneven. Overall, climate variability destabilizes food production: one review notes that climate change is acting as a "hunger risk multiplier" by lowering food availability through lost harvests and altered land suitability.

Water Resources and Fisheries

Altered rainfall and melting snowpack are stressing water supplies. In South Asia's deltas, sea-level rise is pushing saltwater into freshwater aquifers and croplands, reducing irrigable area (Tran and Le 2025). The CSE reports show sea level in the Indian Sundarbans rising at 5.4–12 mm/year (1993–2010), well above the global average (Islam et al. 2019). This intruding salinity already makes soils unproductive in many coastal fields, compelling farmers to abandon them. In parallel, glacier retreat in the Himalayas threatens long-term flows of major rivers (Indus, Ganga-Brahmaputra) that feed irrigation systems.

Water scarcity is amplified by more frequent droughts: UNICEF (2015) finds 68% of India's districts experience repeated water shortages, impacting 200 million people. Rural water scarcity hits women hardest (longer fetching times) and limits irrigated acreage for smallholders. Overexploitation of groundwater is common in response, but this is unsustainable: models suggest global freshwater deficit may reach 5600 km^3/year by 2050 without interventions.

Coastal fisheries are also at risk. Mangrove wetlands like the Sundarbans support small-scale fisheries (e.g. shrimp, crabs) that are sensitive to salinity and storm surges. Cyclones (e.g., Aila 2009) have repeatedly devastated fishing boats and shoreline settlements. Changing estuarine salinity can disrupt fish breeding cycles. For example, communities dependent on Sundarbans fisheries report severe income loss after repeated extreme events, with limited means to cope (Hossain et al. 2024).

Disaster and Health Vulnerability

Rural South Asia faces one of the world's fastest-growing disaster risks. Historical data show that climate disasters in Asia rose from ~100 per year in 1980 to >300 per year by 2012, affecting ~230 million people annually (Krishnamurthy et al. 2023). Events like floods and cyclones destroy crops and infrastructure, causing up to $60 billion in losses per year (2002–2011 average) (ADB 2012; Guha-Sapir et al. 2013). Coastal areas in West Bengal are a "hotspot": an estimated 70–80% of rural coastal residents face regular storm-surge flooding (Mondal et al. 2023). These inundations not only ruin standing crops but leave behind salinized soil (e.g., affecting 20,200 ha in Myanmar after Cyclone Nargis) and drinking-water contamination.

Health impacts compound livelihood losses. Unreliable water supplies and flooding promote water-borne diseases; heat and vector changes increase illnesses among farmworkers (e.g., malaria in India's plains). In tropical areas like India, even small temperature rises (<1 °C) lead to more frequent heat stress in livestock and poultry, reducing productivity. Nutrition vulnerability is also linked: as crop yields fall, poorer households, which spend a large share of income on food, become more food insecure.

Adaptation Strategies for Sustainable Rural Livelihoods

Climate-Smart Agriculture

Climate-Smart Agriculture (CSA) is a holistic approach that transforms agricultural systems to ensure food security and sustainable development in the face of climate change by increasing productivity, enhancing resilience, and reducing greenhouse gas emissions (Kabato et al. 2025). It promotes the use of drought-resistant crop varieties to stabilize yields during dry spells, and organic farming to improve soil health and biodiversity without synthetic inputs. CSA also emphasizes sustainable practices such as crop rotation, conservation tillage, and integrated pest management to maintain ecological balance and resource efficiency (Konfo et al. 2024). Agroforestry and regenerative agriculture further strengthen resilience by integrating trees into farms and restoring soil health, respectively. By combining innovative techniques with traditional knowledge, CSA supports climate adaptation while fostering environmentally and economically sustainable rural livelihoods (Vidadala 2024).

Diversification of Income Sources

Relying on a single income source exposes rural households to significant risks, especially in agriculture, which is vulnerable to climate variability and market fluctuations. In the face of growing climate uncertainty, diversification is essential for spreading risk, stabilizing income, and enhancing resilience (Vernooy 2022). It supports rural development by promoting entrepreneurship, creating job opportunities, and empowering marginalized groups like women and youth, thereby strengthening household and community economies (Mahmood 2009; Ogbari et al. 2024).

Alternative Income Options

Several alternative income-generating options have proven effective in rural contexts, depending on the socio-economic and ecological characteristics of the region. Notable among these are agritourism, handicrafts, and small-scale industries:

Agritourism

This involves integrating tourism with agricultural activities, allowing farmers to generate income through farm stays, guided tours, local cuisine, and cultural experiences. Agritourism not only provides an additional revenue stream but also promotes rural heritage and environmental awareness among visitors. It can invigorate rural areas with increased demand for local products and services (Ammirato and Felicetti 2014; LaPan and Barbieri 2014).

Handicrafts

The production of traditional crafts, including textiles, pottery, woodwork, and other artisanal goods, represents a culturally rooted and economically viable livelihood option. Handicraft production often requires minimal capital investment and can be undertaken alongside agricultural activities (Abisuga-Oyekunle and Fillis 2017). Moreover, the growing demand for ethically sourced and culturally authentic products in national and international markets enhances the viability of this sector (Ghouse 2012).

Small-Scale Industries

Establishing rural-based microenterprises—such as food processing units, dairy farming, poultry, tailoring, and other cottage industries—provides avenues for value addition and employment. These enterprises not only contribute to household income but also stimulate local economies through job creation and increased demand for raw materials and services (Gideon and Georgina 2015).

Sustainable Water Management

Water is a critical resource for sustaining agricultural productivity and rural livelihoods, particularly in regions vulnerable to climate variability and water scarcity (Singh et al. 2018). Sustainable water management encompasses a range of practices and technologies aimed at optimizing water use efficiency, conserving water resources, and ensuring long-term water availability for rural communities. This section explores key strategies, including rainwater harvesting, watershed management, efficient irrigation practices, and community-led water conservation initiatives (Russo et al. 2014; Rastogi et al. 2024).

Rainwater Harvesting

Rainwater harvesting involves the collection and storage of rainfall for agricultural, domestic, and livestock use. It serves as a decentralized and cost-effective method of supplementing water availability in rural areas, especially during dry seasons or in regions with erratic rainfall patterns (Biazin et al. 2012). Techniques such as rooftop harvesting, check dams, percolation tanks, and farm ponds can significantly enhance water accessibility and groundwater recharge (El-Rawy et al. 2024). Rainwater harvesting also contributes to reduced surface runoff and soil erosion, thereby promoting sustainable land and water management (Rastogi et al. 2024).

Watershed Management

Watershed management is an integrated approach that focuses on the conservation and judicious use of land and water resources within a defined drainage area (Heathcote 2009). It involves soil conservation, afforestation, gully plugging, and the construction of water retention structures (Fig. 22.1). The primary objective is to restore ecological balance by improving water infiltration, enhancing soil moisture, and increasing groundwater levels (Singh et al. 2014). Successful watershed management can lead to improved agricultural productivity, reduced vulnerability to droughts, and the sustainable development of rural areas (Rastogi et al. 2024).

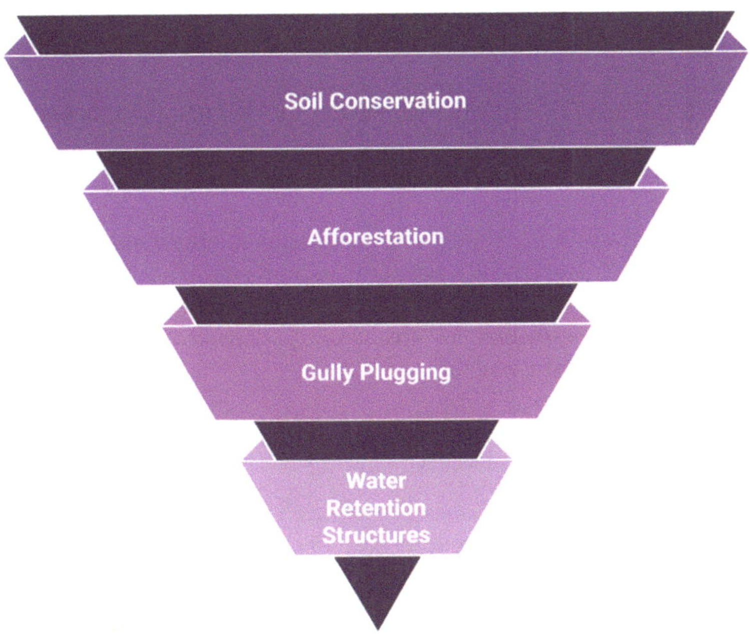

Fig. 22.1 Watershed management process

Efficient Irrigation Practices

Efficient irrigation technologies such as drip and sprinkler systems play a crucial role in minimizing water losses and maximizing crop water productivity. These systems deliver water directly to the root zone, thereby reducing evaporation and runoff (Yang et al. 2023a, b). The adoption of micro-irrigation not only conserves water but also enhances nutrient use efficiency and supports the cultivation of high-value crops (Rastogi et al. 2024).

Community-Led Initiatives for Water Conservation

Community participation is fundamental to the success and sustainability of water management interventions. Community-led initiatives empower local stakeholders to take collective responsibility for water conservation through participatory planning, implementation, and monitoring (Lema 2025; Madhav 2025). These initiatives often involve the formation of water user associations, self-help groups, and village-level water committees that facilitate equitable water distribution and resolve conflicts (Ali et al. 2024). Traditional knowledge systems and indigenous practices, when integrated with modern techniques, can enhance the effectiveness and cultural relevance of water management strategies.

Access to Climate Information and Early Warning Systems

Role of Technology in Improving Climate Preparedness

By improving sensing, data curation and analysis, prediction, and resource optimization, artificial intelligence (AI) technologies have the potential to significantly reduce climate-related security threats. These developments offer a more affordable option to conventional techniques that mainly depend on substantial human resources and physical infrastructure.

Improved Sensing Through Remote Monitoring

Comprehensive Data Curation and Analytical Understanding

Artificial Intelligence (AI) improves environmental monitoring by utilizing cutting-edge remote sensing technologies, such as computer vision and unmanned aerial vehicles (UAVs), which allow for thorough data collection across large, frequently inaccessible areas. While drastically lowering the operational costs connected with conventional monitoring techniques, such in-depth environmental insights are essential for evaluating the complex effects of climate change on ecosystems' provision of food security (Dauvergne 2021). ML systems can automatically sort through and interpret a variety of data streams, including social media feeds, satellite photos, and meteorological data, by utilizing classification and detection techniques. This level of analysis improves our comprehension of the complex connections between security threats and climate change, providing a stronger basis for developing focused policy responses and actions.

Using Predictive Modelling to Forecast Climate Risks

Furthermore, the development and improvement of early-warning systems for a variety of climate dangers and associated security issues rely heavily on machine learning's predictive capabilities. Stakeholders can obtain vital insight into upcoming climate-related challenges by utilizing ML-powered prediction models, which will enable more informed and expedited decision-making.

Optimizing Resources to Increase Climate Resilience

Additionally, ML makes a substantial contribution to resource allocation and management optimization, which is a crucial component of enhancing climate resilience. ML algorithms can suggest more effective and sustainable resource management techniques by analyzing consumption trends and identifying inefficiencies. Machine learning and artificial intelligence facilitate climate risk management via predictive information vital for adaptation and mitigation policy formulation. The technologies combine topographical, climate, and infrastructure data to model climate change impacts, such as sea-level rise, to enable infrastructure resilience (Jain et al. 2023). Projects like Deltares and Coastal AI are such examples. Precision agriculture, which is well developed in the US, China, and the EU, remains underutilized in regions of the Global South (Galaz et al. 2021). AI's transparent and granular analyses allow for more accurate climate risk assessment and more informed decision-making, especially in instances involving huge investments and long-term policy implications.

Examples of Mobile Apps, AI-Driven Weather Forecasting, and Local Adaptation Programs

Agriculture depends heavily on weather factors such as temperature, humidity, dew point, sun and wind data, rainfall, hail, and soil moisture. Crop growth, irrigation requirements, weed and insect management, and harvesting schedules are all impacted by these variables. Farmers can maximize harvesting, support intelligent spraying, increase irrigation efficiency, improve weed and pest control, and make smarter decisions throughout the agricultural supply chain by incorporating this meteorological data. More sustainable farming methods, lower losses, and more productivity result from this. The technology utilized for weather prediction is stated in Table 22.1.

Table 22.1 In depth analysis of Technology by using local weather forecasting for agriculture (Das and Nayak 2024)

Sl. No.	Technology	AI tools	Digital technologies
1	Technology for gathering data	Sensor with low cost	Air temperature and humidity sensor (DHT22), rain sensor (YL83 + LM393), atmospheric pressure sensor (BMP180), and LDR based intensity sensor, node MCU, radio circuit
		Sensor with short range	ESP32, LoRaWAN, Sigfox, NB-IoT, Wi-fi, Bluetooth, HW390 capacitive sensor
		Sensor with long range	Veml7700, NRF24L01, long range radio (LoRa) protocol, wireless sensor- GPRS/3G/4G technology,
		Platform for data collection	Low-cost WSN (wireless electronic modules), ThingSpeak, Zigbee, open-source Arduino, ThingSpeak
		Secondary data	Satellite remote sensing, Data from NASA, Data from IMD
		Autonomous data collection devices	Drones, GIS technology, remote sensing devices, GPS GPS-based receivers
2	Layer of data storage	Block storage	MongoDB NoSQL database
		Storage of a lower-level object	Cloud computing, Amazon's elastic computing cloud (EC2)
		Storage of a higher-level object	Fog and edge computing
		Platform of object storage	Blockchain, AWS
3	Layers for data processing	Artificial intelligence and machine learning	Support vector machine, generative AI -large language models (LLMs), ad boost, naïve bayes, k-means clustering, random Forest, linear regression, decision tree.
		Deep learning	Image recognition techniques, artificial neural network, and Gaussian mixture
4	Layers of application	Layer with UI component	A graphical user-friendly GUI, API
		Layer with UI processes	Watson decision AI platform

Policy Support and Financial Inclusion

Government Policies that Promote Rural Resilience

To fully realize the potential of rural communities and entrepreneurs, innovation policy must be tailored to the needs of rural areas. Using case studies from various nations, the OECD's Enhancing Rural Innovation initiative suggests four key policy areas to foster innovation in rural settings. These consist of:

(a) Extending the concept of innovation beyond technology and science
(b) Customizing programs to fit the distinct firm profiles, community assets, and demographics of rural locations
(c) Streamlining and enhancing the coordination of current governmental services.
(d) Building connections and networks to support rural innovation ecosystems.

These domains are not universally applicable. Examples of the four policy areas are shown in Table 22.2, which also illustrates how these four policy areas vary by rural region.

Access to Credit, Climate Insurance, and Subsidies for Adaptation Efforts

These include the organizations that manage the finances, the sources of funding for climate change adaptation, and the channels or systems that allocate the cash to adaptation initiatives. Governments, international development organizations, and the private sector, which make contributions for commercial or charitable reasons, are the main sources of funding. There is significant overlap among these sources, and it might be challenging to pinpoint the precise source of funding for climate change adaptation. For instance, multilateral development banks (MDBs) finance climate change by using their own funds or special trust funds established by their donors. The main funding under the MDB for adaptation to climate change is listed in Table 22.3.

The Role of Technology and Innovation

In order to reduce the effects of climate change and improve adaptability, rural livelihoods in the future depend on adopting digital technologies and sustainable practices (Balogun et al. 2020). Climate-smart technologies as well as sustainable farming practices can increase resilience and earnings, while digitalisation opens up prospects for trade, information access, and creation of employment in rural areas (Simelton and McCampbell 2021).

Table 22.2 Policy areas to support rural innovation

Sl. No.	Policy domain	Policy action	Example	Accessibility and objectives	Reference
1	Science and technology	Supporting social innovation	Highlands and Islands Enterprise Agency, Scotland, UK	Services and support vary by degree of remoteness	OECD (2025)
		Creating opportunities through public-private partnerships for innovation and public service delivery	Community-based access to digital infrastructure, Scotland, UK IISD-ELA, Ontario, Canada Kamo aquarium, Yamagata, Japan	Joint development for rural remote, but can be more viable in accessible rural	OECD (2025)
		Building a culture for Experimentation	Innovation boosters, Innosuisse, Switzerland; regulatory sandboxes: Fintech in Australia; digital in UK	Different rural remote and accessible opportunities for experimentation	OECD (2025)
2	Rural demographics and firms	Improving access to information in rural areas	National Statistical Services, Scotland rural data portal, Canada rural observatory, EC regular rural impact evaluations in monitoring and evaluation practices	Over-sampling is often needed in more sparsely populated and remote areas	OECD (2025)
		Targeting policies for rural firms	Innovation programmes for small and rural, Regional innovation system, Switzerland Old firms, succession planning, Scottish Enterprise, Scotland	Remote areas have a larger share of smaller and older firms, or firms with strong family ties.	OECD (2025)
		Targeting rural demographics	Women's entrepreneurship programmes in US, Canada, Switzerland black entrepreneurship programmes in US and Canada indigenous entrepreneurship programmes in Canada youth entrepreneurship programme	Remote places may offer different levels of services that may challenge entrepreneurship and innovation for demographics differently. Alternatively, the lack of relevant job opportunities in remote areas may incentivize low growth firms more than in accessible rural areas	OECD (2025)

3	Co-ordination and simplification	Simplification and facilitating access	MGNREGA in India	Provide a leading Example of resilience-building efforts that contribute for the sustainable development in rural areas	Fischer and Ali (2019)
			Locality-based disaster risk reduction (CBDRR) programs and Bangladesh red crescent society (BDRCS) in Bangladesh	Enhance the liveliness Of rural communities to natural disasters and strengthen early warning systems, construct cyclone shelters	Azad et al. (2019)
			CFGORRP in Nepal	Elaborates the additions of resilience-building initiatives with the objectives of sustainable development in rural areas	Gurung et al. (2021)
			PSNP in Ethiopia	Improving the livelihood of rural households related to climate shocks and food insecurity while promoting sustainable development	Tenzing and Conway (2022)
			One-stop shops: Find business support, Scotland; business pathfinder tool (federal, provincial and local), Canada; business promotion guide (federal), Switzerland bottom-up initiatives: EC leader programme	Facilitating access to national resources is often more challenging for rural areas. Bottom-up initiatives can bring a more tailored approach	OECD (2025)

(continued)

Table 22.2 (continued)

Sl. No.	Policy domain	Policy action	Example	Accessibility and objectives	Reference
4	Networks and linkages for rural innovation	Development of entrepreneurial eco-system	Go forward Pine Bluff, Arkansas, USA; Shorefast foundation, Fogo Island, Canada; Brainport Eindhoven, South Holland	Collaboration and multi-usage facilities as a means to deliver services in RR	OECD (2025)
		Building networks with an innovation partner	Interface, Scotland community college, technology transfer institutes; Quebec, Canada nuclear innovation institute, Ontario, Canada	Remote areas do not have easy access to universities or large research conglomerates and need to work with either spin-offs, industry research labs, or make linkages to resources further away	OECD (2025)
		Functional area approaches for building scale	1 billion build Back better regional challenge, USA; regional innovation system, Switzerland; community futures Programme, Canada/EC LEADER; Fogo Island process; inner areas strategy, Italy	Remote areas are more dependent on building links, and as such, programmes should facilitate joint applications	

Table 22.3 Main funding under the MDB for the adaptation of climate change and resilience (Timilsina 2021)

Sl No.	Administrating MDB	Financing instruments	Name of fund
1	World Bank	Weather hedges, contingent financing	World Bank Group catastrophic risk management
2	ADB	Co-financing, Grant, TA	ADB climate change fund (ADB-CCF)
3	AfDB	Grant	Africa climate change fund (ACCF)
4	AfDB	Co-financing, Grant, loan, risk management	Africa water facility (AWF)
5	ADB	Co-financing, loan, TA	ASEAN infrastructure fund (AIF)
6	IDB	Grant, loan, equity, TA	Multilateral investment fund (MIF) of the IDB Group
7	World Bank	Grant	Partnership for market readiness (PMR)
8	World Bank	Grant, loan	Pilot program for climate resilience (PPCR)
9	World Bank	Grant, TA	Public-private infrastructure advisory facility (PPIAF)
10	World Bank	Grant, TA	Korea green growth trust fund
11	MDB	Grant, loan, TA	MDB pilot program for climate resilience
12	IFC	Loan, guarantee	IFC partial credit guarantees
13	World Bank	Different for individual funds	Climate investment funds (CIF)
14	IFC	RSF	IFC risk sharing facility (IFC RSF)
15	World Bank	Grant, loan	International development association
16	World Bank	Grant	Global Facility for Disaster Reduction and Recovery (GFDRR)
17	World Bank and IFC	Worldwide	Global index insurance facility (GIIF)
18	IDB	Grant	IDB regional Fund of Agricultural Technology (FONTAGRO)
19	IDB	Loan, TA	IDB'S infrastructure fund
20	IDB	Grant, TA	IDB'S sustainable energy and climate change initiative (SECCI)

Digital Solutions for Rural Communities

The Potential of Extension Program

By raising farmers' knowledge of land use and climate hazards and encouraging them to adopt adaptation strategies for sustainable livelihoods, agricultural extension programs support regenerative land management techniques. Boosting community involvement in decision-making to promote long-term prosperity, expanding access to extension services, and enhancing the function of extension activities in mitigating and adapting to climate change.

As the effects of climate change on rural environments become more apparent, rural populations are dealing with a number of challenges (Mahmood et al. 2019). Furthermore, putting climate change adaptation plans into action is becoming more and more difficult (Asfaw et al. 2019). This has significant and wide-ranging effects on attaining sustainable development, especially in the agricultural industry (Altieri et al. 2015). Households in many rural regions of the world mostly depend on natural resources for their livelihoods, including agriculture, cattle, and forests (Ullah et al. 2022). Despite the obstacles presented by climate change, rural development initiatives need efficient extension programs to support sustainable growth in the agricultural sector (Iglesias et al. 2012). The goal of extension programs is to give farmers the information and abilities they require to enhance their farming methods and standard of living (Williams et al. 2022). Extension agents offer guidance on forest protection, landscape restoration, and the adoption of creative cropping and diversification techniques through rural development initiatives, which help farming communities adjust to unpredictable weather patterns and changing climates.

Market Access

The ability to sell at greater prices due to improved access to both local and foreign markets encourage rural farmers to create and sell more goods of higher quality, quantity, and diversity. Additionally, food security, poverty and hunger reduction, and regional income growth are all aided by dependable market access. Despite these noteworthy advantages, a number of issues, including remote locations, have limited smallholder farmers' access to markets (Barbier 2023; Muto and Yamano 2009).

For instance, Ogutu et al. (2020) demonstrated that agricultural commercialisation greatly enhanced food security and nutritional quality with regard to calorie, zinc as well, and iron consumption using data from smallholder landowners in Kenya. Participation in the cowpea market raised household income by 0.7% and food expenditure by 1.6% in Nigeria, according to a study by Manda et al. (2020). According to Ojong et al. (2022), Ethiopian households benefit from increased asset and farm ownership and income due to agricultural commercialisation, which also lowers rural poverty and promotes household economic development. Dey and Singh (2023) estimated cross-sectional data from 175 homes in Eastern India and discovered that the market engagement of vegetable growers considerably raises household earnings and per capita consumption.

Rural Advisory Services

Rural Advisory Services (RAS) are key to facilitating climate-smart agriculture through closing the knowledge gap between experts and farmers. They provide specialized, context-relevant advisory services that improve farmers' ability to transition to climate-resilient practices (Simpson and Burpee 2014). Capacity-building for RAS promotes long-term sustainability and adaptive capacities. They also enhance access to finance, shape policy, and support monitoring, all for resilient agricultural systems (FAO 2015). Their community-driven strategy promotes collective adaptation and learning, equipping farmers with the skills and know-how to maintain livelihoods in the context of climate change.

Millions of people are at danger of hunger and poverty as a result of the immediate threat that climate change poses to agriculture and food security. Rising temperatures, more frequent extreme weather events, water shortages, rising sea levels, acidifying the oceans, degradation of land, ecological systems disruptions, and loss of biodiversity are just a few of the predicted effects that could seriously impair agriculture's capacity to feed vulnerable populations and impede efforts to eradicate hunger, malnutrition, and poverty (FAO 2015). Improving farmers' ability to control risks and implement practical adaptation and mitigation measures for climate change is essential. The behaviour, tactics, and farming practices of innumerable producers must fundamentally change in order to advance climate-smart agriculture.

Interventions in Renewable Energy and Rural Livelihoods in India

Indian rural livelihoods are influenced by extreme geographic variation, fluctuating climate, and socioeconomic dynamics. Though the energy-water-food nexus is at the heart of livelihoods, it has generally been inadequately managed (Thomas et al. 2025). Renewable energy programs, based on climate-resonant policies, have in the main taken top-down paths with weak evaluation of their direct effects on livelihoods. The potential for differential, last-mile solutions is underutilized. Access to clean energy is crucial for the conservation of the environment, but rural energy consumption is shaped by availability. With the limitations of grid-based electrification, decentralized technologies—like solar lanterns, biogas plants, microgrids, and solar pumps—have been embraced. Still, ineffectiveness due to factors such as technological immaturity, poor maintenance, limited expertise, poor governance, and absence of need-based design has been hampered despite subsidies (Duić et al. 2008). An integrated, comprehensive renewable energy system, tailored to the local requirements and resources, is necessary for rural development in a sustainable manner.

Conclusion

Climate change represents significant threats to the natural resources and ecological balance rural communities rely on, disorienting traditional livelihoods—most notably in agriculture and agri-related industries, via increased temperatures, unpredictable weather patterns, and environmental degradation. In spite of these increasing threats, most rural communities are proactively seeking adaptation strategies by merging traditional knowledge with new, science-based interventions. These problems are best addressed through a collaborative and inclusive process, wherein researchers create science-based solutions, policymakers formulate enabling frameworks, and rural communities are made capable of spearheading localized adaptation. It is only through such action that context-specific and sustainable solutions are arrived at and scaled up. If there were proper synergy of policy facilitation, knowledge sharing, and community participation, climate-smart, resilient, and diversified rural livelihoods are not only viable but can flourish amidst global change. Yet the absence of proper weather monitoring facilities in rural parts of developing countries impedes the effective monitoring of climatic changes. To fill this gap, the setup of small-scale hydrological stations and training of local school personnel, students, and community-based organizations to gather climate data are suggested. This action would not only produce useful local datasets but also enhance awareness and community capacity to respond effectively to the effects of climate change.

References

Abisuga-Oyekunle OA, Fillis IR (2017) The role of handicraft micro-enterprises as a catalyst for youth employment. Creat Ind J 10(1):59–74. https://doi.org/10.1080/17510694.2016.1247628

ADB [Asian Development Bank] (2012) Addressing climate change and migration in Asia and the Pacific. ADB, Manila

Ali A, Akhtar R, Hussain J (2023) Unveiling high mountain communities' perception of climate change impact on lives and livelihoods in Gilgit-Baltistan: evidence from people-centric approach. Environ Commun 17(6):602–617. https://doi.org/10.1080/17524032.2023.2229044

Altieri MA, Nicholls CI, Henao A, Lana MA (2015) Agroecology and the design of climate change-resilient farming systems. Agron Sustain Develop 35(3):869–890. https://doi.org/10.1007/s13593-015-0285-2

Ammirato S, Felicetti AM (2014) The agritourism as a means of sustainable development for rural communities: a research from the field. Int J Interdiscip Environ Stud 8(1):17–29. https://doi.org/10.18848/2329-1621/CGP/v08i01/53305

Aryal JP, Sapkota TB, Khurana R, Khatri-Chhetri A, Rahut DB, Jat ML (2020) Climate change and agriculture in South Asia: adaptation options in smallholder production systems. Environ Dev Sustain 22(6):5045–5075

Asfaw A, Simane B, Bantider A, Hassen A (2019) Determinants in the adoption of climate change adaptation strategies: evidence from rainfed-dependent smallholder farmers in north-Central Ethiopia (*Woleka* sub-basin). Environ Dev Sustain 21:2535–2565. https://doi.org/10.1007/s10668-018-0150-y

Atta-Aidoo J, Antwi-Agyei P, Dougill AJ, Ogbanje CE, Akoto-Danso EK, Eze S (2022) Adoption of climate-smart agricultural practices by smallholder farmers in rural Ghana: an application of the theory of planned behavior. PLoS Climate 1(10):e0000082

Azad MAK, Uddin MS, Zaman S, Ashraf MA (2019) Community-based disaster management and its salient features: a policy approach to people-centred risk reduction in Bangladesh. Asia-Pac J Rural Dev 29(2):135–160. https://doi.org/10.1177/1018529119898036

Balogun AL, Marks D, Sharma R, Shekhar H, Balmes C, Maheng D, Arshad A, Salehi P (2020) Assessing the potentials of digitalization as a tool for climate change adaptation and sustainable development in urban centres. Sustain Cities Soc 53:101888. https://doi.org/10.1016/j.scs.2019.101888

Banerjee S, Chanda A, Ghosh T, Cremin E, Renaud FG (2023) A qualitative assessment of natural and anthropogenic drivers of risk to sustainable livelihoods in the Indian Sundarban. Sustainability 15(7):6146

Barbier EB (2023) Overcoming digital poverty traps in rural Asia. Rev Dev Econ 27(3):1403–1420. https://doi.org/10.1111/rode.12962

Biazin B, Sterk G, Temesgen M, Abdulkedir A, Stroosnijder L (2012) Rainwater harvesting and management in rainfed agricultural systems in sub-Saharan Africa—a review. Phys Chem Earth A/B/C 47:139–151. https://doi.org/10.1016/j.pce.2011.08.015

Das SK, Nayak P (2024) Integration of IoT-AI powered local weather forecasting: a game-changer for agriculture. arXiv preprint arXiv:2501.14754. https://doi.org/10.48550/arXiv.2501.14754

Dauvergne P (2021) The globalization of artificial intelligence: consequences for the politics of environmentalism. Globalizations 18(2):285–299. https://doi.org/10.1080/14747731.2020.1785670

de Heer JM, Choudhury GA (2025) Climate vulnerabilities and impacts. In: Strategies for adaptation to climate change in a transformative approach: building on the Bangladesh experience 2025 Jul 29. Springer Nature Switzerland, Cham, pp 149–170

Dey S, Singh PK (2023) Role of market participation on smallholder vegetable farmers' wellbeing: evidence from matching approach in eastern India. Agribusiness 39(4):1217–1237. https://doi.org/10.1002/agr.21813

Dey RK, Halder P, Nandy S, Ansary R (2025a) Identifying groundwater potential zones using multi-criteria approach in agriculture-dominated landscapes of Malda district, India. Groundwater Sustain Dev:101497. https://doi.org/10.1016/j.gsd.2025.101497

Dey RK, Ansary R, Mandal S (2025b) Impact of climate variability on mango cultivation: a case study in Malda district of West Bengal, India. In: Livelihoods and Well-being in the era of climate change: risk to resilience across India. Springer Nature Switzerland, Cham, pp 217–232. https://doi.org/10.1007/978-3-031-81132-6

Dogar MM, Kucharski F & Azharuddin S (2017). Study of the global and regional climatic impacts of ENSO magnitude using SPEEDY AGCM. J Earth Syst Sci 126, 30. https://doi.org/10.1007/s12040-017-0804-4

Dogar MMA, & Almazroui M (2022). Revisiting the strong and weak ENSO teleconnection impacts using a high-resolution atmospheric model. Atmospheric Environment, 270, 118866. https://doi.org/10.1016/j.atmosenv.2021.118866

Dogar MM, Kucharski F, Sato T, Mehmood S, Ali S, Gong Z, ... & Arraut J (2019). Towards understanding the global and regional climatic impacts of Modoki magnitude. Global and Planetary Change, 172, 223–241. https://doi.org/10.1016/j.gloplacha.2018.10.004

Duić N, Krajačić G, da Graça Carvalho M (2008) Renew Islands methodology for sustainable energy and resource planning for islands. Renew Sust Energ Rev 12(4):1032–1062. https://doi.org/10.1016/j.rser.2006.10.015

El-Rawy M, Wahba M, Fathi H (2024) Rainwater harvesting for managed aquifer recharge and flood mitigation in the MENA region. In: El-Rawy M, Negm A (eds) Managed aquifer recharge in MENA countries, Earth and environmental sciences library. Springer, Cham. https://doi.org/10.1007/978-3-031-58764-1_3

Fischer HW, Ali SS (2019) Reshaping the public domain: decentralization, the mahatma Gandhi National Rural Employment Guarantee act (MGNREGA), and trajectories of local democracy in rural India. World Dev 120:147–158. https://doi.org/10.1016/j.worlddev.2018.09.013

Food and Agriculture Organization of the United Nations (FAO) (2015) Building resilient agricultural systems through farmer field schools—integrated production and Pest management Programme (IPPM). FAO, Rome

Galaz V, Centeno MA, Callahan PW, Causevic A, Patterson T, Brass I, Baum S, Farber D, Fischer J, Garcia D, McPhearson T (2021) Artificial intelligence, systemic risks, and sustainability. Technol Soc 67:101741. https://doi.org/10.1016/j.techsoc.2021.101741

Ghosh S, Mistri B (2023) Cyclone-induced coastal vulnerability, livelihood challenges and mitigation measures of Matla–Bidya inter-estuarine area, Indian Sundarban. Nat Hazards 116(3):3857–3878

Ghouse SM (2012) Indian handicraft industry: problems and strategies. Int J Manag Res Rev 2(7):1183–1199

Gideon E, Georgina E (2015) Analysis of employment creation and income generation potentials of small and medium scale enterprise in Abia state, Nigeria. Singap J Bus Econ Manag Stud 51(1818):1–9

Guha-Sapir D, Hoyois P, Below R (2013) Annual disaster statistical review 2012. Centre for Research on Epidemiology of Disasters, Louvain

Gurung S, Joshi SD, Parajuli B (2021) Overview of an early warning system for glacial lake outburst flood risk mitigation in Dudh-Koshi Basin, Nepal. Sci Cold Arid Reg 13(3):206–219

Halder P, Satpati L, Dey RK (2022) Water security in the Indian Sundarban under the changing climate. Society, Environment & Development, Raidighi College, Sundarbans, pp 94–123. ISBN 978-8-19-511040-7

Halder P, Dey RK, Mandal S (2023) Long-period trend analysis of annual and seasonal rainfall in West Bengal, India (1901–2020). Theor Appl Climatol 154(1–2):685–703. https://doi.org/10.1007/s0070-023-04577-z

Hamer EA, Oyedotun TD, Charles E, Moonsammy S (2025) Livelihood adaptation to climate-related changes in agroecological zones in developing countries—challenges, prospects, and policy concerns. Front Clim 7:1594747

Heathcote IW (2009) Integrated watershed management: principles and practice. John Wiley & Sons

Hossain MZ, Rahman MA, Rahaman KR, Ha-Mim NM, Haque SF (2024) Investigating critical relationships among vulnerability, livelihoods, and non-migration strategies at the fishing communities in the Sundarbans. Environ Dev Sustain 26(11):29129–29168

Hossain A, Farhad M, Aonti AJ, Kabir MP, Hossain MM, Ahmed B, Haq MI, Azim J (2025) Cereals production under changing climate. In: Challenges and solutions of climate impact on agriculture 2025 Jan 1. Academic, pp 63–83

Iglesias A, Quiroga S, Moneo M, Garrote L (2012) From climate change impacts to the development of adaptation strategies: challenges for agriculture in Europe. Clim Chang 112:143–168. https://doi.org/10.1007/s10584-011-0344-x

Islam MA, Hoque MA, Ahmed KM, Butler AP (2019) Impact of climate change and land use on groundwater salinization in southern Bangladesh—implications for other Asian deltas. Environ Manag 64(5):640–649

Jain H, Dhupper R, Shrivastava A, Kumar D, Kumari M (2023) AI-enabled strategies for climate change adaptation: protecting communities, infrastructure, and businesses from the impacts of climate change. Comput Urban Sci 3(1):1–17. https://doi.org/10.1007/s43762023-00100-2

Kabato W, Getnet GT, Sinore T, Nemeth A, Molnár Z (2025) Towards climate-smart agriculture: strategies for sustainable agricultural production, food security, and greenhouse gas reduction. Agronomy 15(3):565. https://doi.org/10.3390/agronomy15030565

Khan A, Mandal S, Satpati L, Dogar MM, Hasanean H, Salah AM (2025) Analyzing drought: the impact of climate change on precipitation patterns in the Shilabati River basin, West Bengal, India. Earth Syst Environ. https://doi.org/10.1007/s41748-025-00679-y

Köck-Schulmeyer M, Ginebreda A, Petrovic M, Giulivo M, Aznar-Alemany Ò, Eljarrat E, Valle-Sistac J, Molins-Delgado D, Diaz-Cruz MS, Monllor-Alcaraz LS, Guillem-Argiles N (2021) Priority and emerging organic microcontaminants in three Mediterranean river basins: occurrence, spatial distribution, and identification of river basin specific pollutants. Sci Total Environ 754:142344

Kokou K, Syampungani S, Chirwa PW, Makhubele L (2024) Socio-ecological management of multifunctional landscapes. In: Trees in a sub-Saharan multi-functional landscape: research, management, and policy 2024 Oct 1. Springer Nature Switzerland, Cham, pp 267–304

Konfo TR, Chabi AB, Gero AA, Lagnika C, Avlessi F, Biaou G, Sohounhloue CK (2024) Recent climate-smart innovations in agrifood to enhance producer incomes through sustainable solutions. J Agric Food Res 15:100985. https://doi.org/10.1016/j.jafr.2024.100985

Krishnamurthy PK, Lewis K, Kent C, Aggarwal P (2023) Climate impacts on food security and livelihoods in Asia a review of existing knowledge. United Nations World Food Programme Regional Bureau for Asia, pp 1–30

Kumar R, Das A, Mondal S, Upadhyay PK, Bhatt BP, Mishra JS, Singh AK, Choudhary JS, Kumar S, Sundaram PK, Biswas AK (2025) Greening rice-fallow areas: integrating pulses and oilseeds for sustainable cropping in eastern India. Front Sustain Food Syst 9:1502759

LaPan C, Barbieri C (2014) The role of agritourism in heritage preservation. Curr Issues Tour 17(8):666–673. https://doi.org/10.1080/13683500.2013.849667

Lema MW (2025) Sustaining rural livelihoods through participatory water governance: a review of community-driven water resource management models in east and Central Africa. Environ Qual Manag 34(3):e70023. https://doi.org/10.1002/tqem.70023

Madhav MV (2025) Catalysing water governance: empowering communities for sustainable water management. Indian J Public Adm 71(1):119–136. https://doi.org/10.1177/00195561241309140

Mahmood T (2009) Developing entrepreneurship for rural youth and women. In: Fien J, Maclean R, Park MG (eds) Work, learning and sustainable development, Technical and vocational education training: issues concerns prospects, vol 8. Springer, Dordrecht. https://doi.org/10.1007/978-1-4020-8194-1_15

Mahmood N, Arshad M, Kächele H, Ma H, Ullah A, Müller K (2019) Wheat yield response to input and socioeconomic factors under changing climate: evidence from rainfed environments of Pakistan. Sci Total Environ 688:1275–1285. https://doi.org/10.1016/j.scitotenv.2019.06.266

Manda J, Alene AD, Tufa AH, Feleke S, Abdoulaye T, Omoigui LO, Manyong V (2020) Market participation, household food security, and income: the case of cowpea producers in northern Nigeria. Food Energy Secur 9(3):e211. https://doi.org/10.1002/fes3.211

Mandal S, Choudhury BU (2014) Estimation and prediction of maximum daily rainfall at Sagar Island using best fit probability models. Theor Appl Climatol 117:87–97

Mandal S, Choudhury BU, Mondal M, Bej S (2013) Trend analysis of weather variables in Sagar Island, West Bengal, India: a long-term perspective (1982–2010). Curr Sci 105:947–953

Mandal S, Choudhury BU, Satpati LN (2015) Monsoon variability, crop water requirement, and crop planning for kharif rice in Sagar Island, India. Int J Biometeorol 59:1891–1903. https://doi.org/10.1007/s00484-0150995-9

Mandal S, Satpati LN, Choudhury BU, Sadhu S (2018) Climate change vulnerability to the agrarian ecosystem of the small Island: evidence from Sagar Island, India. Theor Appl Climatol 132:451–464. https://doi.org/10.1007/s00704-017-2098-5

Mondal M, Paul S, Mukhopadhyay A, Mandal P, Chatterjee U, Santra S, Bhattacharya S, Paul S (2023) Evaluating rural poverty and livelihood diversification in the context of climate-induced extreme events in coastal area: insights from Indian Sundarban. Reg Stud Mar Sci 67:103191

Muto M, Yamano T (2009) The impact of mobile phone coverage expansion on market participation: panel data evidence from Uganda. World Dev 37(12):1887–1896. https://doi.org/10.1016/j.worlddev.2009.05.004

Muzammal H, Zaman M, Safdar M, Adnan Shahid M, Sabir MK, Khil A, Raza A, Faheem M, Ahmed J, Sattar J, Sajid M (2024) Climate change impacts on water resources and implications for agricultural management. In: Transforming agricultural management for a sustainable future: climate change and machine learning perspectives 2024 Aug 7. Springer Nature Switzerland, Cham, pp 21–45

Ngubane A, Masuku MM, Chamane SC (2025) Decoding market access: a quantitative analysis of challenges faced by small-scale farmers. J Agribus Rural Dev 77(3):299–316

OECD (2025) Reinforcing rural resilience. OECD rural stud. OECD Publishing, Paris. https://doi.org/10.1787/7cd485e3-en

Ogbari ME, Folorunso F, Simon-Ilogho B, Adebayo O, Olanrewaju K, Efegbudu J, Omoregbe M (2024) Social empowerment and its effect on poverty alleviation for sustainable development among women entrepreneurs in the Nigerian agricultural sector. Sustainability 16(6):2225. https://doi.org/10.3390/su16062225

Ogutu SO, Gödecke T, Qaim M (2020) Agricultural commercialisation and nutrition in smallholder farm households. J Agric Econ 71(2):534–555. https://doi.org/10.1111/1477-9552.12359

Rastogi M, Kolur SM, Burud A, Sadineni T, Sekhar M, Kumar R, Rajput A (2024) Advancing water conservation techniques in agriculture for sustainable resource management: a review. J Geogr Environ Earth Sci Int 28(3):41–53. https://doi.org/10.9734/JGEESI/2024/v28i3755

Russo T, Alfredo K, Fisher J (2014) Sustainable water management in urban, agricultural, and natural systems. Water 6(12):3934–3956. https://doi.org/10.3390/w6123934

Shahzad L, Waheed A, Sharif F, Ghafoor GZ, Rafique A (2024) Understanding role of climatic parameters and adaptation strategies in agriculture productivity of south Asian countries. Sustain Environ 10(1):2345453

Simelton E, McCampbell M (2021) Do digital climate services for farmers encourage resilient farming practices? Pinpointing gaps through the responsible research and innovation framework. Agriculture (Basel) 11:953. https://doi.org/10.3390/agriculture11100953

Simpson BM, Burpee G (2014) Agricultural extension and adaptation under the "new Normal" of climate change. In: Leal Filho W (ed) Handbook of climate change adaptation. Springer, Berlin, Heidelberg. https://doi.org/10.1007/978-3-642-38670-1_121

Singh R, Garg KK, Wani SP, Tewari RK, Dhyani SK (2014) Impact of water management interventions on hydrology and ecosystem services in Garhkundar-Dabar watershed of Bundelkhand region, Central India. J Hydrol 509:132–149. https://doi.org/10.1016/j.jhydrol.2013.11.030

Singh C, Osbahr H, Dorward P (2018) The implications of rural perceptions of water scarcity on differential adaptation behaviour in Rajasthan, India. Reg Environ Chang 18:2417–2432. https://doi.org/10.1007/s10113-018-1358-y

Tenzing J, Conway D (2022) Climate discourses as barriers to rights-based adaptive social protection: how historical politics shape Ethiopia's climate-smart safety net. Glob Environ Change 76:102583. https://doi.org/10.1016/j.gloenvcha.2022.102583

Thomas SJ, Sahoo SS, Thomas S, Kumar A, Awad MM (2025) Indian rural livelihoods and renewable energy interventions–a critical analysis for a bottom-up approach for sustainability from an energy-water-food nexus context. Energy Nexus 23:100421. https://doi.org/10.1016/j.nexus.2025.100421

Timilsina GR (2021) Financing climate change adaptation: international initiatives. Sustainability 13(12):6515. https://doi.org/10.3390/su13126515

Tran DQ, Le KN (2025) Sea-level rise and saltwater intrusion: economic estimates of impacts of nature-based mitigation policies under uncertainty. Water 17(9):1355

Ullah A, Zeb A, Saqib SE, Kächele H (2022) Landscape co-management and livelihood sustainability: lessons learned from the billion trees afforestation project in Pakistan. Land Use Policy 115:106034. https://doi.org/10.1016/j.landusepol.2022.106034

UNICEF (2015) Unless we act now: the impact of climate change on children. United Nations Children's Fund, New York

Vernooy R (2022) Does crop diversification lead to climate-related resilience? Improving the theory through insights on practice. Agroecol Sustain Food Syst 46(6):877–901. https://doi.org/10.1080/21683565.2022.2076184

Vidadala R (2024) Reimagining agroforestry: climate-resilient landscapes for regenerative agriculture. In: Kumar S, Alam B, Taria S, Singh P, Yadav A, Arunachalam A (eds) Agroforestry solutions for climate change and environmental restoration. Springer, Singapore. https://doi.org/10.1007/978-981-97-5004-7_8

Yang P, Wu L, Cheng M, Fan J, Li S, Wang H, Qian L (2023a) Review on drip irrigation: impact on crop yield, quality, and water productivity in China. Water 15(9):1733. https://doi.org/10.3390/w15091733

Yang QL, Du T, Li N, Liang J, Javed T, Wang H, Guo J, Liu Y (2023b) Bibliometric analysis on the impact of climate change on crop pest and disease. Agronomy 13:920. https://doi.org/10.3390/agronomy13030920

Chapter 23
Branching the Future: Agroforestry as a Nexus for Climate-Resilient Livelihoods

Pradip Kumar Sarkar, B. U. Choudhury, Rubai Podder, Sujit Das, Bapi Das, Prananath Barman, and Biswajit Das

Abstract The ecological and socio-economic landscape is being rapidly altered by climate change, with a disproportionate impact on rural communities that depend on climate-sensitive resources. The need for sustainable and adaptive land-use strategies has become urgent as rising temperatures, erratic rainfall, soil degradation, and biodiversity loss intensify. The focus of this chapter is on agroforestry, a transformative and integrative approach that not only addresses climate vulnerabilities but also strengthens rural resilience through diversified livelihoods. The deliberate integration of trees with crops and/or livestock, which is known as agroforestry, is a centuries-old land management tradition that is now relevant in the current climate context. The chapter presents different types of agroforestry, including agri-silviculture, silvi-pasture, agri-horti-silviculture, and home gardens, which have great potential for adaptation in both arid and humid ecosystems. We also critically examine agroforestry's role in regulating microclimates, enhancing soil health, conserving water, and mitigating the adverse effects of extreme weather events. By comparing aboveground biomass and belowground root dynamics with conventional agricultural practices, agroforestry systems demonstrate their carbon sequestration capacity. Numerous studies have shown that agroforestry not only stores more carbon than monoculture systems, but it also restores ecosystem stability. In addition, agroforestry has multiple socio-economic advantages. By producing food, fodder, fuel, wood, and fiber, it provides a pathway for income diversification, food and nutritional security, and energy access. It must provide a platform for women's empowerment and inclusive community development, particularly in regions with limited access to resources and opportunities. Both global and national lenses are used to analyze the policy and institutional framework. The institutional mainstreaming of agroforestry has been pioneered by India's National Agroforestry Policy (ICRAF. National Agroforestry Policy—India's pioneering approach. World Agroforestry Centre, Nairobi, 2014), but there are still challenges like fragmented responsibilities, land tenure insecurity, and regulatory hurdles. Success stories and

P. K. Sarkar (✉) · B. U. Choudhury · R. Podder · S. Das · B. Das · P. Barman · B. Das
ICAR Research Complex for NEH Region, Tripura Centre, Lembucherra, West Tripura, India

S. Mandal et al. (eds.), *Climate Change Vulnerability*, Climate Change Management, https://doi.org/10.1007/978-3-032-16783-5_23

field-based innovations from climate-vulnerable regions across India, Africa, and Southeast Asia are documented in the chapter, which demonstrate agroforestry's adaptability and scalability. However, the path to large-scale adoption is not without obstacles. Progress is being hindered by knowledge gaps, weak market linkages, a lack of financial incentives, and limited access to planting materials. The chapter proposes actionable strategies for scaling up agroforestry, including participatory extension models, leveraging climate finance, digital innovations, and ecosystem-based payments. To sum up, agroforestry is at the intersection of ecological restoration, climate resilience, and rural prosperity. Climate-resilient development in rural landscapes can be achieved through strategic policy support, institutional synergy, and community engagement.

Introduction

Climate change has emerged as one of the most defining challenges of the twenty-first century, threatening ecosystems, livelihoods, and economies across the globe (Dogar et al. 2017, 2019, 2023; Dogar and Sato 2018, 2019; Stenchikov and Dogar 2012; Dogar and Almazroui 2022; Mandal et al. 2013; Mandal and Choudhury 2014; Halder et al. 2023; Khan et al. 2025). Although its effects are extensive and profound, they are horrible in rural areas where residents rely primarily on climate-sensitive industries, including forestry, fisheries, and agriculture (IPCC 2021). The rural poor often face two burdens: they are the group least responsible for global greenhouse gas emissions, and they bear the brunt of the effects of climate variability, including unpredictable rainfall, increased frequency of droughts and floods, heat waves, land degradation, and biodiversity loss (FAO 2021; Halder et al. 2022). Delays in the monsoon season, decreased water availability, elevated soil salinity, and decreased crop yields are all signs of climate change in India and other South Asian nations. Among the risks that these shifts exacerbate in rural areas include land fragmentation, poor infrastructure, low technical investment, and limited access to institutional help (Singh et al. 2020; Mandal et al. 2015, 2018). Smallholder farmers, who are the backbone of the rural agricultural economy, are especially at risk. Changes in seasonal weather patterns have a direct impact on agricultural cycles and livestock health, leading to food scarcity, economic loss, and increased migration. The ability of rural populations to adapt is influenced by institutional risk reduction techniques, information accessibility, and the availability of natural resources, according to the Intergovernmental Panel on Climate Change (IPCC). Since traditional agricultural systems often fall short in mitigating climate challenges, it is imperative to seek alternative land-use practices that are both ecologically and socially viable.

One practical way to improve climate resilience in rural areas is through agroforestry, which is the deliberate coexistence of trees with crops and/or cattle on the same land unit (Nair 1993; Dey et al. 2025a). Agroforestry systems' many uses support carbon sequestration, livelihood diversification, enhanced agricultural output, and ecosystem restoration. Because of these qualities, agroforestry is the perfect natural solution to the interconnected problems of rural development and climate adaption. Agroforestry decreases soil erosion, increases soil water retention capacity, and improves soil fertility through nitrogen cycling and organic matter input (Jose 2009). Tree canopies act as windbreaks to protect soil and lower evapotranspiration, while also regulating microclimates to lessen heat stress on crops and animals. Tree roots increase infiltration and decrease surface runoff in areas with irregular rainfall, lowering the danger of drought and flooding. Crucially, agroforestry systems lessen the financial vulnerability of farm households by producing timber, fuel wood, feed, fruits, and medicinal plants, which offer a year-round source of revenue (Leakey 2017). By distributing risk and improving food and income security, agricultural production diversification acts as a natural insurance mechanism. Agroforestry preserves biodiversity and creates ecological corridors, which both increase landscape-level resilience. Governments and international organizations have recognized the potential of agroforestry in mitigating climate change. The first in the world, India's National Agroforestry Policy (2014) highlighted the importance of agroforestry in developing climate-resilient agriculture, strengthening carbon sinks, and boosting rural livelihoods (MoAFW 2014). The significance of ecosystem-based strategies, such as agroforestry, in Nationally Determined Contributions (NDCs) is also emphasized by the Paris Agreement. The following goals will serve as the foundation for the study based on the information above:

1. To analyze the impacts of climate change on rural communities, particularly smallholder farmers, in terms of environmental, economic, and social vulnerability.
2. To examine the principles, functions, and types of agroforestry systems that contribute to climate adaptation.
3. To evaluate the role of agroforestry in enhancing resilience through ecosystem services such as carbon sequestration, biodiversity conservation, and livelihood diversification.
4. To assess the policy landscape, institutional mechanisms, and implementation challenges associated with mainstreaming agroforestry in adaptation strategies.
5. To identify practical pathways and recommendations for scaling up agroforestry interventions in climate-vulnerable regions.

This chapter's scope is both geographical and thematic. Thematically, it discusses the governance elements, socioeconomic advantages, and ecological roles of agroforestry in relation to climate adaptation. Geographically, although the examples are mostly from South Asia, the lessons are applicable everywhere, especially to tropical and subtropical areas that face comparable climate risks. This chapter aims to frame agroforestry as a revolutionary option for sustainable development in an era of climate uncertainty, rather than just as an alternative land use, by connecting scientific knowledge with policy and practice.

Understanding Agroforestry in the Climate Context

Definition and Principles

In order to improve productivity, profitability, diversity, and ecosystem sustainability, agroforestry—a dynamic and ecologically oriented natural resource management system—integrates trees and shrubs into agricultural landscapes (Nair 1993). Three main elements define it: a multifunctional land-use system that offers social, economic, and environmental benefits; purposeful and integrated management techniques; and at least two plant species interacting biologically, one of which is a woody perennial. The fundamental tenets of agroforestry are multi-functionality and synergy. It aims to reduce competition for light, water, and nutrients while maximizing beneficial interactions between the elements—trees, crops, and cattle (Rao et al. 2007). Agroforestry is a naturally sustainable land-use system because of its emphasis on biodiversity, soil health, microclimate regulation, and tolerance to climatic extremes. The idea of climate-smart agriculture, which is being advocated by international organizations such as the FAO and CGIAR, is in line with these concepts.

Historical Evolution and Traditional Systems

Agroforestry is not a modern invention but a revival and formalization of traditional practices that have sustained communities for centuries. Historical records and indigenous knowledge systems demonstrate that farmers across Asia, Africa, and Latin America have long incorporated trees into their farming systems. Keralan home gardens, jhum (shifting farming) in combination with tree crops in the Northeast, and silvo-pastoral systems in dry and semi-arid regions are a few examples from India (Kumar and Nair 2004). These traditional methods maintained soil fertility and water cycles, generated food, fodder, lumber, medicine, and were based on ecological knowledge and social necessity. For example, Southeast Asian home gardens feature vertical vegetation layering that replicates forest ecosystems and supports a wide variety of species. Crop yields and soil fertility have increased in Africa thanks to parkland systems that include sporadic trees on cropland, as Faidherbia albida in the Sahel (Garrity et al. 2010). However, these systems were marginalized as a result of colonial agricultural policy, which favored extractive forestry and monoculture crops. Agroforestry was reintroduced into scientific and policy debates in the 1970s and 1980s, which led to the creation of organizations such as the World Agroforestry Centre (ICRAF). Since then, agroforestry has served as a link between rural development and environmental preservation.

Agroforestry Typologies Relevant to Climate Adaptation

Agroforestry systems are highly diverse and adaptable to agro-climatic zones, soil types, and socio-economic conditions. For climate adaptation, specific typologies have shown particular promise:

1. *Agri-silviculture systems* involve integrating trees and crops on the same land unit. They help reduce crop failure risks due to extreme weather, as the trees provide shade, wind protection, and soil enrichment. Species like *Azadirachta indica* (neem), *Dalbergia sissoo* (shisham), and *Gliricidia sepium* (quickstick) are commonly used.
2. *Silvopastoral systems*: These combine trees with pasture and livestock. In drought-prone regions, trees such as *Leucaena leucocephala* and *Acacia nilotica* provide animal fodder and shade, improving livestock productivity while conserving soil moisture.
3. *Agri-horticultural systems*: They provide food and marketable produce through the use of fruit trees and crops. In these systems, citrus, guava, and mango species are combined with cereals or legumes to provide economic resilience and dietary diversity.
4. *Boundary plantations and shelterbelts*: In addition to serving as windbreaks, trees placed along field boundaries shield crops from heat stress and desiccation. Additionally, they act as living fences and habitat corridors for beneficial insects and pollinators.
5. *Home gardens*: Often seen in humid and tropical areas, home gardens include a multilayered arrangement of trees, shrubs, herbs, and vines. These systems provide year-round food and income while maintaining high levels of biodiversity.
6. *Alley cropping and contour hedgerows*: Especially useful in hilly and sloping terrains, these systems reduce soil erosion, enhance water infiltration, and maintain slope stability. Species like *Vetiveria zizanioides* and *Sesbania grandiflora* are used as hedgerows.

Each of these typologies serves distinct functions in climate adaptation—whether through carbon storage, microclimate regulation, livelihood diversification, or risk buffering. The selection of appropriate systems depends on biophysical, economic, and cultural contexts, and successful implementation often requires participatory planning involving local stakeholders.

Agroforestry as a Climate Adaptation Strategy

Agroforestry is increasingly recognized not only as a sustainable land-use practice but also as a powerful tool for climate adaptation. The integration of trees within farming systems strengthens ecological functions and enhances the resilience of rural communities to climate shocks. This section explores how agroforestry contributes to climate adaptation through the regulation of microclimates, improvement of ecosystem stability, conservation of water resources, enrichment of soils, and the diversification of livelihood strategies—particularly in vulnerable rain—fed areas.

Microclimate Regulation and Ecosystem Stability

Agroforestry systems have a profound impact on microclimates, which play a critical role in determining the health and productivity of crops and livestock. Tree canopies reduce temperature extremes by providing shade, lowering soil surface temperatures, and moderating the diurnal temperature range. This is particularly valuable in hot and arid climates, where high temperatures can lead to heat stress in crops and livestock, reduce pollination, and accelerate evapotranspiration (Jose et al. 2004). Shading from trees in agroforestry systems helps reduce photosynthetic stress, especially during the hottest hours of the day. This moderation in temperature supports sustained crop yields even during erratic weather conditions. Moreover, trees act as windbreaks, reducing wind velocity and minimizing mechanical damage to crops, which is especially beneficial in cyclone-prone and semi-arid areas (Nair et al. 2009). Agroforestry also contributes to ecological stability by promoting biodiversity and enhancing natural pest control mechanisms. The multi-strata design of many agroforestry systems supports a greater diversity of flora and fauna compared to monoculture systems. Birds, insects, and other beneficial organisms that reside in agroforestry landscapes contribute to pollination and natural pest control, thereby reducing the dependence on chemical pesticides (Torquebiau and Penot 2006). Further, these systems enhance ecological connectivity by serving as corridors or buffer zones between fragmented forest patches. This structural diversity contributes to habitat conservation and supports the ecological integrity of landscapes, thus enhancing overall resilience to environmental stressors.

Water Conservation and Soil Enhancement

Water scarcity is one of the most pressing challenges for agricultural sustainability under changing climatic conditions. Agroforestry systems contribute significantly to water conservation through a combination of physical and biological mechanisms. Tree roots improve soil structure by creating channels that facilitate water infiltration, thereby reducing surface runoff and enhancing groundwater recharge (Ong et al. 2002; Dey et al. 2025b). The litters that fall from trees increases soil organic matter, which enhances the soil's water-holding capacity and contributes to greater drought tolerance. Mulching from leaf litter and pruned biomass creates a protective layer over the soil surface, reducing evaporation losses and maintaining soil moisture (Young 1997). Trees also play a role in reducing the frequency and severity of flash floods by stabilizing soil and promoting gradual water infiltration during heavy rains. Furthermore, agroforestry improves soil fertility through nitrogen fixation (in the case of leguminous trees), cycling of deep-soil nutrients, and the addition of organic matter. Species like *Gliricidia sepium*, *Leucaena leucocephala*, and *Sesbania grandiflora* are known for their nitrogen-fixing abilities, which support soil enrichment without the need for synthetic fertilizers (Sanchez 1995).

Agroforestry systems help prevent land degradation by controlling erosion. Trees planted on contours or as hedgerows act as physical barriers, slowing down runoff and trapping sediments. Root systems bind the soil, reducing susceptibility to landslides on sloped terrain. These soil conservation measures are critical in mountainous and semi-arid regions where soil erosion is exacerbated by intense rainfall and sparse vegetation cover. Additionally, biochar derived from tree residues can be used as a soil amendment to improve soil pH, nutrient availability, and carbon retention. Such practices not only enhance productivity but also contribute to long-term soil health and carbon sequestration (Lehmann and Joseph 2009).

Diversified Risk-Buffering in Rainfed Agriculture

Rainfed agriculture, which supports over 60% of farmers globally, is particularly vulnerable to climate variability. In India, nearly 50% of net sown area is under rainfed conditions, where low and erratic rainfall severely constrains crop yields and farm income (ICAR 2015). Agroforestry offers an integrated approach to mitigate these risks through diversification of production, year-round income opportunities, and enhanced ecological resilience. One of the key benefits of agroforestry in rainfed areas is diversification. By integrating multiple species—trees, crops, and livestock—farmers can reduce their dependency on a single crop. This diversification acts as a natural insurance mechanism against crop failure due to drought, pests, or market fluctuations. For instance, when a cereal crop fails due to delayed rains, income can still be generated from fruit trees or fodder trees that are less sensitive to short-term rainfall anomalies (Garrity 2004). Fodder trees in silvo-pastoral systems ensure that livestock can be maintained even during dry seasons when natural pastures are scarce. Fruit trees like *Aonla* (Indian gooseberry), *Ber* (*Ziziphus mauritiana*), and Custard apple (*Annona squamosa)* are resilient to dry conditions and provide supplementary income. Timber and bamboo-based systems offer long-term investments that can be harvested during periods of financial distress. Furthermore, agroforestry systems improve farm productivity and economic returns over time. Studies have shown that integrated farming systems with trees tend to have higher net income and lower yield variability compared to mono-cropping systems. This is due to the multi-layered harvesting schedule and the availability of products throughout the year (Saxena 2000a, b).

In addition to economic diversification, agroforestry enhances food and nutritional security. Trees provide edible fruits, leaves, and seeds rich in micronutrients, which supplement the diet of rural households. This nutritional diversification is crucial during lean periods when conventional crops are unavailable or unaffordable. Importantly, agroforestry also increases the adaptive capacity of smallholder farmers by reducing their exposure to market and climatic risks, while enhancing their access to diversified livelihood options and improved ecosystem services. This makes agroforestry not just a climate adaptation strategy, but also a pathway for inclusive and sustainable rural development.

Carbon Sequestration Potential of Agroforestry

As the global community intensifies its efforts to mitigate climate change, carbon sequestration through nature-based solutions has emerged as a central focus. Agroforestry, a land-use system that integrates trees with crops and/or livestock on the same unit of land, stands out as a particularly effective method of sequestering atmospheric carbon dioxide (CO_2) in both biomass and soils. This section discusses the role of agroforestry systems as carbon sinks, explores aboveground and belowground carbon dynamics, and offers a comparative perspective vis-à-vis conventional agriculture in terms of carbon sequestration potential.

Role of Agroforestry Systems as Carbon Sinks

Agroforestry systems are uniquely positioned to serve as efficient carbon sinks due to their multilayered vegetation structure and prolonged biomass retention. Unlike monoculture agriculture, which typically involves short cropping cycles and little woody biomass, agroforestry integrates perennial components such as trees and shrubs that accumulate carbon over extended periods. These perennial elements contribute to the long-term capture and storage of carbon in plant biomass and soils, mitigating the impacts of greenhouse gas emissions (Montagnini and Nair 2004a, b). The carbon sequestration potential of agroforestry varies depending on climatic zone, species composition, soil type, management practices, and system age. Tropical agroforestry systems, in particular, have been noted for their high carbon accumulation rates due to faster growth rates and greater biodiversity (Albrecht and Kandji 2003). For example, multistate systems like home gardens in Kerala or coffee-based agroforestry in Central America have demonstrated high levels of carbon storage, rivaling even secondary forests in some cases. Agroforestry also promotes a circular nutrient economy where leaf litter, pruned biomass, and root turnover contribute to soil organic carbon (SOC) enhancement. By minimizing land degradation and providing year-round ground cover, these systems reduce erosion and maintain soil structure—further supporting carbon retention in the landscape (Pandey 2007) (Fig. 23.1).

In recognition of these benefits, agroforestry has been integrated into various climate mitigation programs and carbon offset schemes, including REDD+

Fig. 23.1 Flow chart of circular nutrient economy (CNE)

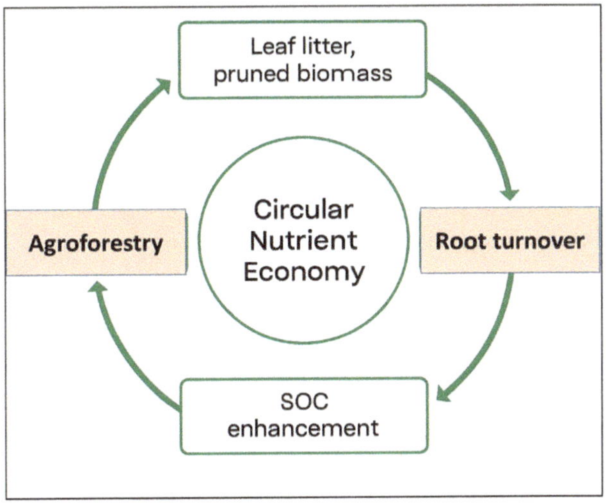

(Reducing Emissions from Deforestation and Forest Degradation), the Clean Development Mechanism (CDM), and voluntary carbon markets.

Aboveground and Belowground Carbon Dynamics

Carbon storage in agroforestry occurs in both aboveground biomass—such as tree trunks, branches, foliage and belowground components like roots and soil organic matter. Aboveground biomass constitutes the most visible form of carbon sequestration, especially in tree-dense systems. Fast-growing species like *Eucalyptus*, *Albizia*, *Dalbergia*, and *Leucaena* can accumulate substantial biomass within a few years, contributing significantly to aboveground carbon stocks (Nair et al. 2010). Quantifying aboveground biomass typically involves allometric equations based on measurements of tree diameter, height, and wood density. Estimates suggest that agroforestry systems can sequester between 2–9 tons of carbon per hectare per year, depending on species and management intensity (Kaonga and Bayliss-Smith 2009). However, belowground carbon dynamics are equally crucial but often underappreciated. Tree roots enhance soil structure, stimulate microbial activity, and contribute to soil carbon through root turnover and exudates. The symbiotic relationship between plant roots and mycorrhizal fungi further supports carbon stabilization by promoting aggregate formation and enhancing soil carbon retention (Six et al. 2002).

Litter fall from trees, combined with minimal tillage practices in agroforestry systems, adds a steady stream of organic material to the soil. This organic matter decomposes slowly, contributing to the build-up of stable soil organic carbon pools. Over time, these systems increase soil carbon stocks, which play a vital role in maintaining soil health, water retention, and nutrient cycling. A study by Kumar and Nair (2011) in the Himalayan region demonstrated that agroforestry plots had significantly higher soil organic carbon compared to adjacent croplands, with increases of up to 45% over a 10-year period. Similarly, experiments in central India under tropical dry climates showed that agroforestry systems enhanced soil microbial biomass and reduced CO_2 efflux from soil, reflecting improved carbon use efficiency (Chaturvedi and Raghubanshi 2015).

Study Sites Multipurpose AFS Models

The present study was conducted in the nine different multi-purpose tree-based agroforestry models of ICAR farm land area in Cocotilla of Lembucherra, West Tripura. The nine different tree-based agroforestry models were such as; acacia based-AFS (*Acacia auriculiformis*), shisham based-AFS (*Dalbergia sisso*), neem based-AFS (*Azadirachta indica*), eucalyptus based-AFS (*Eucalyptus maculata*), teak based-AFS (*Tectona grandis*), gamar based-AFS (*Gmelina arborea*), white siris based-AFS (*Albizia procera*), champaca based-AFS (*Michelia champaca*), and rain-tree based-AFS (*Samanea saman*),. These plots were geographically lies between 23.9022–23.9026 N to 91.3228–91.3232 E and situated in the altitude of 46 m to 54 m asl respectively. Cocotilla farm land area is the large farming area of ICAR Tripura Centre, which had an area of 51.04 acres and fall under Lembucherra region (Debendrachandranagar) of West district of Tripura. The area boasts a clump canopy comprising different AFS-models, fruit orchards, rubber plantations, grasslands, barren lands along with some mixed species croplands *viz.*, coconut + banana + intercrops, jackfruitr + lemon + intercrops, jackfruit + banana + coconut + intercrops, jackfruit + lemon + arjun + wood-apple + parkia + intercrops, mango + elephant-foot-yam, lichi + elephant-foot-yam, mango + elephant-foot-yam + intercrops, jackfruitr + colocasia + intercrops, jackfruitr + tea + bamboo + intercrops, mulberry + areca + intercrops, and some other seasonal crop species. The research site encounters a blend of tropical monsoonal weather and subtropical humid conditions. The monsoon contributes abundant rainfall, effectively blanketing the ground. The average temperature varies from 22.1 °C to 35.2 °C in the summer and from 9.5 °C to 28.1 °C in the winter, while the yearly rainfall ranges from 0 to 714 mm.

Plot Selection and Data Collection

We utilized a non-destructive ground sampling approach to evaluate the vegetation and above-ground tree biomass in these eight different agroforestry systems. To enhance the evaluation of tree diversity and biomass, each substantial plot was subdivided into two smaller sub-plots, each measuring 20 m × 20 m (Singh and Dadhwal 2009). To prevent any overlap, we placed four tiny quadrants (1 m × 1 m) at a randomized position of the sub-plots to collect standing litter biomass. This method allowed us to survey the greatest amount of area without any sub-plot overlap.

Tree Sampling and Vegetational Studies

The data records for the vegetation research assessment were obtained using both qualitative and quantitative analysis methods. The quadrates were randomly set up for this purpose. During the vegetation survey, considerable portions of the tree species thriving in the wild were successfully identified. The diameter at breast height (DBH) measurements were recorded for all tree specimens boasting a diameter surpassing 10 cm, captured at a height of 1.37 m above the ground. These DBH measurements were subsequently harnessed to analyze key variables including basal area and tree density. The basal area at each specific site was defined as the cumulative cross-sectional area of all stems within 1 hectare of land, while stem density represented the count of individual trees with a diameter of >10 cm or less at breast height (DBH).

Biomass and Carbon Stock Estimations

Employing a non-destructive random sampling methodology, the study undertook the task of quantifying the biomass and carbon reservoir of numerous tree species. The computation of above-ground tree biomass (AGTB) and below-ground biomass (BGB) were accomplished through the species specific allometric equations each and individual species (Naik et al. 2018a, b, 2019; Brahma et al. 2017; Jha 2015; Kumar et al. 2016; Giri et al. 2015; Lodhiyal and Lodhiyal 2003; Rao et al. 2007). The computation of leaf litter biomass (LB) on the AFS plots was executed using the phytomass formula formulated by Ajtay et al. (1979). The cumulative aboveground biomass (AGB) encompassed the aggregation of both above-ground tree biomass (AGTB), and litter biomass (LB), the biomass of a particular tree-based agroforestry system. The overall biomass, denoted as TB, was calculated as the summation of AGB and BGB. Pursuant to the procedure outlined by Brown et al. (1989), 50% of the total biomass was acknowledged as the carbon stock representation. Through the summation of biomass carbon attributed to diverse components (AGB + BGB + AFS-plots LB), an evaluation of the vegetation's carbon stock was conducted.

Statistical Analyses

The study recorded the biomass productions, total carbon stock of various tree species, and their components, such as roots, litter, and below-ground soil organic carbon (SOC) stock. For assessing the statistical significance between the mean values, the ANOVA test was carried out utilizing the Microsoft Excel 2016 package developed by Microsoft Corporation, USA. Based on these, a long-term assessment of multipurpose tree species (MPTs) based agroforestry models, established in 1987 at the Cocotilla farm unit of ICAR-NEH Region, Tripura

Centre, underscores their vital role in mitigating climate change through effective carbon sequestration. Over a span of 35 years, significant variations were observed among species in terms of growth performance and carbon storage potential. Among the tested species, *Eucalyptus hybrid* emerged as the most promising, exhibiting the maximum trunk diameter at breast height (50.27 ± 1.75 cm), tallest height (27.33 ± 1.45 m), and greatest volume (3.64 ± 0.40 m^3 per tree). It also recorded the highest total dry biomass (3.00 ± 0.26 t per tree) and annual carbon storage potential (37.24 ± 3.26 kg per tree per year). In contrast, *Azadirachta indica* showed the lowest values across all parameters, including trunk diameter (21.53 ± 1.93 cm), height (13.33 ± 0.88 m), volume (0.34 ± 0.08 m^3), biomass (0.34 ± 0.07 t), and carbon storage (4.10 ± 0.86 kg per tree per year). *Tectona grandis* and *Gmelina arborea* demonstrated intermediate but consistent performance, with notable carbon sequestration capacities of 20.13 ± 2.03 and 17.37 ± 2.30 kg per tree per year, respectively.

These findings highlight the superior ecological performance of *Eucalyptus hybrid*, followed closely by *T. grandis* and *G. arborea*, under the agro-climatic conditions of Tripura. Their robust growth and substantial carbon accumulation make them ideal candidates for integration into agroforestry systems aimed at climate change mitigation. Thus, the strategic incorporation of these high-performing MPTs in agroforestry models can serve as a sustainable solution for enhancing carbon sinks, restoring ecological balance, and building climate-resilient rural landscapes in the North Eastern Himalayan region (Fig. 23.2; Table 23.1).

Fig. 23.2 (**a**) Teak + Pineapple-based agroforestry system. (**b**) Champaca + black paper-based agroforestry system

Table 23.1 Biophysical performance of different multipurpose tree species in the agroforestry plots of ICAR Tripura

SI. No.	Species	Common name	Planting spacing	Trunk DBH (cm)	Tree height (m)	Tree volume (m^3 $tree^{-1}$)	Non-destructive method, following the standard allometric equation				Annual carbon storage potential (kg $tree^{-1}$ $year^{-1}$)
							Dry biomass (t $tree^{-1}$)			Total Carbon storage (t $tree^{-1}$)	
							AGB	BGB	Total biomass		
1.	*Acacia auriculiformis*	Earleaf acacia	2 m × 2 m	34.05 ± 1.57	19.67 ± 0.89	1.21 ± 0.15	0.72 ± 0.08	0.26 ± 0.03	0.98 ± 0.11	0.42 ± 0.05	11.96 ± 1.33
2.	*Dalbergia sissoo*	Shisham	2 m × 2 m	32.30 ± 1.41	17.17 ± 0.60	0.94 ± 0.05	0.69 ± 0.05	0.25 ± 0.02	0.94 ± 0.07	0.40 ± 0.03	11.48 ± 0.83
3.	*Azadirachta indica*	Neem	3 m × 3 m	21.53 ± 1.93	13.33 ± 0.88	0.34 ± 0.08	0.24 ± 0.05	0.10 ± 0.02	0.34 ± 0.07	0.14 ± 0.03	4.10 ± 0.86
4.	*Michelia champaca*	Champaca	3 m × 3 m	36.27 ± 0.73	23.33 ± 0.90	1.61 ± 0.12	0.83 ± 0.06	0.29 ± 0.01	1.12 ± 0.06	0.48 ± 0.04	13.70 ± 0.76
5.	*Eucalyptus hybrid*	Eucalyptus	3 m × 3 m	50.27 ± 1.75	27.33 ± 1.45	3.64 ± 0.40	2.28 ± 0.20	0.72 ± 0.06	3.00 ± 0.26	1.30 ± 0.11	37.24 ± 3.26
6.	*Tectona grandis*	Teak	4 m × 4 m	42.32 ± 1.46	24.33 ± 2.20	2.26 ± 0.10	1.22 ± 0.13	0.41 ± 0.04	1.64 ± 0.16	0.70 ± 0.07	20.13 ± 2.03
7.	*Gmelina arborea*	Gamhar	4 m × 4 m	39.56 ± 2.19	24.00 ± 2.60	1.99 ± 0.36	1.05 ± 0.14	0.36 ± 0.04	1.42 ± 0.19	0.61 ± 0.08	17.37 ± 2.30
8.	*Albizia procera*	White siris	4 m × 4 m	27.36 ± 1.15	17.50 ± 0.76	0.68 ± 0.03	0.41 ± 0.04	0.16 ± 0.01	0.56 ± 0.06	0.24 ± 0.02	6.81 ± 0.69
9.	*Samanea saman*	Rain tree	4 m × 4 m	25.98 ± 1.78	15.50 ± 1.04	0.54 ± 0.04	0.35 ± 0.08	0.14 ± 0.03	0.49 ± 0.10	0.21 ± 0.04	5.91 ± 1.25
L.S.D. ($P \leq 0.05$)				***4.83***	***4.35***	***0.60***	***0.32***	***0.10***	***0.42***	***0.18***	***5.25***

Note: *DBH* Diameter at breast height, *AGB* above ground biomass, *BGB* below ground biomass

Comparative Analysis with Conventional Agriculture

Conventional agriculture, characterized by intensive tillage, mono-cropping, and chemical inputs, tends to have low carbon sequestration potential and often contributes to carbon loss through soil erosion and biomass burning. In contrast, agroforestry offers a more balanced and regenerative approach that actively sequesters carbon while delivering a suite of ecological co-benefits. The comparative carbon balance between agroforestry and conventional systems underscores the superior sequestration capacity of the former. Studies indicate that agroforestry systems can store up to five times more carbon per hectare than conventional croplands (Mutuo et al. 2005). For instance, in sub-Saharan Africa, maize-based systems integrated with *Faidherbia albida* trees sequestered significantly more carbon and increased crop productivity, as compared to mono-cropped maize fields. Moreover, agroforestry systems reduce the emission of nitrous oxide and methane - other potent greenhouse gases - by improving nitrogen use efficiency and integrating organic nutrient cycles. In livestock-dominated landscapes, silvo-pastoral systems reduce methane emissions by improving forage quality and animal health, while simultaneously increasing tree-based carbon stocks (Montagnini 2017).

Conventional agricultural practices often involve periodic clearing, ploughing, and application of synthetic fertilizers, all of which disrupt soil structure and microbial communities. These disturbances lead to rapid decomposition of organic matter and loss of soil carbon. In contrast, agroforestry maintains permanent vegetation cover, minimizes disturbance, and fosters ecological stability - conditions conducive to long-term carbon storage. From a policy perspective, agroforestry holds the potential to contribute significantly to national and global climate goals. India's Nationally Determined Contributions (NDCs) under the Paris Agreement recognize agroforestry as a means to create an additional carbon sink of 2.5–3 billion tonnes of CO_2-equivalent through forest and tree cover by 2030 (MoEFCC 2015). This acknowledgment has resulted in supportive policies such as the Sub-Mission on Agroforestry (SMAF), which incentivizes tree planting on farmlands. Agroforestry also supports climate-resilient development pathways by synergizing mitigation and adaptation. The additional income from timber, fruits, and non-timber forest products provides financial resilience to farmers, while the improved ecosystem services support long-term sustainability. In conclusion, agroforestry systems present a scientifically robust, economically viable, and ecologically sustainable approach to sequestering carbon in terrestrial ecosystems. Their integration into agricultural landscapes offers a practical and scalable strategy to combat climate change while supporting livelihoods and enhancing ecosystem resilience.

Socio-Economic co-Benefits of Agroforestry

Agroforestry is not only an ecologically sustainable land use practice but also a potent socio-economic tool capable of transforming rural livelihoods. Its multidimensional nature allows it to address several pressing challenges facing rural populations, particularly in the context of climate change. This section explores the socio-economic co-benefits of agroforestry, focusing on livelihood diversification, income stability, food-fodder-fuel-fiber security, and the role of agroforestry in promoting women's participation and broader community empowerment.

Livelihood Diversification and Income Stability

One of the most significant socio-economic advantages of agroforestry is its ability to diversify rural livelihoods. In regions where climate variability and market volatility threaten agricultural sustainability, agroforestry provides an insurance mechanism by allowing farmers to derive income from multiple sources. This diversification includes the sale of timber, fruits, nuts, medicinal plants, honey, and other non-timber forest products (NTFPs), which collectively enhance household income security (Saxena 2000a, b). Agroforestry systems such as silvopastoral (trees with livestock), agri-silviculture (trees with crops), and agrihorticulture (trees with fruit crops) allow farmers to buffer risks associated with single-commodity failure. For instance, in semi-arid regions of India, integration of *Emblica officinalis* (amla) and *Azadirachta indica* (neem) with arable crops like pigeon pea and millets has enabled farmers to earn sustained incomes even in drought years (Kumar and Nair 2011). Moreover, agroforestry promotes the concept of delayed returns through long-gestation species like teak, mango, and bamboo, ensuring future income streams. These systems enhance long-term economic resilience, especially in regions where off-farm employment opportunities are limited. The presence of perennial trees also contributes to land value appreciation, offering an additional economic benefit (Mercer and Miller 1998).

Food, Fodder, Fuel, and Fiber Security

Agroforestry systems address food and nutritional security by incorporating a range of edible species that provide fruits, nuts, leafy vegetables, and legumes. Many indigenous agroforestry models include multipurpose trees that yield nutrient-dense produce. For example, *Moringa oleifera* is widely recognized for its high vitamin and mineral content and is commonly integrated into smallholder farms in South Asia and Sub-Saharan Africa. In addition to food, agroforestry provides year-round fodder for livestock, an essential input for mixed farming systems. Species such as *Leucaena leucocephala*, *Gliricidia sepium*, and *Sesbania grandflora* are commonly used as fodder trees. These species not only enhance animal productivity but also reduce pressure on natural forests for grazing. Fuel wood and fiber needs are also addressed through agroforestry. Rural households often rely on biomass for cooking, and fast-growing species like *Prosopis juliflora* and *Casuarina equisetifolia* can supply these needs sustainably. Bamboo and other fibrous plants integrated into agroforestry systems serve as raw materials for construction and cottage industries, generating additional income (Pandey 2007). By catering to these multiple needs—food, fodder, fuel, and fiber—agroforestry reduces the dependency of rural communities on external inputs and enhances their self-sufficiency, particularly during environmental and economic shocks.

Women's Participation and Community Empowerment

Agroforestry plays a pivotal role in advancing gender equity and community empowerment. Women, especially in rural areas, are primary collectors and users of fuel wood, fodder, and food. The proximity of tree-based resources in agroforestry systems reduces the drudgery and time burden of resource collection, thereby freeing up time for other productive activities (Kumar and Ramachandran 2005). In many parts of India and Africa, women's self-help groups (SHGs) and cooperatives actively participate in the management of agroforestry nurseries, seedling production, and value-added enterprises such as leaf plate making, beekeeping, or herbal product processing. These activities not only enhance household income but also promote leadership and decision-making roles among women (Agarwal 2001). Agroforestry-based interventions can also foster collective action and strengthen local institutions. Community-based agroforestry projects, such as joint forest management (JFM) and participatory watershed programs, have demonstrated success in promoting environmental stewardship and equitable benefit-sharing. The participatory nature of these initiatives ensures that marginalized groups, including women and landless laborers, have a voice in land-use decisions (Sanchez 1995). Furthermore, agroforestry promotes intergenerational equity by ensuring that natural resources are managed sustainably and productively. The knowledge-sharing culture embedded in traditional agroforestry systems—often transmitted orally through generations—preserves indigenous practices and strengthens community identity and cohesion. In fragile ecosystems such as the hills of Northeast India or the tribal belts of Odisha and Jharkhand, agroforestry provides a viable development model that aligns ecological integrity with social inclusion. Institutions such as ICAR and ICRAF (now CIFOR-ICRAF) have implemented several successful community agroforestry initiatives that have resulted in increased income, better nutrition, and improved gender dynamics (ICRAF 2013).

Case Example: Tribal Women in Jharkhand

In the Gumla district of Jharkhand, tribal women have adopted agroforestry practices under the guidance of ICAR's tribal sub-plan. These practices include planting *Dalbergia sissoo*, *Amla*, and *Tamarindus indica* along with traditional crops. The produce is either sold in local markets or processed into value-added products such as pickles and herbal powders. The economic independence gained from these practices has led to better health, education, and status for the women in their communities. Another initiative in Odisha's Koraput district introduced agroforestry-based kitchen gardens supported by SHGs. These gardens included a variety of fruit trees, vegetables, and medicinal herbs. Women not only ensured household food security but also generated income by selling surplus produce.

Policy and Institutional Support

Recognizing the socio-economic benefits of agroforestry, governments and international organizations have formulated policies to support its adoption. India's National Agroforestry Policy (2014) specifically highlights the role of agroforestry in empowering rural women, enhancing farm incomes, and promoting inclusive growth. Supportive measures such as decentralized nursery schemes, community seed banks, microcredit facilities, and training programs further promote the uptake of agroforestry. Institutions like NABARD, ICAR, and various state agricultural universities are actively involved in promoting agroforestry as part of rural development programs. In summary, agroforestry is a multidimensional land-use approach that delivers tangible socio-economic benefits, especially for smallholders and marginalized groups. It enhances livelihood security, strengthens rural economies, promotes gender equity, and fosters community resilience. These co-benefits position agroforestry as a key strategy not only for climate adaptation and mitigation but also for sustainable rural development.

Policy Landscape and Institutional Frameworks

The effective promotion and mainstreaming of agroforestry practices in response to climate change depend significantly on a conducive policy landscape and a robust institutional framework. Over the last few decades, both international and national policy instruments have increasingly acknowledged agroforestry as a sustainable land-use practice with the potential to address ecological, economic, and social challenges. This section explores the evolution and impact of policy measures related to agroforestry, focusing particularly on India's pioneering National Agroforestry Policy (2014), its progress, and the broader institutional arrangements that support or hinder agroforestry development.

Overview of National and International Agroforestry Policies

At the international level, agroforestry gained recognition primarily through its contributions to sustainable development, biodiversity conservation, and climate change mitigation. Organizations such as the Food and Agriculture Organization (FAO), World Agroforestry Centre (ICRAF), and International Union for Conservation of

Nature (IUCN) have actively advocated agroforestry as part of integrated land-use strategies (FAO 2013). The Paris Agreement under the United Nations Framework Convention on Climate Change (UNFCCC) has indirectly supported agroforestry by encouraging land-based carbon sinks and nature-based solutions (NBS) for climate mitigation and adaptation. Agroforestry fits squarely within these frameworks, contributing to nationally determined contributions (NDCs) of several countries including India, Brazil, and Kenya (Nair et al. 2009).

In Africa, countries like Ethiopia and Rwanda have incorporated agroforestry into their national adaptation programs of action (NAPAs). Similarly, in Latin America, Costa Rica and Nicaragua have developed agroforestry-specific legal frameworks to support sustainable farming practices, access to extension services, and incentives for ecosystem services. National policies in several countries have begun aligning land use and agricultural strategies with forestry objectives to promote tree integration on farms. However, there remains a considerable variation in policy effectiveness due to administrative, legal, and institutional challenges.

India's National Agroforestry Policy 2014 and beyond

India emerged as a global leader by launching the National Agroforestry Policy (NAP) in 2014, the first of its kind globally. The policy was a milestone in integrating trees into farming systems at a national level. Its formulation was the result of a multi-stakeholder consultative process involving the Ministry of Agriculture, ICAR, ICRAF, civil society organizations, and farmers' groups.

The NAP 2014 laid out seven strategic objectives:

1. Establishing institutional support for agroforestry
2. Simplifying the regulatory mechanism related to tree felling and transit
3. Providing financial and institutional incentives
4. Enhancing research, extension, and capacity building
5. Improving access to quality planting materials
6. Strengthening market access for agroforestry products
7. Promoting agroforestry through convergence with other rural development programs

The policy was aligned with other national missions such as the National Mission for Sustainable Agriculture (NMSA) and the National Adaptation Fund for Climate Change (NAFCC). Importantly, it supported decentralized nurseries and agroforestry-based models for carbon sequestration and ecosystem services. The implementation of the policy saw significant progress in states like Uttar Pradesh, Haryana, and Karnataka, where agroforestry was integrated into state action plans. For instance, the Har Medh Par Ped initiative in Haryana encouraged planting trees on farm bunds, while Uttar Pradesh created simplified transit rules for commercial tree species. In 2018, the Ministry of Agriculture and Farmers' Welfare launched the Sub-Mission on Agroforestry (SMAF) under NMSA to provide financial incentives to farmers for tree planting. SMAF targeted small and marginal farmers, tribal communities, and ecologically sensitive regions. The mission supports plantation activities, capacity-building programs, and convergence with MNREGA and watershed development schemes (GoI 2018). Despite these achievements, challenges remain. Several states have yet to notify tree species exempt from transit permits. Bureaucratic delays, limited extension support, and a lack of awareness continue to hinder the policy's full-scale adoption. Moreover, private sector involvement in agroforestry markets remains weak due to price volatility and the absence of farmer-producer organizations (FPOs) dedicated to tree crops.

Institutional Convergence and Gaps

Effective agroforestry promotion requires coordination across multiple ministries, departments, and levels of government. In India, key institutions involved in agroforestry include:

- *Ministry of agriculture and farmers' welfare*—the nodal ministry for agroforestry programs under NMSA.
- *Indian council of agricultural research (ICAR)*—implements agroforestry research through its Division of Agroforestry and All India Coordinated Research Project on Agroforestry (AICRPAF).
- *Forest departments*—regulate tree felling, transit, and promote social forestry.
- *Ministry of environment, forest and climate change (MoEFCC)*—coordinates climate change mitigation efforts.
- *State agricultural universities (SAUs)*—engage in education, extension, and region-specific agroforestry development.
- *National bank for agriculture and rural development (NABARD)*—provides credit and project finance under rural infrastructure schemes.

Despite these institutions' involvement, convergence is often limited by siloed functioning, duplication of efforts, and differing mandates. For instance, forestry departments emphasize conservation and regulation, while agricultural departments focus on productivity and income generation. This leads to mismatches in implementation priorities. Furthermore, the lack of a national database on agroforestry practices, tree cover outside forests (TOF), and the contribution of agroforestry to GDP makes planning and monitoring difficult. Extension services related to agroforestry are underdeveloped and poorly equipped to advise farmers on long-gestation tree crops. To address these gaps, there is a need for a dedicated National Agroforestry Mission with inter-ministerial coordination. Such a mission should bring together experts in forestry, agriculture, rural development, and climate science to provide integrated solutions. This would also help align agroforestry goals with India's commitments under SDGs, UNFCCC, and the Bonn Challenge. Moreover, strengthening public-private partnerships (PPP) and developing value chains for agroforestry products can boost farmer interest. Policy support in terms of Minimum Support Price (MSP) for key tree crops, carbon credit trading, and digital platforms for market access are also crucial.

India's proactive approach in recognizing agroforestry's potential through the National Agroforestry Policy (2014) has set a benchmark. However, to translate policy into practice, strong institutional convergence, capacity building, regulatory reforms, and market linkages are essential. Globally, agroforestry has gained policy traction, yet its integration into mainstream agricultural and environmental strategies remains a work in progress. By addressing institutional fragmentation and leveraging synergies across sectors, agroforestry can be mainstreamed as a cornerstone of sustainable and climate-resilient rural development.

Case Studies from Climate-Vulnerable Regions

Agroforestry systems are context-specific and vary significantly across regions, shaped by climatic conditions, socio-economic factors, and traditional knowledge systems. In climate-vulnerable regions, the adoption of agroforestry has demonstrated tangible benefits in enhancing resilience, food security, and carbon sequestration. This section presents selected case studies from the Indian subcontinent, Africa, and Southeast Asia to highlight the diversity, adaptability, and success of agroforestry interventions in some of the world's most climate-sensitive regions.

Indian Subcontinent: Silvi-Pastoral Systems in Drought-Prone Areas

Drought-prone regions in India, particularly in semi-arid zones like Bundelkhand (Uttar Pradesh and Madhya Pradesh), Vidarbha (Maharashtra), and parts of Telangana, have historically grappled with water scarcity, soil degradation, and agrarian distress. Silvi-pastoral systems—integrating trees, shrubs, and pasture grasses—have emerged as effective strategies in mitigating the impact of recurrent droughts. A notable case is the adoption of silvi-pastoral systems in Jhansi district (Bundelkhand region), facilitated by ICAR-Indian Grassland and Fodder Research Institute (IGFRI). The system typically includes tree species such as *Leucaena leucocephala*, *Hardwickia binata*, and *Albizia lebbeck*, combined with grasses like *Cenchrus ciliaris* and *Pennisetum pedicellatum*. These combinations stabilize the soil, improve organic matter content, and provide fodder even during dry spells (Patil et al. 2019). Such systems are crucial for livestock-based livelihoods, which are a major source of income in marginal farming households. Moreover, the integration of leguminous species improves nitrogen content in the soil, reducing dependence on chemical fertilizers. An impact assessment by IGFRI showed that land productivity increased by 25–30% under silvi-pastoral systems, with enhanced fodder availability and soil moisture retention. These systems also served as carbon sinks, with aboveground biomass sequestration rates ranging between 3.5 to 5.2 Mg C ha^{-1} $yr.^{-1}$ (Dhyani et al. 2016).

Africa: Farmer-Managed Natural Regeneration (FMNR)

Farmer-managed natural regeneration (FMNR) is a low-cost, sustainable agroforestry practice that involves the protection and management of naturally occurring trees and shrubs in agricultural landscapes. It has gained widespread recognition for its transformative impact in the Sahel region of Africa, particularly in Niger, Burkina Faso, and Mali. FMNR was popularized in the 1980s by Tony Rinaudo and is currently supported by international organizations such as World Vision. In southern Niger, over five million hectares of degraded land have been regenerated through FMNR, leading to the growth of over 200 million trees (Reij et al. 2009). The approach typically involves selective pruning, protecting rootstock, and managing natural coppice regrowth. Indigenous species like *Faidherbia albida*, *Ziziphus mauritiana*, and *Acacia Senegal* are commonly regenerated. These trees provide fuel wood, fodder, and fruits, contributing to household nutrition and income. A major advantage of FMNR is its cost-effectiveness; it requires minimal external inputs and leverages traditional ecological knowledge. Moreover, FMNR improves microclimatic conditions, reduces wind erosion, and enhances soil fertility. Studies have shown that millet and sorghum yields in FMNR fields increased by 30–100%, depending on rainfall and management (Garrity et al. 2010). FMNR has also improved social capital through community-led governance structures, with a strong emphasis on women's participation and land tenure security. The model is now being replicated in Kenya, Ethiopia, and Malawi.

Southeast Asia: Home Gardens and Carbon Offset Models

Southeast Asia, particularly countries like Indonesia, Thailand, and the Philippines, has a long-standing tradition of home gardens—complex, multistory agroforestry systems that mimic forest structure and offer ecological and economic resilience. In West Java, Indonesia, home gardens - locally known as *pekarangan*—consist of layered vegetation including timber trees (*Artocarpus* spp., *Albizia* spp.), fruit trees (*Durio zibethinus*, *Mangifera indica*), shrubs, herbs, and annual crops. These systems are maintained around homesteads and provide year-round access to food, medicines, spices, and small timber. Home gardens are particularly effective in carbon sequestration. A study by Hairiah et al. (2011) showed that multistory home gardens can store up to 92 Mg C ha^{-1} in aboveground biomass. These systems also improve biodiversity, reduce household expenses, and support nutritional security. Southeast Asian countries have begun integrating home gardens into REDD+ (Reducing Emissions from Deforestation and Forest Degradation) and carbon offset schemes. For example, in the Philippines, the Watershed Rehabilitation Project in Mindanao incentivizes farmers to adopt home garden agroforestry under carbon credit agreements. This not only improves ecosystem services but also offers a new stream of income through the voluntary carbon market. In northern Thailand, agroforestry coffee cultivation under shade trees (*Inga edulis*, *Gliricidia sepium*) has been recognized under international certification schemes like Rainforest Alliance and Fair Trade. These initiatives have increased farmers' earnings while promoting sustainable land use and reducing the carbon footprint of cash crops (Roshetko et al. 2007).

Comparative Lessons and Policy Implications

The case studies from different regions reveal some common themes:

1. *Local adaptation:* Agroforestry systems thrive when adapted to local biophysical and socio-cultural contexts.
2. *Low-cost and scalable:* Practices like FMNR demonstrate that effective climate resilience does not always require high capital investment.
3. *Multi-functionality:* Whether it is fodder production in silvi-pastoral systems or carbon storage in home gardens, agroforestry provides multiple co-benefits.
4. *Institutional support matters:* Successful models often have backing from government policies, NGOs, or community-based institutions.

However, challenges such as land tenure insecurity, lack of extension services, and absence of market incentives remain significant barriers. Therefore, a comprehensive approach combining policy, capacity building, and community participation is essential for scaling agroforestry in climate-vulnerable regions.

Challenges and Constraints

While agroforestry offers a multitude of benefits for climate adaptation and rural livelihood resilience, its widespread adoption is often hindered by a range of challenges. These barriers are multidimensional—spanning legal, technical, economic, and institutional domains. A clear understanding of these constraints is crucial to design enabling frameworks and policy interventions that can facilitate the scaling of agroforestry systems.

Land Tenure and Legal Barriers

One of the most critical constraints to agroforestry adoption is land tenure insecurity. Many smallholders, especially in developing countries, operate on lands without formal titles or with uncertain legal rights. This insecurity discourages long-term investments in tree-based systems, which typically require several years to yield benefits. In India, for instance, tenant farmers often lack the legal standing to plant and manage trees, since trees are considered long-term assets and their ownership may default to landowners (Kishwan et al. 2014). Furthermore, restrictive forestry regulations often limit the harvesting, transportation, and marketing of tree products, even if the trees are planted on private lands. Despite India's National Agroforestry Policy (2014), several states still have complex permit systems for felling and transporting species such as teak (*Tectona grandis*) or sandalwood (*Santalum album*), discouraging farmers from tree planting (Dhyani et al. 2013). In sub-Saharan Africa, customary land tenure systems can also pose barriers. Communal ownership may lead to conflicts over rights to tree resources, while women and marginalized groups often face exclusion from land access, curbing their participation in agroforestry practices (FAO 2018a, b).

Knowledge Gaps and Technical Limitations

Another key challenge is the limited awareness and technical knowledge among farmers, extension agents, and policy-makers regarding agroforestry systems. Unlike conventional monoculture agriculture, agroforestry requires integrated knowledge of tree-crop-livestock interactions, species compatibility, and long-term resource management. In many regions, formal agricultural extension systems do not sufficiently address agroforestry. Extension personnel are often more familiar with annual cropping systems and lack the training to advise on appropriate agroforestry models, species selection, or ecological design. A survey by the World Agroforestry Centre (ICRAF) found that over 60% of farmers in Kenya and Uganda were unaware of the full range of agroforestry options suitable for their agro-climatic zones (ICRAF 2015). Moreover, there is a lack of localized research and demonstration plots, especially in ecologically fragile zones. Research has tended to focus on biophysical parameters like yield and biomass, while neglecting social and gender dimensions of adoption. The absence of real-time advisory services and decision-support tools limits farmers' ability to adopt adaptive agroforestry systems in the face of climate uncertainty. Pest and disease management in agroforestry also remains under-researched. Interactions between perennial and annual species can sometimes exacerbate pest cycles, necessitating integrated pest management strategies tailored to diverse systems (Kumar et al. 2016).

Market Access, Price Volatility, and Value Chain Bottlenecks

Agroforestry products—ranging from fruits and nuts to fuel wood, medicinal plants, and timber -face significant market-related constraints. The lack of organized markets, unpredictable price fluctuations, and inadequate value chains can disincentivize farmers from adopting agroforestry systems. In India, for example, farmers cultivating high-value tree crops like *Melia dubia* or *Ailanthus excelsa* face difficulties in accessing markets due to the absence of aggregation centers, quality standards, and buyer linkages. In many states, the sale of timber is constrained by bureaucratic hurdles requiring multiple permits and inspections (Dhyani et al. 2013). Transport infrastructure and post-harvest facilities for perishable products like fruits and flowers are often lacking in remote rural areas. Additionally, value addition through processing is limited by poor access to credit, technology, and business development services. Women entrepreneurs, in particular, face barriers in entering the agroforestry value chain due to social norms and restricted mobility. Price volatility is a major issue, especially for non-timber forest products (NTFPs) such as gums, resins, or medicinal plants, which often lack minimum support prices or insurance mechanisms. The absence of forward and contract markets for agroforestry commodities increases farmers' exposure to market risks.

Institutional and Policy Gaps

Beyond physical and economic barriers, there are also institutional constraints. Agroforestry does not fall neatly under a single institutional jurisdiction, often lying at the intersection of forestry, agriculture, rural development, and environment departments. This leads to fragmentation in planning, funding, and implementation. Coordination between ministries is often weak, and agroforestry programs are frequently underfunded. While the National Agroforestry Policy in India laid a foundation for institutional convergence, the pace of its implementation has been uneven across states. Few states have formulated state-specific agroforestry policies or established Agroforestry Missions to integrate efforts. Capacity development at all levels - farmers, extension agents, and local governments, is inadequate. Incentive mechanisms like payment for ecosystem services (PES) are still in the pilot stage, and few regions have formalized frameworks for recognizing and rewarding ecosystem benefits of agroforestry.

Way Forward

Addressing these challenges requires a multi-pronged approach:

1. *Legal reforms:* Simplifying tree felling and transit regulations, particularly for farm-grown species, and ensuring tenure security through legal recognition of user rights.
2. *Capacity building:* Strengthening extension services, promoting farmer field schools, and investing in research and development focused on local agroforestry models.

3. *Market development:* Establishing aggregation centers, developing farmer cooperatives, and supporting value chain development with financial and technical assistance.
4. *Policy integration:* Creating inter-ministerial coordination platforms and mainstreaming agroforestry into national climate action plans and rural livelihood strategies.
5. *Incentive frameworks:* Expanding carbon finance, PES, and agroforestry insurance schemes to reward ecological stewardship.

Without overcoming these systemic barriers, the transformative potential of agroforestry for climate adaptation and livelihood security will remain underutilized.

Pathways for Scaling up Agroforestry for Adaptation

Scaling up agroforestry for climate adaptation requires an integrated, inclusive, and innovative approach that addresses institutional, financial, and technological enablers. While smallholder farmers have long practiced various forms of agroforestry, elevating these traditional systems into mainstream climate adaptation strategies calls for a coherent framework built on participation, incentives, and smart technologies. This section explores key pathways for scaling up agroforestry, focusing on the roles of extension models, climate finance, and digital innovations.

Role of Extension and Participatory Models

Agricultural extension systems are central to translating research into practice and fostering adoption of climate-resilient agroforestry. However, conventional top-down extension approaches often fall short in addressing the complexity and site-specificity of agroforestry systems. Instead, participatory models - such as Farmer Field Schools (FFS), Participatory Rural Appraisal (PRA), and innovation platforms, which have emerged as more effective mechanisms to engage communities and co-create knowledge (Franzel et al. 2014). Participatory models facilitate experiential learning, encourage local experimentation, and foster horizontal knowledge exchange among farmers. They are particularly crucial in agroforestry, where ecological and socio-economic conditions vary widely. In the Sahel, the success of farmer-managed natural regeneration (FMNR) has been largely attributed to local leadership, peer-to-peer networks, and collective action (Reij and Winterbottom 2015). Extension services must be retooled to provide long-term support, facilitate market linkages, and promote climate-smart practices. This includes training extension agents in agroforestry principles, developing context-specific agroforestry packages, and integrating gender and social inclusion into advisory systems. Public-private partnerships can enhance reach, while leveraging civil society organizations can ensure last-mile delivery. In India, the National Agroforestry Mission and Krishi Vigyan Kendra's (KVKs) play a pivotal role in promoting tree-based models. However, their impact can be scaled further by strengthening the capacity of personnel, improving funding, and promoting decentralized planning (Kishwan et al. 2014).

Climate Finance and Ecosystem Service Payments

Scaling agroforestry also hinges on financial incentives and policy mechanisms that recognize and reward ecosystem services. Agroforestry systems provide multiple environmental benefits—carbon sequestration, biodiversity conservation, soil health improvement, and microclimate regulation—that are often undervalued in conventional economic accounting. Climate finance, such as the Green Climate Fund (GCF), Adaptation Fund, and National Adaptation Plans (NAPs), offers opportunities to mainstream agroforestry into climate investment frameworks. These funds can support both infrastructure (nurseries, irrigation) and capacity-building initiatives that facilitate agroforestry adoption (FAO 2018a, b). Payment for Ecosystem Services (PES) is a promising mechanism that can internalize the public benefits of agroforestry. Under PES schemes, farmers receive financial compensation for practices that enhance ecosystem functions. While PES has shown success in Indonesia, its implementation in Asia and Africa remains limited due to institutional and monitoring challenges (Fripp 2014). India's experience with carbon trading through Clean Development Mechanism (CDM) projects offers insights into monetizing carbon sequestration in agroforestry. However, simplifying verification processes and establishing baselines remain critical for smallholder inclusion. Integration of agroforestry into voluntary carbon markets, such as Gold Standard and Verra, is gaining traction but requires strong technical backstopping. Access to credit is another constraint. Agroforestry is often perceived as a long-gestation investment by financial institutions. Designing tree-based loan products, offering interest subventions, and promoting crop-tree insurance can de-risk farmer participation. Public finance instruments like MGNREGA (in India) can be tapped for initial investments in community agroforestry.

Leveraging Digital Tools and Innovation

The digital revolution provides unprecedented opportunities to accelerate the adoption of agroforestry through improved access to information, decision support, and value chains. Mobile applications, Geographic Information Systems (GIS), remote sensing, and machine learning are transforming how agroforestry is planned, implemented, and monitored. Mobile-based advisory services, such as the mKisan and e-Choupal platforms in India, have demonstrated how real-time, localized information can empower farmers. These platforms can be extended to include tree-crop interaction guidelines, weather-based advisories, and pest alerts tailored to agroforestry systems. Geospatial tools enable the identification of suitable zones for agroforestry expansion, assessment of land-use changes, and monitoring of biomass and carbon stocks. Projects like India's National Agroforestry Decision Support System (NAFDSS) integrate GIS-based planning with socio-economic parameters to assist in landscape-level agroforestry planning (Dhyani et al. 2013). Blockchain and digital traceability solutions are also emerging as tools to enhance transparency in agroforestry product value chains. For example, blockchain can authenticate the source of agroforestry-based coffee, spices, or medicinal plants, thereby fetching premium prices and promoting sustainable practices (Rai et al. 2020). Agri-startups and innovation incubators can play a vital role in co-developing farmer-centric solutions—ranging from low-cost tree planting kits to mobile carbon calculators. Crowd-sourcing platforms can connect tree growers with funders, researchers, and buyers, creating a dynamic innovation ecosystem.

An Integrated Roadmap

To truly scale agroforestry for adaptation, a synergistic framework is required—one that:

1. Empowers local actors through participatory governance and inclusive extension services.
2. Incentivizes ecosystem stewardship via PES, climate finance, and accessible credit mechanisms.
3. Harnesses digital technologies for planning, monitoring, and market linkage.
4. Strengthens policy coherence by mainstreaming agroforestry into national and sub-national climate strategies.

Global alliances such as the AFR100 (African Forest Landscape Restoration Initiative) and India's Agroforestry Mission offer scalable platforms for cross-learning, funding, and replication. Regional cooperation and knowledge sharing, supported by institutions like FAO, ICRAF, and CGIAR, can foster South-South exchange and innovation. By integrating scientific, financial, and social dimensions, agroforestry can become a cornerstone of climate adaptation strategies in rural landscapes, securing not only ecosystems but also livelihoods in the face of mounting climate risks.

Conclusion and Future Outlook

As climate change continues to reshape global ecosystems and socio-economic dynamics, the search for sustainable, inclusive, and resilient agricultural systems has never been more urgent. Agroforestry, a time-tested practice rooted in ecological wisdom and traditional land-use systems, emerges as a key nature-based solution that bridges environmental sustainability with rural livelihood security. This chapter examined the multifaceted potential of agroforestry as a climate adaptation and mitigation strategy, while exploring its socio-economic co-benefits, policy frameworks, and pathways for upscaling.

Summary of Key Findings

Agroforestry systems are deeply intertwined with climate resilience mechanisms. By enhancing microclimatic conditions, conserving soil and water, and reducing exposure to climate-related shocks, these systems provide a robust safety net for smallholder farmers, especially in vulnerable dry land and rainfed regions. Agroforestry serves as a natural shield against the harsh impacts of rising temperatures, erratic precipitation, and soil degradation. Additionally, it supports biodiversity, increases resource-use efficiency, and diversifies agricultural production—creating a more stable and adaptive agricultural matrix. From a mitigation perspective, agroforestry holds significant carbon sequestration potential. Both aboveground biomass and belowground root systems contribute to long-term carbon sinks, making it an effective strategy to offset greenhouse gas emissions when compared to conventional mono-cropping systems. Several studies have highlighted agroforestry's capacity to store 25–50 tons of carbon per hectare over a few decades (Jose 2009; Montagnini and Nair 2004a, b). Beyond ecological functions, agroforestry contributes to livelihood diversification, food and nutritional security, and gender-inclusive community empowerment. It integrates the production of food, fuel wood, fodder, timber, and medicinal plants—ensuring that farmers are less reliant on single income streams and more resilient to market and climate variability. Women often play a critical role in the maintenance of home gardens and processing of agroforestry products, emphasizing the gender-responsive nature of these systems. Policy support at both nationally and globally has been pivotal in mainstreaming agroforestry. India's National Agroforestry Policy (2014), integration into State Action Plans on Climate Change (SAPCCs), and international frameworks like the Paris Agreement and Sustainable Development Goals (SDGs) provide the institutional backing for its promotion. However, implementation gaps, fragmented institutional responsibilities, and weak market linkages persist. Case studies from India, Africa, and Southeast Asia illustrate how context-specific models—ranging from silvi-pastoral systems and FMNR to home gardens and carbon offset programs—can be tailored to local ecologies and socio-economic realities. These examples reaffirm the scalability and adaptability of agroforestry across diverse agro-climatic zones.

Policy and Research Recommendations

To unlock the full potential of agroforestry in climate adaptation, a combination of enabling policies, financial incentives, and research innovations is essential.

1. *Strengthen policy coherence and institutional convergence:* Agroforestry often falls at the intersection of agriculture, forestry, and rural development. Creating unified implementation frameworks across these sectors, backed by adequate funding and decentralization, is critical. Integration of agroforestry into flagship programs like MGNREGA, PMKSY, and FPO promotion schemes can catalyze wider adoption.
2. *Expand climate finance and payment for ecosystem services (PES):* Establish dedicated climate-resilient agroforestry funds at state and national levels, supported by international mechanisms such as GCF, GEF, and carbon markets. PES schemes that reward tree growers for ecosystem services (carbon sequestration, watershed protection, etc.) must be simplified, scaled up, and targeted at marginal farmers.
3. *Invest in extension and capacity building:* Training of extension agents, promotion of Farmer Field Schools (FFS), and development of agroforestry curriculum in agricultural universities can bridge the knowledge gap. Participatory approaches and community-led planning should be institutionalized for greater ownership and context specificity.
4. *Enhance research and innovation*: Research should focus on developing location-specific agroforestry models, understanding long-term carbon dynamics, evaluating gender impacts, and mapping value chains for agroforestry products. Digital tools—such as AI-enabled monitoring systems, remote sensing, and mobile-based advisories—can revolutionize planning and dissemination.
5. *Secure land tenure and reduce bureaucratic barriers*: Unclear land rights and complex tree-felling regulations often discourage farmers from adopting agroforestry. Policy reforms that ensure tree ownership, simplify felling and transit rules, and recognize agroforestry in land use classifications are essential.
6. *Foster market linkages and value chain development*: Creating aggregation platforms for agroforestry produce, setting up minimum support prices (MSP) for select tree products, and investing in post-harvest infrastructure and branding will improve profitability and attracts youth to tree-based farming.

Vision for Climate-Smart Agroforestry Ecosystems

A future-oriented vision of agroforestry must transcend its current role as a peripheral land-use practice and evolve into a central pillar of climate-resilient development. Climate-smart agroforestry ecosystems are integrated landscapes that optimize ecological integrity, economic productivity, and social equity. These systems are not only adaptive and productive but also regenerative—restoring degraded lands, reversing biodiversity loss, and rebuilding community resilience.

Such ecosystems will be characterized by:

- **Ecological Intelligence**: Designed to harness local biodiversity, manage water efficiently, and sequester carbon.
- **Technological Integration**: Powered by digital platforms for decision support, monitoring, and precision planning.
- **Economic Inclusion**: Linked to diversified markets, credit access, and circular bio economy models.
- **Social Empowerment**: Built on gender equity, youth participation, and indigenous knowledge.

Achieving this vision requires a concerted effort from governments, civil society, academia, private sector, and farming communities. Collaborative platforms, such as agroforestry consortia, inter-ministerial task forces, and regional knowledge hubs, can play a pivotal role in knowledge exchange, innovation, and investment mobilization.

Ultimately, agroforestry represents a convergence of ecological necessity and developmental opportunity. It embodies the essence of sustainable adaptation—where mitigation co-benefits, resilient livelihoods, and environmental stewardship align. By mainstreaming agroforestry into rural landscapes, we not only respond to the climate crisis but also co-create a regenerative and equitable future for generations to come.

Acknowledgements The authors express their gratitude to the Director, ICAR Research Complex for NEH Region, Meghalaya, and the Head of ICAR Tripura Centre for all-around support in carrying out the study. The authors duly acknowledge the sincere efforts of the retired Scientists of the Institute in developing and maintenance of the long-period agroforestry systems (35 years old) in the research farm of the Agroforestry Division of the Institute.

Funding The research was not supported by any funding.

Author Contributions All authors have contributed equally to this work. All authors contributed to the final draft and approved the manuscript.

Data Availability The datasets during and/or analyzed during the current study are available from the corresponding author on reasonable request.

References

Agarwal B (2001) Participatory exclusions, community forestry, and gender: an analysis for South Asia and a conceptual framework. World Dev 29(10):1623–1648

Albrecht A, Kandji ST (2003) Carbon sequestration in tropical agroforestry systems. Agric Ecosyst Environ 99(1–3):15–27

Ajtay GL, Ketner P. Duvigneaud P (1979) Terrestrial primary production and phytomass. pp. 129–182. In: The Global Carbon Cycle Bolin, B., Degens, E., Kempe, S. and Ketner, P. (Eds.), SCOPE edition. 13, John Wiley, Chichester.
Brahma B, Sileshi GW, Nath AJ, Das AK (2017) Development and evaluation of robust tree biomass equations for rubber tree (Hevea brasiliensis) plantations in India. For. Ecos.:15–24. https://doi.org/10.1186/s40663-017-0101-3
Brown S, Gillespie AJ, Lugo AE (1989) Biomass estimation methods for tropical forests with applications to forest inventory data. For. Sci. 35(4): 881–902.
Chaturvedi RK, Raghubanshi AS (2015) Carbon density and accumulation in agroforestry systems of a tropical dry climate. Agrofor Syst 89(3):435–445
Dey RK, Ansary R, Mandal S (2025a) Impact of climate variability on mango cultivation: a case study in Malda District of West Bengal, India. In: Livelihoods and Well-being in the era of climate change: risk to resilience across India. Springer Nature Switzerland, Cham, pp 217–232. https://doi.org/10.1007/978-3-031-81132-6
Dey RK, Halder P, Nandy S, Ansary R (2025b) Identifying groundwater potential zones using multi-criteria approach in agriculture-dominated landscapes of Malda district, India. Groundwater Sustain Dev:101497. https://doi.org/10.1016/j.gsd.2025.101497
Dogar MM, Kucharski F, Azharuddin S (2017) Study of the global and regional climatic impacts of ENSO magnitude using SPEEDY AGCM. J Earth Syst Sci 126(2):30
Dogar MM, Kucharski F, Sato T, Mehmood S, Ali S, Gong Z et al (2019) Towards understanding the global and regional climatic impacts of Modoki magnitude. Glob Planet Chang 172:223–241
Dogar MM, Sato T (2018) Analysis of climate trends and leading modes of climate variability for MENA region. J Geophys Res Atmos 123(23):13–074
Dogar MM, Sato T (2019). Regional climate response of Middle Eastern, African, and South Asian monsoon regions to explosive volcanism and ENSO forcing. Journal of Geophysical Research: Atmospheres, 124, 7580–7598. https://doi.org/10.1029/2019JD030358
Dogar MM, Hermanson L, Scaife AA et al. (2023). A Review of El Niño Southern Oscillation Linkage to Strong Volcanic Eruptions and Post-Volcanic Winter Warming. Earth Syst Environ 7, 15–42. https://doi.org/10.1007/s41748-022-00331-z
Dogar MMA, Almazroui M (2022) Revisiting the strong and weak ENSO teleconnection impacts using a high-resolution atmospheric model. Atmos Environ 270:118866
Dhyani SK, Handa AK, Uma (2013) Agroforestry: a strategy for climate change adaptation and mitigation. Indian J Agrofor 15(2):1–9
Dhyani SK, Handa AK, Uma (2016) Agroforestry: a land use for enhancing productivity, carbon sequestration and soil health. Indian J Agrofor 18(2):1–9
FAO (2018a) Scaling up climate finance in agriculture. Food and Agriculture Organization of the United Nations, Rome
FAO (2018b) The state of the world's forests 2018—forest pathways to sustainable development. Food and Agriculture Organization of the United Nations, Rome
FAO (2013) Advancing agroforestry on the policy agenda: a guide for decision-makers. Food and Agriculture Organization of the United Nations, Rome
FAO (2021) The state of food and agriculture 2021: making agrifood systems more resilient to shocks and stresses. Food and Agriculture Organization of the United Nations, Rome
Franzel S, Coe R, Cooper P (2014) Participatory approaches for scaling agroforestry. Curr Opin Environ Sustain 6:66–72
Fripp E. (2014). Payments for ecosystem services (PES): a practical guide to assessing the feasibility of PES projects. Bogor, Indonesia: CIFOR., 26 p
Garrity DP (2004) Agroforestry and the achievement of the millennium development goals. Agrofor Syst 61(1):5–17
Garrity DP, Akinnifesi FK, Ajayi OC, Weldesemayat SG, Mowo JG, Kalinganire A, Larwanou M, Bayala J (2010) Evergreen agriculture: a robust approach to sustainable food security in Africa. Food Secur 2(3):197–214
GoI (2018) Operational guidelines for sub-mission on agroforestry. Ministry of Agriculture & Farmers' Welfare, Government of India

Giri BG, Negi GCS, Dhyani PP, Bargali SS, Saxena R (2015) Ecosystem services of forests: carbon stock in vegetation and soil components in the watershed of Kumaun Himalaya, India. Intern J Ecol Environ Sci 41(3–4):177–188

Hairiah K, Dewi S, Agus F, Velarde S, Ekadinata A, Rahayu S, van Noordwijk M (2011) Measuring carbon stocks across land use systems: a manual. World Agroforestry Centre (ICRAF), Bogor, Indonesia

Halder P, Dey RK, Mandal S (2023) Long-period trend analysis of annual and seasonal rainfall in West Bengal, India (1901–2020). Theor Appl Climatol 154(1–2):685–703. https://doi.org/10.1007/s00704-023-04577-z

Halder P, Satpati L, Dey RK (2022) Water security in the Indian Sundarban under the changing climate. Society, Environment & Development, Raidighi College, Sundarbans, pp 94–123. ISBN 978-8-19-511040-7

ICAR (2015) Vision 2050: Indian Council of Agricultural Research—central research Institute for Dryland Agriculture. CRIDA, Hyderabad

ICRAF (2015) Trees for resilience: agroforestry in the context of climate change. World Agroforestry Centre, Nairobi, Kenya

ICRAF (2013) Agroforestry for improved livelihoods and food security. World Agroforestry Centre, Nairobi

ICRAF (2014) National Agroforestry Policy—India's pioneering approach. World Agroforestry Centre, Nairobi

IPCC (2021) Climate change 2021: impacts, adaptation, and vulnerability. In: Contribution of working group II to the sixth assessment report of the intergovernmental panel on climate change. Cambridge University Press, p 31

Jha KK (2015) Carbon storage and sequestration rate assessment and allometric model development in young teak plantations of tropical moist deciduous forest, India. J. For. Res. 26: 589–604. https://doi.org/10.1007/s11676-015-0053-9

Jose S (2009) Agroforestry for ecosystem services and environmental benefits: an overview. Agrofor Syst 76(1):1–10

Jose S, Gillespie AR, Pallardy SG (2004) Interspecific interactions in temperate agroforestry. Agrofor Syst 61(1):237–255

Kaonga M, Bayliss-Smith TP (2009) Carbon pools in tree biomass and the soil in improved fallows in eastern Zambia. Agroforest Syst 76(1):37–51

Khan A, Mandal S, Satpati L, Dogar MM, Hasanean H, Salah AM (2025) Analyzing drought: the impact of climate change on precipitation patterns in the Shilabati River basin, West Bengal, India. Earth Syst Environ. https://doi.org/10.1007/s41748-025-00679-y

Kishwan J, Dhyani SK, Kumar A (2014) National Agroforestry Policy of India 2014: critical analysis and way forward. Indian Forester 140(4):390–398

Kumar BM, Nair PKR (2004) The enigma of tropical homegardens. Agrofor Syst 61(1):135–152

Kumar BM, Nair PKR (2011) Carbon sequestration potential of agroforestry systems: opportunities and challenges. Springer. https://doi.org/10.1007/978-94-007-1630-8

Kumar M, Sirohi C, Rawat P (2016) Integrated pest management in agroforestry: strategies and prospects. J Appl Nat Sci 8(1):48–56

Kumar S, Ramachandran N (2005) Women's role in agriculture and agroforestry in India: a review. J Sustain For 21(2–3):87–109

Leakey RRB (2017) Multifunctional agriculture: achieving sustainable development in Africa. Academic Press Food Secur 9(6):468

Lehmann J, Joseph S (2009) Biochar for environmental management: science and technology. Earthscan, London. 448 p

Lodhiyal N, Lodhiyal LS (2003) Biomass and net primary productivity of Bhabar Shisham forests in central Himalaya, India. For. Ecosys. Manag. 176(1–3):217–235.

Mandal S, Choudhury BU (2014) Estimation and prediction of maximum daily rainfall at Sagar Island using best fit probability models. Theor Appl Climatol 117:87–97

Mandal S, Choudhury BU, Mondal M, Bej S (2013) Trend analysis of weather variables in Sagar Island, West Bengal, India: a long-term perspective (1982–2010). Curr Sci 105:947–953

Mandal S, Choudhury BU, Satpati LN (2015) Monsoon variability, crop water requirement, and crop planning for kharif rice in Sagar Island, India. Int J Biometeorol 59:1891–1903. https://doi.org/10.1007/s00484-0150995-9

Mandal S, Satpati LN, Choudhury BU, Sadhu S (2018) Climate change vulnerability to the agrarian ecosystem of the small Island: evidence from Sagar Island, India. Theor Appl Climatol 132:451–464. https://doi.org/10.1007/s00704-017-2098-5

Mercer, D. E., & Miller, R. P. (1998). Socioeconomic research in agroforestry: progress, prospects, priorities. Agrofor Syst, 38(1–3), 177–193

MoAFW (2014) National Agroforestry Policy. Ministry of Agriculture and Farmers Welfare, Government of India

MoEFCC (2015) India's intended nationally determined contribution: working towards climate justice. Ministry of Environment, Forest and Climate Change, Government of India. 38 p

Montagnini, F. (2017). Integrating landscapes: agroforestry for biodiversity conservation and food sovereignty. Springer. https://doi.org/10.1007/978-3-319-69371-2

Montagnini, F., & Nair, P. K. R. (2004a). Carbon sequestration: an underexploited environmental benefit of agroforestry systems. Agrofor Syst, 61(1), 281–295

Montagnini F, Nair PKR (2004b) Carbon sequestration: an underexploited environmental benefit of agroforestry systems. Agrofor Syst 61–62(1):281–295

Mutuo PK, Cadisch G, Albrecht A, Palm CA, Verchot L (2005) Potential of agroforestry for carbon sequestration and mitigation of greenhouse gas emissions from soils in the tropics. Nutr Cycl Agroecosyst 71(1):43–54

Naik SK, Sarkar PK, Das B, Singh AK, Bhatt BP (2018a) Biomass production and carbon stocks estimate in mango orchards (Mangifera indica L.) of hot and subhumid climate in eastern region, India. Carbon Manag 10(5):477–487

Naik SK, Sarkar PK, Das B, Singh AK, Bhatt BP (2018b) Predictive models for dry biomass and carbon stock estimation in Litchi chinensis under hot and dry sub-humid climate. Arch Agron Soil Sci 64(10):1366–1378

Naik SK, Sarkar PK, Das B, Singh AK, Bhatt BP (2019) Biomass production and carbon stocks estimate in mango orchards of hot and sub-humid climate in eastern region, India. Car Manag 10(5):1–11.

Nair PKR (1993) An introduction to agroforestry. Springer Science & Business Media. 499 p

Nair PKR, Kumar BM, Nair VD (2009) Agroforestry as a strategy for carbon sequestration. J Plant Nutr Soil Sci 172(1):10–23

Nair PKR, Kumar BM, Nair VD (2010) Carbon sequestration in agroforestry systems. Adv Agron 108:237–307

Ong CK, Black CR, Wilson J (2002) Tree-crop interactions: manipulation of water use and root function. Agrofor Syst 53(1):1–5

Pandey D (2007) Multifunctional agroforestry systems in India. Curr Sci 92(4):455–463

Patil NG, Suresh G, Bhadoria PBS (2019) Silvi-pastoral systems for arid and semi-arid regions. J Soil Water Conserv 18(4):311–318

Rai RK, Tiwari S, Pathak H (2020) Blockchain technology for sustainable agroforestry supply chains: opportunities and challenges. Agric Syst 181:102821

Rao MR, Nair PKR, Ong CK (2007) Biophysical interactions in tropical agroforestry systems. Agrofor Syst 71(1):1–12

Reij, C., & Winterbottom, R. (2015). Scaling up regreening: six steps to success. World Resources Institute, Washington D.C., 66 p

Reij C, Tappan G, Smale M (2009) Agroenvironmental transformation in the Sahel: another kind of green revolution. IFPRI Discussion Paper 00914. 43 p

Roshetko JM, Nugraha E, Tukan JCM, Fay C (2007) Agroforestry for livelihood enhancement and enterprise development. In: World Agroforestry Centre working paper series, ICRAF SEA, pp. 07395

Sanchez PA (1995) Science in agroforestry. Agrofor Syst 30(1–2):5–55
Saxena NC (2000a) Farm forestry in India: evolution and sustainability. World Dev 28(2):307–321
Saxena NC (2000b) Farm forestry in India: evolution, constraints, and potentials. Oxford & IBH Publishing Co. Pvt. Ltd.
Singh RK, Singh A, Kumar P (2020) Climate change and agriculture: adaptation strategies and future perspectives. Curr Sci 118(9):1354–1361
Singh S, Dadhwal VK (2009) Manual on spatial assessment of vegetation carbon pool of India. Indian Institute of Remote Sensing (National Remote Sensing Centre), ISRO, Government of India, Dehradun, pp 1–32
Six J, Conant RT, Paul EA, Paustian K (2002) Stabilization mechanisms of soil organic matter: Implications for C-saturation of soils. Plant and Soil (241):155–176.
Stenchikov, G., & Dogar, M. (2012, April). Regional climate response to volcanic radiative forcing in Middle East and North Africa. In EGU General Assembly Conference Abstracts (p. 6373).
Torquebiau, E., & Penot, E. (2006). Ecology versus economics in tropical multistrata agroforests. Agrofor Syst, 66(1), 53–61
Young A (1997) Agroforestry for soil management, 2nd edn. CAB International, Wallingford. 320 p

Chapter 24
Climate-Smart Villages: A Community-Driven Blueprint for Rural Resilience

Soumik Das, T. S. Sindhushree, N. K. Sinha, Hanumantappa Meti, and S. Dharumarajan

Abstract Land degradation affects 24% of the world's land surface and 1.5 billion of its people. It is the result of human activities, exacerbated by natural processes which are closely linked to climate change and loss of biodiversity. From erratic monsoons to declining biodiversity, South Asian rural populations are increasingly at risk from climate change. This chapter examines Climate-Smart Villages (CSVs) approach, a revolutionary strategy that combines community-led governance, state--of-the-art technology and local knowledge to increase resilience. Using case studies from Bangladesh, Nepal and India, we show how CSVs enable farmers to use solar--powered irrigation, drought-resistant crops and participatory climate-risk mapping, increasing yields by 20–50% while lowering GHGs emissions. However, difficulties still exist. Notwithstanding the achievements in Vietnam's Mekong Delta and India's NICRA program, resolving financing shortages, gender inequality, and policy fragmentation are necessary for growing CSVs. We provide a critical analysis of the growth of CSV in South Asia, comparing the swift adoption in India with the slower adoption in its bordering nations. The chapter ends with practical recommendations, such as blended financing models and AI-driven Agro-advisories, asking policymakers to incorporate CSV ideas into National Adaptation Plans (NAPs).

S. Das · N. K. Sinha (✉)
ICAR-Indian Institute of Soil Science, Nabi Bagh, Bhopal, Madhya Pradesh, India

T. S. Sindhushree · H. Meti
Department of Soil Science and Agricultural Chemistry, UAS, GKVK, Bangalore, Karnataka, India

S. Dharumarajan
ICAR-National Bureau of Soil Survey and Land Use Planning, Regional Centre, Hebbal, Bangalore, Karnataka, India

S. Mandal et al. (eds.), *Climate Change Vulnerability*, Climate Change Management, https://doi.org/10.1007/978-3-032-16783-5_24

Abbreviations

CCAFS	CGIAR Research Program on Climate Change, Agriculture and Food Security
CDKN	Climate and Development Knowledge Network
CGIAR	Consultative Group on International Agricultural Research
CIAT	International Center for Tropical Agriculture
CSA	Climate-Smart Agriculture
CSV	Climate-Smart Village
DSR	Direct Seeded Rice
FAIFA	Federation of All India Farmer Associations
FAO	Food and Agriculture Organization
FPO's	Farmer Producer Organizations
GHGs	Greenhouse gases
GRFC	Global Report on Food Crises
ICAR	Indian Council of Agricultural Research
ICRISAT	International Crops Research Institute for the Semi-Arid Tropics
IPCC	Intergovernmental Panel on Climate Change
IUCN	International Union for Conservation of Nature
LI-BIRD	Local Initiatives for Biodiversity, Research and Development (LI-BIRD)
NAPs	National Adaptation Plans
NGOs	Non-Governmental Organizations
NICRA	National Innovations in Climate Resilient Agriculture
SDG's	Sustainable Development Goals
SHG's	Self Help Groups
UNDP	United Nations Development Programme
UNFCC	United Nations Framework Convention on Climate

Introduction

Globally, a large portion of the population faces insufficient food access, with millions experiencing chronic hunger and food insecurity. According to the 2025 Global Report on Food Crises (GRFC), acute hunger has increased to 295 million people facing high levels of acute food insecurity-three times as many as in 2016 and twice as many as in 2020. Conflict and insecurity, weather extremes and economic shocks continue to drive acute food insecurity and to compound vulnerabilities, pushing millions into Crisis, Emergency and Catastrophic levels of food insecurity. Meanwhile, climate extremes are growing in intensity-wreaking havoc on global food security, crippling harvests and breaking supply chains. Agricultural production systems across the world are expected to change in response to climate change, posing major challenges to the livelihoods and food security of millions of people (IPCC 2019a, b).

Climate change has become a challenge not only for the environment but also as a significant socio-political issue, particularly for South Asia, land of more than 1.8 billion people, is among the most vulnerable regions to climate (Dogar and Almazroui 2022; Adnan et al. 2021). The rural livelihood of them depends largely on climate-sensitive activities such as agriculture, livestock, fisheries, and forestry (Shunsuke 2025). Undoubtedly, climate change poses a significant threat to smallholder farmers (Salmanicheraghabadi et al. 2021). Strengthening farmers' adaptive capacity and resilience has been widely recommended as a key strategy to safeguard rural communities and ensure sustainable livelihoods (Fallah Haghighi et al. 2018). In response to this growing threat, there is a global consensus that farming systems must evolve to become more "Climate Smart." Numerous international and national organizations have taken initiatives to support this and aimed to enhance the resilience of rural communities. Some of them are: CCAFS (CGIAR Research Program on Climate Change, Agriculture and Food Security), CGIAR (Consultative Group on International Agricultural Research), FAO (Food and Agriculture Organization of the United Nations), World Bank, ICRISAT (International Crops Research Institute for the Semi-Arid Tropics), ICAR (Indian Council of Agricultural Research), National Governments & Research Institutes and Non-Governmental Organizations (NGOs).

Number of ways and options are there to reduce the negative impacts of climate change on agricultural systems and rural livelihoods. Options range from change in crop management, such as sowing time, modified irrigation technologies and sustainable land use (Aggarwal et al. 2018; Khatri-Chhetri et al. 2016; Mandal et al. 2018). Along with reduction in GHG emissions, these technologies and practices may increase crop yields, farm income and status of rural population. Among the recent adaptation technologies, Climate-smart agriculture (CSA) focuses on food security and national development goals and, where possible, it also aims to reduce or remove GHG emissions (Aggarwal et al. 2018). Recent analyses indicate that although several CSA programs and on-farm studies were successful, there is a weak uptake of many innovative CSA practices and technologies in the local level (Westermann et al. 2015). For example, new water management practices and technologies in India were adopted by only 12% of farmers in the last 40 years (Palanisami et al. 2015).

Farmers, especially those managing diverse and resource-constrained farms, need more than just tools—they need clarity on how to balance multiple interventions, create synergies, and avoid trade-offs in their everyday practices. Without this contextual understanding, even the most promising solutions risk remaining unused. Climate change complicates this because its impacts will vary across locations (Dogar et al., 2017a, 2017b, 2019, 2020, 2024; 2025; Azharuddin 2016; Dogar 2020). Effective implementation, therefore, requires an integrated approach in which science, technology and decision making interact with local socioeconomic conditions and cultures (Aggarwal et al. 2018). To address the need for context-specific, evidence-based solutions, the CGIAR Research Program on Climate Change, Agriculture and Food Security (CCAFS) initiated the Climate-Smart Village (CSV) approach across Asia, Africa, and Latin America (Aggarwal et al.

2018). This approach aims to generate field-level evidence on the efficacy of climate-smart options by integrating scientific research with local knowledge and farmer-led experimentation. As a cornerstone of the broader agricultural research-for-development (AR4D) agenda, the CSV model seeks to develop scalable pathways for climate-resilient farming systems that enhance productivity, build resilience, and reduce emissions (Campbell et al. 2016). Within this context, the present chapter offers a critical analysis of the emergence and evolution of the CSV approach in South Asia, with a particular focus on its implementation in India and neighbouring countries. It explores the enabling conditions, institutional frameworks, and community participation that have shaped its trajectory.

Impact of Climate Change on Agriculture

Climate change is significantly impacting global agricultural production and crop yields, threatening the world's food supply and economy. Global warming has increased variability in crop yields, leading to increased frequency of severe weather events like floods, droughts and cyclones. Rising temperatures can decrease harvest yields, disrupt planting and harvest schedules. Unpredictable rainfall patterns can cause water logging or drought, affecting crops like rice, wheat and lentils. Extreme weather events also cause physical crop damage, soil erosion and fertility loss (Dimple 2023; Fig. 24.1).

Direct Impacts

Climate warming has led to a general reduction in the growing period of crops (Tan et al. 2021), accompanied by increased temperature fluctuations, which may result in inadequate nutrient accumulation in crops. Crop growth cycles are also

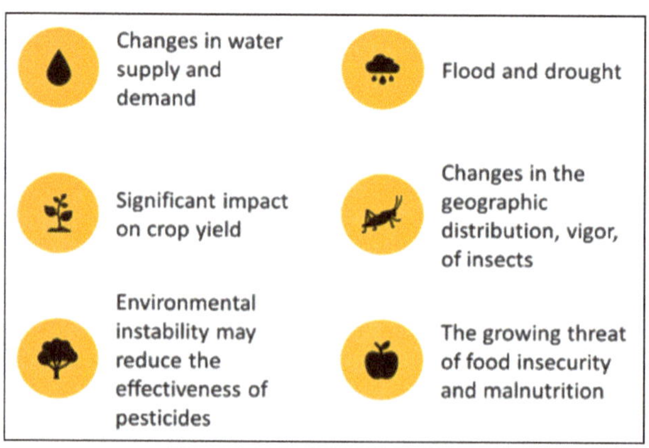

Fig. 24.1 Impact of climate change on agriculture (Source: Kaushik et al. 2023)

influenced by changes in sunshine hours, and reduced sunlight will restrict crop growth (Ozkaynak 2013) and has wide-ranging and complex effects on crop yield. Elevated temperatures increase evapotranspiration rates, exacerbating water evaporation and leading to soil dryness and water scarcity, which adversely affects crop yield (Yuan et al. 2024). A rise in CO_2 concentration and temperature may lead to a decrease in protein content and a decline in quality of staple crops (Janmohammadi and Sabaghnia 2023). As a result, consumer demand for certain agricultural products may be decreased, thereby affecting agricultural production and farmers' income to some extent.

Indirect Impacts

The ecosystem changes resulting from climate change, such as the increased frequency and intensity of extreme weather events, decline in soil fertility, water scarcity and changes in the occurrence patterns of pests and diseases, have profound implications for agricultural production.

Higher Frequency of Extreme Climate Conditions

Increased frequency and intensity of extreme weather events, potentially leading to extensive crop damage or even complete crop failure, may also disrupt agricultural infrastructure, increasing the cost of agricultural production (Newman and Noy 2023).

Deteriorated Soil Health

Climate change-induced temperature rise and altered rainfall patterns disrupt soil organic carbon reserves, threaten soil fertility and biodiversity and worsen salinization (Lal et al. 2021; Hassani et al. 2021). Extreme events like floods and droughts degrade soil fertility, while higher temperatures speed up organic matter decomposition, leading to rapid nutrient loss and reduced soil quality, ultimately hindering crop growth.

Variation in the Water Cycle

Climate change alters precipitation patterns and increases extreme rainfall events, disrupting agricultural water supplies and affecting crop growth, soil quality and productivity (Yuan et al. 2024; Thackeray et al. 2022). Rising temperatures also accelerate evaporation and reduce soil moisture, lowering crop water use efficiency.

Occurrence of New Pests and Diseases

Climate change has complex and multifaceted effects on the occurrence patterns of pests in crops by altering temperature. Temperature fluctuations can accelerate pests' consumption, development, mobility and reproduction, thus enlarging their population sizes and geographical distribution ranges. This may lead to increased crop loss and negatively impact agricultural production (Yuan et al. 2024).

South-Asian Region-Specific Discussion

The south and south-east Asian countries are amongst the most affected countries by climate change, mainly due to poverty and complex socio-economic-demographic challenges. Almost 70% of the total population lives in rural areas, and 75% of the rural population are poor and most at risk due to climate change, particularly in arid and semi-arid regions (Yadav and Lal 2018; Population of Asia 2019). South Asia is projected to see a 2.8 °C maximum temperature increases by 2040–2069, with erratic monsoon rainfall increasing by 10–15% in intensity (Ghaffar et al. 2022).

Rice and wheat are major contributors to food security in Asia. There is a big challenge to increase wheat production by 60% by 2050 to meet ever-increasing food demands (Rezaei et al. 2018). Climate change reduces Asian agricultural productivity, with rice and wheat yields projected to decline by 10–15% by 2050 due to heat stress and monsoon shifts (Lobell et al. 2021). Climate variability has induced floods and droughts, which have decreased the rice yield in South Asia and several other parts of Asia (Mottaleb et al. 2017). In South Asia, rice yield in rain-fed areas has already decreased, and it might reduce by 14% under the RCP 4.5 scenario, while 10% under the RCP 8.5 scenario by 2080 (Chun et al. 2016). A meta-analysis of data in 52 publications shows a projected mean reduction in maize and sorghum yields across South Asia-16% for maize and 11% for sorghum-by the 2050s, but no mean changes for rice yields (Knox et al. 2012). In Bangladesh, cyclone Sidr (2007) damaged 30% of cropland, a trend worsened by rising sea levels (Huq et al. 2015).

India-Specific

Rural India is extremely susceptible because 70% of the population lives in rural areas and heavily depends on agricultural activities for their livelihoods (Parry et al. 2007). High rural poverty (30.9%) is worsening vulnerability, as climate change degrades rural ecosystem services, pushing poor and marginalized communities into greater survival challenges and reducing overall well-being (Raveendran 2016). India's millet cultivation, supporting 60% of livestock, faces 20% yield losses due to drought (Lone et al. 2020). In India, the changing climate is projected to reduce monsoon sorghum yield by 2–14% by 2020 and more by 2050 and 2080 (Srivastava et al. 2010). A large reduction in wheat yield is projected for the Indo-Gangetic plains unless appropriate cultivars and crop management practices are adopted (Ortiz et al. 2008). Climate variability could decline agricultural incomes by 25% in the coming decades, leading to rural indebtedness, food insecurity and distress migration (ICRIER 2019).

Nepal-Specific

Nepal's unique ecology makes it vulnerable to climate change impacts. Nepal has lost 25% of its glacier volume in the last 30 years, putting downstream irrigation at risk and increasing the likelihood of glacial lake outburst floods (GLOFs) (Himalayan Monitoring Report 2020). Traditional cropping systems are collapsing due to unreliable water from glacial-fed sources, affecting rural smallholder farmers. Underinvestment in rural public goods, limited hydrological data and ineffective governance and accountability mechanisms exacerbate the situation of the country.

Bangladesh-Specific

Bangladesh, with its low elevation and dense population, is highly vulnerable to climate change impacts such as sea-level rise, cyclones, and increasing salinity. Around 60% of its population lives in rural areas, heavily dependent on agriculture for livelihood. Coastal regions face increasing salinity due to sea-level rise and tidal surges from frequent cyclones like Sidr (2007) and Amphan (2020), which have devastated farmland, fisheries, and ecosystems. According to IUCN (2020), over 20% of coastal agricultural land is now saline, causing declines in crop yield, freshwater availability, and ecosystem health. Areas like Satkhira and Khulna face chronic inundation and displacement, severely destabilizing rural livelihoods.

Climate Change Mitigation Strategies

Climate change mitigation involves actions to reduce or prevent greenhouse gas emissions from human activities or to enhance carbon sinks that remove them from the atmosphere. Mitigation efforts include transitioning to renewable energy sources, enhancing energy efficiency, adopting regenerative agricultural practices and protecting and restoring forests and critical ecosystems. A comprehensive approach incorporating resilient crop varieties, efficient irrigation, sustainable practices, infrastructure development, financial support, and insurance is essential. Modern farming techniques can help mitigate risks and reduce the impact of climate change on agriculture. If we can slow down the rise in greenhouse gases, we can slow down the pace of climate change and avoid its worst consequences. Mitigation strategies to reduce GHG's include-.

Shifting Away from Fossil Fuels

Fossil fuels are the biggest source of greenhouse gases, so transitioning to modern renewable energy sources like solar, wind and geothermal power and advancing sustainable modes of transportation is crucial.

Improving Energy Efficiency

Using less energy overall—in buildings, industries, public and private spaces, energy generation and transmission and transportation—helps reduce emissions. This is by using thermal comfort standards, better insulation and energy efficient appliances and by improving building design, energy transmission systems and vehicles.

Changing Agricultural Practices

Certain farming methods release high amounts of methane and nitrous oxide, which are potent greenhouse gases. Regenerative agricultural practices—including enhancing soil health, reducing livestock-related emissions, direct seeding techniques and using cover crops—support mitigation, improve resilience and decrease the cost burden on farmers.

The Sustainable Management and Conservation of Forests

Forests act as carbon sinks by reducing the overall concentration of greenhouse gases in the atmosphere. Measures to reduce deforestation and forest degradation are key for climate mitigation and generate multiple additional benefits.

Restoring and Conserving Critical Ecosystems

In addition to forests, ecosystems such as wetlands, peat lands and grasslands, as well as coastal biomes such as mangrove forests, contribute significantly to carbon sequestration, while supporting biodiversity and enhancing climate resilience.

Creating a Supportive Environment

Investments, policies, and regulations that encourage emission reductions, such as incentives, and carbon pricing are crucial to drive climate change mitigation.

Beyond Mitigation Practices: The Need for Adoption

Adaptation is the best way to handle climate variability and change as it has the potential to minimize hazardous climate change effects for sustainable production (IPCC 2019a, b). Innovative technologies and defensive adaptation can reduce the uncertain and harmful effects of climate on agricultural productivity. Therefore, to survive the harmful climate change effects, the development and implementation of adaptation strategies are crucial. Integration of adaptation and mitigation strategies may have immense potential to drive sustainable and productive agricultural systems (Habib-ur-Rahman et al. 2022). Climate resilient agricultural production systems can be developed and diversified with the integration of land, water, forest biodiversity, livestock and aquaculture (Meena et al. 2019).

Climate-smart agriculture (CSA) presents a viable solution by integrating productivity enhancement, adaptation and greenhouse gas (GHG) emissions reduction. A global review spanning 2013–2023 affirmed CSA's effectiveness in increasing productivity, building resilience and reducing GHG emissions across diverse farming systems (Campbell et al. 2014). Practices like cover cropping and catch crops can sequester 4–6 t CO_2 ha^{-1} $year^{-1}$ while improving crop yields by up to 65% through enhanced soil fertility (Poeplau and Don 2015). No-till farming especially when combined with cover crops, enhances carbon storage and reduces nitrous oxide emissions by 40–70% (Six et al. 2004). Agroforestry systems capture carbon in both biomass and soil while also boosting farmers' incomes, as demonstrated in Kenyan agro-ecological projects (Mbow et al. 2014). In Asia, especially in South and Southeast Asia, CSA strategies emphasize alternate wetting and drying (AWD) in paddy fields to curb methane emissions by maintaining yields. Solar-powered pumps, drip irrigation and precision farming and climate-resilient seed varieties help in mitigating climate change. Enhanced weathering methods, like spreading basalt rock dust in U.S. cornfields, have sequestered up to 15 t CO_2 ha^{-1} over 4 years while increasing maize and soybean yields by 12–16% (Beerling et al. 2020).

In addition to environmental benefits, CSA offers substantial socio-economic advantages. Diversified crop rotations and intercropping systems improve biodiversity by up to 23–28% over 20 years and enhance economic returns (Tamburini et al. 2020). According to the Federation of All India Farmer Associations (FAIFA), climate-resilient farming reduces costs and opens access to premium markets (FAIFA 2023). Corporate actors such as Mars and McCain have committed to sustainable sourcing and soil health initiatives, recognizing the economic and reputational value of mitigation (Forbes 2023).

However, barriers such as lack of financing, limited technical knowledge and insecure land tenure constrain the adoption of mitigation practices. Addressing these barriers through education, incentives, and public–private partnerships is essential. Mitigation efforts must also be inclusive. Empowering women through targeted CSA interventions enhances adaptive capacity and decision-making power in low- and middle-income countries (World Bank 2015). Besides, climate-smart practices align with global goals such as the Sustainable Development Goals (SDGs), particularly SDG 2 (Zero Hunger) and SDG 13 (Climate Action). Experts at UN climate summits have repeatedly emphasized the role of agriculture as both a significant emission source and a key opportunity for mitigation (IPCC 2022). With the global population expected to rise by 65% by 2050, sustainable agricultural practices that increase yield while reducing emissions are indispensable (FAO 2021a, b).

In the above context, the imperative to adopt climate change adaptation strategies in agriculture is driven by multiple factors: reversing productivity declines, increasing resilience, reducing emissions, enhancing nutrition, improving livelihoods and aligning with global climate goals. A vast body of evidence affirms that integrated mitigation strategies-such as no-till, cover cropping, agroforestry and precision farming-provide tangible, multifaceted benefits. The time to scale these strategies is now, through coordinated efforts involving research institutions, governments, private actors and local communities. Our collective future depends on proactive mitigation to ensure a sustainable and food-secure planet.

Climate Change Adaptation Strategies-Important Ones

Below are some potential adaptation options to resilient agriculture in general and specific to Indian agriculture (Srinivasarao et al. 2024; Table 24.1).

Table 24.1 Area coverage and yield performance under different adopted technologies of Climate Smart Agriculture (Source: Singh et al. 2021)

Technology	Year	Area (Acre)	Yield (Quintal)	Extra Profit in Indian Rupees (compared to conventional farming)
Direct seeded rice	2019	1	20	₹2356.00
ZT-wheat	2019	1	18	₹4065.80
ZT-mung	2020	1	4.80	₹3256.40
Direct seeded rice	2020	2	44	₹4895.24
ZT-wheat	2020	2	38	₹8566.25
Total				**₹23,139.69**

Smart Agro-Advisory

Ensuring that farmers, particularly smallholder farmers, are well-informed about imminent climate abnormalities is crucial for successful adaptation. This can be achieved through:

(a) ***Swift dissemination of weather information:*** South-Asian countries like India, Bangladesh experience various climatic extremes, making timely weather updates is essential. Well-equipped weather stations and efficient extension services are needed to provide farmers with tailored advice on crop planting and input application. Developing advanced decision support systems (DSS) can further aid this process.

(b) ***Early warning systems:*** Reliable and timely warnings about impending climatic risks help in identifying areas and communities at risk of food insecurity due to climate change.

(c) ***Contingent crop planning***: Contingency plans should be developed for different regions, utilizing modern information and space technologies to monitor pest movements (Srinivasarao et al. 2020).

(d) ***Crop insurance:*** Weather insurance, supported by standardized weather data, offers farmers an alternative to manage increased climatic risks.

(e) ***Community fodder and seed banks:*** Events like droughts, dry spells, and floods can devastate crops and livestock. To cope, farmers, especially smallholder farmers in India, can establish community reserves of food, forage and seeds. Support from the government and development partners is crucial, especially in the initial stages.

Adoption of New-Generation Technologies

Utilizing artificial intelligence, machine learning and mobile app-based agro-advisory services can enhance farmers' decision-making abilities in the face of climate change.

Adoption of Best Management Practices

(a) ***Adoption of climate-smart varieties and climate-resilient crops:*** Millets are gaining importance as they are climate-resilient and can ensure nutritional security for smallholder farmers. Development of stress-tolerant crop varieties with higher yield potential and resistance to multiple abiotic (drought, heat, flood, salinity) and biotic (pests and diseases) stresses is essential to mitigate the adverse impacts of climate risks. Additionally, promoting new and unconventional crops like dragon fruit, quinoa, and chia, which are tolerant to various abiotic stresses, is crucial.

(b) ***Soil management:*** Effective soil management is vital for climate change adaptation in rainfed agriculture, as extreme weather events can accelerate soil erosion. Practices like tree and hedgerow planting, contour ploughing, terrace gardening and water harvesting help reduce erosion. Zero tillage with residue retention enhances resilience to water stress and excess rainfall, while also boosting soil organic carbon.

(c) ***Crop diversification:*** Crop diversification, both in type and timing, is a cost-effective way to support essential ecosystem services like nutrient cycling, carbon storage, erosion control and water regulation, reducing pests and disease risks caused by climate variability, thereby stabilizing crop production under stress. Leguminous crops (red grams, mung beans and peanuts) are used to supplement nitrogen loss due to soil erosion or flooding. In cool and humid climates, legumes are mixed with main crops to protect fallow land.

(d) ***Water management:*** Water harvesting offers a potential solution to manage irrigation water scarcity throughout the year. Adopting micro-irrigation systems such as sprinklers and drip irrigation can save significant water use. Laser land levelling (LLL) significantly improves water and nutrient use efficiency, making agriculture more resilient to water stress. Comprehensive climate-smart water management key components (Sikka et al. 2018) include community tanks, farm ponds (widely adopted in NICRA villages), check dams, in situ soil moisture conservation practices, etc.

(e) ***Watershed management:*** Participatory integrated watershed management approaches ensure resilience to climate variability and increase food production, improve livelihoods, protect the environment, generate employment, address gender and equity issues, and conserve biodiversity.

(f) ***Energy-smart technologies:*** Energy-smart technologies like zero tillage and minimum tillage, along with resource conservation technologies, have the potential to increase crop productivity with less water and energy use. Integrating bioenergy, biodiesel and biomaterial production into the cropping calendar can contribute significantly to achieving energy security. Additionally, the deployment of windmills and solar collectors, as well as the installation of photovoltaic panels, can effectively harness renewable energy sources.

(g) ***Low carbon technology:*** Several carbon-negative technologies are imperative, include integrated nutrient management, Biochar application, Conservation agriculture, Organic farming, and integrated farming system (IFS).

(h) ***Embracing regenerative agriculture:*** It represents a holistic approach to farming that seek to improve soil health, enhance biodiversity, bolster water retention and mitigate greenhouse gas emissions. Key principles include: minimal soil disturbance, cover crops and crop rotation, diverse agroecosystems, Livestock integration, soil health management, water management and Agroforestry.

All the above-mentioned adaptation strategies strive to cultivate agri-food systems that are productive, profitable, environmentally sustainable and resilient to climate change which emphasizes collaboration with nature to leave the land healthier and more productive for future generations. However, these strategies face challenges related to input availability, high production costs, limited access to agricultural extension services, and the need for adequate adaptation knowledge and experience.

Notable Adaptation Strategies to Climate Change: A Global Perspective

There are a great number of countries leading the way in climate change adaptation, many of them showing outsized ambition and innovation, despite limited resources (Fig. 24.2). Some of them are:

(a) In the Pacific, the small island state of Tuvalu has drawn on the best available science—and around 270,000 cubic meters of sand—to reclaim a 780 m-long, 100 m-wide strip of land to protect against sea level rise and storm waves beyond 2100. This is an important initiative for a low-lying atoll country comprised of only around 26 square kilometres of land.
(b) Other countries such as Malawi and Pakistan are modernizing the capture and use of climate data and early warning systems, equipping communities, farmers and policy makers with the information they need to protect lives and livelihoods.

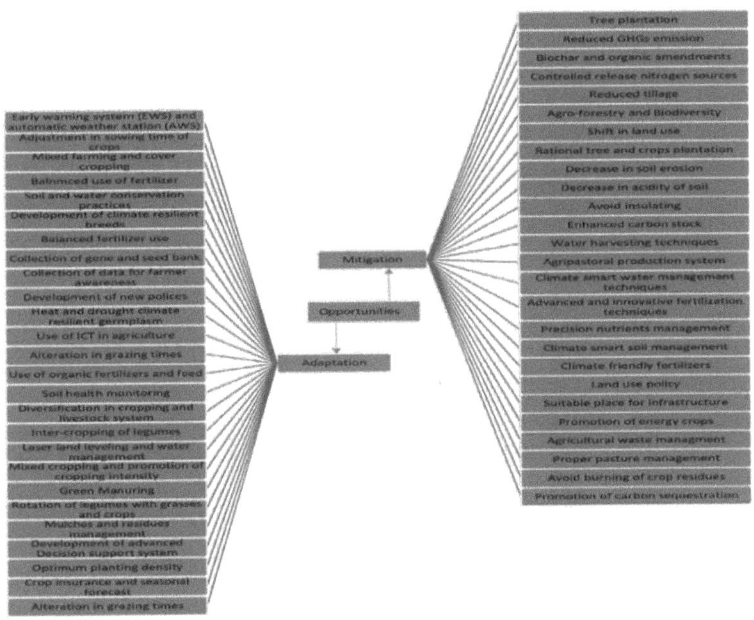

Fig. 24.2 Overview of opportunities, including adaptation and mitigation strategies for sustainable agriculture production systems in Asia (Source: Habib-ur-Rahman et al. 2022)

(c) Bhutan, the world's first carbon-negative country and Chad are among the world's Least Developed Countries (LDCs) to finalize National Adaptation Plans. The result of years of meticulous planning and rigorous consultation, the plans are crucial roadmaps for adaptation in the years ahead. In Bhutan's case, the plan is deeply rooted in the country's unique ethos of Gross National Happiness.

National Adaptation Plans (NAPs) are comprehensive medium and long-term strategies that outline how a nation will adapt to the changing climate and reduce its vulnerability to climate-related risks. NAPs are also crucial because they enable countries to systematically assess their vulnerability to climate change, identify adaptation needs and design effective strategies to build resilience.

Climate-Smart Agriculture (CSA)

Climate-smart agriculture (CSA) is proposed as a solution to transform and reorient agricultural systems to support food security under the new realities of climate change. It is not a one-size-fits-all set of practices that can be adopted by every farmer. In each location, its form needs to be defined by the context (i.e., extent of vulnerability to climate change, varying community risk profiles, availability of resources, and livelihood options). CSA strives to involve all local, regional, and national stakeholders in decision-making. Only by doing so is it possible to identify the most appropriate interventions and form the partnerships and alliances needed to enable sustainable development. CSA consists of co-achieving three objectives or pillars, defined as: (**a**) **Productivity**: Sustainably increasing agricultural productivity to support equitable increases in incomes, food security and development; (**b**) **Adaptation**: Adapting and building resilience to climate change from the farm to national levels; and (**c**) **Mitigation**: Reducing or removing GHG emissions wherever and whenever possible.

CSA Practices

CSA incorporates practices like boosting soil carbon absorption and reducing greenhouse gas emissions to mitigate climate change's detrimental effects and boost agricultural production and profitability. CSA practices are very similar to precision agriculture in their focus on understanding the elements that influence crop output, soil health, air pollution and other farming outcomes.

Fundamental Practices of CSA Include

- Cultivating Climate-Resilient Crop Varieties
- Conservation Agriculture: No-till and reduced-tillage cultivation
- Agroforestry
- Water Management: Precision irrigation, drip irrigation, rainwater harvest
- Integrated Pest Management
- Nutrient Management: Organic fertilizer inclusion along with chemical fertilizer, precision fertilizing and crop rotation
- Livestock Management: Rotational grazing, specialized feed formulations, and manure management (Lowering methane emissions)
- Digital Climate Services: Weather forecasting, climate advisory and customized farming information availability through websites and mobile applications

- Some of the successful CSA practices and cropping systems in India are represented in Table 24.2.

Leading Indian Institutes and Centres of Excellence in CSA

The role of institutions in improving CSA in India has been expanding, starting with the farmers at community-level institutions to policymaking bodies.

Public Institutions (Government and Quasi-Government)
- National Initiative in Climate Resilient Agriculture (NICRA), New Delhi
- National Institute of Rural Development and Panchayati Raj (NIRD & PR), Hyderabad
- National Institute of Agricultural Extension Management (MANAGE), Hyderabad
- National Academy of Agricultural Research Management (NAARM), Hyderabad
- Indian Institute of Millets Research (IIMR), Hyderabad
- Institute of Rural Management Anand (IRMA), Gujarat

Private Institutions (NGOs, For-Profit Companies, Incubation Centres etc.)
- The Energy and Resources Institute (TERI), New Delhi
- CropIn, Bengaluru
- Digital Green, New Delhi
- M. S. Swaminathan Research Foundation (MSSRF), Chennai

Conception About CSV and Its Components

Climate-smart village (CSV) is an AR4D (Agricultural research for development) approach to scale up CSA to rural development that integrates climate-resilient agricultural practices, sustainable natural resource management and community

Table 24.2 Successful CSA technologies and cropping systems in India (Source: The Asia Foundation 2022)

Successful CSA technologies and cropping systems	Primarily climate smart	Productivity enhancement/farm management contributing to climate resilience	Both
Land shaping	●		
Drought-tolerant paddy varieties	●		
Rainwater harvesting	●		
Paddy nursery		●	
Direct seeding rice		●	
Drum seeding technique		●	
Broad bed and furrow		●	
Zero-till sowing of wheat		●	
Rotavator		●	
Crop diversification (intercropping)			●
Integrated farming systems			●
Custom hiring canters (farm machineries)			●
Furrow irrigated raised bed			●

empowerment to enhance adaptation to climate change while contributing to its mitigation. This concept was introduced by the CGIAR (Consultative Group on International Agricultural Research) Research Program on CCAFS (Climate Change, Agriculture and Food Security) around 2011 as a way to test, refine, and scale out **CSA** technologies and institutional innovations within local communities and as an innovative response to the growing threats of climate change on agriculture and rural communities (Aggarwal et al. 2013; CCAFS 2021). It was first piloted in South specifically India, Nepal and Bangladesh-with the aim of testing, refining and scaling CSA practices in vulnerable rural settings.

CSVs serve as learning hubs where local stakeholders-including farmers, extension workers, researchers, NGOs and government agencies-collaboratively design and implement context-specific CSA practices (Aggarwal et al. 2018). The rationale behind introducing CSVs was to bridge the gap between climate science and rural livelihood realities by providing location-specific, participatory approaches that are scalable and measurable (Aggarwal et al. 2013). In South Asia, particularly in Indian states like Haryana and Bihar, CSVs have demonstrated water savings of up to 50% through technologies like laser land levelling and drip irrigation, along with yield improvements from drought-tolerant and flood-tolerant crop varieties (Jat et al. 2016). These changes have significantly improved rural livelihoods, promoted gender-inclusive participation and reduced vulnerability to climate shocks.

With the leadership of CCAFS, the CSV approach has been implemented in more than 20 countries in Asia, Africa and South America (Aggarwal et al. 2018). This includes multiple sites in the three South Asian countries: Andhra Pradesh,

Fig. 24.3 Types of climate-smart options that may be tested as part of context-specific portfolios in the CSV sites (Source: CCAFS, 2016)

Maharashtra, Madhya Pradesh, Bihar, Punjab, and Haryana states of India, six districts of Nepal and Barisal, Khulna and Sylhet divisions of Southern Bangladesh. Globally, the CSV approach has expanded to over 30 countries, tailoring solutions to regional contexts. CSVs are now recognized not only as field-based solutions but also as policy-informing platforms that align community practices with National Adaptation Plans and international climate agreements like the Paris Accord (Fig. 24.3).

The Key Components of a CSV (CCAFS 2020a, b)

(a) ***Climate-Smart Technologies and Practices***: Adoption of techniques like conservation tillage, crop diversification, agroforestry, improved irrigation systems (e.g., drip irrigation), use of climate-resilient seeds and integrated pest management (FAO 2013).
(b) ***Climate Information and Early Warning Systems***: Dissemination of real-time weather forecasts, seasonal climate predictions and decision-support tools to guide farmers' planning and risk management (Hellin et al. 2018).
(c) ***Capacity Building and Knowledge Sharing***: Continuous training programs, farmer field schools and participatory research to enhance awareness, skills and confidence in adopting CSA innovations (Wollenberg et al. 2016).
(d) ***Risk Management Mechanisms***: Implementation of insurance schemes, savings and credit facilities and social safety nets to buffer households against climate-related shocks (Vermeulen et al. 2012).

(e) ***Sustainable Energy and Infrastructure***: Integration of renewable energy sources like solar pumps, biogas plants and off-grid energy systems to promote environmental sustainability (IEA 2020).
(f) ***Institutional and Policy Support***: Strong coordination with local governance bodies, policy makers and development programs to ensure scalability and alignment with national climate strategies (Harvey et al. 2014a, b).
(g) ***Monitoring and Evaluation***: Participatory assessment of outcomes to refine interventions and facilitate replication in other villages (Aggarwal et al. 2018).
(h) ***Social Inclusion and Gender Equity***: Active involvement of women, youth and marginalized groups in decision-making processes to ensure equitable benefits and inclusive development (Jost et al. 2016).

Climate-Smart Interventions for CSV

The options tested as part of the CSV research agenda for dealing with climate change and variability include:

(a) *Weather-smart activities*: weather forecasts, climate-informed agro-advisories, weather insurance, climate analogues as a tool for forward planning, strategies to avoid mal-adaptation.
(b) *Water-smart practices*: aquifer recharge, rainwater harvesting, community management of water, laser-land levelling, micro irrigation, raised-bed planting, solar pumps.
(c) *Seed/ breed smart*: adapted varieties and breeds, seed banks, including community-based activities.
(d) *Carbon/nutrient-smart practices*: agroforestry, minimum tillage, land use systems, livestock management, integrated nutrient management, and biofuels.
(e) *Institutional/market smart activities*: cross-sectoral linkages; local institutions including learning platforms or farmer-to-farmer learning and capacity development, contingency planning, financial services, market information, gender equitable approaches, and off-farm risk management strategies (Fig. 24.4).

There is no fixed package of CSA interventions to be tested, however or a one-size-fits-all approach. Interventions selected differ based on the region, its agro-ecological characteristics, level of development, capacity, and interest of farmers and the local government.

Key Steps of the CSV Approach Implementation

In establishing a CSV AR4D site, the very first step is to build trust and partnerships amongst diverse stakeholders, and to get agreement and buy-in to a common approach. Once partners have agreed on the establishment of a CSV site, the major steps include:

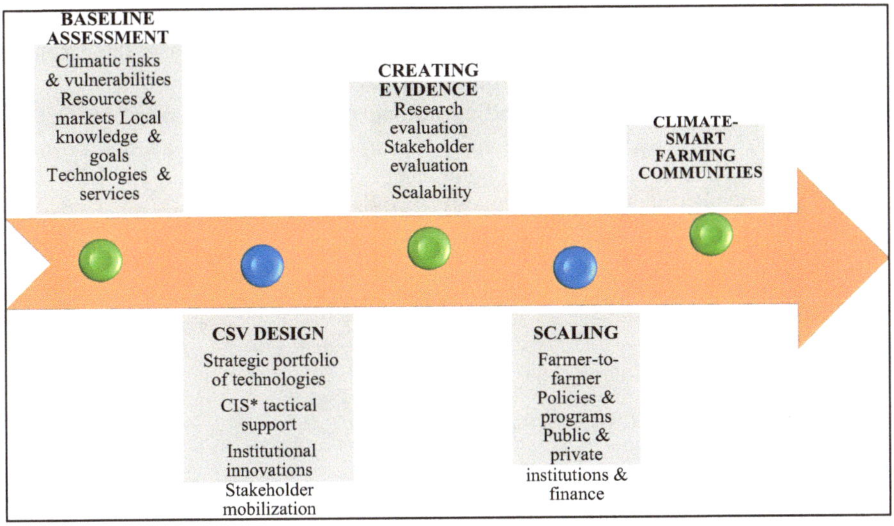

Fig. 24.4 Steps for the implementation of the CSV AR4D approach (Source: CCAFS, 2016)

(a) *Baseline assessment:* Includes climate risk analysis and gender and social inclusion analysis;
(b) *CSV design:* Identification and prioritization of climate-smart technologies, practices and services based on biophysical, socio-economic, gender, policy and institutional context; also considering possible synergies and trade-offs amongst individual activities;
(c) *Creating evidence:* Evaluation and development of portfolios of climate-smart interventions (e.g. providing value-added weather services to farmers, promoting weather-based insurance, building capacity in climate change adaptation and facilitating community partnerships for knowledge sharing);
(d) *Scaling:* This involves scaling up through policies and institutions, and scaling out to large areas through farm-to-farm and ICT-based approaches (Fig. 5).

Importance and Relevance of the CSV Approach on the South-Asian Context

According to Mishra et al. (2016) review of the national-level climate policy documents for most of the South Asian countries reveals a common goal of building people's adaptive capacity by providing livelihood security in the face of climate change risks. However, the level of livelihood-focused adaptation initiatives in practice appears inadequate. Out of 21 adaptation projects reviewed, only three local-scale initiatives in two countries (India and Bangladesh) had an explicit focus on livelihoods. An examination of the national policies and adaptation programmes of South Asian countries clearly shows that the countries have not lagged in terms of policy formulation and launching of action on the ground, but there is widespread concern regarding the effective implementation of measures. In general, limited access to climate change information, knowledge and services across the extended Himalayan region is a major constraint to effective adaptation.

There is also a limited understanding among policymakers of what indicates a good adaptation practice, how to identify and undertake suitable adaptation interventions and how to replicate and scale up successful interventions. The role of the private sector is important in scaling up adaptation, especially for leveraging finance for technology transfer. However, most businesses in the South Asia have yet to take a proactive approach to adaptation. Few have assessed the likely effects of climate change on their own operations.

CSV Case Studies from South Asia

Converting conventional villages into CSVs has emerged as a globally relevant strategy to tackle the dual challenges of food insecurity and climate vulnerability. Southeast Asian countries like Vietnam incorporated CSV models within the national New Rural Development program, establishing pilot CSVs in Yen Bai province using a "One Commune One Product" (OCOP) strategy to build community-based resilience (CCAFS 2019). The Philippines initiated 21 CSVs under the AMIA villages program for making additional focus on climate-resilient agriculture, women's empowerment and disaster preparedness. In Myanmar, CSVs introduced climate-smart nutrition-sensitive agriculture, improving dietary diversity among households, though further monitoring is ongoing (IIRR et al. 2019).

In the case of South Asia, CSVs in India, Nepal, and Bangladesh have improved farmers' income and climate resilience. Haryana state of India integrated CSVs into policy through a convergence approach involving laser land leveling, precision irrigation and agro-advisory systems. Total farm income rose by ₹194,074 in Kurukshetra and ₹92,648 in Sirsa (districts of Haryana) when compared with non-CSV villages (Aryal et al. 2020). Villages such as Anjanthapur in Andhra Pradesh of India benefited from Rs. 2.48 crore investment in CSA techniques like farm ponds, drought-resistant seeds and water-saving practices (ICRISAT 2021a, b). India's Phayeng village in Manipur, declared the first carbon-positive village, adopted forest restoration, solar energy, eco-tourism, and clean cooking initiatives under a Rs. 10 crore climate-resilience project (India Times 2019). Study of Jat et al. (2020) strengthens the fact that CSV practice portfolios produced 5–37% higher crop yields, increased net incomes by 28–40% and reduced GHG emissions by 16–25% through resource-conserving practices such as zero-tillage and residue incorporation. Additionally, CSV efforts were backed by major financing schemes such as PM-KUSUM, Atal Bhujal Yojana and the Per Drop More Crop initiative. These programs collectively funded interventions like solar irrigation pumps, micro-irrigation systems and canal rehabilitation. Although detailed economic data is still emerging, field reports highlight lower labor and input costs, better yields and enhanced resilience due to diversified cropping and crop insurance (CCAFS 2019; Aryal et al. 2020).

In Nepal, the NGO LI-BIRD, in collaboration with CCAFS, launched CSVs across Dang, Bardiya and Gorkha, with baseline assessments mapping climate sensitivity and adaptive capacity. These data-driven interventions helped develop tailored CSV action plans (LI-BIRD and CCAFS 2018). On the other hand, the youth-led Gangarampur CSV, integrating smart agriculture, solar water purification and community disaster plans, created a localized model of resilience for coastal zones in Bangladesh (Youth Climate Lab 2020). These real-world success stories emphasized CSV's potential towards context-specific, inclusive climate action from local to global. CSVs have shown a 15–40% increase in yields, 30% improvement in water-use efficiency and substantial gains in household income in regions like Africa and India (CCAFS 2020a, b). These models are now being embedded in national policies-such as ADB's biodiversity corridors in Cambodia or cross-learning platforms in East Africa-demonstrating potential for regional up scaling (ADB 2018).

Decision Support Framework in CSV Approach

An ideal CSV AR4D site typically provides five types of decision support to farmers and other stakeholders, and allows for the research that assesses the effectiveness of such support:

(a) Village/community agricultural land use plans and contingency plans considering current and future climate risks, soil and socio-economic conditions, and markets.
(b) CSA portfolios that do not become maladaptive in future climate and market scenarios—assessed using models.
(c) Strategic guidance before the planting season, where feasible based on seasonal forecasts, on most suitable CSA practices, technologies, services, processes and institutional options. This is done in a participatory mode with local farmer groups and due consideration is given to institutions in the region such as farmers' self-help groups, water-users associations, markets and the availability of climate and agriculture development finance.
(d) Tactical guidance to farmers on using real-time weather forecasts and value-added ICT based agro-advisories; on accessing good quality inputs and technologies for improving water/nutrient/energy use efficiencies and on risk transfer through insurance mechanisms in case of crop and livestock losses.
(e) Policy level guidance on policy barriers and policy options to unlock CSA and local and national development. This includes consideration of the financial needs to drive scaling.

Success Stories from the Field: Impact of the CSV Approach

(a) *Engaging State Leadership: The Case of Haryana, India*

The state of Haryana presents a compelling example of how proactive government involvement can accelerate the adoption and scaling of Climate-Smart Village (CSV) initiatives. From the very inception of the CSV-AR4D sites, the Government of Haryana has played an active role, fostering a strong sense of ownership and institutional commitment. This engagement has not only led to the formal endorsement of climate-resilient approaches and village-level plans but has also influenced policy shifts in support of climate-smart agriculture (CSA) portfolios. Building on this foundation, the Haryana government has gone on to promote hundreds of CSVs across the state. While distinct from the original CCAFS-AR4D pilot sites, these state-led CSVs emphasize integrated, participatory climate action tailored to local needs. This case clearly underscores the importance of early and sustained engagement of government stakeholders in scaling climate-resilient agricultural interventions.

(b) *Scaling up climate-smart agriculture through mainstreaming Climate Smart Villages (CSV) in Bihar, India*

In 2017, the National Adaptation Fund on Climate Change (NAFCC) launched a project in Bihar to scale up climate-smart agriculture by mainstreaming the Climate-Smart Village (CSV) approach. Targeting climate-vulnerable regions in Patna (Daniyawan, Fatuha) and Nalanda (Chandi, Nagar Nausa, Noorsarai), the project introduced direct-seeded rice, zero tillage wheat, and improved pulses (lentil and moong). These interventions aimed to strengthen farmers' adaptive capacity through locally relevant, participatory strategies.

(c) *Case study: Solar-based irrigation system in Nepal*

In Nepal's Terai region, where irrigation relies on groundwater, farmers traditionally used expensive and polluting diesel pumps. To address this, LI-BIRD, with support from CCAFS and CDKN, piloted 12 solar-powered irrigation systems in Bardiya, Dang, Nawalparasi, and Mahottari districts. Each unit—capable of pumping 120,000 L/day—was co-funded (75% project, 25% community) with added local government support. These systems not only ensured timely rice nursery establishment, reducing crop failure, but also enabled a shift to commercial vegetable farming, raising cropping intensity by 200–300%.

Systemic Modifications for Climate-Smart Village Workflow

For the Climate-Smart Village (CSV) approach to be effectively scaled and sustained, several systemic modifications are necessary to ensure inclusivity, relevance, and long-term impact across diverse agro-ecological regions.

Institutional Strengthening and Decentralised Governance

Effective CSV implementation requires empowering local institutions—particularly Panchayats and village committees—with autonomous planning, budgeting, and execution authority. This decentralization fosters community ownership while reducing implementation inefficiencies through improved coordination across agriculture, water, and rural development sectors.

Land Tenure Security and Asset Access

Secure land tenure emerges as a critical precondition for farmer investment in climate-smart interventions. Evidence from South Africa demonstrates that tenure insecurity significantly limits adoption rates of long-term practices like agroforestry. Formalizing land rights through participatory mapping and simplified titling processes, particularly for women and tenants, substantially increases adoption rates.

Inclusive Financial Mechanism Design

Conventional financing often fails to address smallholder needs due to high upfront costs of climate-smart technologies. Tiered subsidies, low-interest loans, and viability-based financing models reduce adoption barriers. Community-based funding mechanisms, including Community Innovation Funds and Payments for Ecosystem Services (PES), offer scalable alternatives for collective resource access.

Enhanced Extension and Information Systems

Localized, gender-sensitive extension services prove essential for CSV success. Strengthening farmer field schools, lead farmer networks, and ICT-based advisory systems ensures practical, real-time guidance. Integration of female extension agents addresses gender-specific barriers to technology adoption.

Participatory Planning and Adaptive Management

Community-driven, co-created interventions demonstrate superior sustainability outcomes. Participatory planning processes, as evidenced in Lushoto, Tanzania, ensure technology alignment with local contexts. Long-term planning cycles with continuous monitoring, evaluation, and adaptation mechanisms accommodate the delayed benefits of interventions like agroforestry.

Policy Alignment and Subsidy Restructuring

Current subsidy structures often reinforce unsustainable practices. Redirecting subsidies from high-emission inputs toward climate-smart technologies—such as drip irrigation and agroforestry kits—can catalyse behavioural change. Complementary instruments, including conditional cash transfers and carbon-based incentives, further promote equitable transitions.

Ecosystem-Based Solution Integration

Nature-based solutions provide co-benefits for productivity and climate resilience. Practices including agroforestry, terracing, silvopasture, and rainwater harvesting enhance ecosystem services while improving farm incomes. Ethiopian case studies demonstrate 17% yield increases and 7% income gains through agroforestry adoption, highlighting the potential of integrated green-grey infrastructure approaches.

Conclusions

Climate-Smart Villages (CSVs) offer more than technical solutions-they represent a lifeline for rural South Asia. By weaving together ancestral wisdom, cutting-edge tools like solar irrigation and drought-resistant seeds, and community-driven governance, CSVs have turned vulnerability into resilience. Farmers from Vietnam to Bangladesh now harvest hope alongside higher yields (20–50%), using less water and emitting fewer greenhouse gases. Yet this journey is far from complete. Persistent gaps in funding, gender equity, and fragmented policies threaten to leave the most vulnerable behind. The path forward demands courage: blending finance models to empower communities, amplifying women's leadership, and weaving CSV principles into national climate plans. As monsoon patterns fracture and temperatures rise, these villages stand as beacons-proof that when science walks hand-in-hand with tradition, even the smallest hamlet can nurture a climate-proof future.

References

Adnan M, Khan F, Rehman N, Ali S, Hassan SS, Dogar MM, Hasson S (2021) Variability and predictability of summer monsoon rainfall over Pakistan. Asia-Pacific Journal of Atmospheric Sciences, 57(1):89–97.

Aggarwal PK, Jarvis A, Campbell BM, Zougmoré RB, Khatri-Chhetri A, Vermeulen SJ, et al. (2013) Climate-smart villages: an approach for scaling up climate-smart agriculture. CCAFS Policy Brief No. 17. https://cgspace.cgiar.org/handle/10568/33326

Aggarwal PK, Jarvis A, Campbell BM, Zougmoré RB, Khatri-Chhetri A, Vermeulen SJ et al (2018) The climate-smart village approach: framework of an integrative strategy for scaling up adaptation options in agriculture. Ecol Soc 23(1):14. https://doi.org/10.5751/ES-09844-230114

Aryal JP et al (2020) Impacts of climate-smart agriculture on productivity and income in South Asia. J Environ Manag 261:110261

Asian Development Bank (ADB) (2018) Final completion report—biodiversity corridors project, Cambodia. https://www.adb.org/projects/documents/cam-42305-013-tcr

Azharuddin S (2016) Decadal climate variability and predictability. Current Science, 110(8):1397.

Beerling DJ, Kantzas EP, Lomas MR, Wade P, Eufrasio RM, Holden PB et al (2020) Potential for large-scale CO_2 removal via enhanced rock weathering with croplands. Nature 583:242–248

Campbell BM, Thornton P, Zougmore R, van Asten P, Lipper L (2014) Climate-smart agriculture: building resilience to climate change. CCAFS Report, CGIAR Research Program on Climate Change, Agriculture and Food Security

Campbell BM, Vermeulen SJ, Aggarwal PK, Corner-Dolloff C, Girvetz E, Loboguerrero AM et al (2016) Reducing risks to food security from climate change. Glob Food Sec 11:34–43. https://doi.org/10.1016/j.gfs.2016.06.002

CCAFS (2016) Climate-smart villages: an AR4D approach to scale up climate-smart agriculture. CGIAR Research Program on Climate Change, Agriculture and Food Security (CCAFS), Copenhagen, Denmark. www.ccafs.cgiar.org

CCAFS (2019) Climate-smart villages in Southeast Asia. CCAFS, CGIAR Research Program on Climate Change, Agriculture and Food Security

CCAFS (2020a) Climate-smart villages: India and beyond. CCAFS, CGIAR Research Program on Climate Change, Agriculture and Food Security

CCAFS (2020b) CSV implementation guide: climate-smart villages as platforms for scaling climate-smart agriculture. CCAFS, CGIAR Research Program on Climate Change, Agriculture and Food Security

CCAFS (2021) Climate-smart villages as a platform for scaling out climate-smart agriculture. CCAFS, CGIAR Research Program on Climate Change, Agriculture and Food Security

Chun JA, Li S, Wang Q, Lee WS, Lee EJ, Horstmann N et al (2016) Assessing rice productivity and adaptation strategies for Southeast Asia under climate change through multi-scale crop modeling. Agric Syst 143:14–21. https://doi.org/10.1016/j.agsy.2015.12.001

Dogar MM, Kucharski F, Azharuddin S, (2017a). Study of the global and regional climatic impacts of ENSO magnitude using SPEEDY AGCM. J Earth Syst Sci 126(30). https://doi.org/10.1007/s12040-017-0804-4

Dogar MM, Stenchikov G, Osipov S, Wyman B, Zhao M (2017b) Sensitivity of the regional climate in the Middle East and North Africa to volcanic perturbations. Journal of Geophysical Research: Atmospheres, 122(15):7922–7948.

Dogar MM, Kucharski F, Sato T, Mehmood S, Ali S, Gong Z, Arraut J (2019) Towards understanding the global and regional climatic impacts of Modoki magnitude. Global and Planetary Change, 172, 223–241.

Dogar MM, Fujiwara M, Zhao M, Ohba M, Kosaka Y (2024) ENSO and NAO linkage to strong volcanism and associated post-volcanic high-latitude winter warming. Geophysical Research Letters, 51(1), e2023GL106114.

Dogar MMA (2020) Study of the regional climatic impacts of tropical explosive volcanism in the Middle East and North Africa region (Doctoral dissertation, Hokkaido University).

Dogar MMA, Almazroui M (2022) Revisiting the strong and weak ENSO teleconnection impacts using a high-resolution atmospheric model. Atmospheric Environment, 270, 118866.

Dogar MM, Watanabe S, Fujiwara M, Abid MA, Habeebullah TM, Khan B (2025) Tropical volcanic impacts on MENA climate via ENSO and NAO dynamics in a high-top model. npj Climate and Atmospheric Science, 8(1):330.

Dogar MM, Sato T, Liu F (2020) Ocean sensitivity to periodic and constant volcanism. Scientific Reports, 10(1):293.

Fallah Haghighi N, Hajihoseini H, Ramezanpour NG, Bijani M (2018) Gap analysis of current & desired states of entrepreneurship development components in the field of ICTs in Iran. Technol Soc 54:101–110. https://doi.org/10.1016/j.techsoc.2018.03.003

FAO (2013) Climate-smart agriculture sourcebook. FAO Report, Food and Agriculture Organization of the United Nations

FAO (2021a) Climate-smart agriculture policies and practices. FAO Report. https://www.fao.org/publications

FAO (2021b) Energy-smart food for people and climate. FAO Report, Food and Agriculture Organization of the United Nations

Forbes. (2023) Sustainability in agriculture: business and environmental benefits. Forbes Media. https://www.forbes.com

Ghaffar A, Habib MHR, Ahmad S, Ahmad I, Khan MA, Hussain J et al (2022) Adaptations in cropping system and pattern for sustainable crops production under climate change scenarios. CRC Press, Boca Raton, FL. https://doi.org/10.1201/9781003286417-1

Habib-ur-Rahman M, Ahmad A, Raza A, Hasnain MU, Alharby HF, Alzahrani YM et al (2022) Impact of climate change on agricultural production; issues, challenges, and opportunities in Asia. Front Plant Sci 13:925548. https://doi.org/10.3389/fpls.2022.925548

Harvey CA, Chacon M, Donatti CI, Garen E, Hannah L, Andrade A et al (2014a) Climate-smart landscapes: opportunities and challenges for integrating adaptation and mitigation in tropical agriculture. Conserv Lett 7(2):77–90

Harvey CA, Rakotobe ZL, Rao NS, Dave R, Razafimahatratra H, Rabarijohn RH et al (2014b) Extreme vulnerability of smallholder farmers to agricultural risks and climate change in Madagascar. Philos Trans R Soc B Biol Sci 369(1639):20130089. https://doi.org/10.1098/rstb.2013.0089

Hassani A, Azapagic A, Shokri N (2021) Global predictions of primary soil salinization under changing climate in the 21st century. Nat Commun 12:6663

Hellin J, Rosenstock TS, Cairns JE, van Loon J (2018) Climate-smart agriculture and smallholder adoption: building resilience to climate change. World Dev 108:285–297

Himalayan Monitoring Report (2020) The state of Himalayan glaciers: implications for water resources in Nepal. ICIMOD, Kathmandu. https://icimod.org

Huq N, Huge J, Boon E, Gain AK (2015) Climate change impacts in agricultural communities in rural areas of coastal Bangladesh: a tale of many stories. Sustainability 7(7):8437–8460. https://doi.org/10.3390/su7078437

ICRIER (2019) Climate change and rural incomes: challenges for India's smallholder farmers. https://icrier.org/

ICRISAT (2021a) Climate-smart villages in India: lessons from implementation. ICRISAT Report. https://www.icrisat.org

ICRISAT (2021b) Climate-smart villages in India: success stories and pathways. ICRISAT Report. https://www.icrisat.org/wp-content/uploads/2021/03/Climate-Smart-Villages-in-India.pdf

ICRISAT, FAIFA (2023). Climate resilience and crop production in semi-arid tropics. ICRISAT Report

IEA (2020) Renewable 2020. International Energy Agency Report

IIRR, ICRAF, CCAFS (2019) Nutrition-sensitive climate-smart villages in Myanmar. International Institute of Rural Reconstruction Report. https://hdl.handle.net/10568/101017

India Times (2019) Phayeng: Manipur's first carbon-positive village. https://www.indiatimes.com/news/india/phayeng-manipur-s-first-carbon-positive-village-365478.html

IPCC (2019a) Global warming of 1.5°C. Summary for policy makers. World Meteorological Organization, Bern

IPCC (2019b) Refinement to the 2006 IPCC guidelines for national greenhouse gas inventories. In: Buendia CE, Tanabe K, Kranjc A, Baasansuren J, Fukuda M, Ngarize S, Osako A, Pyrozhenko Y, Shermanau P, Federici S (eds). IPCC, Switzerland

IPCC (2022) Climate change 2022: impacts, vulnerability, and adaptation. In: Contribution of working group II to the sixth assessment report of the Intergovernmental Panel on Climate Change

IUCN (2020) Salinity and climate impacts on agriculture in Bangladesh. https://www.iucn.org/

Janmohammadi M, Sabaghnia N (2023) Strategies to alleviate the unusual effects of climate change on crop production: a thirsty and warm future, low crop quality. A review. Biologija 69:121–133

Jat ML et al (2016) Climate-smart villages in South Asia: evidence from indo-Gangetic plains. Int J Agric Sustain

Jat ML et al (2020) Climate-smart soils for securing food and nutrition security. Science 367(6476):eaaz7432

Jost C, Kyazze FB, Naab JB, Neelormi S, Kinyangi J, Zougmore RB, Kristjanson P (2016) Understanding gender dimensions of agriculture and climate change in smallholder farming communities. Clim Dev 8(2):133–144

Kaushik N, Siddique KHM, Ahmed MS, Hakimi SS (2023) Climate-smart agriculture strategies for South Asia to address the challenges of climate change: identification of climate-resilient agriculture practices for India, Bangladesh, and Afghanistan. APN Sci Bull 13(1). https://doi.org/10.30852/sb.2023.2349

Khatri-Chhetri A, Aryal JP, Sapkota TB, Khurana R (2016) Economic benefits of climate-smart agricultural practices to smallholder farmers in the indo-Gangetic Plains of India. Curr Sci:1251–1256

Lal R, Monger C, Nave L, Smith P (2021) The role of soil in regulation of climate. Philos Trans R Soc Lond B Biol Sci 376:20210084

LI-BIRD, CCAFS (2018) Climate smart village baseline report—Nepal. https://cgspace.cgiar.org/handle/10568/91232

Lone FA, Maheen M, Ul Shafiq M, Bhat MS, Rather JA (2020) Farmer's perception and adaptation strategies to changing climate in Kashmir Himalayas, India. GeoJournal 87:1743–1757. https://doi.org/10.1007/s10708-020-10330-0

Mbow C et al (2014) Agroforestry solutions to address food security and climate change challenges in Africa. Curr Opin Environ Sustain 6:61–67

Meena RK, Verma TP, Yadav RP, Mahapatra SK, Surya JN, Singh D et al (2019) Local perceptions and adaptation of indigenous communities to climate change: evidences from high mountain Pangi Valley of Indian Himalayas. Indian J Tradit Knowl 11:876–888

Mishra A, Hussain A, Rasul G (2016) Climate change haunts south Asian livelihoods. Trade Insight—SAWTEE Newsl 12(4):10–13

Mottaleb KA, Rejesus RM, Murty MVR, Mohanty S, Li T (2017) Benefits of the development and dissemination of climate-smart rice: ex ante impact assessment of drought-tolerant rice in South Asia. Mitig Adapt Strateg Glob Change 22:879–901. https://doi.org/10.1007/s11027-016-9705-0

Newman R, Noy I (2023) The global costs of extreme weather that are attributable to climate change. Nat Commun 14:6103

Ortiz R, Sayre KD, Govaerts B, Gupta R, Subbarao GV, Ban T, Hodson D, Dixon JM, Ortiz-Monasterio JI, Reynolds M (2008) Climate change: can wheat beat the heat? Agric Ecosyst Environ 126(1–2):46–58

Ozkaynak E (2013) Effects of air temperature and hours of sunlight on the length of the vegetation period and the yield of some field crops. Ekoloji 22:58–63

Palanisami K, Kumar DS, RPS M, Raman S, Kar G, Monhan K (2015) Managing water management research: analysis of four decades of research and outreach programmes in India. Econ Polit Wkly 50(26/27):33–43

Parry ML, Canziani OF, Palutikof JP, Van Der Linden PJ, Hanson CE (2007) Climate change 2007: impacts, adaptation and vulnerability. Contribution of working group II to the fourth assessment report of the intergovernmental panel on climate change. Cambridge Univ Press, Cambridge, UK

Poeplau C, Don A (2015) Carbon sequestration in agricultural soils via cultivation of cover crops. Agric Ecosyst Environ 200:33–41

Population of Asia (2019) Demographics: density ratios, growth rate, clock, rate of men to woman. Available online at: www.populationof.net (accessed June 2, 2019)

Rao CS, Baral K, Chanadana VM, Jagadesh M, Karthik R (2024) Climate change adaptation and mitigation in Indian agriculture. J Agrometeorol 26(2):137–148. https://doi.org/10.54386/jam.v26i2.2582

Raveendran G (2016) A review of Rangarajan committee report on poverty estimation. Indian J Hum Dev 10(1):85–96. https://doi.org/10.1177/0973703016648033

Rezaei EE, Siebert S, Huging H, Ewert F (2018) Climate change effect on wheat phenology depends on cultivar change. Sci Rep 8:4891. https://doi.org/10.1038/s41598-018-23101-2

Salmanicheraghabadi A, Poursaeed AR, Bayramzadeh V, EshraghiSamani R (2021) Social network analysis of sustainable forest management actors in Zagros region. Arab J Geosci 14:2701. https://doi.org/10.1007/s12517-021-09032-9

Shunsuke S (2025) The impact of climate change on rural livelihoods in South Asia: a comparative study of India, Nepal, and Bangladesh. Authorea, May 15. https://doi.org/10.22541/au.172979332.28241985/v2

Sikka AK, Islam A, Rao KV (2018) Climate-smart land and water management for sustainable agriculture. Irrig Drain 67(1):72–81

Singh NR, Mukherjee A, Sundaram PK, Manisha T, Rao KK, Kumar AK, Mishra JS, Choudhary AK (2021) Climate-smart farmers success stories from Chandi block of Nalanda compressed. Food Sci Rep 2(6):62–65

Six J et al (2004) The role of soil organic matter in soil carbon sequestration. Soil Biol Biochem 36(9):1359–1372

Srinivasarao C, Srinivas T, Rao RVS, Srinivasarao N, Vinayagam SS, Krishnan P (2020) Climate change and Indian agriculture: challenges and adaptation strategies. ICAR-National Academy of Agricultural Research Management, Hyderabad, India, p. 584

Srivastava A, Naresh Kumar S, Aggarwal PK (2010) Assessment on vulnerability of sorghum to climate change in India. Agric Ecosyst Environ 138(3–4):160–169

Tamburini G et al (2020) Agricultural diversification promotes multiple ecosystem services without compromising yield. Sci Adv 6(45):eaaz0739

Tan Q, Liu Y, Dai L, Pan T (2021) Shortened key growth periods of soybean observed in China under climate change. Sci Rep 11:8197

Thackeray CW, Hall A, Norris J, Chen D (2022) Constraining the increased frequency of global precipitation extremes under warming. Nat Clim Chang 12:441–448

Vermeulen SJ, Aggarwal PK, Ainslie A, Angelone C, Campbell BM, Challinor AJ (2012) Options for support to agriculture and food security under climate change. Environ Sci Pol 15(1):136–144

Westermann O, Förch W, Thornton PK (2015) Reaching more farmers: innovative approaches to scaling up climate-smart agriculture. CCAFS working paper no. 135. CGIAR research program on climate change, agriculture and food security (CCAFS), Copenhagen, Denmark

Wollenberg E, Campbell BM, Holmgren P, Seymour F, Sibanda L, von Braun J (2016) Actions needed to halt deforestation and promote climate-smart agriculture. CIFOR Policy Brief No. 3

World Bank (2015) Gender in climate-smart agriculture: module 18 for gender in agriculture sourcebook

Yadav SS, Lal R (2018) Vulnerability of women to climate change in arid and semi-arid regions: the case of India and South Asia. J Arid Environ 149:4–17. https://doi.org/10.1016/j.jaridenv.2017.08.001

Youth Climate Lab (2020) Youth-led climate smart village in Bangladesh: Gangarampur initiative. https://www.youthclimatelab.org

Yuan X, Li S, Chen J, Yu H, Yang T, Wang C, Huang S, Chen H, Ao X (2024) Impacts of global climate change on agricultural production: a comprehensive review. Agronomy 14(7):1360. https://doi.org/10.3390/agronomy14071360

FSC
www.fsc.org
MIX
Papier | Fördert gute Waldnutzung
FSC® C083411

Kolibri 360 GmbH
Ferdinand-Jühlke-Straße 7
99095 Erfurt, Deutschland
produktsicherheit@kolibri360.de